Acrodermatitis enteropathica = zinc sulphate +

OXFORD MEDICAL PUBLICATIONS

AN INTRODUCTION TO HUMAN PHYSIOLOGY

Respiratory centre in Medulla.
Vasomotor " in Medulla (b.p)

bone marrow —< W.b.c
R.b.c
platelets.

Cant inject protein directly into vein ∴ anaphylactic Rea
∴ give A Acids.

AN INTRODUCTION TO HUMAN PHYSIOLOGY

J. H. GREEN

M.A., M.B., B.Chir. (Cantab.)
Ph.D. (Lond.), A.R.I.C.

*Professor of Physiology, University of London
at the Middlesex Hospital Medical School*

FOURTH (S.I.) EDITION

LONDON
OXFORD UNIVERSITY PRESS

Oxford University Press, Ely House, London W.1

GLASGOW   NEW YORK   TORONTO   MELBOURNE   WELLINGTON
CAPE TOWN   IBADAN   NAIROBI   DAR ES SALAAM   LUSAKA   ADDIS ABABA
DELHI   BOMBAY   CALCUTTA   MADRAS   KARACHI   LAHORE   DACCA
KUALA LUMPUR   SINGAPORE   HONG KONG   TOKYO

ISBN 0 19 263328 7

© Oxford University Press 1963, 1964, 1966, 1968, 1970, 1972, 1974, 1976

First Edition 1963
Fourth (S.I.) Edition 1976

Text set in 9/10 pt. Monotype Times New Roman, printed by photolithography, and bound in Great Britain at The Pitman Press, Bath

CONTENTS

v

S. I. UNITS

This edition has been produced to cover the transitional period when medical and scientific literature is changing rapidly from the use of old familiar units to the newer, and now generally accepted, *Système International d'Unités* (S.I. units) based on the kilogram, metre, and second and using the *newton* as the unit of force.

The major changes affecting human physiology may be summarized as follows:

	OLD UNITS	NEW UNITS
AMOUNT:	mg.	→ *millimole* (mmol)
CONCENTRATION:	mg per 100 ml.	→ *mmol per litre of molecules*
	mg. per 100 ml.	→ *mg per litre* (if molecular weight is not known)
	milliequivalents per litre	→ *mmol per litre of ions*
HEAT and ENERGY:	Calorie	→ *kilojoule* (kJ) =joule
GAS TENSION:	mm. Hg	→ *kilopascal* (kPa)
BLOOD PRESSURE:	mm. Hg	→ *kilopascal* (kPa)

In this edition the old and new units are wherever possible given alongside one another to facilitate this transition.

The Millimole

The S.I. unit of quantity is the *mole* (mol.) It is based on the number of atoms in 0·012 kg. (12 g.) of carbon. 12 g. of carbon has been chosen since the internationally agreed table of relative atomic weights [p. 221] now uses exactly 12 as the atomic weight of carbon-12 instead of taking exactly 1 as the atomic weight of hydrogen or 16 as the atomic weight of oxygen. Latest estimates indicate that there are $6 \cdot 0220943 \times 10^{23}$ atoms in 12 g. of carbon (Avogadro's number). The atomic weight in grams of the other elements will contain this number of atoms, and the molecular weight (or ionic weight) in grams of a substance will contain the same number of molecules (or ions).

It is convenient to consider the *mole* simply as a number, albeit a very large number, approximately equal to 6×10^{23}. Thus 6×10^{23} molecules represent a *mole of molecules*, 6×10^{23} atoms represent a *mole of atoms*, 6×10^{23} ions represent a *mole of ions* and 6×10^{23} electrons represent a *mole of electrons*. A mole, in theory, could be applied to 6×10^{23} of anything, but its use is usually restricted to elementary particles such as atoms, molecules, ions, electrons, photons, etc.

It should be noted, however, that when using the term *mole* it is important to state clearly which elementary particle is being considered since, for example, *one mole* of sodium chloride *molecules* (6×10^{23} molecules) will give rise to *two moles of ions*: one mole of sodium ions (6×10^{23} Na$^+$) plus one mole of chloride ions (6×10^{23} Cl$^-$). Furthermore two grams represent *one* mole of hydrogen molecules but *two* moles of hydrogen ions.

Since the *mole* is a large unit, sub-divisions are frequently used. The *millimole* is 1 000 times smaller ($=6 \times 10^{20}$ particles), the *micromole* is 1 000 000 times smaller than the mole ($=6' \times 10^{17}$). The *nanomole* is one thousand-millionth of a mole ($= 6 \times 10^{14}$) and the *picomole* is one million-millionth ($=6 \times 10^{11}$).

Millimoles per Litre

Since the cubic metre is too large for practical use, the unit of volume has been standardized as the cubic decimetre (dm^3). The litre (which was changed in 1964 from $1 \cdot 000027$ dm^3 to 1 dm^3 exactly) may be used as an acceptable alternative. Concentrations are thus expressed in *moles per litre* (mol/l), *millimoles per litre* (mmol/l), *micromoles per litre* (μmol/l), *nanomoles per litre* (nmol/l) or *picomoles per litre* (pmol/l). Such solutions may be referred to as *molar* (M), *millimolar* (mM), *micromolar* (μm), *nanomolar* (nM) and *picomolar* (pM) respectively, the volume of one litre being implied.

Since ions, as well as molecules, may be expressed in millimoles, the term milliequivalent is no longer needed and has become obsolescent. Fortunately most of the ions considered in human physiology (sodium, potassium, chloride, bicarbonate) are monovalent and the numerical value of their concentration in millimoles per litre is the same as that in milliequivalents per litre. The conversion is therefore comparatively easy. Even carbon dioxide reacts with water in the body to form a single hydrogen ion and a monovalent bicarbonate ion. It may therefore be considered to be monovalent. (Divalent carbonate ions are not normally present in body fluids.) Confusion is more likely to occur with the divalent ions, such as calcium and magnesium, where the new concentration in millimoles per litre will be half that in the old milliequivalents per litre.

With the milliequivalent system the presence of an equal number of positively and negatively charged ions in body fluids can be demonstrated by a simple addition [p. 131]. With the millimoles per litre of ions system this equality following summation no longer applies if divalent ions are present. However, the fact that divalent ions carry a double charge (which is allowed for in the milliequivalent system but not in the millimole of ions system) could be allowed for by considering *millimoles of charges* instead of *millimoles of ions*. Direct summation would once again then be possible.

The cells of the body require a constant environment. It is interesting to note that expressing the concentration of the ions in the surrounding fluid in *millimoles per litre* gives a direct indication of the number of ions per unit volume in the fluid surrounding the cell, irrespective of the ionic weights. Thus x mmol/l of sodium chloride gives the same number of ions per litre as x mmol/l of sodium bicarbonate or x mmol/l of potassium chloride. This did not apply when the concentration was expressed as y mg. per 100 ml.

One disadvantage of the molar system is that it can only be applied to substances where the molecular or ionic weight is known. Plasma proteins, for example, and particularly the globulin fraction, have ill-defined molecular weights. Their concentration will continue to be expressed in grams per volume although the volume will now be the *litre*, instead of 100 ml. as previously.

The Kilojoule

The medical Calorie (or more accurately the kilocalorie) is the heat unit traditionally used in metabolic and dietary studies. Unfortunately it is not part of the S.I. unit system. The Calorie is equivalent to approximately 4·2 kJ. of energy, and under the S.I. system it is proposed that only the kilojoule be used. The joule has the advantage that it is common to both electrical units (=1 watt-second) and mechanical units (=1 newton-metre). However, so much of the literature in this field has in the past used the Calorie that it will probably be a long time before the term ceases to be used.

The Kilopascal

$$120 \text{ mm. Hg} = 16 \text{ kPa}$$
$$110 \text{ mm. Hg} = 14 \cdot 7 \text{ kPa}$$
$$100 \text{ mm. Hg} = 13 \cdot 3 \text{ kPa}$$
$$90 \text{ mm. Hg} = 12 \text{ kPa}$$
$$80 \text{ mm. Hg} = 10 \cdot 7 \text{ kPa}$$
$$40 \text{ mm. Hg} = 5 \cdot 3 \text{ kPa}$$

The *kilopascal* (kPa) replaces the mm. Hg as the unit of pressure and gas tension in the S.I. unit system. It is easier to say and write than the S.I. alternative, *kilonewton per square metre* (kN m^{-2}). Since the density of mercury is 13·595.

$$1 \text{ mm. Hg} = 13 \cdot 595 \times 9 \cdot 81 \text{ N m}^{-2} = 0 \cdot 133 \text{ kPa}.$$

Hence 7·5 mm. Hg equals 1 kPa and 300 mm. Hg equals 40 kPa. The millimetre of mercury may be converted accurately to the kilopascal by dividing by 7·50. This may be carried out in one's head by the simple rule '*double it, double it again, divide by 3, divide bv 10.*' Thus 150 mm. Hg becomes 300, 600, 200, *20·0 kPa.*

Unfortunately dividing by 3 often leads to either 0·3̇ (0·3333 . . .) or 0·6̇ (0·6666 . . .).

Expressing gas partial pressures and tensions in kilopascals, instead of in mm. Hg, as in the past, has an unexpected advantage. The barometric pressure in kilopascals is close to 100.

$$100 \text{ kPa} = 1000 \text{ millibars} = 750 \cdot 062 \text{ mm. Hg.}$$

This means that the percentage composition of gases at sea-level will be approximately equal to their partial pressure in kPa. Thus room air contains 21 per cent oxygen at a oxygen partial pressure of approximately 21 kPa. A gas cylinder containing 7 per cent. carbon dioxide will, when released, give carbon dioxide at a partial pressure of 7 kPa. For accurate work a correction factor can be applied to allow for water vapour pressure and for variation in barometric pressure, but this correction factor will probably be so close to unity that it may often be ignored.

Since blood pressure has traditionally been recorded using the height of a column of mercury as the indication of pressure, the use of the *mm. Hg of mercury* as a unit of blood pressure is likely to continue alongside the kilopascal for some time to come.

Low pressures are often recorded in cm. H_2O, using a water manometer.

$$1 \text{ cm. Hg} = 10 \times 9 \cdot 81 = 98 \cdot 1 \text{ N m}^{-2} \simeq 0 \cdot 1 \text{ kPa.}$$

Thus a pressure of 10 cm. H_2O is very close to 1 kPa.

In this edition mm. Hg has been retained on the diagrams since it would appear that the change-over to S.I. units for blood pressure will be a much slower process than for the other units referred to above and that mm. Hg will continue to be used for many years to come.

In addition to the S.I. unit changes a number of amendments have been made to the text in the light of recent developments.

J.H.G.

The Middlesex Hospital Medical School,
London, W.1.
May 1975

EXTRACT FROM PREFACE TO THE FIRST EDITION

This book is intended to fill the need for an introductory book on human physiology. It has been written, in the first instance, for medical students who are just starting their course, and who already have a knowledge of chemistry, physics and biology to 1st M.B. standard. But it is hoped that the book will also prove useful to dental students, physiotherapy students, and as a textbook for a short course in the subject.

My intention has been to give, in a clear and concise manner, the basic concepts of human physiology, unobscured by controversy. It is hoped that this will provide a *framework* to which additional knowledge may be added by the attendance at systematic lectures, or by consulting the larger textbooks (such as *Samson Wright's Applied Physiology*), reviews and original papers. Some guide to this additional reading is given at the end of each chapter.

A brief mention of the apparatus commonly employed in practical classes has been included at the appropriate point in the text so as to integrate the practical class experiments with the theory. This applies particularly to the sections on blood, the circulation, respiration, kidney, and nerve-muscle which lend themselves readily to class experimentation. Where appropriate, these sections have been expanded to enable the student to understand the theory underlying any practical work which is undertaken before further systematic lectures in the subject have been given, and to give the student an indication of the future clinical relevance.

Such a book must of necessity have its limitations. It is not possible, for example, to give full credit to all the authors whose papers have contributed to the present understanding of the subject; reference to such works will be found in the larger volumes. Nor does space permit the detailed discussion of the finer nuances of differing theories, and I have, in such cases, given what appears to me to be the most plausible interpretation of the experimental results. I realize that this gives the impression that the subject is a more clear-cut one than in fact it is, but I feel that in an introductory book this is justified in the interests of clarity.

I have concentrated in this book on the physiology of the higher vertebrates with special reference to man. It has not been possible to include the very interesting consideration of the comparative physiology of the lower species.

Recent techniques have enabled some of the findings, which previously could only be obtained by animal experimentation, to be confirmed in man. In such cases, the classical animal experimental tracings have been replaced in this book by original tracings using man as the subject.

A brief introduction to the subject of biochemistry has been included.

J. H. G.

The Middlesex Hospital Medical School
London, W.1
May 1963

EXTRACT FROM PREFACE TO THE SECOND EDITION

The period since the first edition has been one of continuing transition in the teaching of physiology. In many fields the trend has been for the smoked-drum kymograph to be gradually replaced, as the standard method of recording, by more modern methods. This trend has shown itself in the increased complexity of the apparatus, mainly electronic, which is being made available to students in the practical class. This applies particularly to the field of cardio-vascular physiology. In respiratory physiology the many hours previously spent by students trying to master such techniques as the use of the Haldane gas analysis apparatus are being saved by the use of more modern devices, such as the paramagnetic oxygen meter, the infra-red CO_2 analyser, oxygen electrodes and carbon dioxide electrodes, all of which give almost instantaneous answers. The tendency now is for the students to carry out practical work in small groups rather than individually. Since one of the objects of this book has been to integrate the practical class experiments with the theory, accounts of these newer methods have been incorporated in the text which has been expanded, where necessary, bearing these new trends in mind.

J. H. G.

The Middlesex Hospital Medical School
London, W.1
May 1968

EXTRACT FROM PREFACE TO THE THIRD EDITION

In the interval since the last edition the trend in physiology referred to in the 'Preface to the Second Edition' has continued. In addition, in many medical schools and colleges there has been a move towards more integrated teaching which is breaking down the old rigid boundaries which previously existed between physiology, anatomy, biochemistry, pharmacology, pathology, medicine and surgery, whilst, at the same time, the amount of time available in the course for the study of these subjects is being reduced by the introduction of new academic disciplines based on the social and community aspects of medicine. These changes have been borne in mind in the preparation of this new edition.

J. H. G.

The Middlesex Hospital Medical School
London, W.1
May 1972

S. I. CONVERSION TABLE

	Old Units	S.I. Units
Arterial blood pressure	$\frac{120}{80}$ mm. Hg	$\frac{16}{11}$ kPa
Pulmonary artery blood pressure	$\frac{25}{8}$ mm. Hg	$\frac{3\cdot3}{1}$ kPa
Capillary blood pressure	$32 \rightarrow 12$ mm. Hg	$4\cdot3 \rightarrow 1\cdot6$ kPa
Osmotic pressure of plasma proteins	25 mm. Hg	$3\cdot3$ kPa
Barometric pressure	760 mm. Hg	$101\cdot3$ kPa
	750 mm. Hg	100 kPa
Room air P_{O_2}	160 mm. Hg	21 kPa
P_{CO_2}	0 mm. Hg	0 kPa
Alveolar air P_{O_2}	100 mm. Hg	13 kPa
P_{CO_2}	40 mm. Hg	5 kPa
Arterial blood P_{O_2}	95–100 mm. Hg	13 kPa
P_{CO_2}	40 mm. Hg	5 kPa
Venous blood P_{O_2}	40 mm. Hg	5 kPa
P_{CO_2}	46 mm. Hg	6 kPa
Saturated water vapour pressure at 37°C	47 mm. Hg	6 kPa
Haemoglobin	$14\cdot5$ g/100 ml	$2\cdot2$ mmol/l
Arterial blood O_2	19 ml O_2/ 100 ml	$8\cdot5$ mmol/l
CO_2	48 ml CO_2/100 ml	22 mmol/l
Venous blood O_2	14 ml O_2/100 ml	$6\cdot0$ mmol/l
CO_2	52 ml CO_2/100 ml	24 mmol/l
Blood: glucose	60–100 mg/100 ml	$3\cdot3$–$5\cdot5$ mmol/l
cholesterol	200 mg/100 ml	5 mmol/l
creatinine	1 mg/100 ml	90 μmol/l
urea	30 mg/100 ml	5 mmol/l
uric acid	3 mg/100 ml	$0\cdot2$ mmol/l
Physiological saline	$0\cdot9$ g NaCl/100 ml	150 mmol/l Na^+
		150 mmol/l Cl^-
Plasma: Na^+	145 mEq./l	145 mmol/l
Cl^-	110 mEq./l	110 mmol/l
K^+	5 mEq./l	5 mmol/l
Ca (total)	10 mg/100 ml	$2\cdot5$ mmol/l
Ca^{++}	$2\cdot5$ mEq./l	$1\cdot25$ mmol/l
Mg^{++}	2 mg/100 ml	$1\cdot0$ mmol/l
Proteins	6–8 g/100 ml	60–80 g/l
Metabolism of: 1 g. carbohydrate	$4\cdot1$ Calories	17 kJ
1 g. fat	$9\cdot3$ Calories	38 kJ
1 g. protein	$4\cdot1$ Calories	17 kJ
Basal metabolic rate: Males	40 Cal/m²/hr	170 kJ m^{-2} hr^{-1}
Females	37 Cal/m²/hr	150 kJ m^{-2} hr^{-1}
Daily requirements	2800 Calories	11 500 kJ

1. INTRODUCTION

The term 'physiology' was first used in its modern sense by the French physician Jean Fernel in 1552. It is derived from the Greek *physis*, *logos* which mean 'inquiry into nature'. Physiology is the study of the functions of healthy living organisms and the changes which occur during activity. In its widest sense it embraces all the animal and plant kingdoms, but in this book it will be restricted to the animal kingdom with special reference to man.

Physiology is an experimental science, that is, our knowledge is based on direct experimental evidence. However, where such direct evidence is not available, the function of a certain organ in man has to be deduced from the behaviour of that organ in the higher mammals and other animals closely related to man.

This is, in fact, a book about a *standard* man and how his body functions. However, since all human beings are different, a standard man, as such, does not exist. It is convenient, nevertheless, to consider such a hypothetical individual who is the average of all. This will enable numerical values to be given to many of the physiological constants which will act as a guide to the 'order of magnitude' of the value expected in any normal individual.

The metric system will be used, and we will start by saying that our standard man weighs 70 kg.

It is convenient, when considering function, to divide the body into a number of closely interrelated systems.

Heart and Circulatory System

cells on outer layer of outer layer of skin are dead + flattened

Man is built of cells which require food and nourishment to live. Blood is the main transport system of the body, and by means of it, nourishment arrives at the cells. At the same time their waste products are removed. But just as the transport system of a large city (the bus or train system) does not run all the way to the front door of every house, so the blood does not come in contact with the cells. It runs in the vicinity of the cells, but stays in the blood vessels. The smallest and usually the nearest blood vessels are the **capillaries.** The gap between the blood capillaries and the cells is bridged by **tissue fluid,** which is also known as **interstitial fluid.** The nourishment diffuses from the capillary into this tissue fluid, and from this tissue fluid to the cells.

The blood is pumped round the body by the **heart.** The same blood circulates round and round [FIG. 1], as was first shown by William Harvey in the early seventeenth century.

The heart acts as two pumps. The left side, consisting of the left **atrium** and **ventricle,** pumps the blood through the aorta and the **arteries** to the tissues. The blood, having passed through the tissues, returns via the **veins** to the right side of the heart, which consists of the right atrium and ventricle [FIG. 2]. The blood is pumped by the right side of the heart through the pulmonary artery to the **lungs** and it returns to the left side of the heart again via the four pulmonary veins. *8 pints!*

Standard man has 5 litres of blood in his body. The volume of blood which leaves each side of the heart and enters the arteries per minute is known as the *cardiac output*. It is 5 litres per minute, so that the blood is circulating on the average once a minute. However, not all the blood is circulating at this rate, some is stagnating, particularly in the lower parts of the body, and some is circulating more rapidly.

In exercise the cardiac output may increase to 30 litres per minute. This indicates that the circulation has speeded up, and that the blood is now circulating, on the average, once every 10 seconds. It will be noted that just as a motor car can travel at 30 miles per hour without needing a road 30 miles long (it could do so round a 5 mile circular track) so the heart can pump out 30 litres of

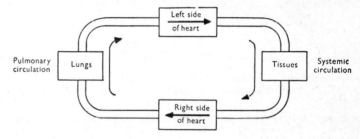

FIG. 1. The basic plan of the circulation. The left side of the heart pumps oxygenated blood, which is bright red in colour, to the tissues. The deoxygenated dark blood, returning from the tissues, is pumped by the right side of the heart to the lungs and round to the left side of the heart again.

blood from each side per minute when there are only 5 litres of blood in the circulation. The 30 litres per minute is not a quantity but a 'rate of flow'.

Respiratory System

Oxygen from the air is absorbed by the blood as it passes through the lungs and is transported to the tissues. The gaseous waste product carbon dioxide (CO_2), produced by the tissues is carried to the lungs and is breathed out in the expired air. Respiratory movements (breathing) ensure that the gas in the depth of the lungs (*alveolar air*), which is in contact with the blood, is maintained at a constant composition by the replacement of the oxygen as it is used up, and by the removal of the surplus carbon dioxide.

The blood changes colour as it passes through the lungs. It arrives dark in colour having given up its oxygen, and changes to a bright red colour as it becomes oxygenated.

Digestive System

The food taken in by the mouth is used for the growth and repair of tissues, and for the production of heat and energy. After appropriate breakdown of the large molecules in the food by biological catalysts known as *enzymes* in the digestive tract, the food is absorbed into the blood stream and passes via the portal vein to the **liver** and thence to the tissues [FIG. 2].

The unabsorbed food together with bacteria and cells shed from the lining of the intestines are passed out via the rectum and anal canal as faeces.

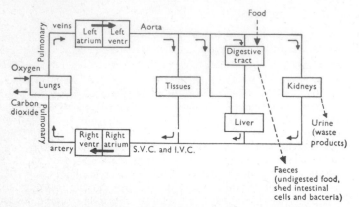

FIG. 2. The circulation of the blood acts as the transport system of the body. Oxygen is carried from the lungs to the tissues. Carbon dioxide is carried from the tissues to the lungs. Food, after absorption, is carried to the liver and then to the general circulation. Waste products are transported to the kidneys for excretion in the urine.

Left ventr = left ventricle

Right ventr = right ventricle

Body tissue is formed of **protein**. The food protein (meat, fish, etc.) is broken down by digestion to its constituent **amino acids.** These amino acids circulate round the body, and the cells remove the amino acids they require for growth and repair. The amino acids are built up again by the cells into body protein. Thus the skin cells join the amino acids to form skin protein, the bone marrow cells use the amino acids to form blood cells, the muscle cells make muscle protein, etc., all from the **amino acid pool.** It is rather like a man who, wishing to have a house of a given size and shape, and being unable to find such a house, buys any house, knocks it down to the individual bricks, transports these bricks to the correct site and then uses these bricks to build his house. The food proteins are different from the body proteins, but they contain the correct 'building bricks', the amino acids.

Body proteins consist of long chains built up of many hundreds of these amino acids often with cross-links between chains or between two parts of the same chain. Twenty different amino acids are employed in the formation of body protein. Each may be incorporated at several places in the protein molecule.

The names of the amino acids (with their abbreviations) are:

Alanine (ala)	Leucine (leu)
Arginine (arg)	Lysine (lys)
Asparagine (asn)	Methionine (met)
Aspartic acid (asp)	Phenylalanine (phe)
Cysteine (cys)	Proline (pro)
Glutamic acid (glu)	Serine (ser)
Glutamine (gln)	Threonine (thr)
Glycine (gly)	Tryptophan (try or trp)
Histidine (his)	Tyrosine (tyr)
Isoleucine (ile or ileu)	Valine (val)

The chemical formulae of these amino acids will be considered later [p. 114]. It will be noted that the abbreviation is the first three letters of the name with the exception of asparagine, glutamine and isoleucine. Using these abbreviations the amino acid sequence in a protein may be represented as follows:

$$\overset{1}{ala}-\overset{2}{gly}-\overset{3}{his}-\overset{4}{phe}\ldots$$

All the amino acids contain nitrogen (in addition to carbon, hydrogen and oxygen) but **cysteine** and **methionine** also contain sulphur. Cross-links between chains are formed by sulphur-bonds (—S—S—) between cysteine units.

$$\ldots\overset{5}{ala}-\overset{6}{gly}-\overset{7}{his}-\overset{8}{cys}-\overset{9}{leu}\ldots$$

$$\ldots\underset{27}{arg}-\underset{28}{glu}-\underset{29}{val}-\underset{30}{cys}-\underset{31}{pro}\ldots$$

A pair of cysteine units joined in this way may alternatively be referred to as cystine [p. 114]. The cross-links may be between two separate chains, as in insulin [p. 168], or between two parts of the same chain forming loops as in the human growth hormone [p. 161]. Glutamic acid also exists in pyro form [p. 209] (pyroglu or pyr).

Kidneys

The waste products of the cells, other than carbon dioxide, are excreted from the body, dissolved in water, by the kidneys and after temporary storage in the bladder pass out via the urethra as **urine.**

Control Systems

The functions of the organs of the body are controlled in two ways. The first system of control is by sending 'chemical messengers' known as **hormones** via the blood stream from an endocrine gland, where they are produced, to modify the activity of some distant organ. Hormones play a particularly important part in the control of the reproductive system.

The second system of control is by means of the **nervous system.** Messages are transmitted along nerves in the form of coded nerve impulses from the central nervous system (C.N.S.)—the brain and spinal cord—to the organ concerned to modify its function. The nerves conveying this information away from the C.N.S. are termed *efferent* or motor nerves.

Voluntary movement is brought about by activity in the motor nerves which run to the muscles. This activity causes shortening of the voluntary (striated) muscle fibres. Information is fed back into the C.N.S. from the muscles concerned so that the movements can be co-ordinated.

The C.N.S. is kept aware of the outside world by the sense organs such as those of the skin which respond to touch, pain and temperature, and by the **special senses** of sight, smell, hearing and taste. This information, together with that from the muscles, is transmitted to the C.N.S. along the *afferent* or sensory nerves, once again in the form of coded nerve impulses.

Peripheral Nerves

The spinal cord develops on a segmental basis. It is protected by the vertebral column in which it lies.

A pair of spinal nerves, one on each side, leave the cord between each of the vertebrae through the intervertebral foramina. The first cervical nerve (C.1) leaves between the first cervical vertebra and the skull. The last cervical nerve (C.8) leaves between the seventh cervical vertebra and the first thoracic vertebra. Thus, there are eight cervical nerves although there are only seven cervical vertebrae. The other nerves are each named according to the vertebra lying immediately above them. There are twelve thoracic nerves (T.1 to T.12), five lumbar nerves (L.1 to L.5) and five sacral nerves (S.1 to S.5) and a coccygeal nerve making thirty-one nerves on each side [FIG. 3].

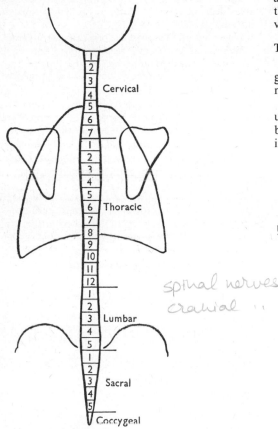

FIG. 3. Spinal nerves leave the spinal cord on each side through the foramina between the vertebrae. The first cervical nerve (C.1) leaves between the skull and the first cervical vertebra (the atlas). Including a coccygeal nerve there are thirty-one spinal nerves on each side.

In addition to the spinal nerves, twelve pairs of nerves originate from the brain. These are known as **cranial nerves.** Each has a name, but they are also numbered from one to twelve using roman numerals:

I	Olfactory
II	Optic
III	Oculomotor
IV	Trochlear
V	Trigeminal
VI	Abducent
VII	Facial
VIII	Vestibulocochlear (auditory)
IX	Glossopharyngeal
X	Vagus
XI	Accessory
XII	Hypoglossal

Each spinal nerve has a ventral (anterior) and dorsal (posterior) root attaching it to the spinal cord. The motor or efferent nerves run in the ventral nerve root and the sensory or afferent nerves run in the dorsal nerve root. This fact was shown independently by Bell and Magendie at the beginning of the last century and is known as the Bell–Magendie Law.

The muscles employed in voluntary movement are controlled by a single motor nerve system and these motor nerves are excitatory, that is, their activity causes the muscles to contract [FIG. 4(A)]. A voluntary muscle relaxes when the activity in its motor nerve ceases.

The Autonomic Nervous System

The part of the nervous system which controls the heart, secreting glands and the involuntary muscle is termed the autonomic nervous system.

Involuntary muscle, sometimes known as smooth, plain or unstriated muscle, is found in the digestive tract, the air passages, the bladder, the blood vessels and in the eye. The control system for involuntary muscle differs from that for voluntary muscle.

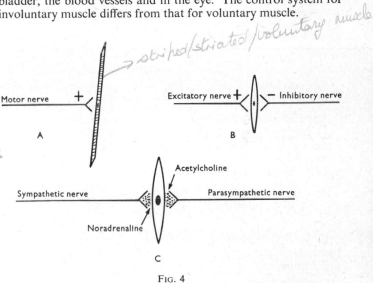

FIG. 4

A. The motor nerve to a voluntary muscle fibre is excitatory. Its activity causes the muscle to contract.
B. Involuntary muscle (plain or smooth muscle) often has a double nerve supply with an excitatory nerve (+) which causes the muscle to contract, and an inhibitory nerve (−) which causes relaxation.
C. The autonomic nerve supply has two divisions termed sympathetic and parasympathetic. When the involuntary muscle has a double supply one nerve is sympathetic and the other parasympathetic. The sympathetic is excitatory to some organs and inhibitory to others. The chemical transmitter at the sympathetic nerve termination is usually noradrenaline. That at the parasympathetic termination is acetylcholine.

Involuntary muscle usually has a double motor nerve supply. One nerve is excitatory and its activity will cause the involuntary muscle fibre to contract. The other nerve is inhibitory and when active will cause the muscle fibre to relax [FIG. 4(B)]. We can consider the excitatory nerve as a plus nerve, and the inhibitory nerve as a minus nerve. If both nerves are active simultaneously, algebraic summation will take place, the pluses and minuses will tend to cancel out, and the resultant state of the muscle will depend upon which activity is the greater.

The autonomic nervous system is subdivided into two divisions termed the **sympathetic nervous system** and the **parasympathetic nervous system.** When an involuntary muscle fibre has a double nerve supply, the excitatory nerve will belong to one division and the inhibitory nerve to the other [FIG. 4(C)].

The two divisions usually act in a complementary manner to one another [TABLE 1]. Thus when the activity of one system supplying an organ increases, the activity of the other usually decreases.

TABLE 1

Involuntary muscle causing	Excitatory nerve (plus nerve)	Inhibitory nerve (minus nerve)
Constriction of the pupil of the eye	Parasympathetic	Sympathetic
Constriction of the air passages (bronchi and bronchioles)	Parasympathetic	Sympathetic
Contraction of the alimentary tract	Parasympathetic	Sympathetic
Contraction of the sphincters of the alimentary tract	Sympathetic	Parasympathetic
Constriction of blood vessels	Sympathetic	Nil*

[See also CHAPTER 20]

* As will be seen later, some blood vessels are supplied with *vasodilator* parasympathetic or sympathetic nerves.

The two systems differ anatomically. The sympathetic nerves leave the central nervous system between the first thoracic and the second lumbar segments of the spinal cord. It is a thoracolumbar outflow. The parasympathetic nerves, on the other hand, leave the central nervous system in the spinal region only at the second, third and fourth sacral segments, but in addition there is an important outflow in some of the cranial nerves which arise from the brain itself. The parasympathetic system thus has a craniosacral outflow.

By far the most widely distributed of the cranial nerve outflows is that of the tenth cranial nerve, the vagus, which supplies the contents of the thorax and most of the abdomen with parasympathetic activity.

Although the terminal ramifications of the sympathetic nervous system are widely distributed throughout the body, the parasympathetic nervous system is restricted to the trunk and skull. It should be noted therefore that *there is no parasympathetic supply to the limbs* and thus the autonomic nervous system supply to the arms, legs, hands and feet is restricted to the sympathetic nervous system only.

Nerve activity leads to the release of small amounts of a chemical substance at the nerve termination. The systems differ in the chemical transmitter released at the junction between each nerve fibre and the muscle fibre. This chemical bridges the gap between the nerve and muscle and its release by the nerve causes the muscle to contract or, maybe, to relax depending on whether we are dealing with a plus or minus nerve. The chemical transmitter released at the sympathetic nerve terminations is **noradrenaline** and such sympathetic nerves are said to be *noradrenergic*. The chemical transmitter released at the parasympathetic nerve termination is **acetylcholine.** Such parasympathetic fibres are said to be *cholinergic* [FIG. 4(C)].

Noradrenaline is also released as a hormone by the central part (medulla) of the suprarenal (adrenal) endocrine gland, and noradrenaline circulating in the blood from this source will augment the activity of the sympathetic nervous system. No endocrine gland has been found that produces acetylcholine to augment the activity of the parasympathetic system.

FURTHER READING

Bernard, Claude (1855) *Introduction to the Study of Experimental Medicine* (English translation, 1949, New York).
Foster, M. (1901) *Lectures on the History of Physiology*, Cambridge.
Franklin, K. J. (1949) *A Short History of Physiology*, London.
Gordon-Taylor, G., and Walls, E. W. (1958) *Sir Charles Bell: His Life and Times*, Edinburgh.
Singer, C., and Underwood, E. A. (1962) *A Short History of Medicine*, 2nd ed., Oxford.

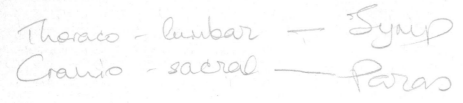

Thoraco - lumbar → Symp
Cranio - sacral — Paras

2. THE BLOOD

The first system to be considered in more detail will be the circulatory system, but before discussing the action of the heart as a pump which circulates the blood, the blood itself will be considered.

Blood is the transport medium of the body. The circulating blood carries oxygen from the lungs to the tissues; food from the digestive tract to the tissues; waste products from the tissues to the kidneys for excretion; carbon dioxide from the tissues to the lungs for excretion; hormones from the endocrine glands to the tissues on which they act; and heat from the site of production to the cooler parts of the body.

Haematocrit or Packed Cell Volume

Blood is a red syrupy fluid which has a specific gravity of 1·055 and a viscosity of two and a half times that of water.

If blood is allowed to stand, and clotting is prevented, it separates out into layers under the influence of gravity [FIG. 5]. This

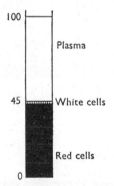

FIG. 5. Haematocrit. If blood clotting is prevented, the blood separates out under gravity into plasma and cells. The percentage of red cells by volume is known as the *haematocrit*.

process may be speeded up by spinning the blood in a centrifuge. The denser **red cells** sink to the bottom. The slightly less dense **white cells** form an intermediate layer (the 'buffy coat'). The top layer consists of a straw-coloured fluid known as **plasma.** If a calibrated tube is used, it is seen that 45 per cent. of the blood volume is made up of red cells. The figure of 45 is known as the **haematocrit** or **packed cell volume.** It varies with the number and size of the red cells in the blood and with the plasma volume. The normal range for the haematocrit is 40–47.

A high haematocrit reading indicates either an increase in the number of circulating red cells, or a reduction in the circulating plasma volume. The high haematocrits (65 or more) seen in cases of *cholera* are due to the large loss of water in the stools and the resultant reduction in plasma volume.

A low haematocrit usually indicates a reduction in the number of circulating red cells, but the reduction will also occur if the plasma volume increases.

Determination of Haematocrit using Microcentrifuge

Haematocrit may conveniently be determined using a small blood sample (50 microlitres) with the capillary tube and microcentrifuge technique [FIG. 6(*A*)].

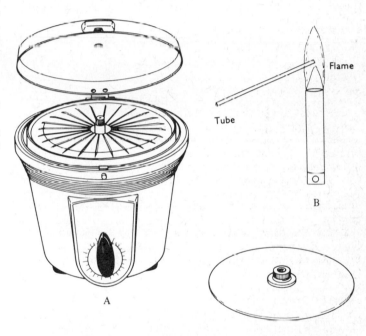

FIG. 6A. Microcentrifuge for use with heparinized capillary tubes for determination of haematocrit.

FIG. 6B. Flame sealing. When flame sealing, the dry end of the tube must be used and care taken not to haemolyse the blood by heat.

An uncalibrated capillary tube 7·5 cm. long and 1 mm. in diameter is used. It is coated internally with heparin to prevent blood clotting. The skin of the finger or lobe of the ear is first cleaned and then pricked with a sterile lancet. The end of the capillary tube is placed in the drop of blood. The blood enters by capillary attraction. The tube is removed when it is two-thirds full with blood. One end is sealed by a hot flame [FIG. 6(*B*)] or alternatively a sealer compound may be used. The tube is then placed with the sealed end outward in the centrifuge.

Since the tubes are so light no balancing is necessary and about 20 tubes can be centrifuged at the same time. The high speed rotation of the tubes gives a centrifugal force of 12,000 times that of gravity, and with this force the separation of the blood into cells and plasma is complete in 5 minutes.

Since the tubes are uncalibrated it is necessary to measure the length of the columns to calculate the haematocrit.

A suitable measuring device, based on the principle of similar triangles, may be constructed using graph paper as in FIGURE 6(C).

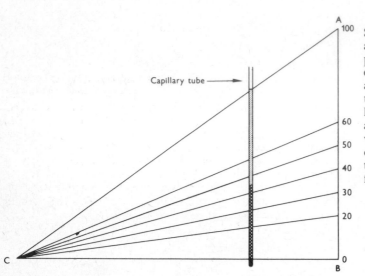

FIG. 6C. Measuring device for haematocrit tubes. Tubes are placed parallel to AB with ends of blood column on BC and AC.

Line **AB** is divided into 100 units. Suitable points on **AB** are joined to a point **C.** The uncalibrated capillary tube is placed parallel to **AB** in such a position that the ends of the blood in the tube lie on **BC** and **AC.** The haematocrit can then be read off. A line from **C** to the cell-plasma junction extended to **AB** gives the haematocrit.

Blood Volume

A 70 kg. adult man has 5–6 litres of blood in his circulation. At any given time, assuming a blood volume of 5 litres, about 1 litre of this blood will be in the lungs, 3 litres in the veins of the systemic circulation, and the remaining 1 litre will be in the heart and systemic arteries, arterioles and capillaries.

As a rough guide, the blood volume may be taken as being approximately one-twelfth (8 per cent.) of the body weight. Thus a 7 lb. baby (=3·2 kg. weight) will have a much smaller blood volume of only $\frac{1}{12} \times 3·2$ litres = 270 ml. This is an important point to remember when a blood transfusion is given to such a baby.

For an accurate determination of blood volume the *dilution technique* is employed. The basis of this technique is that a known quantity of a harmless substance is added to the circulation. It is diluted by the blood, and after thorough mixing has occurred, its dilution is determined. From the known amount of substance injected, the quantity of blood needed to dilute it to the final concentration may be deduced. This is the blood volume.

The quantity of a substance in a solution is equal to the product of the volume of the solution and its concentration. Let the solution injected have a volume v_1 and let the concentration of the substance in it be C_1. After thorough mixing has occurred, let the concentration of the substance in the blood be C_2, and the unknown

blood volume be v_2. Then since the quantity of the substance is the same both before and after the injection:

$$v_1 \times C_1 = v_2 \times C_2$$

or the blood volume,

$$v_2 = \frac{v_1 \cdot C_1}{C_2}$$

Suitable substances for the determination are harmless dyes such as Evans Blue, Rose Bengal, Vital Red and Indigo Carmine, or plasma labelled with radio-active iodine (^{132}I or ^{131}I). ^{131}I for example (with a half life of 8 to 14 days), combines with the amino acid tyrosine in the plasma albumin. These substances measure the plasma volume. Alternatively a suspension of red cells that have been labelled with radio-active chromium (^{51}Cr) or radio-active phosphorus (^{32}P) may be used to measure the red cell volume. Unfortunately no substance has yet been found that is diluted by both the cells and plasma and would thus measure the total blood volume directly. The total blood volume is calculated from the plasma or cell volume using the haematocrit:

Since,

$$\frac{\text{Red cell volume}}{\text{Blood volume}} = \frac{\text{Haematocrit } (H)}{100} \text{ (by definition)}$$

Therefore,

$$\text{Blood volume} = \text{Red cell volume} \times \frac{100}{H} \quad . \quad . \quad . \quad (1)$$

And,

$$\text{Blood volume} = \text{Plasma volume} \times \frac{100}{100 - H} \quad . \quad . \quad (2)$$

For an accurate determination of the blood volume simultaneous determinations of plasma and cell volumes must be employed.

Dye Dilution Method for the Determination of Cardiac Output (Stewart-Hamilton Dye Method)

A modification of the method, whereby the dilution is timed, may be used to determine blood flow, and if we are dealing with the whole circulation, this will determine the cardiac output. A known amount of the dye is injected into a peripheral vein and samples of the arterial blood are taken continuously. A graph is plotted connecting the dye concentration in this arterial blood with time. From the graph the time during which the dye appears and its mean concentration are determined. The cardiac output is then calculated as follows:

If the dye appears in the arterial blood for time T and the average concentration of the dye in this blood per litre is C, then, since the quantity of dye is a constant,

Quantity of dye injected (Q)

$$= \text{Concentration} \times \text{volume}$$
$$= \text{Concentration in arterial blood } (C)$$
$$\times \text{blood flow} \times \text{time } (T)$$

Thus,

$$\text{Blood flow (cardiac output)} = \frac{Q}{C \times T} \text{ litres per minute}$$

A correction has to be applied if recirculation of dye occurs.

Modifications of this technique include the injection of cold saline into a jugular vein and its detection by a small thermistor thermometer at the tip of a catheter in the pulmonary artery, the measuring of the dye concentration using an oximeter attached to the ear, and the use of radio-active substances which may be detected by means of a Geiger counter without cannulating an artery.

Radio-active Isotopes

Radio-active isotopes are detected by their nuclear disintegration which results in the release of α, β or γ-radiation. β-radiation may be detected by a Geiger-Müller counting tube. The gas in this tube is ionized by the radiation and converted into electrical pulses. γ-radiation is detected by a scintillation counter. The radiation acts on a phosphor producing light flashes (scintillations) which are detected by a sensitive photo-electric cell (a photomultiplier). The radio-activity is measured in *curies* which is a unit based on the disintegrations per second. One micro-curie of a radio-active substance produces $3\cdot7 \times 10^4$ detectable disintegrations per second.

RED BLOOD CELLS (ERYTHROCYTES)

Blood cell counts are traditionally expressed as the number of cells per cubic millimetre of blood. A cubic millimetre is equivalent to a microlitre (10^{-6} litre).

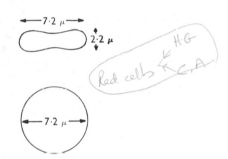

FIG. 7. The red blood cell (erythrocyte) is a biconcave disc.
1 micron (μ) = 10^{-3} mm. = 10^{-4} cm. = 10^{-6} m.

There are 5 million red cells in every cubic millimetre (1 mm.³) of blood, that is, in a drop of blood about the size of a pin's head. Five million is a very large number, but do we realize how large it is? It takes 3 months to reach 5 million, counting at the rate of one a second for 16 hours every day!

Each red cell is a biconcave disc having a diameter of $7\cdot2$ microns (μ) and a thickness of $2\cdot2$ microns [FIG. 7]. A micron is one-thousandth of a millimetre (10^{-3} mm.) or one-millionth of a metre (10^{-6} m.). In S.I. units the micron is known as the micrometre (μm). These cells contain the pigment *haemoglobin* (Hb) which enables them to transport oxygen round the circulation. The cells also contain the enzyme *carbonic anhydrase*, which, as will be seen later, plays an important part in the carriage of carbon dioxide.

Haemoglobin is a compound of the ferrous-iron–containing pigment *haem* combined with the protein *globin*. It has a molecular weight of 67,000. (Globin should not be confused with the protein 'globulin' found in plasma which has a much higher molecular weight of about 140,000.)

The detailed structure of the haemoglobin molecule has been studied by the technique of X-ray crystallography. It consists of four haem groups each with an associated protein chain having a molecular weight of about 17,000. Two of these protein chains are α chains with 141 amino acids. The other two are β chains with 146 amino acids. Each protein chain is bent in several places, and the chains fit into one another to form a compact molecule [see p. 80].

Each haemoglobin molecule contains 4 atoms of ferrous iron, one in each haem group, and is able to unite with 4 molecules of oxygen.

Haemoglobin combines loosely with oxygen, the process being known as 'oxygenation'. The iron remains in the ferrous state. Should 'oxidation' of the haemoglobin molecule occur, and the ferrous iron be converted to ferric iron, *methaemoglobin* is formed. Methaemoglobin, unlike haemoglobin, does not transport oxygen and the oxygen previously combined with the haemoglobin is given off. This principle is used in the Van Slyke method for the determination of oxygen in the blood [p. 86].

1 g. haemoglobin will combine with 1·34 ml. oxygen. This follows from the fact that:
1 molecule of haemoglobin will carry 4 molecules of oxygen.
1 mole (gram-molecule) of haemoglobin = 67,000 grams. 1 mole of oxygen has a volume of 22·4 litres.
Thus 67,000 grams of haemoglobin will carry $4 \times 22,400$ ml. O_2.

$$\therefore \text{1 gram of haemoglobin will carry } \frac{4 \times 22,400 \text{ ml. } O_2}{67,000}$$
$$= 1\cdot34 \text{ ml. } O_2$$

A normal person has about 14·5 g. haemoglobin in every 100 ml. of his blood (2·16 milli-moles/litre = 2·16 mmol/l). Thus, every 100 ml. of blood can carry as a maximum $14\cdot5 \times 1\cdot34 = 20$ ml. oxygen combined with the haemoglobin.

Cyanosis

Haemoglobin without oxygen is known as **reduced haemoglobin**, and is dark blue, almost black, in colour. Haemoglobin combined with oxygen is known as **oxyhaemoglobin** and is bright red in colour.

The blood passing through the lungs changes colour. It arrives blue and leaves with 95–100 per cent. of the haemoglobin in the form of oxyhaemoglobin and bright red in colour. The blood passes through the left side of the heart and on through the arteries still bright red in colour to the tissue capillaries. Here the oxygen is given up to the tissues and the blood's colour darkens. The blood returns to the right side of the heart via the veins and passes on to the lungs with at least 30 per cent. of the haemoglobin in the reduced form. It is the presence of this reduced haemoglobin that gives the venous blood its characteristic dark blue coloration.

If the blood does not absorb oxygen during its passage through the lungs then even the arterial blood will be blue. When the oxygen uptake in the lungs is impaired, and the blood in the capillaries contains more than 5 g. haemoglobin in the reduced form per 100 ml. of blood (0·75 mmol/l), the subject becomes visibly blue and is said to be *cyanosed*. The condition is known as **cyanosis**.

The Red Cell Count

To determine whether there is an adequate number of red cells in the circulation, a red cell count is carried out. Since it would take several months to count all the red cells in only 1 mm.³ of blood, we have to be content, for practical purposes, with a count of the number of cells in a much smaller volume of blood. However, the smaller the volume used the greater the care that has to be taken to ensure that this sample is representative of the whole blood volume. The red cell pipette is employed to dilute a small known quantity of blood. It is graduated at 0·5, 1 and 101 [FIG. 8(A)].

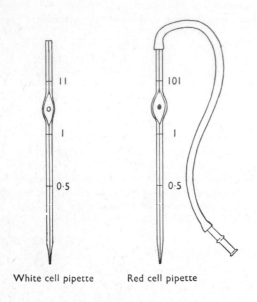

White cell pipette   Red cell pipette

FIG. 8A. Red and white cell pipettes for diluting the blood.

The bulb of this pipette has a volume of 100 units and the stem a volume of 1 unit. Blood from the finger or lobe of the ear is sucked up to the 0·5 mark, followed by diluting fluid up to the 101 mark. A suitable diluting fluid has the following composition:

Sodium chloride	.	. 0·6 g.
Sodium citrate	.	. 1·0 g.
Formalin 1 %	.	. 1·0 ml.
Distilled water	.	. 99 ml.

Thorough mixing is carried out, with the aid of a bead in the bulb, by rotating the pipette. The 0·5 units of blood in the bulb have now been diluted by 99·5 units of diluting fluid, i.e. it has been diluted 200 times. The 1 unit of fluid in the stem contains no blood. It is simply diluting fluid, and is discarded.

The counting takes place in a counting chamber or *haemocytometer*. A glass cover-slip is pressed tightly on to the haemocytometer and forms a chamber $\frac{1}{10}$ mm. deep [FIG. 8(B)]. The dilute blood is run into this space. Small squares, visible under the microscope, have been etched on the glass in the counting area. In the 'Improved Neubauer' type of haemocytometer [FIG. 8(C)], the side of each square is $\frac{1}{20}$ mm. and, therefore, the volume of diluted blood in it is $\frac{1}{4000}$ mm.^{3}. The number of red cells in this square is counted using a microscope. To avoid counting the same cell twice, a convention is adopted in connexion with cells touching the

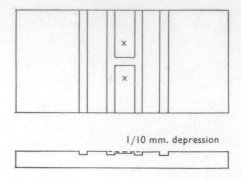

1/10 mm. depression

FIG. 8B. The counting chamber (haemocytometer). Counting areas marked X. A glass cover-slip placed across this area makes a chamber $\frac{1}{10}$ mm. deep.

boundary line. One such convention is to count all cells on or touching the top and left-hand side of a square, but to disregard any cell touching the right-hand side or bottom of the square [FIG. 8(D)]. It is usual to count the cells in 80 such squares distributed across the counting area, so that the cells in a volume of $\frac{1}{50}$ mm.^{3} ($= 80 \times \frac{1}{4000}$) will have been counted. Suppose that the answer is Y. The number of cells in 1 mm.^{3} of dilute fluid will be $50 Y$. The number of cells in 1 mm.^{3} of undiluted blood will be $50 \times 200 Y = 10,000 Y$. So the number of cells in 80 squares is multiplied by 10,000 to give the red cell count per mm.^{3}.

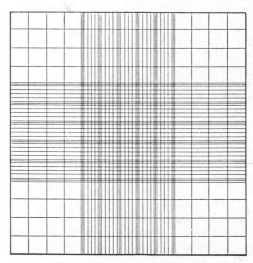

Counting area   (improved Neubauer)

FIG. 8C. Area X enlarged (Improved Neubauer haemocytometer). Eighty small squares distributed across the central counting area are used for the red cell count. The sparsely ruled areas at the four corners are used for the white cell count.

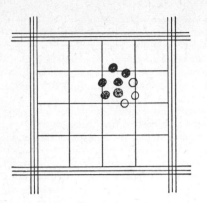

Shaded cells are counted as in,
unshaded cells as out

Fig. 8D. Part of the central counting area enlarged. Cells in the square or touching the top and left-hand side are counted (shaded cells). Cells touching the right-hand side and bottom (unshaded cells) are not counted for this square.

FORMATION AND REPLACEMENT OF RED CELLS

The red cell has a life of 120 days after which time it is broken down and replaced by a new cell. This means that $\frac{1}{120}$ or nearly 1 per cent. of the total body red cells are replaced each day.

Since 5 million red cells per mm.³

$$= 5 \times 10^9 \text{ cells per ml.}$$
$$= 5 \times 10^{12} \text{ cells per litre}$$
$$= 2 \cdot 5 \times 10^{13} \text{ cells per 5 litres}$$

$\frac{1}{120}$ of this figure will be 2×10^{11} cells.

Thus two hundred thousand million (200,000,000,000) new red cells need to be made each day to replace losses. This prodigious activity takes place in the bone marrow which is situated in the medullary cavity in bones. The bone marrow also manufactures the platelets and many of the white blood cells.

The red cells are formed from reticulo-endothelial cells which develop into *proerythroblast* cells. The succeeding stages are as follows, the process being known as *erythropoiesis*:

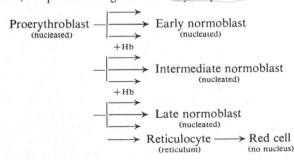

The rate of cell multiplication (mitosis) in the bone marrow required to maintain the red cell count is probably higher than anywhere else in the body (with the possible exception of the intestinal

mucosal cells). It is not surprising, therefore, that these marrow cells are among the first to be affected by an overdose of ionizing radiation such as may be caused by a nuclear explosion, a mishap in an atomic power station or an overdose of X-rays. Strontium-90 is a potential hazard since it is radio-active and may be deposited in bone in close proximity to the red bone marrow. A deficiency of red cells, and, therefore, haemoglobin, in the circulating blood is known as **anaemia** (G. *a-* no; *haima*, blood). This type of anaemia is called *aplastic anaemia*. A similar type of anaemia may be produced by an overdose of some drugs particularly certain benzene derivatives.

In early foetal life, there are no bones and hence no bone marrow. The blood is made first by the blood vessels. In later foetal life, red cells are made in the liver, kidneys, spleen and muscles as well as in the bone marrow. In a child all the marrow is engaged in red cell formation, whereas in the adult only the bones of the trunk and skull have red marrow in their medullary cavities. The marrow of the long bones of the limbs, with the exception of the upper ends, contains *yellow fatty marrow* which is not engaged in red cell formation. The yellow marrow may revert to red marrow if the demand for more red cells arises. A sample of bone marrow for haematological examination in the adult is taken by introducing a hollow needle into the marrow cavity of the sternum or the crest of the ilium.

Iron and the Formation of Red Cells

The structure of the red cell, like that of all cells in the body, is built of protein which is derived from the amino acids of dietary protein. The formation of new red cells thus requires an adequate diet. In addition, a dietary intake of iron is required to replace losses. The iron required is not as great as at first seems necessary as, to a large extent, the iron from broken down red cells is re-used in the formation of new ones.

An adult man has 4·5 g. iron (80 mmol) in his body, of which 2·5 g. (45 mmol) is in the blood haemoglobin, and the rest in the myohaemoglobin of muscles and enzymes.

Blood iron consists of:

(*a*) *Plasma iron*—0·1 mg. per cent. (18 μmol/l)—as ferrous iron combined with a β-globulin known as **transferrin**. The total amount of iron that can be carried by this protein (iron binding capacity) is about 0·3 mg. per cent. (54 μmol/l). The transferrin is thus normally about 30 per cent. saturated.

(*b*) *Red cell iron*—in haemoglobin. The percentage of iron in haemoglobin is 0·33 per cent. Thus, 100 ml. blood which contains 14·5 g. Hb contains 50 mg. iron (0·9 mmol).

The 50 ml. blood destroyed daily yields 25 mg. iron (0·45 mmol), and a similar amount is needed for the formation of new cells. The losses in urine, faeces and from the skin amount to about a milligram a day.

Since normally only about 10–15 per cent. of the ingested iron is absorbed, a daily intake of about 5–10 mg. (90–180 μmol) of iron is required for a man. Iron from liver and meat is well absorbed whereas that from leafy vegetables and eggs is poorly absorbed.

Children need iron during growth for the increase in red cell volume. The start of menstruation in girls increases the requirement still further since the menstrual flow represents an absolute loss of iron in blood shed. The daily requirements of women are thus higher than those of men.

With a normal menstruation about 50 ml. blood are lost per month. This contains 25 mg. iron (0·45 mmol). Heavy periods may result in a loss of 240 ml. blood = 120 mg. iron (2·15 mmol), but replacement of this iron loss works out at only 4 mg (72 μmol) per day spread over the whole month. The recommended daily intake for a woman is 18 mg. per day.

The requirements of pregnancy are greater. The baby and its birth represents a loss of 700 mg. iron (12·5 mmol) to the mother and she needs an additional 200 mg. (3·6 mmol) for lactation.

The need is mainly in the second half of the pregnancy and a saving results from the absence of menstrual periods during the pregnancy (amenorrhoea).

A deficiency of iron in the diet leads to *iron deficiency anaemia* with pale red cells deficient in haemoglobin. It is a *hypochromic* anaemia. A trace of copper in the diet is thought to aid the utilization of the iron.

An iron deficiency anaemia is associated with a low plasma iron, an absence of stainable iron in the bone marrow and a low iron content of the red cells [see **M.C.H.C.** p. 14].

Simple iron salts are absorbed better than more complex iron compounds. The body is able to utilize small quantities of metallic iron as is shown by the fact that iron deficiencies were rarer when iron pots and pans were in common use for cooking. Bread has a higher iron content than the constituent wheat, presumably from the steel grinding mills used to grind the corn.

Vitamin B$_{12}$

A form of anaemia exists which does not respond to treatment with iron. If untreated it is often fatal and for this reason is known as *pernicious anaemia*. The red cells are reduced in number and are irregular in size and shape, but on the average are larger than normal. It is a form of *macrocytic* (*megaloblastic*) anaemia.

In 1926 it was found that patients with this condition responded to treatment with liver and liver extracts. Castle put forward the theory that for the development of red cells two factors are needed —the extrinsic factor in the diet and an intrinsic factor produced by the stomach. These two combine to form the *haematinic principle* or *haemopoietic factor*. This factor is stored in the liver, and is sent from there to the bone marrow to aid the normal maturation of the blood cells.

The extrinsic factor has now been isolated and is termed vitamin B$_{12}$ (*cyanocobalamin*). It contains the element cobalt, an element not previously known to be present in the body. The intrinsic factor is a glycoprotein (amino acids + sugars) and has not yet been synthesized. Its presence appears to be necessary to facilitate the absorption of vitamin B$_{12}$ from the intestines.

Macrocytic anaemias are usually due to a failure of absorption of vitamin B$_{12}$ rather than a failure of intake. Patients with pernicious anaemia itself, have an inherent failure on the part of the stomach to produce the intrinsic factor,[1] and without it the absorption of vitamin B$_{12}$ is insufficient. It is, however, effective if absorption from the digestive tract is by-passed by injecting the vitamin B$_{12}$ intramuscularly. The dose required is very small, of the order of 25 micrograms (≡ one millionth of an ounce) per week. A similar quantity is all that needs to be absorbed by a normal person, with the help of the intrinsic factor, to prevent pernicious anaemia.

The liver appears to contain a large store of vitamin B$_{12}$ in a normal person. As a result symptoms of pernicious anaemia may not appear for several years following the operative removal of

[1] Recent work suggests that some patients with pernicious anaemia have developed a specific auto-antibody to their own gastric mucosa [see p. 45].

the stomach (gastrectomy) although such an operation will remove the source of the intrinsic factor.

A similar macrocytic anaemic develops in patients with the fish tape worm, *Diphyllobothrium latum*, which grows to a length of 30 ft. and occupies the whole of the small intestine. Such a worm presumably absorbs the ingested vitamin B$_{12}$ leaving the host with a deficiency.

In cases of sprue and tropical diarrhoea the speed of movement of food through the digestive tract may be too rapid for adequate vitamin B$_{12}$ absorption and, if this is so, a macrocytic anaemia will ensue.

Vitamin B$_{12}$ appears to be necessary for red cell formation because of its ability to make another B vitamin—folic acid—available for use by the bone marrow. Macrocytic anaemias of short duration (e.g. pregnancy) can often be treated by giving folic acid by mouth as an alternative to injecting vitamin B$_{12}$. But vitamin B$_{12}$, in its own right, is essential for the adequate nutrition of nerves, and in this case, folic acid is not an effective substitute.

A deficiency of vitamin B$_{12}$ leads to a degeneration of both the sensory and motor columns in the spinal cord with loss of sensation and paralysis. The condition is known as *subacute combined degeneration of the cord*. A patient with pernicious anaemia may first complain of neurological symptoms.

Oxygen Lack and Red Cell Formation

The stimulus for red cell formation is oxygen lack. Thus, at high altitudes where the partial pressure of oxygen in the inspired air is reduced [p. 78], an increase in the red cell count occurs as part of the process of acclimatization. Similarly, children with congenital heart disease in which the blood in part by-passes the lungs, have arterial blood that is not fully saturated with oxygen. These children often have an increased red cell count due to the oxygen lack. They may also be cyanosed, hence the expression 'blue baby' which is applied to them.

An excess of red cells in the blood is known as *polycythaemia*.

Erythropoietin

Erythropoietin is a humoral agent produced by the kidneys which stimulates red cell formation. It is a glyco-protein in the α-globulin class. It seems likely that the increase in red cell count associated with oxygen lack is brought about by an increase in circulating erythropoietin since irradiation of the kidneys causes anaemia, and an animal with its kidneys removed (nephrectomized) shows no response to oxygen lack.

Destruction of Red Blood Cells

At the end of their life, broken-down red blood cells are removed from the circulation by the reticulo-endothelial system. This system consists of cells scattered throughout the body, which are *phagocytic* and have the property of engulfing foreign particles, such as Indian ink injected into the circulation. They are found in the bone marrow itself, in the liver and in the spleen.

The protein of the cell and haemoglobin molecule is broken down to amino acids which enter the general *amino acid pool* of the body. The iron from the *haem* portion of the molecule is removed and stored as *ferritin* for re-use in future cells. The rest of the haem molecule is converted to the yellow pigment **bilirubin** and the green pigment **biliverdin.**

The colour changes from the dark blue, almost black, colour of reduced haemoglobin through to the green colour of biliverdin to the yellow colour of bilirubin may be seen to be occurring under the skin in a bruise.

The bilirubin molecule is not used in the formation of new haemoglobin for new cells. A fresh haem molecule is synthesized for this purpose. The bilirubin is excreted from the body. The excretory pathway is unusual and of medical importance. Most waste substances of the body's metabolism are excreted via the kidneys in the urine. Bilirubin, and its oxidation-product biliverdin, however, pass via the blood to the liver. They do so combined with protein (pre-hepatic bilirubin). In the liver, they are transferred by the liver cells to the bile-duct as glucuronides (post-hepatic bilirubin) and enter the duodenum. They pass throughout the small and large intestines and are excreted in the faeces.

RETICULO-ENDOTHELIAL SYSTEM

via blood
as BILIRUBIN and BILIVERDIN combined with protein

LIVER

via bile-duct
as BILE PIGMENTS combined with glucuronic acid

some re-absorption

DUODENUM

SMALL INTESTINE

LARGE INTESTINE

Bacterial action

FAECES
STERCOBILINOGEN
STERCOBILIN (brown colouring matter of faeces)

Bacterial action changes the bilirubin to stercobilinogen and stercobilin which are brown in colour and give the stools their characteristic brown coloration. Some reabsorption of the bilirubin glucoronide takes place during the passage along the intestines and is excreted in the urine as urobilinogen. The yellow colour of urine is, however, due to a different pigment, urochrome.

Jaundice

Any blockage or failure of the bilirubin excretory pathway will result in the accumulation of bilirubin in the blood. When this occurs, the patient turns yellow. The condition is known as *jaundice* or *icterus*.

Jaundice occurs when the liver cells are overloaded by the excessive breakdown of red cells such as occurs in malaria. At birth a *physiological jaundice* may occur owing to the relative immaturity of the liver coupled with the breakdown of red cells as the blood count falls from 7,000,000 per mm.³ to 5,000,000 in the first few weeks of life.

The breakdown of red cells associated with a *Rhesus factor mismatch* [p. 15] gives rise to a much more serious form of jaundice (icterus gravis neonatorum).

Jaundice occurs in *infectious hepatitis* [p. 156]. When the liver cells are being attacked by the virus they are apparently unable to transfer the bilirubin from the blood to the bile-ducts.

Jaundice also occurs when the bile-ducts become blocked due to any cause. The commonest cause is a gall-stone which has moved

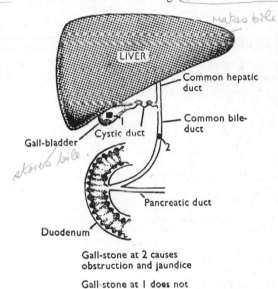

Gall-stone at 2 causes obstruction and jaundice

Gall-stone at 1 does not

FIG. 9. Bile is formed in the liver, stored in the gall-bladder, and passes down the bile-duct to the duodenum. An obstruction to the bile-duct gives rise to jaundice.

into the bile-duct [FIG. 9]. Gall-stones are formed in the gall-bladder following attacks of inflammation (cholecystitis) and are the result of precipitation of one of the bile constituents, usually cholesterol. Whilst in the gall-bladder or in the cystic duct, they do not cause jaundice [FIG. 9 (1)]. This occurs if the stone moves to the common bile-duct [(2)].

Diffusion and Osmosis

If a strong salt solution is separated by a membrane permeable to salt and water from a weak salt solution [FIG. 10], salt passes

FIG. 10. Diffusion. The two salt solutions are separated by a membrane permeable to both salt and water. The salt passes from the strong solution to the weak solution until both solutions have the same strength.

from the strong solution to the weak, and in time the solutions on the two sides of the membrane will have the same strength. This is known as **diffusion.**

If the membrane is now replaced by a second membrane that is permeable to water and not to salt (often called a semi-permeable membrane although *selectively permeable* is more accurate), once again it is found that in time the solution on the two sides will have the same strength. This obviously cannot be due to the passage of salt from the strong solution to the weak since the membrane is impermeable to salt. It is due to the passage of water from the weak solution to the stronger, thus making the weak solution stronger and the strong solution weaker [FIG. 11(*A*) and (*B*)]. Such a phenomenon is known as **osmosis.**

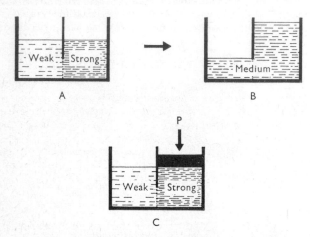

FIG. 11. Osmosis. The solutions are separated by a membrane permeable only to water (A). Water passes from the weak solution to the strong solution until both solutions have the same strength (B). The movement of water may be prevented by applying a hydrostatic pressure **P** (C). This pressure equals the osmotic pressure difference between the two solutions.

In both *diffusion* and *osmosis* the direction of movement is such that the concentrations throughout the solution tend to become equal.

This movement of water in osmosis may be prevented by applying an opposing hydrostatic pressure [FIG. 11(*C*)]. The pressure that is just sufficient to prevent the movement gives the magnitude of the **osmotic pressure.**

The osmotic pressure of a solution depends on the number of *undissociated molecules* or, alternatively, *ions* per litre.

The molecular or ionic weight in grams in 1 litre (1 osmole [p. 140]) gives an osmotic pressure of

$$22\cdot4 \text{ atmospheres} = 22\cdot4 \times 760 \text{ mm. Hg.} \qquad (1)$$

The same solution has a freezing point of $1\cdot85$ °C. below that of water. Depression of the freezing point provides a simple way of calculating the osmotic pressure of a solution.

$$1\cdot85 \text{ °C. depression} \equiv 22\cdot4 \text{ atmospheres osmotic pressure.} \qquad (2)$$

Two solutions which produce no resultant flow through a semi-permeable membrane are said to be **isotonic.** Solutions which are isotonic with a given solution are isotonic with each other. Solutions with a higher osmotic pressure are said to be **hypertonic,** and those with a lower osmotic pressure are termed **hypotonic.**

Haemolysis of Red Cells (Laking)

When red cells are placed in distilled water they swell, burst, and the haemoglobin is released. This breakdown of the cell and the liberation of the haemoglobin is known as **haemolysis.** It is due to osmosis. The cell membrane acts as a semi-permeable membrane, allowing the passage of water into the cell, but not the movement of salts out of the cell. The cell is hypertonic with respect to the distilled water (it behaves as if it contained a strong salt solution), and, therefore, water enters the cell, making it at first more spherical (spherocytosis). Finally, the cell membrane ruptures.

If, on the other hand, red cells are placed in a solution of 0·9 per cent. sodium chloride (NaCl), they do not change their shape. This is because 0·9 per cent. NaCl is isotonic with the cells. Such a solution is termed 'physiological saline' or 'normal saline'. The term 'normal' in this connexion means 'isotonic' with blood. It should not be confused with the term 'a normal solution' which is used to mean *the equivalent weight of a substance dissolved in 1 litre.*

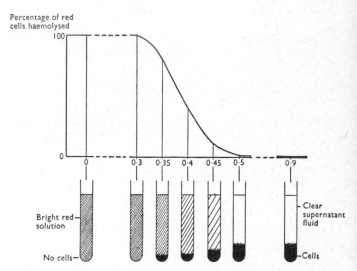

FIG. 12. Fragility of red cells. The percentage of red cells haemolysed increases as the NaCl solution becomes more hypotonic. Haemolysis of normal cells starts at 0·5 per cent. NaCl and is complete by 0·3 per cent. NaCl. The cells become more 'fragile' and haemolyse more easily after storage. An increased fragility is a characteristic feature of the red cell in *acholuric jaundice.* The degree of haemolysis is seen more clearly (*lower*) after centrifugation, or allowing the tubes to stand overnight. (Abscissa = NaCl concentration in g. per cent.)

The osmotic pressure of 0·9 per cent. saline solution may be calculated as follows:

The molecular weight of NaCl is $23 + 35\cdot5 = 58\cdot5$.

An 0·9 per cent. solution = 9 g. NaCl in 1 litre of water.

Thus, from equation 1 above, it exerts an osmotic pressure of

$$2 \times 22\cdot4 \times \frac{9}{58\cdot5} \times 760 \text{ mm. Hg} = 5,100 \text{ mm. Hg.}$$

The factor '× 2' is necessary because the sodium chloride in the solution is completely dissociated into Na^+ and Cl^- ions. Each ion contributes towards the osmotic effect and, therefore, the solution has twice the osmotic pressure of that calculated from the molecular concentration of NaCl.

A 5 per cent. solution of glucose (molecular weight 180) will have a similar osmotic pressure. Both these solutions, if sterile, may be run directly into the veins of a patient.

Hypotonic solutions (less than 0·9 per cent. NaCl) cause the red cells to swell and become more like spheres (spherocytosis). If the hypotonicity is low enough, i.e. less than 0·4 per cent. NaCl, the majority of red cells will be haemolysed [FIG. 12]. This forms the basis of the red cell fragility test.

Hypertonic solutions have the opposite effect on red cells. The cells shrivel up and become dehydrated. This is known as *crenation*.

The cell membrane is a lipoprotein and is destroyed by **fat solvents.** Red cells are haemolysed by soap, saponin, detergents and fat solvents such as ether, chloroform, acetone, carbon tetrachloride and bile salts. Care should therefore be taken not to place blood in vessels that have been cleaned with a detergent, unless all traces of the detergent have been removed.

Freezing causes the disruption of the blood cells by ice crystals and haemolysis on thawing. Blood that has become frozen must never be used for a blood transfusion. The reason for this is partly that potassium ions are released when the cells break down and partly that the haemoglobin released has a deleterious effect on the kidneys. It produces alterations in the blood flow in the glomeruli and blockage of the kidney tubules due to the formation of acid haematin so that complete failure of urine production (anuria) may ensue. Haemoglobin outside the red cell in the plasma is thus a highly dangerous substance.

Determination of the Haemoglobin Content of the Blood

The haemoglobin content may be determined by haemolysing the red cells and then comparing the colour density of the haemoglobin solution obtained with a standard. However, since the colour of haemoglobin depends upon its oxygen content, a more stable haemoglobin derivative is first formed. In the Haldane method, this compound is the cherry-pink carboxyhaemoglobin made by passing carbon monoxide, or more conveniently, coal-gas through the blood. In its simplest form [FIG. 13] colour matching may be carried out by diluting the unknown blood in a calibrated tube until it matches the standard. 100 on the Haldane haemoglobinometer tube represents 100 per cent. haemoglobin with reference to a standard which is 14·8 g. Hb/100 ml. blood (2·2 mmol/l).

Sahli's method is similar but the acid haematin derivative is formed by the addition of hydrochloric acid.

More complex methods of measuring the colour density using photo-electric cell colorimeters, and haemoglobin derivatives, such as cyan-methaemoglobin, acid and alkaline haematin are employed in routine laboratories. Cyan-methaemoglobin is formed by adding the blood to a diluting fluid containing potassium cyanide and ferricyanide.

HAEMATOLOGICAL INDICES

Once the haematocrit, red cell count and haemoglobin content are known, it is possible to deduce further information concerning the red blood cells. We are able, for example, to determine the amount of haemoglobin in each cell, the volume of the cells, and the haemoglobin concentration in the cells.

In iron deficiency anaemia the amount of haemoglobin in each cell is reduced. If such a condition is suspected, it is important to know the haemoglobin content of a single red cell.

Colour Index (C.I.)

$$\text{Colour Index} = \frac{\text{Haemoglobin (per cent. of normal)}}{\text{Red cell count (per cent. of normal)}}$$

This ratio gives a rough indication of the haemoglobin content of the red cells.

Normal for haemoglobin is arbitrarily taken as 14·5 g./100 ml. blood and normal for the red cells as 5 million per mm.³ blood.

The range of colour index is normally 0·9 to 1·1.

Mean Corpuscular Haemoglobin (M.C.H.)

Mean corpuscular haemoglobin is the haemoglobin content of a single red cell measured in absolute units instead of with reference to an arbitrary haemoglobin content and cell count.

The mean corpuscular haemoglobin is calculated by dividing the amount of haemoglobin in 100 ml. blood by the number of red cells in the same volume of blood.

Thus:

$$\text{M.C.H.} = \frac{\text{Haemoglobin in grams per 100 ml. blood}}{\text{Number of red cells per 100 ml. blood}}$$

The top line is equal to approximately 15 g. Hb per 100 ml. blood in a normal person. The bottom line is obtained by multiplying the red cell count per mm.³ by 10^5 since there are 100,000 mm.³ per 100 ml. This works out at approximately $5,000,000 \times 10^5 = 0.5 \times 10^{12}$ cells/100 ml. blood. The quotient or M.C.H. is thus 30×10^{-12} g. Hb per cell = 30 pg. (since 10^{-12} g. = 1 pg.).

The result is expressed in micromicrograms of Hb per cell.

The normal range = 27−32 pg.

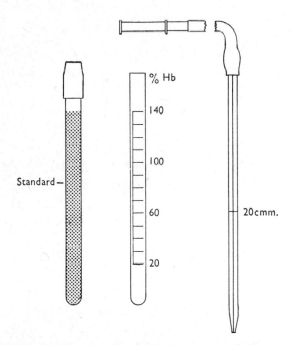

FIG. 13. Haldane haemoglobinometer. Blood is sucked up to the 20 cmm. mark on the pipette. The outside is wiped clean and the contents carefully expelled below the diluting fluid in the calibrated tube. Coal-gas is bubbled through to form carboxyhaemoglobin. Further diluting fluid is added until the colour matches the standard.

The diluting fluid contains 0·4 per cent. ammonia to haemolyse the red cells and to prevent cloudiness due to the precipitation of protein.

Mean Corpuscular Haemoglobin Concentration per cent. (M.C.H.C.)

Both the colour index and the mean corpuscular haemoglobin suffer from the limitation that they take no account of cell size. Thus the value may be low either because the cells are small or because the haemoglobin concentration is low.

The M.C.H.C. is a measure of the haemoglobin in a 100 ml. of packed red cells (not blood). It is expressed as a percentage.

$$\text{M.C.H.C.} = \frac{\text{Haemoglobin in grams per 100 ml. blood}}{\text{Volume of red cells in ml. per 100 ml. blood}}$$
$$\times 100 \text{ per cent.}$$

$$= \frac{\text{Haemoglobin content}}{\text{Haematocrit}} \times 100 \text{ per cent.}$$

The top line is about 15 g. Hb per 100 ml. blood in a normal person. The bottom line of this equation is given by the haematocrit, i.e. 45 ml. The quotient is, therefore, $\frac{15}{45} \times 100 = 33$ per cent.

The normal range is 32–38 per cent. Less than 30 per cent. usually indicates an iron deficiency anaemia.

Mean Corpuscular Volume (M.C.V.)

In anaemias of the pernicious anaemia type, the red cells, although they are fewer in number and more irregular in size, are on the average larger than normal. It is a macrocytic anaemia. If such an anaemia is suspected, we wish to know the volume of an average red cell. The mean volume of a single red cell is termed the mean corpuscular volume (M.C.V.). It is expressed in cubic microns (μ^3).

$$\text{M.C.V.} = \frac{\text{Volume of red cells in ml. per 100 ml. blood}}{\text{Number of red cells per 100 ml. blood}}$$

The top line is the haematocrit, i.e. about 45 in a normal person. The bottom line is of the order of 0.5×10^{12} as seen above. The quotient is, therefore, of the order of 90×10^{-12} ml. $= 90\mu^3$

$$\left[1\mu = \frac{1}{1,000} \text{ mm.} = \frac{1}{10,000} \text{ cm. Therefore } 1\mu^3 = \frac{1}{10^{12}} \text{ ml.} = 10^{-12} \text{ ml.} \right]$$

The normal range is 80–94 μ^3.

BLOOD GROUPS

When a moderate haemorrhage occurs (up to 1 litre loss) the red cells are replaced during the subsequent weeks and all that is needed is that the diet should provide an adequate intake of iron. With a larger haemorrhage, and particularly when the haemoglobin percentage has fallen to below 40 per cent., a blood transfusion is indicated.

If blood of an incompatible group is transfused, the cells of the given blood agglutinate, that is, they stick together to form clumps of red cells. Such an agglutination may have severe consequences. The clumps of agglutinated red cells block capillaries and other small blood vessels and the patient complains of violent pains. These clumped red cells are haemolysed releasing a large amount of haemoglobin into the plasma. As has been seen above, anuria may follow from kidney failure.

Until 1900 human blood transfusions often had fatal results. Then Landsteiner introduced the concept of blood groups, which forms the basis on which blood transfusions are now given.

The principal blood group system is based on the presence or absence of two mucopolysaccharides known as agglutinogens which are carried on the surface of all the red cells in an individual. These agglutinogens are termed A and B. Anyone with agglutinogen A on all his red cells is classified as belonging to blood group A. Those with agglutinogen B belong to Group B. Those with both A and B belong to Group AB. Those with neither A nor B belong to Group O (usually pronounced as the letter O rather than the figure nought or zero) [FIG. 14]. These blood groups are

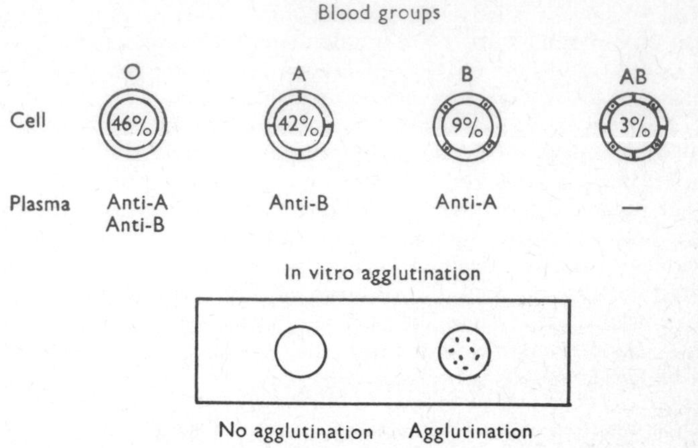

FIG. 14. The four main blood groups are based on the presence of the agglutinogens O, A, B and AB on the red cells. The agglutinins Anti-A and Anti-B, when present, are found in the plasma.

inherited from one's parents according to the Mendelian laws. A and B are dominants,[2] O is recessive.

When cells with agglutinogen A are transfused into a recipient whose plasma contains the agglutinin Anti-A, agglutination of the cells occurs. It is the cells given that are agglutinated.

Anti-A is found in the plasma of Group O and Group B subjects. Anti-B is found in the plasma of Group O and Group A. These agglutinins appear shortly after birth. They decrease with advancing age. The agglutinins are immunoglobulins [p. 22].

The distribution of the four groups amongst the population is therefore:

Blood group	O 46 per cent.	A 42 per cent.	B 9 per cent.	AB 3 per cent.
Cells	—	A agglutinogen	B agglutinogen	A and B agglutinogens
Plasma	Anti-A and Anti-B agglutinins	Anti-B agglutinin	Anti-A agglutinin	—

Agglutinogen A + Agglutinin Anti-A → agglutination
Agglutinogen B + Agglutinin Anti-B → agglutination

To safeguard against agglutination it is desirable **in a transfusion to give the patient blood of the same group as his own.** If this is not possible then blood may be used that will not lead to an agglutination reaction between the donor's cells and the recipient's plasma.

[2] The units of heredity or *genes* occur in pairs. One gene is obtained from the father and one from the mother. If one gene suppresses the effect of the other, it is described as *dominant*.

Thus blood transfusion may be given along the lines indicated by the arrows:

Thus, if the correct group is not available, Group O blood may be given to any of the other three groups. Group O is known as 'universal donor blood'. A and B may be given to AB, but not to O. AB blood can only be used for a transfusion to an AB recipient. An AB patient can receive blood from any group and is known as a 'universal recipient'.

With a limited transfusion, the effect of the donor's plasma on the recipient's red cells may be ignored as the dilution of the donor's plasma by the recipient's plasma will lower the concentration (titre) of the agglutinins to a level below which they become inoperative. In other words, if Group O blood is given to a Group A patient, the Anti-A agglutinin present in the Group O plasma will not reach a high enough titre in the recipient's circulation to affect the recipient's own Group A cells.

ABO system agglutination will take place *in vitro* at room temperature and this enables the blood group of unknown blood to be readily determined. The unknown red cells are added to serum containing Anti-A and to serum containing Anti-B agglutinins. These are obtained from Group B and Group A donors respectively. Gentle rocking causes the red cells to clump if the corresponding agglutinogen and agglutinins are present. There are four possible results for this test, one for each of the possible blood groups. These are:

	Anti-A serum	Anti-B serum
Unknown blood if:		
O	–	–
A	+	–
B	–	+
AB	+	+

+ = Agglutination.

Rhesus Factor

In 1940 another important blood group was discovered and has been termed the *Rhesus factor* or *Rh system*.

In addition to agglutinogens A and B, three further agglutinogens C, D and E occur in association with the red cells. D is the chief and when present, the cells are said to be Rhesus positive. 85 per cent. of the population have agglutinogen D. The remaining 15 per cent. of the population are without D and are Rhesus negative. All combinations of O, A, B, AB grouping with Rhesus positive and Rhesus negative occur.

Unlike the previous ABO system, no preformed Rhesus agglutinin (Anti-D) is found. However, a Rhesus negative person, and only a Rhesus negative person, can make Anti-D following sensitization with Rhesus positive blood. Thus, during the first transfusion of a Rhesus negative patient with Rhesus positive blood no external signs of a mis-match are seen. The production of Anti-D slowly occurs and once Anti-D has been formed any such Rhesus

positive blood will be agglutinated by the Rhesus negative recipient even though the ABO grouping is correct. From then onwards only Rhesus negative blood of the correct ABO grouping must be used.

The Rhesus positive individuals cannot form Rhesus agglutinins and for them it is immaterial whether they receive Rhesus positive or negative blood.

Thus, the following transfusion is permissible:

$$Rh- \rightarrow Rh+$$

The situation with the Rhesus system may be summarized as follows. Rhesus positive patients can receive blood of either group. Rhesus negative patients will become sensitized if given the wrong group, namely Rhesus positive. In the case of a man, the only untoward effect will be that he may develop Anti-D, in which case his correct group must be used for future transfusions. The position with regard to a Rhesus negative woman is very different.

In the case of the female of child-bearing age or younger every care must be taken to prevent sensitization and the formation of Anti-D. This means that no Rhesus negative woman should ever be given Rhesus positive blood. The reason for this is that the formation of Anti-D in the plasma may prevent the woman from having a viable child.

Rhesus factor is an inherited dominant and a person may be a homozygous Rhesus positive D.D. or a heterozygous Rhesus positive D.d. Rhesus negative is only possible with complete absence of D (i.e. homozygous d.d.). (d. is the allelomorphic gene which replaces the gene carrying the D factor.)

If we label the paternal genes 1, 2 and the maternal genes 3, 4, then the possible combination of genes in the child is given by:

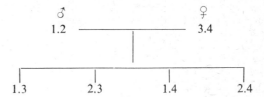

Applying this principle to the inheritance of the Rhesus factor by the child of a Rhesus negative mother:

Homozygous Rhesus positive father

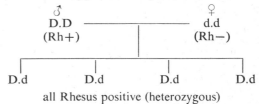

all Rhesus positive (heterozygous)

Heterozygous Rhesus positive father

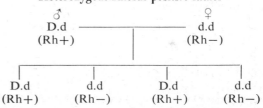

Thus if a Rhesus negative woman marries a homozygous Rhesus positive man all their children will be Rhesus positive. This is a comparatively common occurrence, and is in itself of no significance. The baby *in utero* has its own heart and circulation and its blood cells are independent of the maternal blood cells [FIG. 15(*A*)].

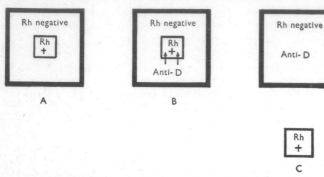

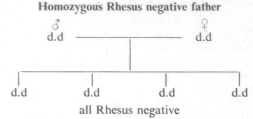

FIG. 15. A Rhesus negative mother may have a Rhesus positive baby *in utero* without untoward consequences since the two circulations are separate (A). Should the mother become sensitized to Rhesus positive blood and develop **Anti-D** in her plasma, this will pass across the placental barrier into the foetus and destroy the red cells of the foetus (B). Once the baby has been born no further **Anti-D** will enter its circulation (C). An exchange transfusion with Rhesus negative blood is given as a temporary measure to stop further haemolysis. Later the baby makes its own Rhesus positive cells, but by that time all the **Anti-D** will have disappeared. If the baby is Rhesus negative its cells are unaffected by the **Anti-D** in the mother.

The placenta allows the passage of food substances to the baby and the removal of waste products, but does not, under normal conditions, allow the passage of red cells from one circulation to the other. However, if the maternal circulation contains **Anti-D** (which has no effect on the mother's cells which are without **D**), the **Anti-D** can pass through the placenta into the baby's circulation where it will meet **D** cells, which will be haemolysed and destroyed [FIG. 15(*B*)]. If sufficient **Anti-D** is present, the foetus will not survive. If it is born alive it will be severely jaundiced—*icterus gravis neonatorum*. Once born the baby is removed from the sources of **Anti-D** [FIG. 15(*C*)]. An exchange transfusion with Rhesus negative blood is usually carried out. The baby's **ABO** grouping is determined and the correct **ABO** group blood is used. At birth the **ABO** agglutinogens are present on the red cells, but the **Anti-A** and **Anti-B** agglutinins in the plasma are not detectable until later in the first year of life. If it not possible to establish the baby's **ABO** grouping, Group O Rhesus negative blood is employed. This Rhesus negative blood will be unaffected by any remaining **Anti-D**. Over the course of the next few weeks the baby's blood will revert to cells which will be Rhesus positive, but, by then the **Anti-D** will have disappeared from the circulation.

Anti-D may develop in Rhesus negative women following pregnancies in which the babies are Rhesus positive, without any obvious Rhesus positive transfusion having been given at any time. In these cases slight placental damage may have led to some of the baby's Rhesus positive blood entering the maternal circulation.

[**Anti-D** has been given to Rhesus negative women immediately after child-birth to destroy any possible Rhesus positive cells from the baby which may have entered the mother's circulation during the separation of the baby's placenta from the mother's uterus. If these Rhesus positive cells are not destroyed, the mother will herself develop **Anti-D** which will persist and affect a subsequent

pregnancy whereas the injected **Anti-D** will have disappeared before then.]

With a heterozygous Rhesus positive father and a Rhesus negative mother then according to the Mendelian laws of inheritance, half the babies will be Rhesus positive, but half will be Rhesus negative and these will, therefore, be unaffected by any maternal **Anti-D**.

Homozygous Rhesus negative father

♂ d.d ——————— ♀ d.d

d.d   d.d   d.d   d.d

all Rhesus negative

With a Rhesus negative father and a Rhesus negative mother all the children will be Rhesus negative, and sensitization is then unimportant. Ideally, a sensitized girl should find a Rhesus negative husband. Unfortunately, he may not be so easy to find as only one man in seven is Rhesus negative.

WHITE BLOOD CELLS (LEUCOCYTES)

White blood cells in the circulation are not white in the sense that a sheet of white paper is white, but in the sense that they are transparent and not coloured. White cells are fewer in number than red cells. There are 600 red cells to every 1 white cell, giving a white cell count of 8,000/mm.³ (range 4–11,000).

The white cells have the power of amoeboid movement. They can pass through the capillary walls into the tissue spaces.

In blood vessels the white cells are seen to be moving slowly along the sides of the vessels out of the main axial stream which contains the red blood cells.

Unlike red cells, white cells are not necessarily the 'end of the line'. Under suitable conditions, they can grow, divide and produce offspring cells. They can be grown in tissue-culture.

The White Cell Count

White cells are counted in a similar manner to red cells, using a haemocytometer, but since they are less numerous the blood is only diluted to 1 in 20 instead of 1 in 200 as was the case with the red cells. A white cell counting pipette has graduated marks at 0·5 and 1·0 on the stem and another at 11 above the bulb [see FIG. 8(*A*), p. 8]. The bulb has a volume of 10 units. Blood is sucked up to the 0·5 mark followed by fluid to make a total of 11 units. The 0·5 units of blood in the bulb have been diluted by 9·5 units of diluting fluid giving a dilution of 1 in 20. The diluting fluid in the stem is discarded.

The diluting fluid contains acetic acid to haemolyse and destroy the red cells, and a stain to show up the otherwise transparent white cells.

A typical diluting fluid is composed of:

Glacial acetic acid . . .	1·5 ml.
Malachite Green (saturated solution)	0·5 ml.
Water	98·0 ml.

The white cell 'counting area' is employed. Each square in this area has a side of 1 mm. Since the depth is $\frac{1}{10}$ mm. the volume of fluid in this area is $\frac{1}{10}$ mm.³ Four such areas are counted. If the total number of white cells counted is z, this is the number of cells in $4 \times \frac{1}{10}$ mm.^{3} of diluted blood. The number of cells in 1 mm.^{3} of undiluted blood will be $20 \times \frac{10}{4} \times z = 50z$. The number of cells counted is multiplied by 50 to give the white cell count per cubic millimetre.

It will be noted that since white cells can leave the circulation and enter the tissue spaces, a white cell count, unlike a red cell count, does not give an indication of the total number of cells in the body, but only of the number that are in transit from the bone marrow to the tissues.

Stained Blood Film

If a blood film is stained, five different types of white cell may be distinguished. The stains usually employed are an acidic red dye such as eosin and a basic blue dye such as Methylene Blue. The white cells are divided firstly into *granulocytes*, which have granules in their cytoplasm, and **lymphocytes** and **monocytes** which do not [FIG. 16]. The granulocytes are then further subdivided into

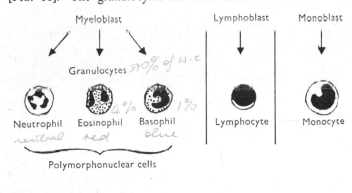

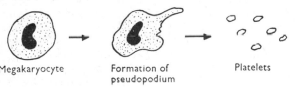

.FIG. 16. White cells and platelets. The granulocyte cells develop from myeloblasts. The lymphocytes develop from lymphoblasts. The monocytes develop from monoblasts. The granulocytes are subdivided by the staining reaction of their granules.

neutrophils, **eosinophils** and **basophils** according to the staining affinity of these granules. In the eosinophil cells the granules take on the bright red appearance of the eosin. In the basophils the granules take on the bright blue of the basic dye. Neutrophils are neutral in appearance as the granules are faintly stained by both dyes.

Granulocytes

The granulocytes are formed in the bone marrow from myeloblast cells which divide into myelocytes and finally into the circulating granulocytes. Seventy-five per cent. of the bone marrow is engaged in myeloid activity (white cell formation) and only 25 per cent. is engaged in erythroid activity (red cell formation). Although white cells are present in fewer numbers than red cells, their life is in general shorter, only a few days in the case of the granulocytes. White cells disappear in a few hours when blood is taken for a transfusion.

The nucleus of the granulocyte cell is divided into lobes or segments. Such cells are termed *polymorphonuclear leucocytes* or

'polymorphs' for short. This lobulation is shown most clearly in the neutrophils in which the lobulation increases throughout life from two lobes to five or more. The nucleus of the eosinophil cell is usually in two lobes, sometimes three. The basophil nucleus is not so clearly lobulated as the other two but shows a constriction across the centre. Thus it is useful to remember, when examining a stained blood film, that a multi-lobulated white cell (with 3, 4 or 5 lobes to its nucleus) is almost certainly a neutrophil and not a basophil, eosinophil, lymphocyte or monocyte.

Granulocytes constitute 70 per cent. of the white cells in the adult. Of the 70 per cent. 1 per cent. will be basophils, 4 per cent. eosinophils and the remainder neutrophils.

A deficiency of granulocytes in the circulating blood is known as *granulocytopenia*, that of neutrophils, *neutropenia* and a complete disappearance of granulocytes is called *agranulocytosis*.

Neutrophils

Neutrophil polymorphonuclear cells are phagocytes, that is they have the power to engulf and digest foreign particles and bacteria. They form one of the first lines of defence of the body against infection. The granules contain digestive enzymes and the cell dies when they have been used up.

The pus of an abscess contains a large number of living and dead polymorphs as well as bacteria and necrotic tissue. The polymorphs will be in a high concentration in the surrounding tissues having migrated there from the nearby blood capillaries.

Eosinophils

The percentage of **eosinophils** is increased in allergic conditions, such as asthma and hay fever, and when parasites are present in the body. The eosinophils congregate round liberated histamine and have antihistamine properties. The level of eosinophils in the blood is regulated by hormones from the adrenal cortex.

The eosinophil granules contain profibrinolysin [p. 20] which may be released at the site of fibrin deposition to dissolve away the clot.

Eosinophils appear to have a very short life, 12 hours in blood, 20 hours in tissues.

Basophils

Basophils are considered to be circulating mast cells. The mast cells, which are also found in the lining of blood vessels, produce heparin and histamine. The histamine may be released as a result of an antigen-antibody reaction. In some species the mast cells also produce 5-hydroxytryptamine but not in man.

Lymphocytes

Lymphocytes are present in the circulation in two forms. Large lymphocytes with a diameter of 10 μ which become small lymphocytes with a diameter of 8 μ. Together they constitute 25 per cent. of the white cells in the adult. In the child the lymphocytes are more numerous and their percentage may exceed that of the granulocytes. The small lymphocyte has the largest nucleus for its size of any cell in the body and is very rich in DNA [p. 116].

In addition to being made in the bone marrow, lymphocytes are also made by the lymphoid tissue of the body in the lymph nodes, tonsils, spleen and in the **thymus**. They proliferate in a chronic infection. Such an infection is of long duration. In tuberculosis for example, after several months' duration the tubercule bacilli are found to be surrounded by lymphocytes.

Lymphocytes recirculate between blood and lymph every few hours. Some (*B*-lymphocytes) have a short life of a few hours. Others (*T*-lymphocytes from the thymus) live 200 days or longer. Obviously, if a lymphocyte divides its life-span could theoretically be infinite.

Lymphocytes play an important part in the formation of protective substances, antibodies, which are released by *B*-lymphocytes (and their offspring, *plasma cells*) into the circulation where they are found mainly in the γ-globulin fraction of the plasma proteins [see pp. 22, 45 and thymus, p. 157].

Monocytes

Five per cent. of the circulating white cells are large cells which have a kidney-shaped nucleus. These cells are called monocytes. They are formed in the bone marrow alongside the granulocytes, lymphocytes and red cells. The monocytes are actively phagocytic.

Differential White Cell Count

A blood smear is first obtained by placing a drop of blood near one end of a glass slide. Another slide, held at an angle of 45 degrees, is moved towards the drop from the other end until it just touches and the blood spreads along its end; it is then pushed along the length of the other slide to the end, carrying a film of blood with it [FIG. 17]. This film is now allowed to dry.

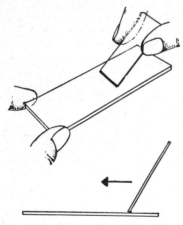

FIG. 17. Making a blood smear for a differential white cell count.

For a differential white cell count the blood smear is fixed and stained. Several hundred white cells are examined microscopically by traversing the slide in a systematic fashion, and classified according to their type. The percentage of each type is then calculated. If the absolute number of each type of white cell per mm.[3] is required, it is calculated by multiplying this percentage by the total white cell count. Thus if the white cell count is 8,000 per mm.[3] then,

			Absolute Figures *per* mm.[3]
Neutrophils	65 per cent.	=	5,200
Eosinophils	4 per cent.	=	320
Basophils	1 per cent.	=	80
Lymphocytes	25 per cent.	=	2,000
Monocytes	5 per cent.	=	400
Total white cell count		=	8,000

PLATELETS—THROMBOCYTES—AND THE CLOTTING OF BLOOD

Platelets are made from the large megakaryocyte cells (40 μ diameter) in the marrow [FIG. 16]. These cells are thought to extend a pseudopodium into the blood sinusoid. The tip is nipped off and circulates in the blood as a non-nucleated platelet. Platelets are 2–3 μ in diameter and there are 250,000 platelets per mm.[3]. A deficiency in the circulating blood is called *thrombocytopenia*.

Platelets aggregate and break down at a site of injury or when they come in contact with a foreign surface. The released 5-hydroxytryptamine (serotonin) causes vasoconstriction. The platelet aggregation may arrest bleeding by forming a platelet plug. This is important in the arrest of capillary bleeding.

Clotting of Blood (Coagulation)

Blood clots or coagulates when the soluble plasma protein *fibrinogen* changes into insoluble *fibrin*. Fibrin forms a network of long strands which entrap blood cells. These strands are sticky so that the clot adheres to the surrounding tissues. At first the clot is soft and jelly-like. Gradually it retracts (contracts down) and exudes a fluid known as **serum.** If exposed to the air the clot hardens. Clot retraction starts after about 10 minutes and is complete in 2 hours. Serum has the same composition as plasma without the fibrinogen (and Factors V and VIII).

It will be seen that clotting concerns the plasma. Plasma without cells will clot. The cells present, with the exception of the platelets, play an entirely passive role and are not essential for the clot formation. Care should be taken not to confuse 'clotting' with 'agglutination' or 'clumping' which concerns primarily the red cells.

The fibrinogen is a natural constituent of the plasma. It only changes into fibrin when *thrombin* is formed. [Thrombin is a protein-splitting (proteolytic) enzyme which breaks down the large fibrinogen protein molecule into smaller units which then 'polymerize' into fibrin.] Thrombin does not exist as such in the plasma, but is present as the precursor substance prothrombin which is a plasma protein. The conversion of prothrombin to thrombin takes place in the presence of *calcium ions* and *thromboplastin* (thrombokinase).

Thromboplastin is not present in the circulation, but is formed when the platelets break down and when tissues are damaged.

The classical theory of blood clotting (Morawitz, 1904) is, therefore, that blood clotting occurs in three stages.

Stage 1. The formation of thromboplastin (thrombokinase).
Stage 2. Conversion of prothrombin to thrombin by the thromboplastin in the presence of Ca^{++}.
Stage 3. Conversion of fibrinogen by the thrombin to fibrin.

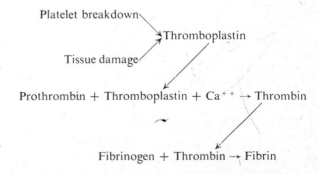

Platelet breakdown
Tissue damage
→ Thromboplastin
Prothrombin + Thromboplastin + Ca^{++} → Thrombin
Fibrinogen + Thrombin → Fibrin

Recent work has shown that the formation of thromboplastin from platelet breakdown and tissue damage are complicated processes involving many additional 'factors'.

The clumping of platelets which precedes their breakdown is facilitated by Factor XI (plasma thromboplastin antecedent, PTA) and Factor XII (Hageman factor). The platelet breakdown produces a *platelet factor* with the aid of Factor VIII (anti-haemophilic globulin, AHG) and the Factor IX (the Christmas factor or plasma thromboplastin component, PTC). AHG is the substance absent in haemophilia, the bleeding disease. The Christmas factor is named after the patient who was first shown to have a deficiency; it also produces a bleeding disease when absent. The platelet factor, with the aid of Factor V (the labile factor or accelerator globulin AcG) and Factor X (the Stuart-Prower factor) form the thromboplastin (blood thromboplastin). This is known as the intrinsic system.

An alternative form of thromboplastin, known as tissue thromboplastin (prothrombinase), is formed when the tissues are damaged. This extrinsic system needs Factors V and X, and another factor, Factor VII (the stable factor or serum prothrombin conversion accelerator, SPCA), but Factors VIII, IX, XI and XII are not required.

Factor XIII (fibrin stabilizing factor) is needed in both systems for the formation of cross-linkages when the fibrinogen is converted into fibrin.

When blood is stored at 4 °C. Factor VIII (antihaemophilic globulin) is destroyed. To obtain Factor VIII, fresh plasma is frozen to −20 °C. and then thawed out at 4 °C. Factor VIII does not redissolve and may be centrifuged off.

The factors associated with blood clotting are given in TABLE 2.

TABLE 2

Factor I	Fibrinogen
Factor II	Prothrombin
Factor III	Tissue thromboplastin
Factor IV	Calcium ions
Factor V	Labile factor (AcG)
	There is no Factor VI
Factor VII	Stable factor (SPCA)
Factor VIII	Antihaemophilic globulin (AHG)
Factor IX	Christmas factor (PTC)
Factor X	Stuart-Prower factor
Factor XI	Plasma thromboplastin antecedent (PTA)
Factor XII	Hageman factor
Factor XIII	Fibrin stabilizing factor

Thrombus Formation

A clot occurring in the circulation is called a *thrombus* and the clinical condition is known as *thrombosis*. A thrombus forms when a change occurs in the smooth inner lining of the vessel wall. Due to a degenerative change or due to injury a roughening occurs and the platelet adheres to this area. Thromboplastin is liberated and a clot is formed.

A thrombus forming in, for example, a coronary vessel supplying the heart muscle (a coronary thrombosis), may have very serious consequences and is one of the causes of sudden death in middle age.

A thrombus which becomes detached is termed an embolus. Thrombosis of the leg veins is one of the complications that may occur after an operation. If such a thrombus becomes detached it will pass through the right atrium and ventricle of the heart to lodge in the blood vessels of the lungs giving rise to a *pulmonary embolus*. Stagnation increases the likelihood of thrombus formation.

To Prevent Clotting—Anticoagulants

When taking blood for a future blood transfusion it is essential to prevent it clotting. Thromboplastin formation is retarded by making sure that the blood is taken from the veins of the donor with the minimum of tissue damage, and that only smooth lined tubing and vessels are used. Glass is much more active than polythene and silicone in promoting platelet breakdown and the formation of thromboplastin. If glass tubing has to be used it can be coated with silicone to reduce the likelihood of clot formation.

Thromboplastin only converts prothrombin into thrombin in the presence of ionized calcium, a point which is made use of when blood is collected for blood transfusion storage. Acid sodium citrate (disodium hydrogen citrate) is added to the blood. This removes the ionized calcium by forming an un-ionized sodium calcium citrate complex. Acid sodium phosphate may also be added as a buffer. Glucose (dextrose) is added as food for the red cells. These combinations are referred to as acid citrate dextrose (ACD) blood and citrate phosphate dextrose (CPD) blood.

The blood is stored at 4° C. Under these conditions the red cells have a usable life of about 3 weeks. After this time so much haemolysis will have taken place (due to red cells disintegrating) that the blood will be unsafe to use. During storage the white cells, platelets and antihaemophiliac globulin disappear but the fibrinogen and Christmas factor remain. Much of the potassium which has leaked out of the red cells during storage, returns to the cells after retransfusion.

When blood is not to be used for transfusion, fluoride or oxalate may be used which will precipitate the calcium as insoluble calcium fluoride or calcium oxalate, and have certain advantages when biochemical determinations are to be made. Fluoride, for example, prevents any change in blood glucose content whilst the sample is awaiting analysis. The di-sodium salt of ethylenediamine tetra-acetic acid (EDTA) may also be used to remove the calcium ions.

$$\begin{array}{ccc}
\text{HOOC.CH}_2 & \text{H} \quad \text{H} & \text{CH}_2.\text{COOH} \\
& | \quad\; | & \\
\text{N} - \text{C} - \text{C} - \text{N} & & \\
& | \quad\; | & \\
\text{HOOC.CH}_2 & \text{H} \quad \text{H} & \text{CH}_2.\text{COOH}
\end{array}$$

EDTA

Heparin prevents the clotting of blood by inhibiting the conversion of fibrinogen to fibrin as well as prothrombin to thrombin. It occurs naturally in the body, being made by the mast cells, but it is doubtful whether the circulating heparin plays any part in the prevention of blood clotting intravascularly. Its function may be that of fat clearing. Heparin injected intravenously every 5 hours will, however, increase the clotting time and it is used for this purpose in patients in the hope of preventing the spread of an intravascular thrombosis. It could be used in place of citrate for a blood transfusion, but is seldom used on the grounds of expense.

Oral Anticoagulants

Prothrombin is made in the liver and vitamin K is necessary for its formation. But a dietary intake of vitamin K is seldom necessary as sufficient of this vitamin is synthesized by the bacterial flora living in the intestines. Vitamin K deficiencies may occur at birth when the intestines are sterile, and when the gut has been sterilized by the use of antibiotics. The deficiency at birth is suspected when bleeding continues from the cut end of the umbilical cord.

O

CH₃

Phytyl

Vitamin K

Substances such as dicoumarol and phenindione (*Dindevan*) inhibit the formation of prothrombin by substrate-competition. They also depress the synthesis in the liver of Factors VII, IX and X. These substances are active when given by mouth, and are used to reduce the likelihood of an intravascular thrombosis.

OH   CH₂   OH

O   O   O   O

Dicoumarol

O

O

Phenindione

To Speed up Clotting

Blood may be made to clot by the addition of the substance thrombin which is available for this purpose. The addition of thrombin by-passes the first two stages of the clotting reaction and, therefore, clotting will occur even if the calcium ions are absent. Moderate heat may be employed to accelerate the clotting process, hence the hot swabs used by surgeons.

Bleeding Time and Clotting Time

If a prick is made, say on the lobe of the ear, and the blood is removed every 15 seconds with filter paper, bleeding ceases in 2–6 minutes. This is not due to clotting, but is due to spasm of the capillaries and the formation of platelet plugs. The time taken for bleeding to cease is known as the *bleeding time*. If on the other hand, blood is placed in a glass tube and rocked in a water bath at 37 °C., the blood stops flowing in less than 10 minutes due to true blood clotting. The *clotting time* is, therefore, a test of the clotting mechanism and particularly of the amount of thromboplastin formed.

Fibrinolysis

In time the fibrin of a thrombus is dissolved away and the circulation in the blood vessel re-established. This is brought about by the enzyme **fibrinolysin (plasmin).**

Fibrinolysin is a protein-splitting enzyme which resembles the pancreatic digestive enzyme trypsin [p. 104]. It breaks the insoluble fibrin down to smaller molecules which are soluble. The activation of fibrinolysin is complex. It is present in the plasma as an inactive precursor **profibrinolysin (plasminogen)** which is converted into fibrinolysin when **plasminogen activator** is present.

Profibrinolysin   <u>plasminogen activator</u>   Fibrinolysin
(Plasminogen)   →   (Plasmin)

Plasminogen activator can be extracted from sweat, tears and urine. The action of any fibrinolysin formed in the normal blood is suppressed by the presence of a further substance **plasmin inhibitor.**

PLASMA

The straw-coloured fluid plasma is composed of water (91 per cent.), plasma proteins (7 per cent.) plus inorganic salts, mainly sodium chloride and bicarbonate (0·9 per cent.) and substances being transported from one part of the body to another. It is a slightly alkaline fluid with a pH of 7·4 [see Chapter 7].

Plasma Proteins

The proteins present in the plasma consist of:

Albumin	.	4·5 g./100 ml. plasma (45 g./l)
Globulin	.	2·7 g./100 ml. plasma (27 g./l)
Prothrombin	.	0·04 g./100 ml. plasma (0·4 g./l)
Fibrinogen	.	0·25 g./100 ml. plasma (2·5 g./l)

Albumin and globulin are proteins that are coagulated by heat in a similar way to the white of an egg which solidifies when the egg is boiled. This property is used to detect the abnormal presence of plasma protein in the urine. The urine is boiled at the correct pH and any protein present is precipitated.

The division between albumin and globulin is based on molecular weight. The albumin molecules are smaller than the globulin molecules. The globulins are further subdivided into α_1, α_2, β and γ-globulin fractions. The γ-globulin fraction contains the antibodies.

Albumin has a molecular weight of 69,000. The globulins have molecular weights of up to 156,000 for the γ-globulin fraction. Fibrinogen has a molecular weight of 400,000.

Separation of Plasma Proteins

The division into albumin and globulin was originally made on the basis of precipitation by salts. The globulin fraction is precipitated by a half-saturated ammonium sulphate solution, whereas a fully saturated ammonium sulphate solution is needed to precipitate the albumin fraction.

The separation nowadays is made in the ultracentrifuge where the different proteins separate out according to their densities, by fractional precipitation at low temperatures, or by electrophoresis.

The basis of the electrophoretic method is as follows:

A protein molecule is built up of amino acids and the molecule has free amino (—NH₂) and carboxyl (—COOH) groups:

COOH
|
PROTEIN
|
NH₂

A protein will combine with an acid such as HCl. The —NH₂ group ionizes and forms —NH₃⁺ Cl⁻. Thus in acid solutions a protein acts as a base and the ionized protein molecules carry a positive charge:

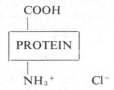

COOH
│
┌─────────┐
│ PROTEIN │
└─────────┘
│
NH_3^+ Cl^-

The protein will also combine with a base such as NaOH. This time the —COOH group ionizes and forms —COO^- Na^+. Thus in alkaline solutions the protein acts as an acid and when ionized carries a negative charge:

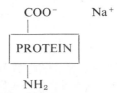

COO^- Na^+
│
┌─────────┐
│ PROTEIN │
└─────────┘
│
NH_2

At an intermediate pH, known as the *iso-electric point*, the molecule carries an equal number of positive and negative charges and is electrically neutral.

The plasma is alkaline and, therefore, the plasma proteins will be acting as acids and will carry a negative charge. If an electric current is passed through a solution of these proteins, the molecules will move towards the positive electrode, the anode.

The rate of movement varies for the different proteins and thus separation may be affected. A modification of this technique is to cause the movement of the proteins to occur along a strip of filter paper, a technique known as *paper electrophoresis*.

Formation of Plasma Proteins

The plasma albumin, α and β-globulin and the blood clotting factors, prothrombin and fibrinogen, are formed in the liver. The globulin fraction is also formed by cells of the reticulo endothelial system and their derivatives, such as plasma cells and lymphocytes.

The albumin/globulin (A/G) ratio is usually 1·7:1. Different methods of separation, however, give slightly different separations of the protein molecules into the two groups and, therefore, different A/G ratios. Electrophoretic methods, for example, give a ratio of 1·2:1.

In liver disease the ratio may be reversed, due to the reduction in albumin formation. Many empirical liver flocculation tests such as colloidal gold, and thymol turbidity are based on this alteration.

An increase in plasma viscosity and an increase in fibrinogen form the basis of the erythrocyte sedimentation test for the progress of an illness.

Erythrocyte Sedimentation Rate (E.S.R.)

If a sample of blood is placed in a standard tube 100 mm. long and left for 1 hour, the red cells commence to settle under gravity. Under normal conditions the rate of sedimentation is no more than 10 mm. per hour, but this rate is increased in many diseases involving inflammation and tissue destruction.

The erythrocyte sedimentation rate is employed clinically to judge the progress of such a disease. The sedimentation rate depends upon the viscosity of the plasma and the ratio of mass to surface area of the red cells. The red cell mass to surface area ratio

increases when the cells pile on top of one another forming rouleaux, like piles of coins. Such rouleaux have the same mass, but a lower surface area than the individual red cells [see Fig. 18].

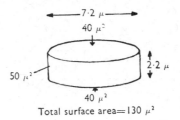

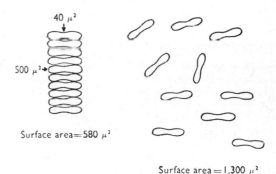

FIG. 18. An increase in fibrinogen leads to increased rouleau formation. Rouleau formation increases the mass to surface area ratio of the red cells. This leads to an increase in the sedimentation rate.

Rouleau formation, and thus the sedimentation rate, is increased when the fibrinogen content increases.

Function of Plasma Proteins

1. Osmotic Pressure

The plasma proteins exert an osmotic pressure of 25 mm. Hg (or 32 cm. blood) which plays an important part in the formation and absorption of tissue fluid [see p. 58]. The osmotic effect depends on the number of molecules present. The albumin molecule is smaller than the globulin. Albumin is also present in greater amounts than globulin. Therefore, the albumin fraction is by far the more important in relation to osmotic pressure. In liver disease there may be a fall in this osmotic pressure, in excess of that expected from the decrease in plasma protein level, due to less albumin and more globulin being formed.

2. Transport

Many substances circulating in the blood are partially or wholly bound to the plasma proteins. Some of these substances and the protein fraction to which they are bound are given below. In certain cases the combination is with only a small subdivision of the plasma fraction which is shown in italic.

Albumin

Calcium   Bilirubin
Sulphonamides   Tryptophan

α-globulin

Cortisol (*Cortisol-binding globulin, CBG*)
Thyroxine (*Thyroxine-binding globulin, TBG*)
Vitamin B_{12}

β-globulin

Cholesterol
Iron (*Transferrin*)
Lipids
Insulin
Vitamin A
Vitamin D
Vitamin K

γ-globulin

Circulating antigens
Histamine

3. Protein Reserve

Plasma proteins given intravenously can, for a limited period, satisfy the body's need for protein. They are probably converted to amino acids by the liver and used as such by the cells of the body. It is unlikely that the plasma proteins are used by the cells directly, since the blood capillaries are impermeable to these proteins, and only the molecules that enter the tissue fluid are available for use by the tissue cells.

4. Viscosity

The viscosity of plasma depends on the size and shape of the protein molecules. The fibrinogen content contributes as much to the viscosity as the albumin, although it is present in very much smaller amounts. The plasma and cells contribute equally to the viscosity of blood.

5. Buffering Effects

Plasma proteins, as has been seen, will combine with both acids and alkalis. In the alkaline plasma, the plasma proteins are partially ionized as acids. They can both donate H^+ to alkalis and accept H^+ from acids. In both cases there is little change in pH, that is they act as a pH buffer [see Chapter 7].

The plasma proteins account for one-sixth of the total buffering power of blood.

6. Antibodies

The antibodies, formed by the lymphocytes (and plasma cells) in response to a foreign protein or other substance which acts as an antigen, are *immunoglobulins* which are found mainly in the γ-globulin fraction of the plasma proteins. [An injection of γ-globulin from a person who has had a virus disease such as measles will give temporary immunity against this infectious disease.] Recent work suggests that the antigen is first taken up by the macrophage cells of the body and is then transferred in a modified form to the antibody-producing cells which manufacture the γ-globulin. Once formed the antibody destroys any further antigen by an antigen-antibody reaction [pp. 45, 158].

The principal γ-globulin antibodies (IgG) have recently been shown to consist of two pairs of cross-linked peptide chains. Two of the chains are short chains containing 214 amino acids (see *myelomatosis*, p. 145) whilst the other two are long chains containing 446 amino acids making a total of 1320 amino acids. The whole molecule has a molecular weight of about 150,000. A difference in the amino acid sequence in the chains probably accounts for the specificity of an antibody molecule for one antigen.

Other γ-globulin antibodies include IgA (m.wt. 160,000) found in saliva, bronchial secretions and intestinal juice (but not in plasma), IgM (m.wt. 900,000) found in plasma, IgD and IgE which are associated with skin reactions.

REFERENCES AND FURTHER READING

Biggs, R., and Macfarlane, R. G. (1971) *Human Blood Coagulation and its Disorders*, 4th ed., Oxford.

Diggs, L. W., Sturm, D., and Bell, A. (1956) *The Morphology of Human Blood Cells*, Philadelphia.

Downey, H. (Ed.) (1938) *Handbook of Hematology*, New York.

Galton, D. A. G. (Ed.) (1959) Haematology, *Brit. med. Bull.*, **15**, No. 1.

Goldsmith, K. L. G. (Ed.) (1959) Blood groups, *Brit. med. Bull.*, **15**, No. 2.

Hamilton, W. F., *et al.* (1929) Studies on the circulation, *Amer. J. Physiol.*, **89**, 322 and 331.

Hamilton, W. F., *et al.* (1948) Comparison of the Fick and dye injection methods of measuring the cardiac output in man, *Amer. J. Physiol.*, **153**, 309.

Laki, K., and Gladner, J. A. (1964) Chemistry and physiology of the fibrinogen-fibrin transition, *Physiol. Rev.*, **44**, 127–160.

Landsteiner, K. (1900) Blood groups, *Zbl. Bakt.*, **27**, 357.

Landsteiner, K., and Weiner, A. S. (1940) Agglutinable factor in human blood recognised by immune sera for rhesus blood, *Proc. Soc. exp. Biol.* (*N.Y.*), **43**, 223.

Lennox, E. S., and Cohn, M. (1967) Immunoglobulins, *Ann. Rev. Biochem.*, Part I, **36**, 365.

Lister, J. (1862) On the coagulation of the blood, *Proc. roy. Soc.*, **2**, 580.

Miller, L. L., and Bale, W. F. (1954) Synthesis of all plasma protein fractions except gamma globulins by the liver, *J. exp. Med.*, **99**, 125.

Ponder, E. (1948) *Hemolysis and Related Phenomena*, New York.

Porter, R. B. (1967) The structure of antibodies, *Scientific American*, **217**, No. 4, 81.

Race, R. R., and Sanger, R. (1968) *Blood Groups in Man*, 5th ed., Oxford.

Riley, J. F. (1959) *The Mast Cells*, Edinburgh.

Rimington, C. (1959) Biosynthesis of haemoglobin, *Brit. med. Bull.*, **15**, 19.

Stewart, G. N. (1921) The output of the heart in dogs, *Amer. J. Physiol.*, **57**, 27.

Whitby, L. E. H., and Britton, C. J. C. (1963) *Disorders of the Blood*, 9th ed., London.

Wintrobe, M. M. (1968) *Clinical Hematology*, 6th ed., Philadelphia.

Wintrobe, M. M., and Landsberg, J. W. (1935) A standardized technique for the blood sedimentation test, *Amer. J. med. Sci.*, **189**, 102.

3. THE HEART AND CIRCULATION: THE HEART

Having considered the fluid in the blood vessel (blood) we now turn to a consideration of the pumping mechanism which enables this fluid to circulate round the body.

The heart is the principal pump which circulates the blood round the body. It will be seen from FIGURE 19 that the left side of the heart transfers blood from the pulmonary veins to the aorta whilst

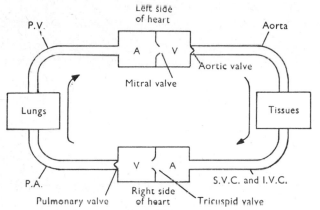

FIG. 19. Schematic diagram of the circulation showing the four heart valves during ventricular diastole.

S.V.C. = superior vena cava; I.V.C. = inferior vena cava; P.A. = pulmonary artery; P.V. = pulmonary veins; A = atrium; V = ventricle.

the right side of the heart transfers blood from the great veins, the superior and inferior venae cavae, to the pulmonary artery. Each side of the heart consists of two chambers, an atrium, previously known as the auricle, and a ventricle.

Valves exist between the atria and the ventricles and between the ventricles and the aorta and pulmonary artery. The left atrioventricular valve has two cusps and is termed the *mitral valve*. The right atrioventricular valve known as the *tricuspid valve* has three cusps; so also have the *aortic* and *pulmonary valves*.

These valves enable the alternate contraction and relaxation of the ventricles to pump blood round the body. If a valve becomes faulty the efficiency of the heart as a pump is greatly impaired.

It should be noted that it is the ventricles which circulate the blood and that the circulation of blood stops when the ventricles (and not the atria) cease to contract.

Historical

Until the seventeenth century the teachings of the second century Greek physician Galen were accepted without question. According to Galen the blood was made from the food, acquired natural spirits in the liver, and was moved to and fro along the veins by the right side of the heart to supply the tissues with nourishment.

Some of the blood was thought to pass through pores in the septum between the ventricles to enter the left side of the heart. Here it received air from the lungs which mixed with the blood to form vital spirits which ebbed and flowed in the arteries.

In 1628 the book *De Motu Cordis* by William Harvey was published. In this treatise Harvey put forward the experimental evidence that led him to the conclusion that the blood circulates round the body. His experiments and observations were so exhaustive and his reasoning so sound that there is very little additional evidence that could be added today.

Among his experiments he demonstrated the pumping action of the heart and the action of the valves. Harvey used quantitative reasoning when he pointed out that if each ventricle holds 2 oz. of blood and pumps out a quarter or even one-eighth of this blood with each beat, then when it has made over 1,000 beats it will have pumped out more blood than is present in the whole body. Since it makes this number of beats in half an hour, Harvey argued that all this blood could not possibly have been made from the food in so short a time as required by Galen's theory. The only possible explanation, Harvey said, was that the blood is moving in a circle.

Harvey showed that systole and not diastole was the active phase of the heart's activity. That the apex beat was due to the heart hitting the chest wall in systole and not due to the active expansion of the heart in diastole as previously thought.

He found that a cut artery bleeds from its *proximal* end (that is, the end nearer the heart) whilst a cut vein bleeds from the *distal* end. He concluded that the blood flows out to the tissues via the arteries and returns via the veins. Harvey did not know how the blood passed from the arteries to the veins. He postulated pores in the flesh. It was left to Malpighi in 1661 to discover the capillaries, using the microscope.

Harvey's teacher in Padua, Fabricius, had been interested in the valves in veins. Harvey explained that the purpose of these valves was to allow a flow of blood in one direction only, namely towards the heart.

Harvey described an experiment that anyone can perform to demonstrate this fact. If a vein is compressed by a finger a short distance below a valve, and the blood is squeezed upwards by a second finger to beyond the valve, then the segment of vein will remain empty as there is no back flow through the valve. But as soon as the first finger is removed the vein fills rapidly from below. This, Harvey said, can be repeated a thousand times and the vein still fills rapidly from below. So much blood has been transmitted in this way through this segment of vein and back to the heart that it must be the same blood moving round in a circle.

The passage of blood through the lungs had previously been described by Ibn-al-Nafis in 1268 and by Michael Servetus in 1553, but these works were unknown to Harvey.

Harvey reported later in correspondence that he had confirmed the suggestion of Columbus in *De Re Anatomica*, published 1559, that blood flowed from the right ventricle through the lungs to the left side of the heart. He showed experimentally that fluid poured into the right ventricle could pass easily via the lungs into the left ventricle. If, however, the pulmonary artery was first tied, no fluid entered the left ventricle, and hence the pores between the two ventricles, postulated by Galen, did not exist.

In *De Motu Cordis* Harvey states: 'Blood circulates sometimes rapidly, sometimes slowly, according to temperament, age, etc., of the individual, external or internal causes, normal or abnormal factors, sleep, rest, exercise, mental state and such like'.

Some of these factors concerning the action of the heart and the circulation of the blood will now be considered in the light of modern knowledge.

The Structure of the Heart

The heart lying in the thorax resembles an inverted cone. The superior aspect of the heart, where the vessels enter, is called its base. The extremity of the ventricles is termed the apex.

The two atria and the two ventricles of the heart lie side-by-side [FIG. 20]. When the heart is beating the atria contract simultaneously, then, after a short pause, both the ventricles contract.

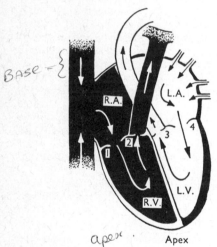

1. Tricuspid valve
2. Pulmonary valve
3. Aortic valve
4. Mitral valve

FIG. 20. The chambers of the heart. The blood passing through the right atrium and ventricle contains reduced haemoglobin and is dark in colour. The blood passing through the left atrium and ventricle is oxygenated and is bright red in colour.

R.A. = Right atrium; L.A. = left atrium; R.V. = right ventricle; L.V. = left ventricle.

There is then a longer pause during which time the whole heart is in a state of relaxation.

The inlet and outlet valves of each ventricle lie alongside one another. All the four valves lie in the same plane, in the fibrous septum or ring which separates the atria from the ventricles [FIG. 21].

The Spread of the Cardiac Impulse

The heart is composed of cardiac muscle, and such muscle has the inherent property of rhythmicity, that is, the muscle fibres contract and relax alternately in a rhythmical manner. This property is developed to the greatest extent in a region known as the sinu-atrial node (S.A. node) which is situated in the wall of the

right atrium near the entrance of the superior vena cava. The S.A. node originates each heart beat. It is known as the *pacemaker*.[3]

Cardiac muscle cells, known as fibres, are cylindrical in shape with central nuclei and faint cross-striations. The fibres branch to form a network or sheet of muscle in which it is difficult to see where one cell ends and another begins. It behaves like a syncytium. (If it were a true syncytium, there would be no cell boundaries at all.) This syncytial-like arrangement enables a contraction wave to spread rapidly from cell to cell until the whole muscle mass is in a state of contraction.

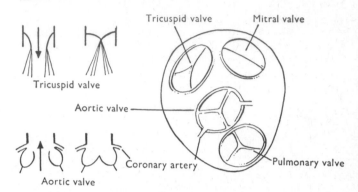

FIG. 21. The four heart valves lie in the same plane. They lie in the fibrous septum separating the atria from the ventricles. The coronary arteries arise from the aorta beyond the aortic valve cusps. Chordae tendineae run from the papillary muscles of the ventricles to the atrioventricular valves.

No nerves are involved in the spread of a contraction wave through cardiac muscle. This may be confirmed by making a series of interdigitating cuts in a piece of atrial muscle. Such cuts would sever any nerves running in the muscle, yet when one part of the muscle is made to contract the contraction wave spreads through the intact parts of the muscle to reach the whole muscle mass. Furthermore, in the embryo the heart beats before the nerves have developed.

Although the two atria are separate chambers the muscle fibres are arranged in rings round both atria. The heart beat, originating as a contraction wave at the S.A. node, spreads rapidly through the atrial muscle causing both atria to contract simultaneously [FIG. 22]. The blood in the atria is forced through the atrioventricular valves into the ventricles.

Rings of cardiac muscle around the entry of the superior and inferior venae cavae, and the pulmonary veins, close off the veins with a sphincter-like action so that blood is not regurgitated back into the veins when the atria contract.

The spread of the contraction wave through the cardiac muscle ceases at the fibrous septum between the atria and the ventricles which contains the four heart valves [plane *A–B* FIG. 22]. The only pathway through this non-conducting septum is from the atrioventricular node (A.V. node) down the atrioventricular bundle (bundle of His) named after the German physiologist Wilhelm His. This bundle runs into the ventricles in the septum between the right and left ventricles. The passage of the contraction wave down this bundle of modified cardiac muscle is not visible from the surface

[3] Recent work using the chick embryo has shown that the pacemaker is initially situated in the right ventricle which develops first. It shifts to the right atrium when this develops and finally settles in the S.A. node (Adolph, 1967).

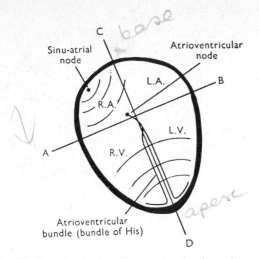

FIG. 22. The spread of the cardiac impulse from the pacemaker (sinu-atrial node) through the atrial muscle to the atrioventricular node. It reaches the ventricular muscle via the atrioventricular bundle. *A–B* is the plane of the fibrous septum between the atria and ventricles. *C–D* is the plane between the right and left chambers of the heart.

R.A. = Right atrium; R.V. = right ventricle; L.A. = left atrium; L.V. = left ventricle.

of the heart, and there appears to be a slight pause following the atrial contraction.

The contraction wave enters the ventricles near the apex and spreads upwards towards the base. The blood in the ventricles is forced upwards towards the base of the heart and out through the aortic and pulmonary valves.

Pericardium

The heart lies in a conical sac known as the pericardium. This consists of an inner serous pericardium and an outer fibrous pericardium. The inner serous pericardium is composed of two smooth layers, the visceral and parietal layers, one attached to the heart and the other to the fibrous sac. These two smooth layers allow the heart to beat in the mediastinum of the thorax with the minimum of friction.

The pericardium sets a limit to the maximum size of the chambers of the heart and prevents excessive stretching of the cardiac muscle fibres due to overfilling with blood.

The pericardium is attached to the diaphragm, and when the heart beats, it behaves as if the apex were relatively fixed. Thus when the ventricles contract, instead of the apex moving upwards towards the base, the base, and particularly the atrioventricular ring, descends towards the apex. This has the effect of increasing the size of the atria at the same time as blood is ejected from the ventricles.

Cardiac Cycle—The Heart as a Pump

The sequence of events may, therefore, be summarized as follows. The heart beat originates in the S.A. node, and shortly afterwards the atria contract. This is followed by a short pause whilst the contraction wave is moving down the bundle of His. Then the ventricles contract, the atrioventricular ring moves downwards and blood is ejected into the arteries. The ventricular muscle relaxes and the atrioventricular ring returns to its initial position. There is then a long pause, when all chambers are relaxed before the next beat occurs.

The ventricular contraction phase is termed systole (pronounced sis'to.lee). It lasts for 0·3 seconds. The ventricular relaxation phase is termed diastole (pronounced dye.ass'to.lee) and this lasts for 0·5 seconds. The complete sequence of events, the cardiac cycle, last for 0·8 seconds, so that there are $\frac{60}{0·8} \simeq 70$ cycles per minute. This is the heart rate.

A table connecting duration of the cardiac cycle in seconds with the heart rate in beats per minute will be found on page 221.

The heart beats continuously for the whole of a person's life and its only period of rest is after each contraction, during diastole. When the heart rate speeds up as it does in exercise or emotional excitement, it does so mainly at the expense of diastole which becomes shorter.

The Heart Sounds

The valves in the heart close passively whenever there is a tendency for the blood to flow in the reverse direction. Since blood

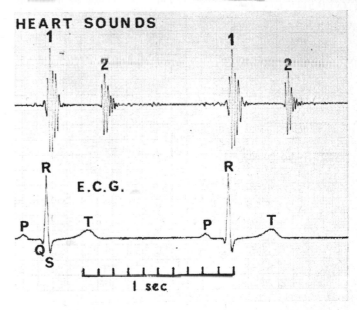

FIG. 23. *Upper trace:* Heart sounds recorded by applying a microphone to the chest wall (phonocardiography). A subject with a slow heart rate has been chosen for clarity. The interval between the start of the first heart sound (1) and the start of the second heart sound (2) corresponds to ventricular systole. The interval between the start of the second heart sound and the start of the next first heart sound corresponds to ventricular diastole. Note that systole is shorter than diastole and that a slow heart rate is associated with a very long diastole. *Lower trace:* Electrocardiogram (Lead II) recorded at the same time. Using the heart sounds as a guide, note that the QRST complex of the E.C.G. also corresponds to ventricular systole.

flows from a region of high pressure to a region of low pressure, it is the relative pressures in the atria, ventricles and arteries that will determine the opening and shutting of the valves.

With the onset of ventricular systole, the pressure in the ventricles starts to rise. As soon as it exceeds that in the corresponding atrium, the atrioventricular valves will shut. This simultaneous closure of the mitral and tricuspid valves with the onset of systole produces the first heart sound [FIG. 23]. It can be heard by applying the ear to the chest wall of a subject. Alternatively a stethoscope

N. L.V.—aorta
R.V.—lungs

may be used. The sound heard may be likened to the word 'lub' spoken very softly. The sound of the valvular closure may be augmented by the impact of the heart against the chest wall and by the noise produced by the contraction of the ventricular muscle fibres. It lasts for 0·15 seconds and the principal frequencies of the sound produced are in the range of 25–45 cycles per second.

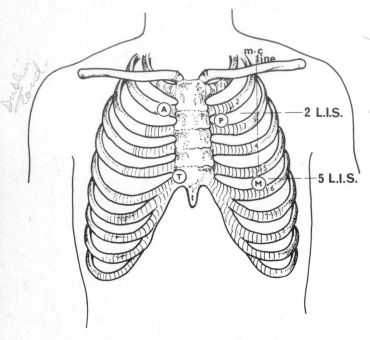

FIG. 24. Diagram of the chest showing ribs, interspace and heart sound areas.

The intercostal spaces are named according to the ribs immediately above. Thus the second left intercostal space (2 L.I.S.) lies between the second and third ribs. The second rib joins the sternum at the sternal angle (manubriosternal junction) which lies about two inches below the jugular (sternal) notch. This landmark can be felt as a ridge when the finger is run down the sternum.

The mitral component of the first heart sound can be heard best at the apex beat in the fifth left interspace (5 L.I.S.) and midclavicular (m–c) line. The tricuspid component is heard best over the sixth costal cartilage at the right border of the sternum.

The pulmonary component of the second heart sound is heard in the second left intercostal space at the left sternal border. The aortic component of the second heart sound is heard over the second costal cartilage at the right sternal border.

All four components of the heart sounds can often be heard in the fourth left interspace at the left sternal border.

The impact of the apex of the heart against the chest wall with each systole can be felt, and frequently seen, in the fifth left intercostal space. It is termed the **apex beat.** The intercostal spaces are named according to the rib that lies immediately above. Thus the fifth interspace lies between the fifth and sixth ribs [FIG. 24]. The apex beat lies about 3–3½ in. from the mid-line. A line through the apex beat parallel to the midline passes through the middle of the clavicle (mid-clavicular line).

As soon as the ventricular pressures exceed those in the aorta, and the pulmonary artery, the aortic and pulmonary valves will open. The opening of the valves does not produce any detectable sound. (A clapping sound is produced when we bring our hands

sharply together, but there is no sound when we take them apart again.)

During the short interval of time between the closure of the mitral valve and the opening of the aortic valve, the left ventricle is a closed chamber. At the same time the right ventricle will be closed off by the closure of the tricuspid and pulmonary valves. Blood is incompressible and although the contraction of the ventricular muscle is increasing the pressure in the ventricles there is no actual change in volume during this phase. It is known as the **isometric contraction phase.** As soon as the aortic and pulmonary valves open, the ventricles decrease in size as the blood is ejected into the aorta and pulmonary artery. *& simultaneously*

At the end of ventricular systole the pressure in the ventricles drops and the aortic and pulmonary valves close since the pressure in these vessels now exceeds that in the ventricles. There is a short **isometric relaxation phase** during which time the ventricles are once again closed chambers, but as soon as the ventricular pressure has fallen to below that in the atria the mitral and tricuspid valves will open.

The closure of the aortic and pulmonary valves gives rise to the second heart sound. This is a shorter and sharper sound and has been likened to the word 'dup'. It lasts for 0·1 seconds. The principal frequency is of the order of 50 cycles per second [FIG. 23].

Since systole is shorter than diastole the rhythm when listening to the heart is:

'LUB'   'DUP'   'LUB'   'DUP'
I   —   II ^{pause} I   —   II ^{pause}

with a shorter interval between the first and next second heart sound than between the second and next first sound.

Under suitable conditions a third heart sound can be heard. It is easier to detect on a phonocardiogram. FIGURE 25 shows a recording made with a microphone applied to the chest wall at the apex. A third heart sound is clearly seen. It occurs about 0·15 seconds after the commencement of the second heart sound, and thus occurs in early diastole. Like the other heart sounds, the amplitude fluctuates with respiration.

The third heart sound is caused by the blood rushing into the ventricles during diastole. The third heart sound is probably due to vibrations of the mitral valve cusps since it is no longer heard in patients with prosthetic mitral valves.

The sound of atrial contraction is sometimes audible and when present is termed the A sound or the fourth heart sound.

The contribution of each valve to the heart sounds is usually heard best at the following sites, known as the valve areas [FIG. 24]:

Pulmonary Valve—2nd left intercostal space at sternal border.
Aortic Valve—2nd right costal cartilage at sternal border.
Mitral Valve—Apex beat (5th intercostal space and mid-clavicular line—about 3½ in. from mid-line).
Tricuspid Valve—6th right costal cartilage at sternal border.

It follows that the first heart sound will usually be heard loudest at the mitral and tricuspid areas, whilst the second heart sound will be heard loudest at the pulmonary and aortic areas.

Since the valves are passive structures and contain no contracting muscle, diseased valves may be replaced by mechanical valves (prosthetic valves) which have been specially designed with smooth surfaces to minimize the risk of blood clotting and red cell haemolysis. Alternatively homografts may be used.

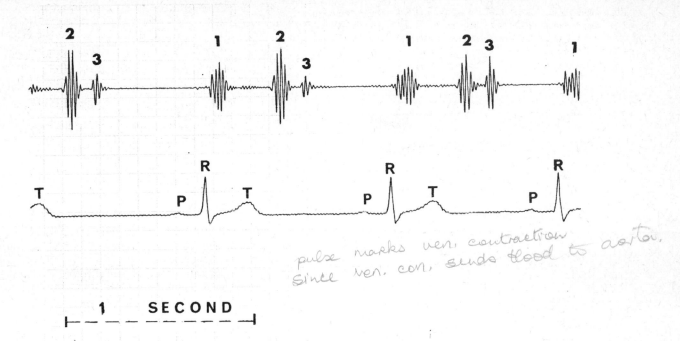

pulse marks ven. contraction since ven. con. sends blood to aorta.

FIG. 25. Third heart sound. *Upper trace:* phonocardiogram recorded at apex. Three heart sounds are present. Third heart sound occurs 0·15 seconds after start of second heart sound. *Lower trace:* electro-cardiogram Lead II.

Quantity of blood ejected with each beat.

Blood Flow Round the Circulation

With each systole 70 ml. of blood are ejected from each ventricle. This quantity is termed the **stroke volume**. The aorta and large arteries are elastic vessels and they accommodate this blood with a relatively small rise in pressure. During the ensuing diastole, there is no output from the heart, and the pressure in the arteries falls. The blood flow in the arteries is pulsatile, and if an artery is cut, the blood shoots out in spurts. The blood flow to the tissues is maintained by the elastic recoil of the arterial walls, and by the time the capillaries have been reached, the flow has ceased to be pulsatile; the steady flow of blood through these vessels shows no variation in systole or diastole.

The blood returns to the heart along the veins, and the venous flow is a steady one. If a vein is cut the blood oozes out from the distal end without any pulsation.

On approaching the heart, the flow again becomes pulsatile as the ventricles are unable to receive the venous blood during systole.

Function of the Atria

The blood returning to the heart during ventricular diastole passes through the atria and enters the ventricles [FIG. 19, p. 23]. The blood returning during systole is unable to do this as the atrioventricular valves are closed. The atria act as a storage reservoir for this blood until the end of systole when the atrioventricular valves open. With the opening of these valves the ventricles fill rapidly with the waiting blood. Further filling occurs as blood returns during diastole. One-tenth of a second before the onset of systole, when the ventricles are already 70 per cent. full of blood, the atria contract and complete the filling of the ventricles by adding the remaining 30 per cent.

Atrial contraction is not essential for life, but the heart is a much more efficient pump when the atria are contracting.

The flow throughout the circulation may be summarized:

	0·3 seconds **SYSTOLE**	0·5 seconds **DIASTOLE**
Left ventricle to arteries	← Rapid flow →	← Nil →
Capillaries	← Continuous flow →	
Veins to right ventricle	← Nil →	← Rapid flow →
Right ventricle to lungs	← Rapid flow →	← Nil →
Lung capillaries	← Continuous but pulsatile flow →	
Pulmonary veins to left ventricle	← Nil →	← Rapid flow →

The flow of blood through the lungs and tissues is continuous, but the flow is intermittent through the heart.

The blood flows through the tissue capillaries at a constant velocity throughout the cardiac cycle, that is, the flow is non-pulsatile. The pulmonary circulation does not contain arterioles and as a result the blood flow through the lungs fluctuates throughout the cardiac cycle being greater in systole than diastole. The pulmonary flow is thus pulsatile.

Pressure Changes in the Heart

FIGURE 26 shows the pressure changes in the heart during a cardiac cycle starting with the onset of the atrial contraction at time 0. The timing of events will refer to this diagram.

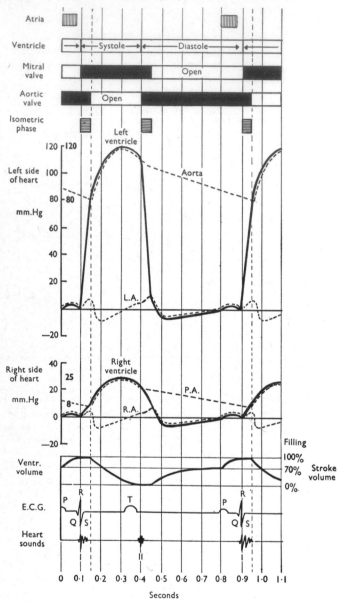

FIG. 26. The sequence of events in the cardiac cycle starting with the onset of atrial systole. Part of the subsequent cycle is shown. P.A. = Blood pressure in the pulmonary artery; R.A. = right atrial pressure; L.A. = left atrial pressure.

It is usual to express blood pressure in mm. Hg, i.e. the height of a corresponding column of mercury measured in millimetres. These pressures may be converted to kilopascals by dividing by 7·5, such as by multiplying by 4 and dividing by 30, or, in the head, by doubling, doubling again, dividing by 3, dividing by 10. Thus 30 mm. Hg becomes: 60, 120, 40, *4·0 kPa*.

These pressures are measured with reference to the atmospheric pressure. If at any time the absolute pressure with reference to a vacuum is needed, the barometric pressure (≃760 mm. Hg ≡ *101 kPa*) must be added. Thus when we say that the left ventricle develops a pressure of 120 mm. Hg (*16 kPa*) during systole, the pressure is 120 mm. Hg above the atmospheric pressure; on the absolute scale it would be:

$$760 + 120 = 880 \text{ mm. Hg } (117 \, kPa)$$

The use of the absolute scale would have the disadvantage that all blood pressure readings would vary from day to day with changes in the barometric pressure.

The pressure in the intrathoracic part of the circulation may be less than the atmospheric pressure. This is due to the elastic recoil of the lungs. The pressure in the atria, for example, may be 755 mm. Hg (absolute) whilst the outside barometric pressure is 760 mm. Hg. Such a pressure is denoted as −5 mm. Hg meaning 5 mm. Hg less than atmospheric. Such a pressure is referred to as a 'negative pressure'.

The atria are thin-walled vessels and the pressure of blood in them is never very far from the pressure outside in the surrounding mediastinum. This pressure is slightly negative (−2 mm. Hg) and it becomes more negative when breathing in (−8 mm. Hg). Respiratory variations in atrial pressure are seen in records of atrial pressure which are obtained with the chest closed by passing a catheter along the veins into the heart and into the right atrium. They disappear when the chest is opened as, for example, in thoracic surgery. For simplicity these respiratory changes will be ignored in this discussion.

When the atrioventricular valves (mitral and tricuspid) are open during diastole (up to 0·1 and from 0·45 to 0·9 seconds [FIG. 26]), the atrial pressure is very slightly greater than the ventricular pressure, otherwise the blood would not enter the ventricles. But to all intents and purposes the atrial and ventricular pressures are identical during this phase.

Similarly during ventricular systole, when the aortic and pulmonary valves are open, the pressures in the ventricles will be equal to those in the aorta and pulmonary artery (from 0·15 to 0·4 and from 0·95 seconds onwards).

Ventricular Pressure Changes

The left ventricle develops a maximum pressure of 120 mm. Hg during systole and therefore the aortic pressure reaches the same peak value. During diastole the ventricular pressure falls to that in the thorax which is approximately 0 mm. Hg (atmospheric pressure). The ventricular changes are thus 120/0 mm. Hg. The pressure in the aorta, however, is maintained by the elastic recoil of the arterial walls. Due to this elastic recoil the pressure in the aorta has only fallen to 80 mm. Hg by the time the next systole occurs. Thus the pressure in the aorta fluctuates between 120 mm. Hg and 80 mm. Hg. These changes are denoted as 120/80 mm. Hg.

The right ventricle pumps out exactly the same quantity of blood as the left ventricle but at a much lower pressure. The right ventricle develops a maximum pressure of 25 mm. Hg. The pressure falls to 0 mm. Hg in diastole and the right ventricular pressure changes are 25/0 mm. Hg. The pressure in the pulmonary artery falls to 8 mm. Hg during diastole and the pressure changes in this vessel are, therefore, 25/8 mm. Hg.

Atrial Pressure Changes

The atrial pressure tracing follows a complex pattern due to the interplay of several factors. Atrial contraction (atrial systole) is associated with a pressure rise in the atria. It lasts for only 0·1 seconds (time 0 to 0·1 and 0·8 to 0·9 seconds [FIG. 26]). For the remainder of the cardiac cycle the atrial muscle is in a state of relaxation (atrial diastole).

The isometric contraction of the ventricles causes the mitral and tricuspid valves to bulge into the atria causing a further increase in atrial pressure (time 0·1 to 0·15 and 0·9 to 0·95 seconds [FIG. 26]). However, as soon as the aortic and pulmonary valves open the ventricular volumes decrease rapidly. The atrioventricular ring descends, increasing the volume of the atria. The atrial pressure falls sharply. Blood returning to the heart cannot now enter the ventricles and the pressure in the atria builds up (time 0·2 to 0·45 seconds [FIG. 26]).

With the relaxation of the ventricles the atrioventricular ring moves upwards towards the base of the heart decreasing the volume of the atria and causing the atrial pressure to rise further (just before 0·45 seconds).

With the opening of the mitral and tricuspid valves at time 0·45 seconds the pressures in the atria and ventricles fall. Blood returning to the heart enters the atria and ventricles and their pressures build up together till time 0·8 seconds is reached when another atrial contraction occurs.

Venous Pressure Changes

The veins near the heart (central veins) show similar pressure waves to those in the right atrium. Since the veins and atria are in communication at all times except during the brief atrial systole, the atrial pressure changes will be transmitted back to the veins.

FIGURE 27 shows a typical venous pressure tracing. There are three maxima in each cardiac cycle lettered a, c and v.

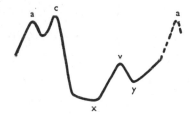

FIG. 27. The pressure waveform in the veins close to the heart during a single cardiac cycle shows three maxima. These are designated a, c and v. The minima on either side of v are designated x and y.

The a wave is caused by atrial systole. The sphincter-like action of the atrial muscle round the entry of the veins prevents this being due simply to a transmission of atrial systole to the veins. The pressure builds up because the veins are unable to empty their blood into the atria.

The c wave is synchronous with the pulse wave in the carotid artery hence the letter c. It corresponds to the atrial pressure peaks occurring at 0·15 and 0·95 seconds [FIG. 26] and is due to the bulging of the A–V valves during the isometric contraction phase.

The v wave corresponds to the peak in the atrial pressure which occurs just before 0·45 seconds. It is due to the filling of the atria whilst the A–V valves are shut and the upward movement of the atrioventricular ring at the end of ventricular systole.

These pressure changes in the veins produce the **venous pulse** which is visible in the neck veins when the subject is lying down. All veins above the sternal angle are usually collapsed because their pressure is sub-atmospheric. No pulsations are therefore visible in the neck veins in the upright posture unless venous congestion is present.

The venous pulsations are seen best in the right external jugular vein which is in direct continuity with the superior vena cava and the right atrium. The venous pressure changes are too small to be palpable. The fact that venous pulsations can be seen but not felt enables them to be readily differentiated from the neighbouring arterial pulsations which are readily felt [FIG. 28].

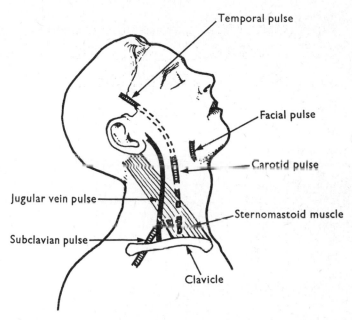

FIG. 28. Arterial and venous pulsation in the head and neck region. The arterial pulsations are palpable and are often visible. The venous pulsations are only present when the venous pressure exceeds atmospheric pressure, that is, with the subject lying down. They are visible but are too feeble to be felt.

A venous pressure tracing may, at first sight, appear so complex with its triple pulsations per cardiac cycle, that difficulty may be experienced in deciding which wave is which. The venous pulsation also shows a respiratory fluctuation owing to the pressure changes that take place in the thorax with breathing [FIG. 29].

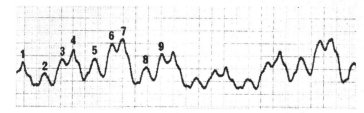

FIG. 29. Venous pulsations recorded from the right external jugular vein. These pulsations show both a cardiac and a respiratory rhythm, making it difficult at first sight to determine which waves are a, c, and v (see FIG. 27). Key in footnote 4 on next page.

However, on either side of the v wave are two pronounced depressions x and y [FIG. 27]. These act as landmarks which enable the v wave to be recognized. Once the v wave has been established at any point on the tracing, it is usually possible to locate it in successive cardiac cycles. The a and c waves which precede the x and v waves lie close together as a pair. The c wave may be absent.

The waves in the venous pulse tracing in FIGURE 29 have not

been labelled so that the reader may see if he can locate the *a*, *c* and *v* waves. The key is given in the footnote[4] below.

The mean value of the *central venous pressure* is low in hypovolaemia (low blood volume) and high in hypervolaemia (high blood volume). It may be used as a guide, like the fuel gauge on a car, when restoring blood to a patient following a haemorrhage.

Ventricular Volume Changes

The changes taking place in the volume of each ventricle during the cardiac cycle may be explained as follows. The left ventricle

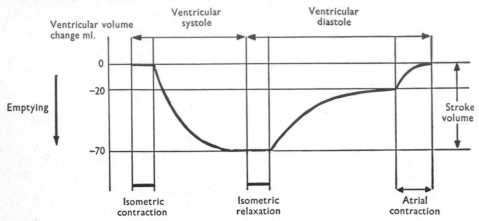

FIG. 30. Ventricular volume changes. During systole blood is ejected at first rapidly and then more slowly. The ventricles fill rapidly when the atrioventricular valves open at the end of the isometric relaxation phase. The filling of the ventricles is completed by the atrial contraction [see also FIG. 26].

will be considered, but simultaneous and identical changes will be occurring in the right ventricle.

The ventricular volume is constant during both the isometric contraction phase and the isometric relaxation phase. Thus with the onset of systole, for the first 0·05 seconds there is no reduction in volume [FIG. 30].

As soon as the aortic valve has opened the ventricular volume decreases as blood is ejected into the aorta. The ejection is rapid at first. Late in systole the ejection rate is reduced.

For the first 0·05 seconds of diastole there is again no ventricular volume change as the ventricular muscle fibres relax isometrically. As soon as the mitral valve opens the blood rushes into the ventricle, the ventricular volume increases rapidly at first and later more slowly. By the time 0·4 of the 0·5 seconds of diastole has elapsed the filling has been such that the ventricular volume has increased by 50 ml. = 70 per cent. of the next stroke volume. Atrial contraction in the last 0·1 seconds of ventricular diastole completes the filling by the addition of a further 20 ml.

Murmurs

Murmurs are sounds heard during the cardiac cycle in addition to the normal heart sounds. Those occurring during systole are known as *systolic murmurs*. Those occurring during diastole are termed *diastolic murmurs*. They are caused by turbulence in the blood as it flows through the valves. Valvular disease may prevent adequate closure of the valves, so that blood regurgitates back

through the partially closed valves. This will occur during systole in a case of mitral or tricuspid incompetence, giving rise to a systolic murmur. It will occur during diastole if the aortic or pulmonary valve is faulty—giving a diastolic murmur.

Murmurs are also caused by a narrowing of the valve orifice which is known as a *stenosis*. The murmur will occur during the rapid passage of blood through the valve. Thus, mitral and tricuspid stenosis will give rise to a diastolic murmur which will reach a maximum during late diastole when atrial contraction forces blood through the valve into the ventricle. Aortic and pulmonary stenosis will give rise to a systolic murmur.

If the turbulence produces vibrations of the chest wall with low frequency components, the vibrations can be felt when the flat of the hand (or the ulnar border of the hand) is applied to the chest wall. Such a palpable murmur is termed a **thrill**.

In general, low frequencies can be felt rather than heard, whilst high frequencies can be heard rather than felt.

ELECTROCARDIOGRAPHY

The invasion of the cardiac muscle by the contraction wave is associated with electrical changes. These potentials may be recorded at points remote from the heart and after electronic amplification may be displayed on a pen recorder or a cathode ray oscilloscope. The waveform obtained is termed an *electrocardiogram*. The various waves present in the waveform are lettered starting with the letter P [FIG. 31].

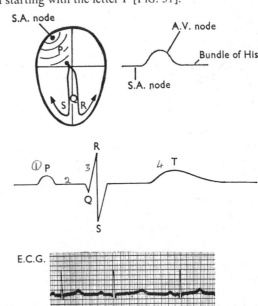

FIG. 31. The electrocardiogram. Time 0·1 seconds. P–R interval = 0·15 seconds. QRST complex = 0·43 seconds.

[4] Venous pulse waves 1, 4, 7 = *c*; 2, 5, 8 = *v*; 3, 6, 9 = *a*.

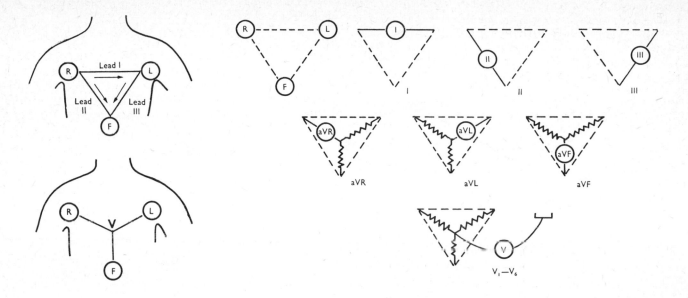

Fig. 32. The standard electrocardiogram leads. Electrodes are applied to the left arm
(L), the right arm (R), and the left leg or foot (F). Leads I, II and III are recorded by
connecting the appropriate pair of electrodes to the amplifier. Leads aVR, aVL and
aVF are recorded by connecting the amplifier between one limb and the junction of
two resistances (100,000 ohms) which are connected to the other two limbs.
    Chest leads are recorded by connecting the amplifier between the chest electrode and
resistances connected to all three limbs. Such a recording is designated V1, V2, V3,
V4, V5 or V6 according to the position of the electrode on the chest.

The P-wave corresponds to the spread of excitation from the
S.A. node over the atrial muscle and therefore represents atrial
systole. The propagation of the contraction wave down the bundle
of His does not produce any detectable external electrical change. It
is denoted on the E.C.G. by the *iso-electric* P–Q interval where there
is no deflection.

The invasion of the ventricles causes electrical changes which to
a large extent cancel each other out. The QRS complex at the
commencement of the ventricular excitation and the T-wave at the
end are all that remain of this algebraic summation.

Should changes occur in the ventricular muscle such as *necrosis*
following a *coronary thrombosis*, then the cancellation will not be so
effective and the S–T segment will no longer be iso-electric.

The relaxation phase T–P is iso-electric.

Leads I, II, III, aVR, aVL and aVF

The detailed waveform depends on the site of the recording
electrodes. To record the *standard limb leads*, electrodes are
applied to the left arm, right arm, and left leg. These electrodes
usually consist of a metal plate which is strapped over the flat part
of the limb. Electrode fluid or jelly is used between the electrode
and the skin to ensure a good electrical connexion. The right leg
is not used for recording the electrocardiogram, but a further

electrode is often applied to this limb to 'earth' the subject, and
thus minimize interference from the mains and electrical apparatus.

In order to make a recording, two input connexions must be
made to the amplifier and there is usually a switch on the electro-
cardiograph apparatus which makes the appropriate connexions
internally when a particular lead is chosen. When the switch is set
at 'Lead I' the electrodes on the right and left arms are connected
to the amplifier. The amplifier then records the potential difference
between the right arm and the left arm [Fig. 32]. When the switch
is set to Lead II, the potential difference between the right arm and
left leg is recorded. When set to Lead III, the potential difference
between the left arm and left leg is recorded. If no switch is present
the appropriate limb electrodes are connected manually to the
amplifier.

An alternative way of recording the electrocardiogram is to
connect one input of the amplifier to a limb electrode, and to connect the other input of the amplifier to the other two limbs through
two resistances. Such an arrangement records the difference
between the potential in one limb and the mean potential in the
other two. If the single limb is the right arm the lead is termed
'aVR', if the left arm it is termed 'aVL', and if the left leg (or foot)
it is termed 'aVF' [Fig. 32].

Leads I, II, III, aVR, aVL and aVF are the six standard limb
leads which are recorded clinically.

Cardiac Vector

FIGURE 33 shows the electrocardiogram recorded using these six limb leads. They have been arranged to correspond to FIGURE 34.

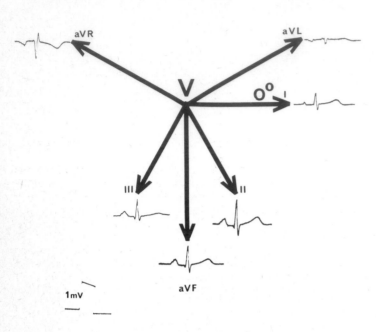

FIG. 33. Recording made using Leads I, II, III, aVR, aVL, and aVF arranged to correspond to FIG. 34. It will be noted that all the recordings are similar but differ in the amplitude of the waves. In this subject the QRS complex was largest in Lead II indicating that the 'electrical axis of the heart' lay in the direction of this lead which is at 60° to the horizontal. Ink jet recording.

largest QRS complex occurs in Lead II (=60°) and hence 60° is the approximate direction of the vector. As a check the minimum QRS complex occurs in aVL (= −30°) and the vector will therefore lie at right angles to this (−30° + 90° = 60°).

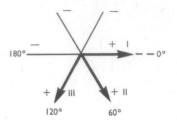

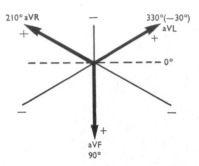

FIG. 34. Leads I, II, III, aVR, aVL, and aVF may be considered to be recordings of the same electrocardiogram vector projected in different directions in the frontal plane. Lead I records the horizontal component (0°). Lead II records the component at 60° to the horizontal. Lead III records the component at 120°. Lead aVF records the component at 90°. Lead aVR records it at 210° and Lead aVL records it at 330° or −30°.
From the relative magnitude of the QRS complexes the direction of the cardiac vector (electrical axis of the heart) may be calculated.

It will be noted that although all six recordings are similar they differ in amplitude. Note also that aVR (and in this case aVL) is upside down compared with the rest. These differences can be explained by considering that the voltage produced by the heart behaves like a vector and has both magnitude and direction and that the different leads are the result of taking the component of this vector in different directions. When a vector is resolved in a plane parallel to the direction of the vector the resultant component is large. When it is resolved in a plane at approximately right angles to the vector the resultant component is small. At more than a right angle it is inverted (negative).

The cardiac vector may be considered as being situated in the centre of the chest and pointing in the general direction of the left foot. Lead I will give the component of this vector in the horizontal direction. This direction is taken as the reference plane 0°. Lead II records the component at 60° to the reference plane and Lead III represents the component at 120° [FIG. 34].

The other limb leads aVF gives the component in the vertical plane (90°), aVR at 210° and aVL at 330° (−30°).

The direction of the cardiac vector may be deduced from the magnitude of the QRS complexes. The vector lies in the same direction as the lead which shows the largest amplitude. Alternatively, it lies at right angles to the lead with the smallest amplitude. Referring to FIGURE 33, it will be seen that, in this case, the

Determining the direction of the cardiac vector is like trying to deduce the direction of the wind from a series of photographs of a flag which have been taken from different viewpoints around the flag-pole. The picture where the flag has the greatest apparent length will give the direction of the wind, and this may be confirmed by finding the picture where the flag has no apparent length because it is viewed end on.

In the normal heart the cardiac vector lies between −10° and +100°. Hypertrophy of the left ventricle will cause the vector to lie at an angle which is more negative than −10°—*left axis deviation*. Hypertrophy of the right ventricle causes the vector to lie at an angle that is more positive than 100°—*right axis deviation*.

Chest Leads

The six limb leads so far discussed give information about the 'electrical axis of the heart' in the frontal plane. To study the activity in a horizontal plane, **chest leads** are used. These usually take the form of a single chest electrode which will adhere to the chest wall by suction, and which is connected to one input of the amplifier. The other input of the amplifier is connected to an electrically neutral point V which is obtained by joining the three limb leads together through resistances [FIG. 32]. It is found that the point V does not change its potential during the cardiac cycle and this potential may thus be used as a reference.

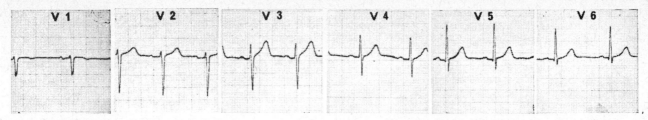

FIG. 35A. Chest Leads V1 to V6 are recorded with the suction electrode applied to the chest positions shown and with the three limbs providing the reference potential. They are each connected via a 100 K ohm resistor to point V (see FIG. 32). Recordings made using Leads V1 and V2 usually show a large S-wave, whilst Leads V5 and V6 usually show a large R-wave. An intermediate lead will show R-waves and S-waves of approximately equal amplitude. Ink jet recording.

As the search electrode is moved across the chest the electro-cardiogram shows a dominant S-wave in chest positions 1 and 2 and a dominant R-wave in chest positions 5 and 6 [FIG. 35].

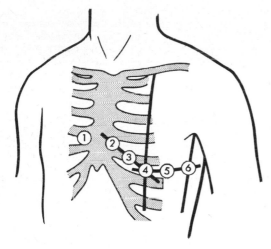

FIG. 35B. Chest lead positions.

1—at the right margin of the sternum in the fourth intercostal space
2—at the left margin of the sternum in the fourth intercostal space
3—midway between positions 2 and 4
4—at the left midclavicular line in the fifth intercostal space
5—at the left anterior axillary line and at the same level as position 4
6—at the left midaxillary line and at the same level as position 4

It will be noted [FIG. 32] that if the chest electrode were placed on the right arm (VR), electrically the arrangement would be very similar to that used when recording aVR. The only difference would be that a resistance would be present across the input of the amplifier and this resistance would simply reduce the amplitude of the recording. If it is removed the VR record will be augmented and is now termed 'augmented VR' or more simply aVR.

Vectorcardiography

The 'electrical axis of the heart' represents the direction of the cardiac vector during the R-wave only. To represent the complete electrocardiogram PQRST a series of continuously changing vectors is required. These can be determined by connecting simultaneously **Lead I** to the **X-plates** of a cathode ray oscilloscope (CRO) and **Lead aVR** (or Lead III if aVR is not available) to the **Y-plates.** The time base is switched off and the CRO spot adjusted

so that it is in the centre of the CRO during the iso-electric (zero voltage) phase of the electrocardiogram.

The pathway taken by the spot consists of a series of three loops which correspond to the P-wave, QRS complex and T-wave of the electrocardiogram respectively [FIG. 36]. If a time scale is required, the trace is alternately switched on and off at a fixed frequency (Z modulation) giving a series of gaps at known time intervals apart in the loops.

At any given instant of time, the direction of the cardiac vector is given by an imaginary line from the iso-electric point to the point reached by the CRO spot. The *electrical axis of the heart* corresponds to the major axis (i.e. general direction) of the **R loop.**

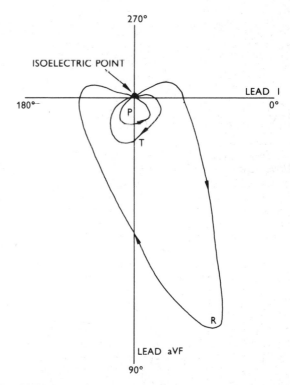

FIG. 36. Vectorcardiography. Direction of movement of the spot of a cathode ray oscilloscope when Lead I is connected to X-plates and Lead aVF (or III) is connected to Y-plates. The general direction of the R loop corresponds to the cardiac vector referred to above.

The axes of the **P** and **T** loops are not always in the same direction as the **R loop.** The directions of the loops change slightly from beat to beat. The **R loop** usually becomes more vertical with a maximum inspiration presumably because the descent of the diaphragm pulls on the pericardium and moves the heart to a more vertical position.

These vectorcardiogram loops are often more difficult to interpret than the standard leads if information other than the direction of axes is required. However, they contain basically the same information, and projecting these loops in the direction of the six frontal planes will regenerate the pattern of the six frontal leads.

CARDIAC RHYTHM ABNORMALITIES

Heart Block and Cardiac Pacemakers

The duration of the P–Q, or as it is often called the P–R, interval is a measure of the conduction time in the bundle. It is normally 0·15 to 0·2 seconds. Any longer time denotes delay in the transmission along the bundle and may be the forerunner of complete failure of conduction along the bundle known as **heart block.** If complete heart block occurs the ventricles will be cut off from the pacemaker and will stop beating. However, they may restart at a slow independent rate due to the inherent rhythmicity of the ventricular muscle. The rate (about 30 beats per minute) is often too slow for an efficient circulation. In all probability the S.A. node and atria will continue at their original rate of 70 beats per minute. Atrial contractions will cease to be effective in aiding the circulation as many will occur whilst the A–V valves are shut. The electrocardiogram will show numerous P-waves but few abnormal QRS complexes and there will be no time correlation between them.

In such a case an electrical stimulator which is either external or implanted in the axilla is connected to the heart via an electrode wire in a fine catheter which is passed into a neck vein and down the superior vena cava and through the right atrium to be embedded in the muscle at the apex of the right ventricle. The tricuspid valve closes satisfactorily around such a catheter. A neutral electrode on the skin in the neck region completes the electrical circuit. Such a device acts as an artificial pacemaker and the ventricular contractions follow at the rate set by the electronic circuits. About 20×10^{-6} joules of energy at 1–3 volts are needed for each pulse. Batteries are available that will supply this energy 70–100 times every minute for several years without renewal.

Ventricular Extrasystoles

If a ventricular beat originates at some ectopic focus an abnormal QRST complex will appear on the E.C.G. [FIG. 37]. This is a

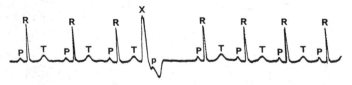

FIG. 37. An electrocardiogram showing a ventricular extrasystole (X). It will be noted that the P-waves are recurring at regular intervals and that each is normally followed by a QRST complex. No P-wave precedes the extrasystole and the P-wave which occurs during the extrasystole (p) does not produce a QRST complex. There is thus electrical silence following the extrasystole until the next successive P-wave occurs. This electrical silence is interpreted by the subject as the heart missing a beat.

common occurrence in young adults and is known as an extrasystole or dropped beat. The term dropped beat is used because the ectopic beat may prevent the next pacemaker beat from contracting the ventricular muscle. There is therefore a pause until the next successive beat arrives [FIG. 38(A)]. Just as we may first

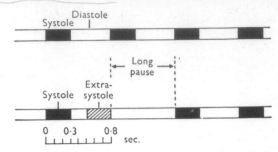

FIG. 38A. Extrasystole (ectopic beat). The long diastole following the extrasystole is noted.

become aware of a clock in a room when it stops ticking, so we may be unaware of the beating of the heart but notice the pause which gives the impression that the heart has missed a beat.

FIG. 38(B) shows the heart sounds associated with an extrasystole. The first heart sound is louder than normal. With a premature beat the valve cusps are still wide open when systole starts. With a normal beat they tend to float towards apposition at the end of diastole. On the other hand, the second heart sound is softer than normal due to the short filling time and absence of an atrial contraction. As a result the stroke output of this beat has been low giving a reduced *valve-closing pressure* in the aorta and pulmonary artery.

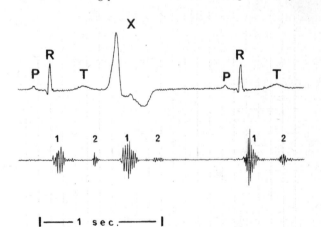

FIG. 38B. Heart sounds associated with ventricular extrasystole.
Upper trace: Electrocardiogram (Lead II).
Lower trace: Heart sounds recorded at third left intercostal space. The first normal beat is followed by a ventricular extrasystole X. After a long diastole it is followed by another normal beat. Because of the short filling time, the stroke volume associated with the extrasystole is smaller than normal. The mitral and tricuspid valves are wide open when this extrasystole starts and their closure gives a louder than normal first heart sound. The low arterial blood pressure associated with the reduced stroke volume gives a soft second heart sound when the aortic and pulmonary valves close.

Atrial Extrasystoles

If cells in the atria, other than those in the sinu-atrial node, originate a heart beat an atrial extrasystole will result. Since the spread across the atrial muscle will be in a different direction to normal, an unusual P-wave will be observed on the electrocardiogram. This P-wave is often inverted.

When the extra beat reaches the A–V node, it will pass down the A–V bundle and spread to the ventricles in the normal manner. The QRST complex of the electrocardiogram and the ventricular contraction will be indistinguishable from a normal beat. As a result the subject is seldom aware of an atrial extrasystole.

FIGURE 39 shows an unusual sequence of three heart beats recorded in a healthy young adult. Each beat has originated from a different part of the heart.

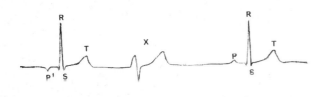

|←—1 SECOND—→|

FIG. 39. Electrocardiogram (Lead II) showing atrial extrasystole (P′) followed by ventricular extrasystole (X) followed by normal sinus rhythm.

The first, on the left, shows an atrial extrasystole. This beat has originated from an ectopic focus in one of the atria. Note the inverted P-wave (P′) followed by a normal QRST complex.

The next beat is a ventricular extrasystole. This beat has originated from an ectopic focus in one of the ventricles. Note the absence of any P-wave and the bizarre QRST complex (X).

The third, on the right, shows the electrocardiogram of normal sinus rhythm. This beat has originated in the S.A. node.

Atrial Fibrillation

Atrial fibrillation occurs when the increased excitability of the atrial muscle leads to abnormal beats arising from *ectopic foci* in the atria. As a result different parts of the atria are contracting and relaxing out of phase with one another. The general appearance of the atria during fibrillation has been described as resembling a bag of wriggling worms. The atria never empty and the atrial wall becomes a quivering sheet of muscle.

In atrial fibrillation the atria cease to pump blood into the ventricles. In addition, the beat originated by the pacemaker is not transmitted to the ventricles. Instead impulses pass down the bundle of His in a haphazard fashion at a rate too fast for the ventricle to respond to each impulse. The ventricular contractions which do occur are irregular both in rate and amplitude. A proportion of these beats fail to develop sufficient pressure to open the aortic valve. As a result the radial pulse is 'irregularly irregular' and the heart rate measured at the heart may be higher than if it is measured by the pulse at the wrist. This is known as a *pulse deficit*.

Ventricular Fibrillation

Should the ventricular muscle fibrillate, the circulation stops immediately. Unless immediate cardiac massage is carried out, life will cease. Under suitable conditions ventricular fibrillation may be stopped by passing an electric current through the heart. The apparatus is termed a *cardiac defibrillator*.

Demand and Sequential Pacemakers

Although external pacemakers can have an adjustable rate, the implanted pacemakers usually work at a fixed rate and produce only ventricular contractions. A sequential form is, however, available. Two electrode wires are used, one to stimulate the right atrium and a second to stimulate the ventricles 0·1 sec. later.

With a pacemaker *in situ*, complications occur when the heart starts beating on its own, since there will now be two 'pacemakers' in the heart. An artificial beat produced at or near an electrocardiogram T-wave (i.e. when the cardiac muscle is highly excitable) may lead to a fatal ventricular fibrillation. For this reason patients with pacemakers are given drugs such as *lignocaine* to suppress natural beatings and as an additional precaution 'demand' pacemakers are used which switch off if the heart starts beating on its own.

The Work Done by the Heart

The heart is a remarkable organ for its size. It weighs only 300 g. and is about the size of a clenched fist, being 12 cm. from apex to base, 8 cm. transversely at its broadest part and 6 cm. anteroposteriorly. It beats seventy times a minute for seventy odd years and each time it beats it pumps out 70 ml. blood from each ventricle. As has been seen, the only rest it takes is for half a second after each beat. In the course of a lifetime it will beat 2,500,000,000 times, and each ventricle will have pumped out 170 million litres of blood. It is not surprising, therefore, that appreciable engineering problems are encountered when designing prostheses for the heart, such as mechanical valves. Even the problem of finding a piece of wire to run to the heart from the artificial pacemaker, that can be flexed this number of times without breaking, has proved to be difficult.

Each time the heart beats it transfers energy to the ejected blood. The potential energy conferred is determined by the product of the volume and the pressure. The kinetic energy transferred is equal to 0·5 multiplied by the mass of blood ejected multiplied by the square of the velocity ($\frac{1}{2} mv^2$). Although the velocity is $\frac{1}{2}$ metre per second in the arteries the potential energy is far in excess of the kinetic energy. With each beat the left ventricle ejects 70 ml. blood at a pressure of 120 mm. Hg. This is equivalent to raising the blood vertically in the air to a height of 150 cm. (1·5 m., 5 ft.). The potential energy gained by the blood is its weight multiplied by this height $= 70 \times 1\cdot5 = 105$ g. metres per beat. The kinetic energy $= \frac{1}{2}.\frac{70}{9\cdot81}.(0\cdot5)^2$ is negligible in comparison (9·81 m. per sec. is the acceleration under gravity). The right ventricle conveys a further one-fifth of this energy to the blood in the pulmonary artery. The energy conveyed would equal the work done by the heart, if the efficiency of the heart were 100 per cent. In practice the efficiency is only 33 per cent. and, therefore, three times this amount of energy must be expended by the heart.

In the course of a lifetime the work done by the heart is over 90,000,000 kg.m. which is equivalent to taking 1 cwt. (=112 lb.) of blood to a height of 1,200 miles.

In exercise this small pump can increase its minute output by six times to 30 litres per ventricle and deliver this blood at the same or even higher ejection pressure. The simultaneous output of the two ventricles is 60 litres per minute. If the blood were not circulating this would be equivalent to a bathful of blood (=26 gallons) every two minutes.

PROPERTIES OF CARDIAC MUSCLE

Amphibian Heart

Some of the more important properties of cardiac muscle may be studied in the amphibian heart. The heart of a frog or toad has two atria and one ventricle. It has no coronary arteries, and no bundle of His like the mammalian heart, and will continue to beat for some time even though no blood is flowing through it. To demonstrate the conduction of the heart impulse from one chamber to another, ligatures may be placed around the junction between the chambers and tightened until the conduction is at first impaired and finally abolished. If such a ligature is placed between the atria and ventricles in the mammalian heart, the coronary arteries will be occluded and the ventricles will fibrillate.

The heart beat in the amphibian heart originates in the sinus venosus which is the homologue of the sinu-atrial node in mammals. If a ligature is placed between the sinus and the atria (1st Stannius ligature) the sinus is isolated [FIG. 40]. The atria and ventricle may stop beating but restart again at a slower rate. The heart beat now originates in the atria.

If the ligature is placed between the atria and the ventricles (2nd Stannius ligature), the atria continue to beat, but the ventricle becomes quiescent. After a long interval the ventricle starts to beat, often at an even slower rate.

During the quiescent phase the ventricular muscle will respond to electrical and mechanical stimulation. If such a ligature is tightened gradually, various degrees of heart block between the atria and ventricle may be produced. When this occurs not every atrial contraction will be followed by a ventricular one.

In these and similar experiments the following properties of cardiac muscle may be demonstrated.

1. Excitability and Contractility

When the quiescent ventricle is stimulated, it responds as a whole with a maximum contraction provided that the stimulus is strong enough. A stronger stimulus has the same effect. A weak stimulus has no effect. This is an 'all or none' response.

The magnitude of the response depends, however, on the physical state of the muscle fibre. It increases for the first few stimuli—the staircase phenomenon. It is decreased if the diastolic pause is inadequate. The force of contraction is increased if the muscle is stretched.

2. Refractory Period

Throughout the entire contraction phase a second stimulus has no effect. During the contraction phase the muscle is *absolutely refractory*. During the relaxation phase only a very strong stimulus is effective. This is a relatively refractory phase. Unlike skeletal muscle [p. 186] cardiac muscle does not go into a state of sustained contraction known as tetanus.

3. Rhythmicity

The Stannius ligatures demonstrate that all three chambers have the power to initiate beats. This property is termed *rhythmicity*. When no ligatures are present, cooling the sinus venosus slows the heart. Warming it, speeds up the heart. This corresponds to the effect of temperature change on the S.A. node in the mammalian heart.

Perfused Mammalian Heart

The mammalian heart muscle receives its blood supply from the coronary arteries. The right and left coronary arteries arise from

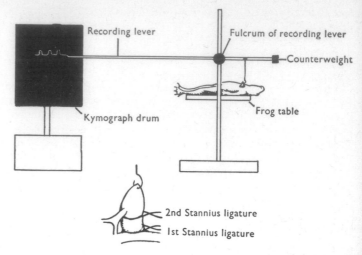

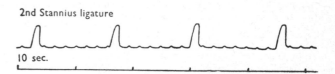

FIG. 40. Recording the activity of the frog heart. The first Stannius ligature is tied between the sinus venosus and the atria. The second Stannius ligature is tied between the atria and the ventricle. Kymograph pen-writer recording.

the aorta immediately above the cusps of the aortic valve [FIG. 21, p. 24]. If we label the three cusps of the aortic valve, the anterior cusp, left posterior cusp and right posterior cusp, then the right coronary artery arises from the sinus beyond the anterior cusp, whilst the larger left coronary artery arises from the sinus associated with the left posterior cusp. The right coronary artery supplies the right atrium and ventricle. The left coronary artery supplies the left atrium and the powerful left ventricle.

Importance of Ions

The perfused mammalian heart requires an adequate oxygen supply to the cardiac muscle in order to continue beating. Provided that these nutritional requirements are met, the mammalian heart will continue to beat for several hours outside the animal. One way of achieving this is to perfuse the aorta with oxygenated fluid in a retrograde manner with a pressure adequate to perfuse the coronary vessels. The aortic valves remain shut and no blood is pumped through the heart, but beating continues. An attachment may be made to the ventricle so that a graphical recording is obtained [FIG. 41].

Such a preparation demonstrates the importance of metallic ions in the perfusion fluid for the continued beating of the heart.

If isotonic NaCl is used as the perfusion fluid, the rhythmicity disappears and the beating ceases. If calcium ions are added to this solution the heart beats for a time and ... in systole. If potassium ions are added the heart st... . If all three ions are present, the beating contin...

A solution which maintains the beating is known as Ringer's solution.

It has the following composition:

NaCl	.	.	0·9 g.
CaCl$_2$	.	.	0·024 g.
KCl	.	.	0·042 g.
NaHCO$_3$	.	.	0·02 g.
Distilled water to	.	100 ml	

The effects of ions on the heart are of practical importance. Firstly, they illustrate the potential dangers of giving an intravenous transfusion or even an intravenous injection containing potassium or calcium ions. The heart may be arrested in both

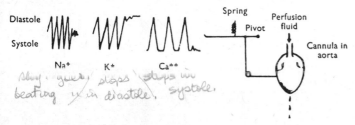

Diastole   Systole   Na⁺   K⁺   Ca⁺⁺   Spring   Pivot   Perfusion fluid   Cannula in aorta

Fig. 41. The isolated mammalian heart. A thread attached to the ventricle passes round a pulley and is attached to a spring-loaded lever. An excess of K⁺ causes the heart to stop in diastole. An excess of Ca⁺⁺ causes the heart to stop in systole.

cases. Further, such a transfusion may be given unintentionally. When blood is stored, potassium leaves the cells and enters the plasma. A blood transfusion with old blood, apart from the dangers associated with haemolysis, will be equivalent to giving a solution containing potassium ions.

On the other hand, the injection of these ions provides a convenient way of intentionally stopping the heart beating. For example, in an operation on the heart, where the pumping action of the heart has been taken over mechanically, the surgeon can inject calcium or potassium salts into the coronary arteries to arrest the heart, so that septal defects and other abnormalities may be repaired. When these salts have been washed out, the heart starts beating again.

Heart Transplants

An extension of the perfused heart technique is the transplantation of the heart from one animal to another and from one human being to another. Work on dogs has shown that provided that the rejection problem can be overcome [p. 157], the transplanted heart is able to maintain an adequate circulation of the blood. Recent experience has shown that the same applies to transplanted hearts in man.

The perfused mammalian heart experiment demonstrates that the heart will continue to beat after all nerves to it have been sectioned. Thus when transplanting a heart it is not necessary to re-establish continuity of the cardiac nerves for beating to continue.

The transplanted heart, even when denervated, is still under the influence of hormones, especially those from the adrenal medulla. It also has its inherent regulation system (see Starling's Law of the Heart, p. 46) and is able to increase its activity in exercise.

Dog experiments have shown that a transplanted heart becomes reinnervated with autonomic nerve fibres after a comparatively short period of time (a few months).

REFERENCES AND FURTHER READING

Adolph, E. F. (1967) The heart's pacemaker, *Scientific American*, **216**, No. 3, 32.

Hamilton, W. F. (Ed.) (1962) Circulation, *Handbook of Physiology*, Section 2, Washington, D.C.

Harvey, W. (1628) *Circulation of the Blood* (*De Motu Cordis*), translated by: Willis, R. (1847) London; Leake, C. D. (1949) 3rd ed., Springfield; Franklin, K. J. (1957) Oxford.

His, W. (1893) Die Thätigkeit des embryonalen Herzens und deren Bedeutung für die Lehre von der Herzbewegung beim Erwachsenen. *Arb. med. Klin. Lpz.*, **14**, 49.

Hutter, O. F., and Trautwein, W. (1956) Vagal and sympathetic effects on the pacemaker fibres in the sinus venosus of the heart, *J. gen. Physiol.*, **39**, 715.

Katz, L. N. (1947) The genesis of the electrocardiogram, *Physiol. Rev.*, **27**, 398.

Keith, A., and Flack, M. (1907) The form and nature of muscular connections between the primary divisions of the vertebrate heart, *J. Anat.* (*Lond.*), **41**, 172.

Kent, A. F. S. (1893) Researches on the structure and function of the mammalian heart, *J. Physiol.* (*Lond.*), **14**, 233.

Locke, F. S. (1895) Towards the ideal artificial circulating fluid for the isolated frog's heart, *J. Physiol.* (*Lond.*), **18**, 332.

Prinzmetal, M., Corday, E., Brill, I., Oblath, R. W., and Kruger, H. F. (1952) *Auricular Arrhythmias*, Springfield.

Ringer, S. (1883) A further contribution regarding the influence of the different constituents of the blood on the contraction of the heart, *J. Physiol.* (*Lond.*), **4**, 29.

Tawara, S. (1906) *Das Reizleitungssystem des Säugetierherzens*, Jena.

Wiggers, C. J. (1928) *The Pressure Pulses in the Cardiovascular System*, London.

FILM

William Harvey and the Circulation of the Blood, 1971, Royal College of Physicians, London.

4. THE HEART AND CIRCULATION: THE CIRCULATION

The circulation consists basically of two pumps and two sets of tubes in series. The two pumps are the right and left sides of the heart. The two sets of tubes are the systemic and pulmonary circulations. In such a system the two pertinent questions are: How much blood is moving round in a circle and how fast is it going? As Harvey said: 'The blood circulates sometimes rapidly, sometimes slowly . . .' The answer to the first question may not be the same as the *blood volume* since some of the blood may be stagnating in some 'cul-de-sac' of the circulation. Dividing the 'volume of blood circulating' by the 'time it takes', will give the volume flow per minute into the aorta or through the lungs. Such a measurement is termed the *cardiac output*.

There is a further question that must be asked when considering the systemic circulation and that is 'How is the blood flow being distributed?' The systemic circulation consists of a large number of pathways in parallel and the blood may be directed away from one organ and redirected to another.

There is no simple way of determining the answer to any of these questions. As yet, no one has invented a blood flow recorder that will measure the blood flow to any organ quickly and easily from outside of the body. The methods that are available for blood

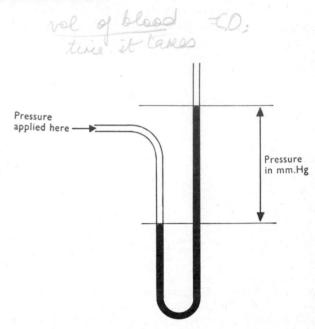

FIG. 42. The mercury manometer introduced by Poiseuille in 1828. When a pressure is applied the mercury in the left limb descends and that in the right limb rises. The pressure in mm. Hg is given by the vertical height between these two levels measured in millimetres. The movement of the mercury in the right-hand limb is only one-half of this distance. 1 mm. Hg = $13 \cdot 595 \times 9 \cdot 81$ Nm^{-2} = $0 \cdot 133$ kPa. To convert mm. Hg to kilopascals divide by $7 \cdot 5$.

flow determination, are complex. The dye dilution method has already been considered [p. 6]. Other methods will be considered later.

In 1828 Poiseuille introduced the *mercury manometer* [FIG. 42] and in 1847 Carl Ludwig introduced the graphical method of recording blood pressure using the *kymograph*. A modern version of this apparatus is shown in FIGURE 43. Since then a great deal of attention has been paid to measurement of systemic blood pressure. Heart rate is also easy to measure, and this too has been considered in detail. But these two are of less importance than blood flow. Unfortunately, it is not possible to deduce the cardiac output from these two measurements alone.

Since the problems associated with the flow of blood through the systemic and pulmonary circulations are different, it is convenient to consider them separately.

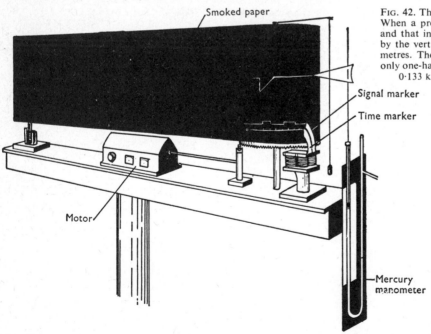

FIG. 43. A modern form of the kymograph first introduced by Carl Ludwig in 1847. Blood pressure is recorded by means of a float and writing point attached to the Poiseuille mercury manometer. The tracing is made on smoked paper which is afterwards varnished to obtain a permanent record.

SYSTEMIC CIRCULATION

The left ventricle pumps blood through the arteries, arterioles and capillaries of the systemic circulation, and the blood returns to the right side of the heart via the veins. Blood is like any other fluid and only flows from a region of high hydrostatic pressure to a region of low hydrostatic pressure. The reason why the blood flows through the systemic circulation is because the pressure at *A* is higher than the pressure at *B* [FIG. 44]. The pressure at *A* is

SYSTEMIC CIRCULATION

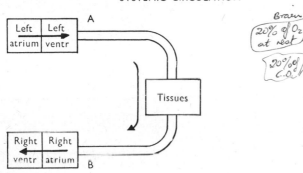

FIG. 44. The systemic circulation. The blood flows from *A* through the tissues to *B* due to the pressure at *A* being higher than the pressure at *B*.

produced by the contractions of the left ventricle. It fluctuates between 120 and 80 mm. Hg (*16 and 10·7 kPa*) and has a mean value of about 100 mm. Hg (*13·3 kPa*) above atmospheric pressure. This pressure is known as the *arterial blood pressure* or more simply the *blood pressure* (B.P.).

The pressure at *B* is the pressure in the great veins and right atrium, and is approximately equal to the *atmospheric pressure*, i.e. 0 mm. Hg. It will be termed the *venous blood pressure*. The pressure difference, equal to the arterial blood pressure minus the venous blood pressure, drives the blood through the systemic circulation.

A blood vessel offers a resistance to the flow which is inversely proportional to the fourth power of the radius of the vessel. The resistance thus increases as the vessels become smaller. In the systemic circulation the main resistance lies in the *arterioles*. The capillaries are even smaller vessels, but each arteriole feeds many capillaries in parallel, and the resistance offered by this capillary network is less than that of the arteriole.

The resistance offered by the blood vessels is termed the **peripheral resistance** (P.R.).

Volume Flow

The volume flow through the systemic circulation will be equal to the pressure difference divided by the peripheral resistance. The rate of blood flow (volume flow) equals the cardiac output (C.O.) from the left ventricle (by definition).

Thus,

$$\text{Cardiac output} = \frac{\text{Arterial blood pressure} - \text{Venous blood pressure}}{\text{Peripheral resistance}} \quad (1)$$

But, the *venous blood pressure* is equal to 0 mm. Hg. Hence equation (1) becomes:

$$\text{Cardiac output} = \frac{\text{Arterial blood pressure}}{\text{Peripheral resistance}} \quad . \quad . \quad . \quad (2)$$

Or,

$$\text{Arterial blood pressure} = \text{Cardiac output} \times \text{Peripheral resistance} \quad (3)$$

This is one of the fundamental equations of circulatory physiology.

Arterial Blood Pressure

The tissues of the body need an adequate blood flow. To provide this an adequate head of pressure must be available in the vessels supplying the tissues.

The brain, in particular, needs a continuous, as well as adequate, blood supply. If its blood flow is impaired, the brain cells cease to function and consciousness is lost. Furthermore, the cells of the brain die if their blood supply is interrupted for more than a very few minutes. (*3*).

The brain, in the sitting or standing position, is situated at a higher level than the heart, and blood has to be pumped 'up hill' from the heart to the brain. Unless an adequate blood pressure is maintained at the level of the heart, the cerebral blood flow will be insufficient to maintain consciousness. Even in a normal person, the blood pressure in the arteries of the brain will be 30–40 mm. Hg less than that at the heart level when in the sitting or standing position [p. 57]. This difference will disappear when in the lying position. A person who faints due to a fall in blood pressure should be placed with his head at the same level as the heart, such as flat on the floor. To sit or stand such a person up will decrease the brain blood flow still further, and may have serious consequences.

The kidneys are organs which need an adequate blood pressure for a different reason. As will be seen later, this pressure is needed to provide the energy for filtering the blood to make urine. Urine production ceases when the blood pressure falls to low levels.

The maintenance of an adequate blood pressure in the aorta will depend on the product of the two factors, the *peripheral resistance* and the *cardiac output* [Equation 3]. These factors will be considered separately in more detail.

PERIPHERAL RESISTANCE

It has been seen that the resistance in the systemic circulation lies mainly in the arterioles. Since we are dealing with a fourth power law, relatively small changes in the radii of arterioles will make a large difference to the resistance. Thus the vessel in FIGURE 45(*A*) has 0·85 the radius of that in (*B*) but double the

FIG. 45. Peripheral resistance is proportional to the fourth power of the diameter of the vessel for laminar flow. The diameters in *A*, *B*, and *C*, are in the ratio of 0·85 : 1·00 : 1·20. The resistance to blood flow is in the ratio of 2 : 1 : ½.

resistance. This is because $0·85^4 = \frac{1}{2}$. The vessel in (*C*) has 1·2 times the radius of that in (*B*) but half the resistance ($1·2^4 = 2$).

REGULATION OF THE SIZE OF ARTERIOLES

The smooth muscle in the arteriolar wall is in a state of partial contraction due to the continuous sympathetic nerve activity, called *sympathetic tone*, or *vasoconstrictor tone*, which is acting upon it. This tone originates in the medulla in a collection of cells known as the **vasomotor centre** (V.M.C.). An increase in activity on the part of the vasomotor centre will cause a decrease in the radius of the arterioles, *arteriolar vasoconstriction*, and an increase in the peripheral resistance. From equation 3 this will lead to a rise in blood pressure, provided that the cardiac output remains constant.

Conversely a decrease in the vasomotor centre activity will cause a relaxation of the vasoconstrictor tone, and an increase in the size of the arterioles. This is termed *vasodilatation*. If the cardiac output is constant, the blood pressure will fall.

Baroreceptors and the Vasomotor Centre Activity

The arterial blood pressure is maintained at an approximately constant level by the baroreceptor mechanism. Certain parts of the arterial tree have receptors in the arterial wall which respond to the pressure in the lumen of the vessel. Such receptors are found in the aortic arch, in the common carotid arteries and at the bifurcation of each of the common carotid arteries into the internal and external carotid arteries—a region known as the carotid sinus

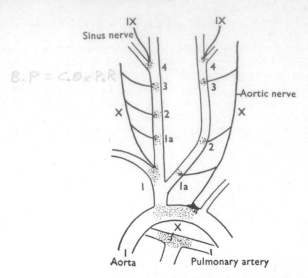

FIG. 47. The baroreceptor areas associated with the systemic and pulmonary arteries (cat).

[FIG. 47]. Such receptors are termed *baroreceptors*. As the pressure rises the nerve impulse activity passing from these areas to the medulla increases. If the pressure falls, the impulse activity decreases. FIGURE 46 shows the impulse activity recorded from a common carotid baroreceptor area [area 3, FIG. 47] with different pressures in the artery.

The nerve impulses run in the vagus nerve (X) and the glosso-pharyngeal nerve (IX). The nerve from the aortic arch baroreceptors is called the *aortic nerve (depressor nerve)* and that from the carotid sinus the *sinus nerve*.

The baroreceptor activity inhibits the vasomotor centre. Due to the continuous baroreceptor activity, the vasomotor centre is

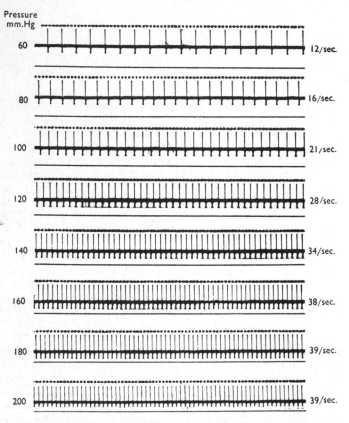

FIG. 46. The activity of a single baroreceptor ending when a steady pressure is applied to the lumen of its artery. Recording made from **the nerve supplying area 3 (FIG. 47).** Photographed from cathode ray oscilloscope.

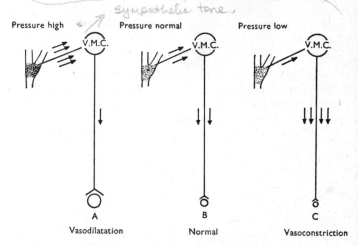

FIG. 48. Baroreceptor activity and arteriolar tone.

B. Normal. Baroreceptor activity partially inhibits the vasomotor centre (V.M.C.) giving normal arteriolar tone.
A. High pressure in baroreceptor region gives increased baroreceptor activity, and a reduction in vasoconstrictor tone. This results in vasodilatation of the arterioles.
C. A low pressure in the baroreceptor regions reduces baroreceptor inhibition of the V.M.C. The arteriolar tone increases giving vasoconstriction.

not so active as it would otherwise be, and, therefore, the arteriolar vasoconstriction is not so intense. This is a feed-back system which tends to maintain a constant blood pressure by minimizing any change. This may be seen by reference to FIGURE 48.

FIGURE 48(*B*) represents the normal state. The baroreceptor activity partially inhibits the vasomotor centre, and the arterioles are in a moderate state of vasoconstriction. Should the blood pressure tend to rise for any reason [FIG. 48(*A*)], the baroreceptor activity will increase and cause inhibition of the vasomotor centre. The vasoconstrictor tone to the arterioles will be reduced. The arterioles will dilate, and by lowering the peripheral resistance, will limit the blood pressure rise. Should the pressure start to fall [FIG. 48(*C*)], the baroreceptor activity will be reduced, there will be less inhibition of the vasomotor centre, more vasoconstriction of the arterioles, and the pressure fall will be checked.

The carotid artery baroreceptors in the region of the carotid sinus of the neck may be *artificially* stimulated by external pressure in the neck. Pressure in this region will increase the activity in the sinus nerve due to mechanical distortion of the receptor endings. The inhibition of the vasomotor centre will lead to a fall of blood pressure. FIGURE 49(*B*) shows the effect on blood pressure of

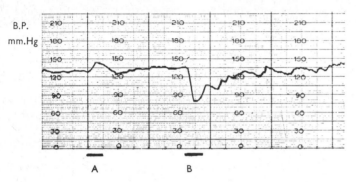

FIG. 49. Blood pressure reflexes from the carotid sinus in man.
A. Pressure on the common carotid artery below the carotid sinus causes a rise in blood pressure.
B. Pressure on the carotid sinus stimulates the baroreceptor nerve endings and causes a profound fall in blood pressure. Heated stylus recordings made using the blood pressure follower.

applying external pressure to the carotid sinus in a subject with a particularly 'sensitive' carotid sinus. In such susceptible subjects, the blood pressure fall may be sufficient to lead to unconsciousness. They may faint when bending their head to one side, particularly if they are wearing too tight a collar. This may be reason why the carotid arteries were so named (Greek *karotikos* = stupefying).

The fact that the fall of blood pressure is not the result of an interruption of the blood supply to the brain may be demonstrated by compressing the same artery a few centimetres below the carotid sinus. This will have a similar effect on the brain blood supply, but the result is completely different; the blood pressure rises [FIG. 49(*A*)]. In this case, the pressure in the carotid sinus falls due to the obstruction to the carotid artery. There will be less inhibition of the vasomotor centre, and, notwithstanding the presence of the other baroreceptors which will tend to keep the blood pressure constant, a pressure rise results.

Although these manœuvres create an abnormal situation in which the stimulus applied to one carotid sinus is different from the natural stimulus, and also different from that applied to the

other baroreceptor areas (the systemic blood pressure), nevertheless they demonstrate the activity of the baroreceptor mechanism in man. An implanted sinus nerve stimulator has recently been used to reduce excessively high blood pressure.

When changing from the lying to the standing position the cardiac output falls, but, due to the baroreceptor mechanism, the blood pressure remains unchanged. After a prolonged stay in bed (several weeks), this mechanism may be lost, and then any sudden change from the horizontal to the vertical position may lead to fainting. A similar loss occurs in astronauts after a period of weightlessness in orbit.

Due to the pulsatile nature of the arterial blood pressure, the baroreceptor impulse activity in the normal circulation is not a steady activity as in FIGURE 47, but occurs in bursts with the maximum activity during systole and early diastole [FIG. 50]. There is

FIG. 50. An electroneurogram (E.N.G.) showing baroreceptor and chemoreceptor activity in a few fibre preparation of the sinus nerve. The large spikes (action potentials) are from a single baroreceptor nerve fibre; the small spikes are action potentials from the chemoreceptor nerve fibres. Each spike corresponds to one nerve impulse.

Photographed from cathode ray oscilloscope. (Time = $\frac{1}{50}$ sec.)

often a pause in late diastole. Any increase in pulse pressure will increase this activity and cause still further inhibition of the vasomotor centre.

Receptors having a similar reflex effect are found in the heart (atrial and ventricular receptors).

Other Factors Influencing the Vasomotor Centre Activity

The vasomotor centre is situated in the floor of the fourth ventricle of the brain at the level of the apex of the *calamus scriptorius* which has been so named because this part of the medulla resembles the point of a pen.

The continuous activity of the cells in this region is modified by many factors in addition to the baroreceptor activity.

1. Carbon Dioxide

An adequate level of carbon dioxide in the blood is essential for the proper functioning of the vasomotor centre. An increase in the carbon dioxide in the blood supplying the centre will increase the centre's activity. Insufficient carbon dioxide, on the other hand, will lead to underactivity of the vasomotor centre and to a fall of blood pressure due to a reduction in the arteriolar tone. Under normal conditions, however, the respiratory centre (a collection of cells in the medulla controlling respiration) adjusts the breathing so that the correct level of carbon dioxide is maintained in the blood. When mechanical methods of respiration are employed, such as the cabinet respirator (iron lung) with a paralysed patient, care has to be taken to see that the patient is not over-ventilated. This would have the effect of washing out too much carbon dioxide, one of the effects of which would be to cause a fall in blood pressure (see also tetany, page 164).

2. Oxygen Lack and Chemoreceptors

A moderate degree of oxygen lack in the blood increases the vasomotor centre activity both by direct action on the vasomotor centre cells and indirectly via the *chemoreceptors* found in the carotid body and the aortic body. The nerve supply of this chemoreceptor tissue is the same as that of the adjacent baro-receptors (IX and X) [FIG. 50]. These cells sample the arterial blood for oxygen tension, and increase the activity in their nerves when the oxygen tension falls. They also respond to an excess of carbon dioxide when this occurs concurrently with an oxygen lack. In addition the chemoreceptors are stimulated by a reduction in their blood flow, such as occurs following a haemorrhage.

The main action of the chemoreceptors is to increase respiration when the oxygen in the arterial blood falls [p. 96], but, when stimulated, they also increase the activity of the vasomotor centre and cause a rise of the blood pressure.

FIGURE 51 shows the increase in activity of the vasomotor centre

CONTROL

B.P.
100 mm. Hg

Sympathetic
E.C.G.
Time

30 ml. HAEMORRHAGE

30
mm. Hg

Sympathetic

BLOOD RESTORED

100
mm. Hg

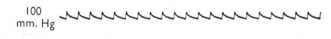

Sympathetic

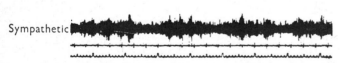

FIG. 51. Sympathetic vasoconstrictor activity, before, during and after a haemorrhage. Photographed from cathode ray oscilloscope.
Time = 0·1 second. (Green, J. H. and Heffron, P. F.)

following a haemorrhage in an animal. The marked increase in sympathetic activity which occurs when the blood pressure has fallen, is, in part, due to the decrease in baroreceptor inhibition of the vasomotor centre, and, in part, due to the chemoreceptor stimulation. In this case the haemorrhage was so severe that the

intense sympathetic activity, vasoconstriction and increased peripheral resistance, could not compensate for the drop in cardiac output, and, as a result there was a profound drop in blood pressure to 30 mm. Hg. On returning the blood the cardiac output and blood pressure were restored, and the sympathetic activity decreased to its original level.

3. Respiratory Centre

The activity in the neighbouring respiratory centre affects the vasomotor centre, and as a result, the vasomotor activity may fluctuate rhythmically. The maximum activity occurs during inspiration.

4. Sensory Nerves

Many sensory nerves, particularly those associated with pain, modify the activity. Mild degrees of pain cause an increase in vasomotor centre activity and a rise of blood pressure. More severe pain may decrease the vasomotor centre activity leading to such a severe drop of blood pressure that fainting occurs.

5. Higher Centres

In the normal conscious human being the part of the brain most likely to affect the vasomotor centre activity is the cerebral cortex, or in more general terms the 'higher centres'. In any form of excitement or emotional stress an increase in vasomotor centre activity occurs. This is often associated with an increase in cardiac output and the two together cause an increase of blood pressure. FIGURE 52 shows the blood pressure changes occurring in a

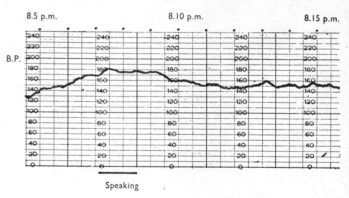

FIG. 52. The effect of stress on the blood pressure in man. At signal marker—subject speaks at a scientific meeting. (Blood pressure follower heated stylus recording.)

speaker at a scientific meeting. It will be seen that his systolic pressure rose from 130 to 170 mm. Hg before he actually spoke. On speaking the pressure rose further to 180 mm. Hg. In such conditions of generalized sympathetic overactivity, the suprarenal endocrine gland will be releasing noradrenaline and other catecholamines which will augment the activity of the sympathetic nervous system on the heart and arterioles. With all these factors in operation the baroreceptors are unable to prevent some rise in blood pressure occurring.

This pressure rise with emotional excitement is shown by patients with complete heart block where the heart is being paced by an electronic pacemaker and the heart rate is fixed.

In some forms of emotional shock the higher centres may have the exact opposite effect on the vasomotor centre and lead to vasodilatation of the arterioles and a fall of blood pressure. The fainting at the sight of blood, which occurs in some people, is an example of inhibition of the vasomotor centre by the cerebral cortex.

The factors affecting the activity of the vasomotor centre are summarized in FIG. 53:

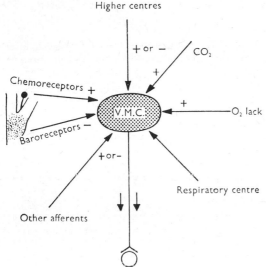

FIG. 53. Factors affecting the vasomotor centre activity.

OTHER FACTORS AFFECTING THE SIZE OF BLOOD VESSELS

Redistribution of Blood

The vasomotor centre controls the overall peripheral resistance of the circulation by its widespread *vasoconstrictor tone* to the arterioles throughout the body. However, when a tissue becomes more active it needs a greater blood flow for its metabolic requirements. To achieve this the vasoconstrictor tone from the vasomotor centre is modified, and sometimes overridden, by local mechanisms to allow vasodilatation to take place in that tissue. If the total peripheral resistance is to remain constant the vasoconstrictor tone elsewhere must be increased by the vasomotor centre. This will result in a redistribution of the blood flow; the blood will be directed away from the inactive tissue and sent to the active tissue.

Metabolites

Metabolites resulting from the metabolic activities of the tissues are vasodilator substances. The carbon dioxide and other metabolites such as adenylic acid and potassium ions released from the active muscle cells in exercise overcome the sympathetic vasoconstrictor tone and dilate the arterioles thereby increasing the muscle blood flow.

Carbon dioxide has now been seen to have two opposite effects on arteriolar tone. Acting centrally on the vasomotor centre it increases the vasomotor activity and vasoconstrictor tone, whilst acting locally it is a vasodilator substance. It is not surprising, therefore, that the effects on blood pressure of increasing the blood carbon dioxide by breathing a high carbon dioxide mixture are very variable. Its vasodilator action is, however, of importance in maintaining an adequate cerebral blood flow and mental changes rapidly occur if the blood carbon dioxide is lowered by voluntary over-ventilation.

In the coronary blood vessels supplying the heart, it is the oxygen lack or the metabolic products such as adenosine compounds resulting from oxygen lack that cause the vessels to dilate and increase the blood supply to the heart muscle when more work has to be done by the heart.

An accumulation of metabolites, resulting from an obstruction to blood flow to a contracting muscle causes *ischaemic pain*. The pain disappears within a few minutes of the circulation being restored.

The ischaemic pain associated with an insufficient blood flow to heart is termed **angina pectoris**. The pain is not localized to the heart itself, but is a referred pain [p. 191]. It has been likened to a band being tightened round the chest. The pain may radiate down the insides of the arms (upper thoracic nerve roots). The pain is relieved when the coronary vessels are dilated by sublingual nitroglycerine tablets. These act by relaxing the smooth muscle in the coronary vessel wall.

Reactive Hyperaemia

If the blood flow to a limb is occluded for a period of 1–5 minutes, vasodilator metabolites are formed. When the circulation is restored the arterioles and capillaries are seen to be dilated. The skin is warm (arteriolar dilatation) and flushed (capillary dilatation), and there is a marked increase in the limb blood flow of up to 20 times due to the simultaneous dilatation of the muscle blood vessels. The effect, termed *reactive hyperaemia*, usually subsides in 2–3 minutes.

Vasodilator Nerves

The sympathetic tone controlled by the vasomotor centre is vasoconstrictor in character. Vasodilatation, in general, is brought about by a decrease in the vasoconstrictor tone. However, certain blood vessels in the body are supplied with nerves which when active cause vasodilatation. Such are termed *vasodilator nerves*.

The parasympathetic nerve fibres supplying the salivary glands, and whose activity cause salivary secretion, also increase the blood flow to the gland. This was demonstrated by Claude Bernard as long ago as 1858 when he stimulated the chorda tympani and noted an increase in blood flow as well as a secretion of saliva. It is now thought that part of this vasodilatation may be due to the formation of the peptide *bradykinin* as a result of the nerve activity.

arg—pro—pro—gly—phe—ser—pro—phe—arg
(Bradykinin)

The parasympathetic fibres to the external genitalia are vasodilator in character and increase the blood flow to the *penis* in the male and the *clitoris* and *labia* in the female in sexual excitement. It is for this reason that the sacral parasympathetic nerves are termed the *nervi erigentes* (=the nerves which cause an erection).

Although most of the sympathetic nerve fibres to voluntary muscle are vasoconstrictor, a few are vasodilator. Special techniques are necessary to show up their presence. For example, constrictor fibres degenerate more rapidly. Bowditch and Warren in 1886 showed that if the sciatic nerve is stimulated 2–3 days after cutting, vasodilatation instead of vasoconstriction is obtained.

At the onset of a faint due to a severe haemorrhage, the profound vasoconstriction changes suddenly to a vasodilatation and the blood pressure falls. This vasodilatation which occurs mainly in the muscles is too great to be accounted for by a reduction in the sympathetic tone. It is due to activation of the vasodilator fibres in muscle, not by the vasomotor centre, but directly from the hypothalamus via the reticulo-spinal pathways which open shunts between the arteries and veins and thus lower the peripheral resistance.

It should be remembered that the increase in muscle blood in exercise is due to metabolites which over-ride the sympathetic vaso-constrictor tone, and is not due to these sympathetic vasodilator fibres.

Axon Reflex and Dorsal Root Nerves

In 1876 Stricker found that the stimulation of the dorsal nerve roots of the spinal cord, which are normally conveying sensory impulses from the sense organs into the central nervous system, resulted in the dilatation of the vessels in the area of skin and muscle supplied by the roots. Impulses generated by the stimulation passed backwards (antidromically) along the sensory nerve. This result was confirmed by Gartner in 1889 and again by Bayliss in 1901. There is, however, no evidence that vasodilator impulses ever leave the spinal cord by this route. At one time it was thought that these results might be due to the stimulation of aberrant motor fibres which were leaving by the dorsal nerve root of the spinal cord instead of by the ventral nerve root. Degeneration experiments, after cutting the nerve at either A or B [FIG. 54], showed that this

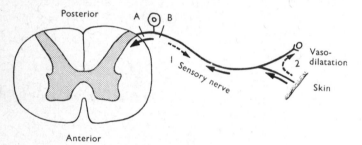

FIG. 54. The axon reflex. Section at B causes the fibres to degenerate. Section at A does not.

was not so. These experiments demonstrated that the cells of origin of these fibres lie in the spinal (dorsal root) ganglia, as is the case with a normal sensory nerve, and not in the spinal cord as is the case with the motor nerves. The nerve fibres degenerated when cut off from their cells of origin by a cut at B but not when the cut was at A. A cut at either A or B would cause degeneration in a fibre having its cell of origin in the spinal cord.

It is now thought that the nerves stimulated are normal sensory nerves, and that they are connected, not only with the sense organs, but also with the neighbouring arterioles as in FIGURE 54. [See also herpes zoster, p. 191.]

Response of the Skin Vessels to Injury

The only apparatus needed to carry out the following simple experiments is a pin.

White Line

If the skin is lightly stroked with a pin, then after a slight delay a white line appears exactly limited to the area stimulated. It lasts for a minute or so and then disappears. It can be elicited after all the nerves to the skin have degenerated and is due to a local capillary contraction. It is not due to arteriolar contraction, because the arterioles supply a larger and more irregular area. It can be

obtained after the arterial blood flow has been obstructed. Con-striction of the arterioles when the arteries are obstructed would not empty the capillaries, but would drive more blood into them.

The capillaries will constrict down even when the blood pressure inside them is as high as 50 mm. Hg. Once shut a pressure of 90 mm. Hg is needed to open them again [see p. 55].

Triple Response

If a greater pressure is used when stroking the skin a sequence of three events occurs [FIG. 55]. First a **red line** appears which is

FIG. 55. The triple response. A red line, flare and wheal develop after injury to the skin due to scratching. The wheal may project 1–2 mm. above the skin surface.

bright red at first, and then becomes bluish indicating stagnation of the blood. This is due to capillary dilatation, and, like the white line, is seen after nerve degeneration.

Secondly, after a delay of 30 seconds a **flare** or flush appears. This is due to arteriolar dilatation. The colour is bright red and it does not become bluish, indicating a rapid blood flow. The area corresponds to the arteriolar distribution. There is a rise of skin temperature due to vasodilatation. This effect is dependent upon an intact sensory nerve supply. It is taking place over the local nerve network [FIG. 54] and is termed an **axon reflex**. Degeneration of nerves following section at A has no effect on the reflex whereas section at B abolishes it. Hence the cell of origin of the nerve is in the spinal ganglion.

Thirdly a **wheal** or blister develops due to the leaking of fluid and protein from the capillaries.

Many agents which mildly injure the skin produce an identical triple response. A few of these are pricking, scratching, freezing, burning, electric currents, acids, alkalis and insect bites. Thus many dissimilar agents produce the same complex response, and the only common factor is that they injure the skin. It is thought that the skin injury liberates a substance such as *histamine*, or a mixture of histamine and polypeptides, collectively known as *H-substance*, and that this H-substance elicits the triple response. Histamine itself injected under the skin produces a typical triple response.

Hypersensitivity and Anaphylaxis

When a foreign protein enters or is injected into the body for the first time, it produces no obvious effect, but hypersensitivity develops. This shows itself when a second injection is given some weeks later. Then the symptoms of histamine poisoning develop. It is thought that the *antigen—antibody reaction* occurring in the body between the foreign protein (the *antigen*) and the *antibody* formed by the body, liberates histamine.

The guinea-pig is very sensitive to histamine and, in this animal, histamine release may cause a fatal constriction of the air-ways (bronchoconstriction).

This state of hypersensitivity is termed *anaphylaxis* (Greek = without protection) or *anaphylactic shock*. The symptoms are spasm of smooth muscle, capillary dilatation, increased capillary permeability and glandular secretion. The triple response shows itself as a skin rash.

The various *allergies* are examples of this hypersensitivity. The antigen is usually a protein, but non-protein substances and even metals such as nickel may occasionally act as antigens. Once sensitized to nickel, for example, the wearer would develop a rash from any nickel fasteners on the clothing that come in contact with the skin.

The effects of the released histamine are, in part, minimized by the antihistamine drugs.

Auto-immunity

Any protein can act as an antigen and bring about the formation of an antibody. The antigen and antibody combine and the antigen is destroyed or inactivated. Such inactivation can often be demonstrated *in vitro* as a precipitation or agglutination.

The body is able to distinguish between its own body protein which it tolerates, and a foreign protein which it rejects. Occasionally the body fails to recognize one of its own body proteins and proceeds to make an antibody against it. This process is termed **auto-immunity,** and the antibody is termed an **auto-antibody.** Patients with Hashimoto's disease (lymphadenoid goitre) destroy their own thyroid tissue by making an antibody against it. Some patients with pernicious anaemia (p. 10) are thought to have made an antibody against the intrinsic factor. *Rheumatoid arthritis ?*

Heat and Blood Vessels

The skin is the only organ of the body whose blood supply is not regulated according to its metabolic need. The blood flow to the skin is regulated by the heat-regulating centre in the hypothalamus in the interest of body temperature regulation [see CHAPTER 14].

The skin of the hands (particularly the fingers) and feet contain numerous vascular connexions between arteries and veins (A–V shunts). These A–V shunts are normally kept closed by the tonic sympathetic vasoconstrictor activity. When the body temperature rises, these shunts open (by a reduction in the sympathetic vasoconstrictor tone) and the blood flow to these parts of the body greatly increases.

The fact that heat dilates the blood vessels of the skin can be readily seen after a hot bath. The skin is red and warm due to the vasodilatation of the underlying vessels. It can be shown experimentally that vasodilatation has also occurred in the deeper vessels such as those in the muscles. Local heat produces vasodilatation by a direct action on the blood vessels and by the release of H-substance. In addition there are remote effects. If one leg is placed in hot water at 40–45 °C., dilatation also occurs in the other leg and throughout the body, especially on the hands and face. This is, in part, due to the warm blood from the heated limb arriving at the temperature regulating centre and, in part, due to nerve impulses arriving from temperature receptors in the warmed limb. This double pathway may be confirmed experimentally in the following manner. Remote vasodilatation still occurs after: (a) obstructing the venous return from the warmed limb. This vasodilatation is brought about by the nervous mechanism. However, it still occurs after, (b) blocking the nerves with a local anaesthetic, but not obstructing the venous return. This vasodilatation is due to the warm blood reaching the hypothalamus. It does not occur after both (a) and (b).

Cold

The effect of cold is more complex. In general it causes vasoconstriction. At a temperature of 10–20 °C. constriction of the arterioles results in blood being trapped in the capillaries of the skin. This blood will lose oxygen and the skin will become blue in colour giving rise to local cyanosis. If the capillaries are also constricted the skin will become white in colour.

If the hands are placed in cold water 5–10 °C. they may become bright red in colour. This phenomenon is termed **cold induced vasodilatation** (C.I.V.D.). It is due to the axon reflex and is probably a protection mechanism against the local effects of cold. The bright colour is also due to the fact that the dissociation of oxy-haemoglobin is slowed down and may cease at low temperatures. Studies in the Antarctic show that this effect is reduced after one year's exposure to a cold environment (Elkington, 1968).

Low temperatures may cause the subcutaneous fat to solidify. At normal body temperature, the body fat is in the liquid state as an oil.

Freezing of tissues causes extensive cell damage because of the formation of ice crystals.

The remote effects of cold are to produce generalized vasoconstriction. This may be sufficient to cause a detectable rise in blood pressure.

The blood vessels in the extremities of some people are more than usually sensitive to the effects of cold. In the hands and feet, even the *arteries* may go into such profound vasoconstriction that the fingers and toes become completely numb when the environmental temperature falls only slightly. This condition is known as *Raynaud's disease.* These patients are often helped by the removal of the sympathetic vasoconstrictor tone to the blood vessels. This is achieved by interrupting the sympathetic nerve pathway by drugs or by cutting the sympathetic nerve fibres at a sympathectomy operation.

Angiotensin   *also ↓ Na⁺*

An inadequate blood flow to the kidneys leads to the release of renin into the blood stream. Renin is a protein-splitting enzyme which acts on angiotensinogen (hypertensinogen) which is present in the α_2-globulin fraction of the plasma proteins to form angiotensin (angiotonin, hypertensin). Angiotensin is a peptide and is first released as angiotensin I with 10 amino acids. This is converted by a plasma enzyme to angiotensin II, which has eight amino acids, by the removal of the two terminal amino acids.

angiotensinogen $\xrightarrow{\text{renin}}$ angiotensin I $\xrightarrow[\text{enzyme}]{\text{plasma}}$ angiotensin II
(decapeptide)　　　　　　(octopeptide)

asp—arg—val—tyr—ileu—his—pro—phe—his—leu
(angiotensin I)

asp—arg—val—tyr—ileu—his—pro—phe
(angiotensin II)

Angiotensin is a powerful vasoconstricting agent which causes a widespread vasoconstriction and an increase in blood pressure. It also brings about the release of aldosterone [p. 166] and stimulates the thirst centre [p. 129].

Myogenic Tone—Bayliss–Folkow Hypothesis

In addition to the central and local control of blood vessels, the inherent *myogenic tone* of these vessels must be considered. In 1902 Bayliss showed that the expansion of the vessel wall brought about by the pulsatile nature of the blood pressure, causes the smooth muscle of the vessel wall to contract and thus to increase its vascular tone. This theme was elaborated by Folkow in 1964 and may be referred to as the Bayliss–Folkow hypothesis. Such a mechanism allows vascular tone to return to blood vessels after all nervous and hormonal influences have been removed.

VISCOSITY

The peripheral resistance depends not only on the size of the blood vessels, but also on the viscosity of the blood. It is directly proportional to the viscosity.

$$\text{Peripheral resistance} = \text{constant} \times \frac{\eta}{r^4}$$

where r is the radius of the vessel and η is the viscosity of the blood.

The viscosity of the blood usually remains constant, but should the viscosity increase, the resistance of the blood vessels will increase even though their size remains unaltered. In animals the contraction of the spleen may force red cells out into the circulation and appreciably increase the viscosity of the blood.

Blood has a viscosity two and a half times that of water. Part of this increase in viscosity is due to the cells and part is due to the plasma. Thus if normal saline were circulating through the arterioles in the place of blood and no alteration in the calibre of the arterioles occurred, then the peripheral resistance would only be one-third. With the same cardiac output, the mean blood pressure would only be 37 mm. Hg. Plasma substitutes such as *dextran* have a viscosity similar to that of blood.

The viscosity of blood increases markedly with an increase in haematocrit [TABLE 3]. Thus polycythaemia (high red cell count) may be associated with a high blood pressure.

TABLE 3

Haematocrit	Viscosity	Haematocrit	Viscosity
20	1·6	60	3·8
30	1·8	65	4·5
40	2·3	70	5·3
45	2·8	80	9·0
50	3·0		

The viscosity of water = 1·0.

CARDIAC OUTPUT

Cardiac output = Stroke volume × Heart rate

(Output per ventricle　(Output per ventricle　(Beats per
per minute)　　　　　per beat)　　　　　minute)

The factors affecting the cardiac output were studied in detail by Starling in the years 1907–15 using his heart-lung preparation. Although Starling's conclusions have come under a great deal of criticism in recent years because they do not adequately explain the changes which occur in exercise, nevertheless they are of great importance when considering the action of the heart as a pump circulating blood round the body.

In the heart-lung preparation the systemic circulation in the dog was replaced by tubing having an adjustable resistance [FIG. 56]. The stroke volume was measured by enclosing the ventricles in an air-tight chamber connected to a volume recorder. Changes in volume with each contraction gave the stroke volume of the two ventricles, and half this quantity multiplied by the heart rate gave the cardiac output per ventricle.

Such studies led to three important, and in some ways remarkable, properties of the heart as a pump:

1. The cardiac output is independent of the peripheral resistance (cardiac output and peripheral resistance).

2. The heart transfers the blood it is given on the venous side to the arterial side (cardiac output and venous return).

3. Changes in heart rate do not *necessarily* change the cardiac output (cardiac output and heart rate).

These properties will be considered in turn:

1. Cardiac Output and Peripheral Resistance

From 1 we see that the heart acts as a constant-flow pump rather than a constant-pressure pump. Until the limit is reached, the heart is able to maintain a constant cardiac output and therefore a constant blood flow through the systemic circulation against an ever-increasing peripheral resistance. It does this by increasing the blood pressure. As the blood pressure rises, more work has to be done by the heart in order to maintain the same output. This is completely different from a constant-pressure system, such as the domestic water supply, where altering the resistance (by turning the tap) changes the rate of flow, the supply pressure remaining constant.

The fact that peripheral resistance changes do not affect the cardiac output means that overall vasoconstriction of the arterioles will lead to an increase in blood pressure whereas vasodilatation will lead to a fall in blood pressure.

Obviously, if the vasoconstriction is such that the arterioles close altogether, the circulation will cease because no blood will return to the heart.

2. Cardiac Output and Venous Return

Under normal conditions each side of the heart is able to transfer all the blood it is given from its venous to its arterial side. If for any reason the blood returning to the heart increases, more ventricular filling will take place during diastole. The ventricular muscle fibres will be stretched and on the next systole will give a more powerful contraction. The resultant increased stroke volume transfers the blood to the arterial side. The fact that there is increase in stroke volume associated with an increase in the initial length of ventricular muscle fibres is known as 'Starling's Law of the Heart'.

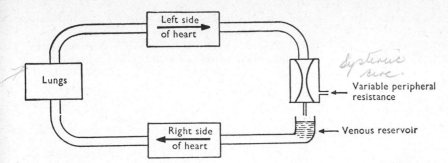

Systemic circ.

FIG. 56. The basic arrangement of the heart-lung preparation. The systemic circulation is replaced by a variable peripheral resistance which consists of a thin-walled rubber tube. The resistance is adjusted by altering the external pressure applied to this tube. The blood is collected in a venous reservoir and returned to the right side of the heart.

This property is important in ensuring that the outputs of the two ventricles are matched. Should the output of one ventricle exceed that of the other, the increased venous return to the weaker ventricle will increase its output.

Heart Failure

If the ventricles fail to deal with the venous return then the blood will accumulate on the venous side. The veins near the heart will become distended and the venous pressure will rise. This condition is known clinically as 'heart failure'. The neck veins will become engorged with blood. Normally all veins above the manubrio-sternal junction are collapsed since the pressure inside them is sub-atmospheric. The high venous pressure in heart failure will lead to oedema which is characteristic of the condition [see p. 59].

Venous Return and Stroke Volume — *amt. ejected / beat*

The stroke volume is determined by the quantity of blood in the ventricles at the onset of systole. During diastole the tricuspid valve is open and the pulmonary valve is shut. The right ventricle is in communication with the veins. The amount of blood in this ventricle will depend upon its size and distensibility compared with that of the veins.

The veins are thin-walled vessels containing at any time about three-fifths of the blood in the circulation. The volume of these veins is the main factor in determining the capacity of the circulation.

When the venous pressure is low a vein has an elliptical cross-section. It becomes more circular as the pressure increases [FIG. 57]. A circle has a larger area than an ellipse of the same circumference. When adopting the erect position the conversion of the

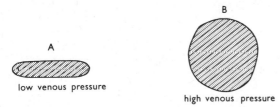

FIG. 57. Change in cross-sectional area of vein with increase in venous pressure.

leg veins from an elliptical to a circular cross-section due to the effect of gravity will greatly increase their capacity. The veins have a muscular coat which receives tonic sympathetic activity (venomotor tone) in much the same way as the arterioles. Alterations in the venomotor tone produce changes in the capacity of the circulation without affecting their resistance to blood flow since they are large diameter vessels.

A very delicate balance is maintained between the capacity of the circulation and the blood volume [FIG. 58]. The blood available to fill the ventricles for the next systole is the difference between these two. Thus if the blood volume is 5 litres and 2×70 ml. are required for filling the ventricles, the capacity of the rest of the circulation must be exactly $5,000 - 140 = 4,860$ ml. An uncompensated increase of only $1\frac{1}{2}$ per cent. in the capacity of the

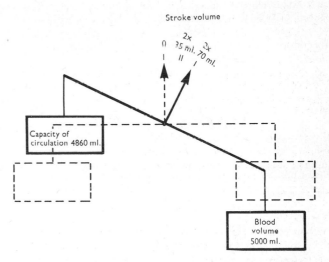

FIG. 58. The venous return to the heart is dependent upon the blood volume being slightly in excess of the capacity of the rest of the circulation. Any increase in the capacity of the circulation will reduce the blood available to the heart and hence cause a reduction in the cardiac output.

An uncompensated increase in the capacity of the circulation from 4,860 ml. to 4,930 ml. will reduce the stroke volume of each ventricle from 70 ml. to 35 ml.

A sudden increase in the capacity of the circulation to 5,000 ml. or more will cause the venous return (and hence the cardiac output) to fall to zero!

circulation would decrease the stroke volume to one half! The capacity would now be 4,930 ml. and, therefore, only 35 ml. blood would be available for each ventricle. With an uncompensated increase of 3 per cent. there would be no cardiac output at all!

The venomotor tone plays an important part in determining the stroke volume. If the veins contract, more blood is forced into the ventricles giving an increased stroke volume, and an increased cardiac output. If the blood volume is reduced, following a haemorrhage for example, the capacity of the circulation is reduced by venoconstriction to compensate for this. Should this compensation be incomplete the blood in the ventricles at the end of diastole will be reduced leading to a small stroke volume and a reduced cardiac output. Notwithstanding the profound arteriolar vasoconstriction, the blood pressure falls. The low cardiac output means a poor blood flow to the tissues which become anoxic leading to the state of *surgical shock*.

Replacement of the blood, lost by a blood transfusion, has a dramatic effect on the cardiac output by making more blood available for filling the ventricles. The length of the ventricular muscle fibres at the end of the filling phase is greater and by Starling's law of the heart, there is an increase in the stroke volume. This is particularly true in the early stages of shock; in the later stages the heart may be unable to deal with this additional blood (irreversible shock).

Blood Volume and Capacity

It has been seen that a delicate balance exists between the capacity of the veins and the blood volume. The sites of the volume receptors which must be regulating the venomotor tone to maintain this equality have not been accurately localized, but they are thought to be in the thorax and may be the atrial receptors.

Changing from the lying to the upright posture leads to pooling of the blood in the dependent parts and even with compensation there is a fall in cardiac output from 5 litres per minute to 4 litres per minute. The blood pressure, however, is maintained at the same level by arteriolar constriction.

Two generalizations may now be made:

1. *Arteriolar vasoconstriction causes an increase in peripheral resistance.*

2. *Venous vasoconstriction (venoconstriction) leads to an increase in stroke volume and cardiac output.*

An increase in blood volume by a blood transfusion after a haemorrhage has the same effect as a reduction in the capacity of the veins since it is balance between the blood volume and capacity that is important.

Muscle Pump

The presence of valves in the veins breaks up the columns of blood and prevents excessive venous distension. Should the valves

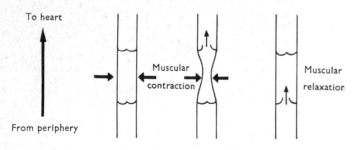

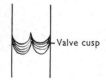

FIG. 59. Valves in veins. The alternate contraction and relaxation of the muscles surrounding the veins helps to return the blood back to the heart. This is known as the 'muscle pump'.

become faulty, such as in varicose veins, then local venous distension may occur particularly in the standing position.

The contraction of skeletal muscles squeezes the segment of veins and the presence of valves forces the blood towards the heart [FIG. 59]. Filling the venous segment occurs from the distal end in the interval between muscular contractions. This forms the muscle pump which speeds up the circulation by aiding the venous return and facilitating the return of blood from the legs and feet.

Gravity and Venous Return

Gravity greatly hinders the return of blood from the dependent parts. The blood tends to collect in the veins below the heart. These veins are then distended by the blood in them. Should the venous tone fail there would be no venous return and the cardiac output would drop to zero. This is exactly what happens when the body is subjected to accelerations several times that of gravity (several g). Unless precautions are taken, a black-out will occur especially if the feet are trailing. This is a problem whenever high g's are encountered such as with astronauts during the take off and during the firing of the booster rockets. The effect is overcome by arranging that the astronaut is on a couch facing the direction of travel but with his body at right-angles to the direction of travel. His legs are bent and raised and he is enclosed in a pressure suit. The acceleration now presses the astronaut's body against the couch and the effect on the circulation will be to force the blood from the anterior to the posterior part of the body. For landing, the space capsule is turned round so that the deceleration will act in a similar manner.

Pilots in high speed conventional planes are seated in the normal sitting position but employ pressure suits over the lower part of the body to prevent the accumulation of blood in the legs when making a tight turn with the feet outward.

Respiratory Pump

So far it has been assumed that the pressure in the great veins is 0 mm. Hg (that is, at atmospheric pressure). However, the heart lies in the thorax, and the thin-walled atria are exposed to a sub-atmospheric pressure that varies between −2 mm. Hg in expiration and −8 mm. Hg in inspiration [see p. 28]. These negative figures mean that if the barometric pressure is 760 mm. Hg then the pressure in the thorax will vary between 758 when breathing out and 752 when breathing in. The pressure gradient along the veins is thus 2–8 mm. Hg greater than anticipated and each inspiration sucks blood towards the heart.

Respiration increases in exercise and the pressure now ranges between −2 and −30 mm. Hg. The mean pressure throughout the respiratory cycle has now increased to −16 mm. Hg which will greatly improve the venous return back to the heart. This is known as the respiratory pump.

If the chest is open, as in thoracic surgery, this respiratory pump action is lost and there is a fall in cardiac output. The atria and other thin walled structures will be exposed to atmospheric pressure (0 mm. Hg).

Valsalva Manœuvre

The venous return may be temporarily modified for experimental purposes by the following manœuvre. A forced expiration is made against a closed glottis. In this way the intrathoracic pressure can be increased to as much as +80 mm. Hg. With this high pressure there is no venous return since no venous blood can enter the thorax. The blood in the lungs and heart will be expelled at a higher pressure than normal. The result is that for a few beats a normal or increased stroke output is obtained from the left ventricle into the arteries. Then the output falls since there is no venous return. In time the venous pressure will build up to the intrathoracic pressure and then some venous return occurs. Such a manœuvre can only be kept up for a relatively short time. At the end a deep breath is taken. The blood enters the thorax and fills the empty lung vessels so that little is available for the left ventricle for several beats. The subject may feel faint until the normal circulation conditions are re-established.

This manœuvre dates back to the seventeenth century when it was introduced by Valsalva as a method of testing the patency of the Eustachian tubes which lead from the nasopharynx to the middle ear. A forced expiration is made as above but with the mouth and nose closed instead of the closed glottis. If the tubes are patent air will be forced into the middle ear (tympanic cavity).

Circulatory changes of a similar, but smaller, nature occur when positive pressure ventilation is employed with the chest closed. The lungs are inflated by forcing air into the air passages for inspiration. The elastic recoil of the lungs is allowed to bring about expiration. The overall increase in thoracic pressure impedes the return of blood to the heart and reduces the cardiac output. It will be noted that owing to the time taken for the blood to pass through the heart and lungs, there will be a delay of several seconds before changes in venous return are reflected in changes in left ventricular output and systemic blood pressure. Rhythmic changes in blood pressure will therefore be 'out of phase' with the rhythmic changes in venous return.

Stroke Volume and Exercise

In exercise the increase in stroke volume is in part brought about by a different mechanism. X-ray evidence shows that the heart is not markedly larger at the end of the filling phase, as postulated by Starling's law of the heart, but may be the same size or even smaller. In this case the increase in stroke volume is brought about by a more complete emptying of the ventricles. Thus the volume at the end of systole (end-systolic volume) is less. This more powerful contraction of the ventricular musculature (*positive inotropic action*) is in response to sympathetic stimulation of the cardiac muscle from higher centres. Sarnoff (1954) has shown that the efficiency of the ventricular muscle increases when the sympathetic nerves are stimulated.

The increase in stroke volume in exercise may thus be seen to be due to the increased *end-diastolic ventricular volume* as a result of the increased venous return (due to the action of the muscle and respiratory pumps and the vasodilatation of the muscular vascular beds due to metabolites) together with a decreased *end-systolic volume* as a result of the increased sympathetic activity. But as will be seen, much of the increase in cardiac output in exercise is due to the increase in heart rate.

Stroke Volume—Summary

Stroke Volume

= (Volume at beginning of systole − Volume at end of systole)
= (End-diastolic volume − End-systolic volume)

      ↑   ↑

Venous return   Sympathetic activity
Capacity of veins

3. Cardiac Output and Heart Rate

For many years following Starling's work it was thought that alterations in heart rate had no effect on the cardiac output, but in recent years there has been a reappraisal of the situation and it is now known that the two are frequently closely connected.

The heart rate is controlled by the cardiac centre in the medulla. This may be subdivided into: (a) a cardiac inhibitory centre which is the dorsal nucleus of the vagus and gives rise to the continuous vagal activity known as vagal tone, and (b) a cardiac accelerator centre which controls the cardiac sympathetic activity.

In a state of mental and physical rest, the sympathetic activity is minimal and the heart rate is controlled by the vagal tone. This is an inhibitory tone which slows down the rate of the pacemaker to 70 beats a minute (*negative chronotropic action*). It was first described by the Weber brothers (Ernst and Eduard) in 1845. The vagus to the heart was one of the first nerves to be shown to have an inhibitory rather than a stimulating action. As a result the Webers' findings did not meet with a ready acceptance at the time. In 1888 Laborde localized the cardiac vagal centre to the floor of the fourth ventricle. Miller and Bowman (1915) identified the centre with the dorsal motor nucleus of the vagus.

FIGURE 60 shows the effect of stimulating the vagus nerve. The heart slows and there is a fall in blood pressure.

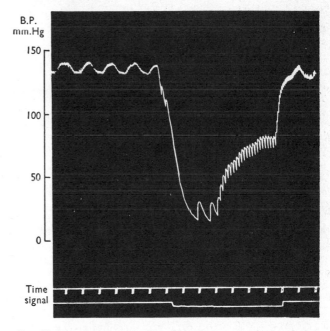

FIG. 60. A smoked-drum tracing of blood pressure using a mercury manometer. The excursions of blood pressure at the commencement of the record are due to respiration.
At the signal the vagus nerve was electrically stimulated causing the heart to slow and the blood pressure to fall. At this slow heart rate the mercury manometer is more able to follow the changes in systolic and diastolic blood pressure and the excursions shown during the period of stimulation are caused by the heart itself. The partial recovery whilst the stimulation is continuing is known as 'vagal escape'. Time = 5 seconds.

The vagal tone acts as a 'brake' on the heart. The greater the degree of vagal activity the slower the heart. Conversely if the vagal tone is reduced the heart speeds up. The small changes in heart rate which are continually occurring are due to the variations in this vagal tone.

If all the vagal tone is removed, the heart rate would be 150 beats per minute. But heart rates of up to 210 beats per minute may occur in very severe exercise and in extreme emotional excitement. Thus there must be some mechanism for stimulating the pacemaker. This is the sympathetic activity which has a *positive chronotropic action*. It is particularly active in emotional excitement and exercise—the 'fight or flight' response as Cannon described it. The sympathetic activity will be augmented by the release of noradrenaline and adrenaline as hormones from the suprarenal medulla. The fact that sympathetic stimulation causes cardiac acceleration was first shown by Vierordt in 1855 and confirmed by von Bezold in 1863 and Cyon in 1866.

Factors Affecting the Cardiac Centre

The cardiac centre like the vasomotor centre is under the influence of many factors both humoral and nervous [FIG. 61]. These

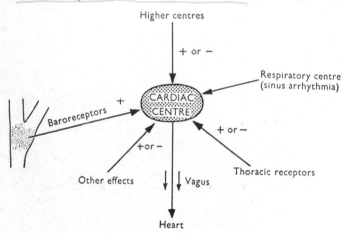

FIG. 61. Factors affecting the activity of the cardiac inhibitory centre.

will be discussed in terms of the cardiac inhibitory centre. They will have the opposite effect on the cardiac accelerator centre.

The baroreceptors stimulate the cardiac inhibitory centre. Thus an increase in blood pressure or an increase in the pulsatility will increase the vagal tone and slow the heart. The relationship between the blood pressure rise and the heart rate fall is sometimes known as Marey's Law, but it is preferable to consider it as a baroreceptor reflex because of the exception seen below.

The rise in blood pressure will thus cause a reduction in the heart rate and a reduction of the cardiac output. This reduction in cardiac output will act in the same direction as the vasomotor changes (vasodilatation) to limit the rise in blood pressure.

Conversely a fall in blood pressure will decrease the vagal tone and speed up the heart. There will be an increase in the cardiac output and with the associated vasoconstriction from the action of the baroreceptors on the vasomotor centre the blood pressure fall will be limited.

It will be noted that these reflexes do not completely prevent the blood pressure changes from occurring. If they did there would be no change in the baroreceptor stimulation. Further, they never reverse the trend. They simply minimize the changes that do occur. Often only a change in the pressure waveform is seen with very little change in the mean blood pressure.

When the blood pressure increase is primarily due to an increase in the heart rate and thus cardiac output, such as occurs in exercise, then Marey's Law does not apply. Obviously the baroreceptor reflexes will not slow the heart to below the resting level in such a condition because if they did they would take away their own drive. They will, however, by their action oppose the heart rate increase so that the increase is less than it otherwise would have been.

In children and young adults deep breathing produces a fluctuation in vagal tone and heart rate. The heart speeds up during inspiration and slows down during expiration. This is called *sinus arrhythmia* and is mainly due to irradiation from the respiratory centre to the cardiac centre. FIGURE 62 shows sinus arrhythmia recorded in a young adult. The heart sounds, expansion of the chest, brachial pulse and electrocardiogram were recorded simultaneously. It will be seen from the intervals between successive R-waves that the heart speeded up during inspiration and slowed during expiration. A study of the intervals between the heart sounds will show that the change is in the duration of diastole (2–1). The duration of systole (1–2) remains constant.

Higher centres affect the activity of the cardiac centre. Emotional excitement speeds up the heart by a reduction in vagal tone and an increase in cardiac sympathetic activity. Heart rates of 150 beats per minute have been found in airline pilots during take-off and landing compared with the normal heart rates of 70 beats per minute when airborne. Even higher heart rates (of over 200 beats per minute) have been reported in Grand-Prix racing drivers not only during the race, but also before the race starts—whilst waiting for the starting flag. On the other hand cardiac slowing due to vagal overactivity may lead to fainting in cases of sudden shock.

The stimulation of sensory nerves produces varying effects on heart rate. Most painful stimuli quicken the heart, but a blow on the abdomen, for example, may cause slowing of the heart.

Oxygen lack and carbon dioxide excess acting on the cardiac centre increase the heart rate. Oxygen lack acting via the chemoreceptors has a complex action. Chemoreceptor stimulation probably slows the heart by a primary action but the resulting increase in respiration causes cardiac acceleration by another reflex mediated from receptors in the lungs. Once again the overall result is cardiac acceleration.

A rise in body temperature acts directly on the pacemaker (S.A. node) and also on the cardiac centre, causing an increase in heart rate. Circulating thyroxine from the thyroid gland [p. 162] accelerates the heart by a direct action on the pacemaker.

Bainbridge Effect

Bainbridge showed in 1915 that an increased venous return to the heart causes an increase in heart rate. Provided that the initial heart rate is slow, this effect may be readily demonstrated. However, no receptors have so far been found in the great veins or right atrium to form the afferent pathway for this reflex. The venous and atrial receptors so far discovered behave like the baroreceptors and when stimulated by an increased venous filling cause cardiac slowing rather than cardiac acceleration. It may be that the Bainbridge effect is not a reflex but is due to the increased filling of the right atrium increasing the rate of firing of the S.A. node which lies in this atrial wall.

Although the Bainbridge effect may play a small part in accounting for the increase in heart rate found in exercise, it is much more

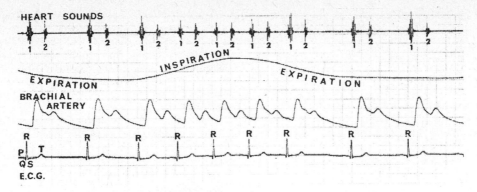

FIG. 62. Sinus arrhythmia in a young adult. The heart speeds up during inspiration and slows during expiration. Time = 0·2 seconds.
Simultaneous recordings of the heart sounds (first heart sound = 1, second heart sound = 2), respiration (recorded as changes in chest circumference), brachial pulse and electrocardiogram (Lead II). Note the changes in the interval between successive R-waves of the electrocardiogram (E.C.G.), and that the interval between the second and first heart sounds (diastole) becomes less as the heart speeds up, whilst the interval between the first and second heart sounds (systole) remains unchanged. Note also the fluctuations in diastolic blood pressure as a result of this alteration in the length of diastole. (Ink-jet recording.)

likely that the higher centres responsible for the voluntary movement are also responsible for the cardiac acceleration. Cardiac acceleration may take place in a competitor in an athletics race whilst waiting for the gun to go off before a single step has been taken.

Influence of Heart Rate on Cardiac Output

At first sight it would appear that since:

$$\text{Cardiac output} = \text{Heart rate} \times \text{Stroke volume}$$

that an increase in heart rate would always give an increase in cardiac output. Starling using his heart-lung preparation on the other hand found that an increase in heart rate had no effect on cardiac output. These two views are diametrically opposite. It now appears that the truth is somewhere between the two.

TABLE 4

Oxygen used (l./min.)	Heart rate (beats/min.)	Stroke volume (ml.)	Cardiac output (l./min.)	Work done (kg.m./min.)
0·25	60	110	6·6	At rest
1	100	120	12	350
1·5	120	120	14·4	550
2	160	120	19·2	900
3	175	120	21	1,250

Based on Asmussen and Nielsen (1952) (simplified)

Within limits the heart can transfer the blood from the venous side to the arterial side either with a slow heart rate and a large stroke volume or a fast heart rate and a small stroke volume, the product being the same in the two cases.

However, if the rate is very fast or very slow then the heart becomes inefficient. If the rate is excessively high, diastole will be too short for adequate filling and for the proper recovery of the heart muscle. If the heart rate is excessively low, there will be an insufficient number of beats per minute to transfer the blood from the venous to the arterial side. The pericardium limits the size of the heart and the maximum stroke volume.

In man an alteration in heart rate may or may not alter the cardiac output. It is unwise to assume that it does unless it is known that the stroke volume is unaffected. But in exercise, up to quite high levels of work, the increase in heart rate may account for practically all the increase in cardiac output with the stroke volume remaining constant [see TABLE 4].

In very severe exercise with cardiac outputs up to 40 litres per minute an increase in stroke volume must occur since such outputs require a stroke volume of 200 ml. with a heart rate of 200 beats per minute:

Cardiac output (40 litres/minute)
= Heart rate (200 beats/minute) × Stroke volume (200 ml.)

CAPILLARY BLOOD PRESSURE

The blood pressure falls from a mean value of 100 mm. Hg to 0 mm. Hg as the blood flows from the arteries through the arterioles and the capillaries to the veins. The greatest pressure drop occurs in the region of greatest resistance. It is a well-known experience that when a large number of people are trying to leave a hall quickly through a narrow doorway, this doorway offers a resistance to their movement, and quite a considerable crush or pressure may be developed on the hall side of the door. Having passed through the door this pressure disappears.

Precapillary Resistance Vessels

As has been seen the main resistance to blood flow in the systemic circulation lies in the arterioles before the capillaries are reached. The arterioles are the *precapillary resistance vessels* and it follows therefore that most of the drop in pressure will occur across these vessels.

The pressure gradient throughout the circulation is shown in FIGURE 63. The pressure falls from 100 mm. Hg (*13·3 kPa*) to

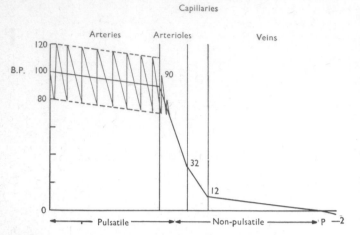

FIG. 63. The blood pressure throughout the circulation. The pressure falls as the blood flows through the arteries, arterioles and capillaries to the veins. The pressure is pulsatile in the arteries, but this pulsation has disappeared by the time the capillaries have been reached. The pressure becomes pulsatile again in the veins as the heart is approached (P).

about 90 mm. Hg (*12 kPa*) as the blood flows from the aorta to the small arteries. As it passes through the arterioles the pressure falls to 32 mm. Hg (*4·3 kPa*). It falls to 12 mm. Hg (*1·6 kPa*) by the time the venous ends of the capillary is reached.

Postcapillary Resistance Vessels

The pressure gradient of 12 mm. Hg to 0 mm. Hg is available for the return of blood from the venous end of the capillaries to the heart. The veins are large diameter vessels and the pressure drop along them is only a few mm. Hg. The remainder of the 12 mm. Hg pressure drop occurs in the small venules between the capillaries and the veins proper. These constitute the *postcapillary resistance vessels*. Although they are unimportant when considering the total peripheral resistance, they are important when considering the pressure in the capillaries. The ratio of precapillary to postcapillary resistance determines the mean capillary pressure. This pressure is a dominant factor in the formation of tissue fluid [p. 58].

INSTRUMENTATION

Measuring and Recording Blood Pressure

Mercury Manometer

Blood pressure may be measured either directly or indirectly. The direct method consists of inserting a cannula or alternatively a hollow needle into the artery or blood vessel and transmitting the pressure of the blood in the lumen of the vessel to a manometer (pressure recorder).

In its simplest form a glass cannula is tied into the blood vessel and the pressure is transmitted via heparinized saline to a mercury U-tube manometer [FIG. 64]. Approximately the correct pressure must be applied to the system from a pressurized saline reservoir to prevent blood entering the recording system. A float and pointer

writing on smoked paper which may be later varnished enables a permanent record to be obtained.

Since a mercury manometer is used, calibration marks can be added later provided that at least one pressure level is known. The scale will be ½ mm. on the record = 1 mm. Hg, since the mercury in the right-hand limb goes down when that in the left limb rises. Thus the movement of the pointer has to be doubled to give the pressure in mm. Hg.

Such a method of recording is reliable and inexpensive, but it gives an indication of only the mean pressure. Due to the high

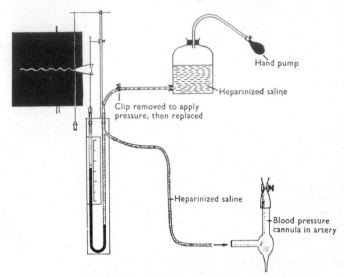

FIG. 64. Recording blood pressure using the mercury manometer. To prevent blood entering the apparatus, a pressure approximately equal to the blood pressure is first applied to the mercury manometer by removing the clip from the rubber tubing leading to the pressure bottle which contains heparinized saline. This clip is replaced and then a connexion is made to the blood pressure cannula in the artery.

inertia of the mercury, the blood pressure changes associated with the cardiac cycle are highly damped. On the other hand the changes in mean pressure with respiration are clearly shown and may at first glance be thought to be due to the heart beat [see FIG. 60]. In many cases a glance at the time scale will show that this is not so.

Optical and Electronic Manometers

In order to obtain a more accurate recording of the blood pressure a diaphragm is used. The arterial blood pressure is transmitted through a column of saline to one side of the diaphragm from a cannula or hollow needle in the artery. By making the diaphragm stiff only a very small amount of blood need enter the cannula for the pressure to be recorded. Clotting can be prevented by adding heparin to the saline in the cannula tip.

The displacement of the diaphragm with the pressure is very small and various methods have been employed to enable this small movement to be recorded. If a small mirror is attached to the edge of the diaphragm, its movement with increased pressure may be employed to deflect a beam of light. This forms the basis of the **optical manometer.** The displacement of the light beam may be recorded photographically.

In the electronic manometer the displacement of the diaphragm is used to change an electronic parameter such as capacitance, inductance or resistance. FIGURE 65 shows three types of 'pressure transducer' commonly employed. In the upper diagram the metallic diaphragm forms one plate of a condenser. The other plate of the condenser is placed in a fixed position very close to, but not touching, the diaphragm. As the pressure increases the plates become closer together and this increases their electrical capacitance.

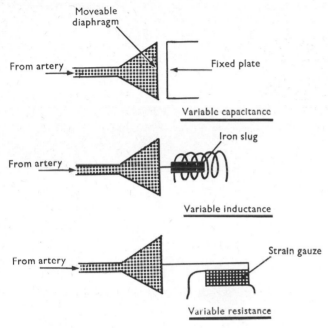

FIG. 65. Electronic manometers. Three types of pressure transducer used to record blood pressure. All employ a diaphragm, one side of which is connected by saline to a blood pressure cannula in an artery. Changes in blood pressure cause movements of the diaphragm which in turn cause changes in the capacitance, inductance or resistance.

In the middle diagram [FIG. 65] the diaphragm is connected to an iron core which moves further into a coil as a pressure rises. This increases the inductance of the coil.

In the lower diagram the diaphragm is connected to a strain gauge which increases in electrical resistance as the pressure increases.

Other methods include connecting the diaphragm to a movable anode valve (R.C.A. 5734) or to a coil which changes the mutual inductance with a fixed coil as the pressure rises.

All these electronic transducers require quite complicated electronic circuitry to convert the small changes in the electrical parameters into signals of sufficient strength for recording.

The recording apparatus is of two main types. In the first the pressure signal is amplified until a voltage is obtained that is sufficient to deflect the beam of a cathode ray oscilloscope. The displacement of this beam is photographed using moving paper so that a pulse tracing is obtained. For display purposes a time-base is employed on the cathode ray oscilloscope and provided that a persistent phosphor is used, the pulse wave will be visible as the spot moves across the screen of the cathode ray tube.

In the second type of recording system the pressure signal is amplified until a current is sufficient to displace a coil in a magnetic field (moving-coil galvanometer). The movement of this coil is employed to produce a permanent record. An ink pen attached to the moving coil will produce such a record. Alternatively a heated stylus moving over heat sensitive paper may be employed [FIG. 66].

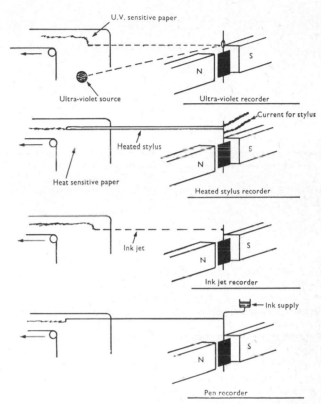

FIG. 66. Four types of recorder used to record blood pressure and other physiological phenomena. All use a moving-coil galvanometer which consists of a coil of wire moving in a magnetic field. The pen and heated stylus recorders have the disadvantage that friction between the paper and pen limits the frequency response to below 100 cycles per second. This disadvantage has been eliminated in the ink-jet and ultra-violet recorders.

In an attempt to eliminate the friction between the pen and the recording paper which limits the frequency response of the recorder, the ink-jet recorder has been introduced. In the apparatus the ink is ejected from the moving-coil at a high velocity and makes a recording on the paper without there being any mechanical contact between the coil and the paper. For the highest frequency response, and therefore the most faithful recording, an ultra-violet recorder may be used. This uses a similar principle to that used in the optical manometer. A mirror attached to the moving-coil is used to reflect, not a beam of light, but a beam of ultra-violet which is allowed to fall on ultra-violet sensitive paper thus giving a pulse tracing.

By using several of these recording galvanometers writing simultaneously on the same paper, a multi-channel recording of other physiological events, in addition to blood pressure, may be made.

Using an optical or electronic manometer the arterial blood pressure is seen to fluctuate between two limits with each cardiac cycle. The maximum pressure reached is the *systolic pressure* [Fig. 67]. The minimum pressure is termed the *diastolic pressure*.

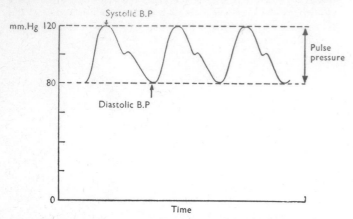

FIG. 67. The arterial blood pressure fluctuates between a maximum (or systolic blood pressure) and a minimum (or diastolic blood pressure). The difference between these two pressures is the pulse pressure. A second peak is seen on the down-stroke. This is termed the 'dicrotic notch'.

The difference between the two is the *pulse pressure*. Typical figures are 120 mm. Hg (*16 kPa*) for the systolic pressure, 80 mm. Hg (*11 kPa*) for the diastolic pressure, and 40 mm. Hg (*5 kPa*) for the pulse pressure.

With this method of recording, the respiratory variation appears insignificant compared with the cardiac variations.

With electronic manometers it is possible to record the pressure through a catheter passed into the chambers of the heart.

A heavily damped optical or electronic manometer will record the *mean pressure*. Alternatively the averaging may be carried out electronically.

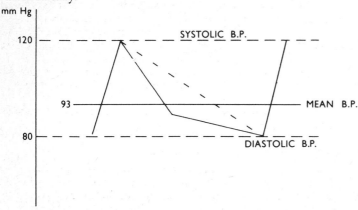

FIG. 68. Mean blood pressure. The shape of the blood pressure waveform gives a mean blood pressure which is nearer to the diastolic pressure than to the systolic pressure.

Since the area under the blood pressure curve [Fig. 68] is usually less than that of the corresponding triangle, the mean pressure throughout the cardiac cycle will be about one third of the way up from the diastolic pressure to the systolic pressure (and not half-way). Thus in FIGURE 68:

Mean pressure =
Diastolic pressure + $\frac{1}{3}$ (systolic pressure − diastolic pressure)
= Diastolic pressure + $\frac{1}{3}$ pulse pressure
= $80 + \frac{1}{3} . 40$
= 93 mm. Hg

(and not $\frac{1}{2}(80 + 120) = 100$ mm. Hg).

Sphygmomanometer

In 1896 Riva-Rocci introduced an inflatable cuff for recording blood pressure in man *indirectly* that is, without cannulating an artery. In 1905 Korotkov described a convenient method for finding the systolic and diastolic pressures by listening to the sounds produced in the artery beyond the cuff as the cuff pressure is reduced. Although the results obtained by this method are inaccurate compared with those given by an electronic manometer, it has become the standard method of blood pressure recording throughout the world.

The cuff is applied to the arm and inflated rapidly to well above the systolic pressure (to say 200 mm. Hg) using a small hand pump [Fig. 69]. The brachial artery blood flow is now obstructed. A

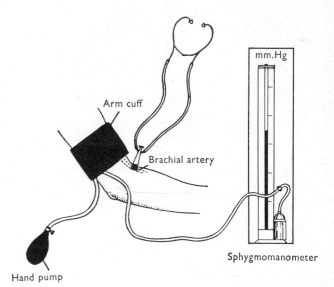

FIG. 69. The sphygmomanometer for recording blood pressure in man without cannulating an artery. The rubber bag is enclosed in a cloth cuff and is wrapped round the arm. It is inflated using a small hand pump and the pressure recorded on a mercury manometer [see FIG. 70].

stethoscope is placed over the brachial artery in the cubital fossa, that is, distal to the cuff. The cuff pressure is slowly released until a tapping sound in rhythm with the heart rate is first heard. The cuff pressure at which this sound is first heard is taken as the systolic pressure [Fig. 70].

As the cuff pressure falls the sounds heard in the stethoscope become louder and more banging in character. Then as the diastolic pressure is reached the sound becomes suddenly muffled. A little later the sounds disappear. In Britain the point where the sounds become muffled is taken as indicating the diastolic pressure. In the United States it is common practice to take the point of disappearance of the sound as the diastolic pressure.

When the cuff pressure exceeds the pressure in the artery, the artery will be closed. When the cuff pressure is less than the pressure in the artery the arterial lumen will be open. It is the turbulence which results from the opening and closing of the artery that produces the Korotkov sounds [FIG. 70].

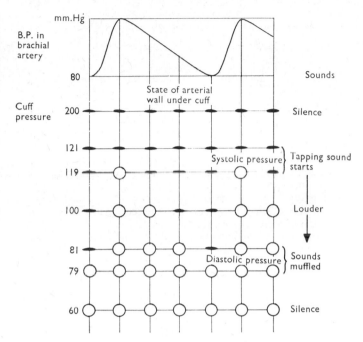

FIG. 70. Korotkov sounds heard in a stethoscope placed over the brachial artery enable the systolic and diastolic blood pressure to be determined. These sounds are produced by turbulence in the intermittent blood flow as the artery opens and shuts with each cardiac cycle. Note the changes in the state of the artery around the systolic and diastolic pressure which gives the changes in the sounds heard. B.P. = 120/80 mm. Hg.

The cuff pressure is measured using a mercury manometer or an aneroid dial manometer.

Systolic Blood Pressure

The systolic blood pressure indicates the maximum pressure produced by the left ventricle during systole [FIG. 26, p. 28].

Diastolic Blood Pressure

The diastolic pressure is the pressure in the artery at the end of diastole. It will depend upon the systolic blood pressure, the elasticity of the aorta and large arteries, the peripheral resistances, the viscosity of the blood, and the length of diastole.

After the aortic valve has closed there is an exponential fall-off in blood pressure in the aorta as the blood flows through the arterioles and capillaries to the veins [curve 1, FIG. 71]. The level to which the pressure has fallen just before the aortic valve opens again with the next ventricular systole [point A, FIG. 71] is the diastolic blood pressure.

If the peripheral resistance is high, the blood pressure will fall along a higher curve—[curve 2, FIG. 71]. The pressure has dropped to a lesser extent (point B) and the diastolic pressure is higher.

If the peripheral resistance is low, the pressure will fall along curve 3. The pressure will fall markedly to point C, giving a low diastolic blood pressure. The interval between the systolic pressure and diastolic pressure will be greater giving a large pulse pressure and, if the difference is great enough, a collapsing pulse. This is found in vasodilatation states such as occur following a hot bath and in patients with overactive thyroid glands.

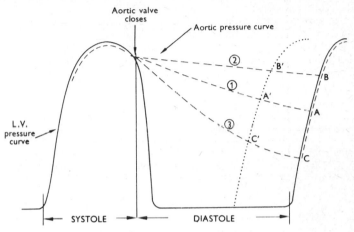

FIG. 71. Factors affecting diastolic pressure. Normally arterial blood pressure falls during diastole along curve ① to point A. When peripheral resistance is high, it falls more slowly along curve ② to a higher diastolic pressure point B. When peripheral resistance is low, it falls more rapidly along curve ③ to point C giving a lower diastolic pressure and a large pulse pressure.
A reduction in the duration of diastole, for a given peripheral resistance, will raise the diastolic pressure (points A′, B′, and C′).

It will be noted in FIGURE 71 that for a given peripheral resistance (e.g. curve 1) the diastolic blood pressure also depends upon the duration of diastole. A short diastole will give a raised diastolic pressure [points A′, B′, and C′, FIG. 71]. A long diastole will give a lowered diastolic pressure.

Since sinus arrhythmia (fluctuation of heart rate with respiration) is principally due to a variation in the length of diastole it follows that, when it is present, there will be a marked fluctuation in the diastolic blood pressure with respiration. This can be clearly seen in FIGURE 62, p. 51.

Other Uses of the Blood Pressure Cuff

If the arm cuff is inflated to a pressure of less than the arterial blood pressure, the cuff will obstruct the veins, but will not obstruct the arteries. Let us apply a pressure of say 60 mm. Hg. Blood will continue to enter the limb but will be unable to leave. The arm will become engorged with blood. Ultimately the venous pressure will build up to 60 mm. Hg and then the blood will be able to pass under the cuff and return to the heart. Not only will the veins be at a pressure of at least 60 mm. Hg, but so also will the capillaries. This provides a simple experimental method for changing the venous pressure and capillary pressure for experimental purposes, such as in studies in the production of the white line.

Venous Occlusion Plethysmography

During the early stages of the occlusion, before venous engorgement has occurred, the blood flow into the limb will continue unchanged. Since no blood can leave, the rate of increase in volume of the limb is a measure of the rate of blood inflow. This is the basis of the venous occlusion plethysmography method of measuring limb blood flow. The limb is placed in an air-tight chamber which is connected by air or fluid to a volume recorder. Changes in limb volume displace air or fluid from this chamber and the rate of volume displacement is measured [FIG. 72].

Chambers of suitable shapes and sizes enable the blood flow to be recorded in the hand, finger, forearm, leg and foot. When recording forearm muscle blood flow, the hand blood flow is eliminated by inflating a wrist cuff to well above the systolic blood pressure shortly before the inflation of the occluding cuff [FIG. 72].

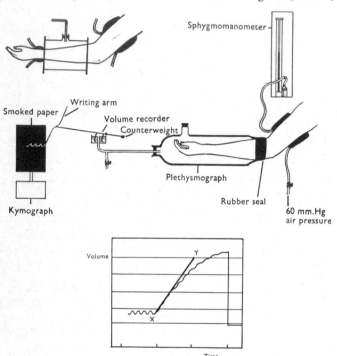

FIG. 72. Recording blood flow in the hand and forearm using the plethysmograph. The cuff is inflated rapidly to 60 mm. Hg, thus occluding the veins. The rate of blood flow into the arm is measured by the rate of air or fluid displacement from the plethysmograph to the volume recorder at the beginning of the occlusion. This is given by the slope of XY. The hand blood flow (mainly skin flow) may be recorded by having only the hand in the plethysmograph. When recording forearm blood flow only (mainly muscle), a wrist cuff may be inflated to 200 mm. Hg to cut off blood flow to the hand (top left).

Muscle Blood Flow

Such determinations at rest and after exercise show that the resting muscle blood flow is 1–3 ml. blood per 100 g. muscle per minute, and that it may increase to 40 or more ml. blood/100 g./minute in exercise.

Since a 70 kg. man has approximately 30 kg. of muscle tissue, a resting blood flow of 2 ml./100 g./minute would represent a muscle blood flow of 600 ml. per minute, that is 12 per cent. of the cardiac output (5 litres/minute). In exercise a flow of 40 ml./100 g./minute would represent a muscle blood flow which would be 12,000 ml. per minute or 40 per cent. of a cardiac output of 30 litres/minute.

It will be noted that the cardiac output has increased by 6 times (from 5 to 30 litres per minute), and that the percentage of this output received by the muscles has increased by 3 times (from 12 per cent. to 40 per cent.). The remaining factor determining the oxygen supply to the muscle (the amount of oxygen extracted from every 100 ml. blood) will be discussed on page 83.

Further Methods of Recording Blood Flow

Venous occlusion plethysmography enables blood flow changes to be studied in such parts of the body as the hand, forearm, leg and foot. Elsewhere blood flow recording is more difficult. It has been seen [p. 6] that the total blood flow (cardiac output) can be determined using an injected dye or radio-isotope. Another method using the Fick Principle is discussed later on page 82, and modifications of this principle may be employed to determine liver blood flow [p. 156] and kidney blood flow [p. 143].

Various methods have been employed to measure the blood flow in arteries in experimental animals. These include the *thermostromuhr* (based on the transference of heat by the blood from a heat source in the blood stream to thermocouples or thermisters up and down stream), rotameters, bristle flowmeters (where the displacement of a bristle in the blood stream is measured) and Venturi systems (where the pressure drop across a resistance is measured). All these methods involve cannulation of the artery.

The most satisfactory system so far developed is probably the electromagnetic flowmeter [FIG. 73] since arterial cannulation is

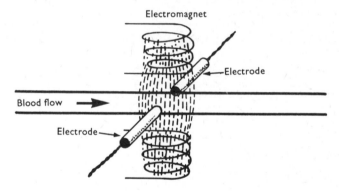

FIG. 73. Electromagnetic flowmeter. Blood acts as an electrical conductor and its movement through a magnetic field produces an electrical potential at right angles to the magnetic field. The magnitude of this potential, which is picked up by the electrodes, is an indication of the rate of blood flow.

not essential. It consists of an electromagnet placed so as to create a magnetic field across the blood flowing in the artery. The blood, being an electrical conductor, produces by its movement an e.m.f. or voltage at right angles to the magnetic field and at right angles to the direction of blood flow. This can be picked up using two electrodes in contact with the wall of the artery and after suitable amplification can be arranged to give a signal related to the velocity of blood flow. From the known cross-sectional area of the artery, the volume flow can be calculated. In order to produce more stable electronic circuitry, a sinusoidal or square-wave current is employed in the electromagnet, thus producing an alternating magnetic field and an alternating voltage at the electrodes.

The ultrasonic flow meter transmits ultrasonic sound along the blood in a blood vessel. Since the sound is carried by the blood itself, the velocity of blood flow can be determined from the increase in the velocity of the sound.

Effect of Gravity on Blood Pressure

All the pressures so far discussed have been assumed to be at the level of the heart. Below the heart the action of gravity on the column of blood in the arterial tree increases the arterial blood pressure by an amount equal to:

$$10 \times h \times \frac{1 \cdot 055}{13 \cdot 6} \text{ mm. Hg}$$

where h is the vertical distance below the heart in cm., $1 \cdot 055$ is the specific gravity of blood, and $13 \cdot 6$ is the specific gravity of mercury.

$$= 0 \cdot 78h \text{ mm. Hg} = 0 \cdot 1h \text{ kPa}$$

Hence for every 10 cm. below the heart the pressure is increased by 8 mm. Hg (1 kPa).

Above the heart the pressure is reduced by this amount.

Thus in the standing position, if the pressure is 120 mm. Hg at the level of the heart, it will be approximately 80 mm. Hg more in the feet, i.e. 200 mm. Hg. The venous pressure when standing perfectly still will also be increased by 80 mm. Hg. When walking about the action of the muscle pump lowers the venous pressure considerably.

Above the heart the pressure is less than that at heart level. The pressure in the head arteries will be approximately 30 mm. Hg less, that is, 120 − 30 = 90 mm. Hg. The pressure in the veins will theoretically be sub-atmospheric (−30 mm. Hg) but in practice the superficial veins collapse and when this occurs the pressure in them remains at atmospheric pressure (0 mm. Hg). Sub-atmospheric pressures, however, will occur if the veins are held open and air may be sucked into the heart and circulation with serious consequences (air embolism). The venous sinuses in the skull have rigid walls and are unable to collapse; as a consequence the pressure in these will be sub-atmospheric.

ARTERIAL PULSES

Radial Pulse

Pulsations in the radial artery at the wrist provide a convenient means of determining the heart rate. When the pulse is taken the number of times the artery impinges on the fingers in a given time is noted. It is the changes in the blood pressure in the radial artery that cause the expansion and contraction of the arterial wall. This pressure change is transmitted from the aorta at a very high velocity (the pulse-wave velocity = 6 metres per second) and takes only one-tenth of a second to reach the wrist. The blood, on the other hand, flows from the aorta to the radial artery at a much slower rate and the journey takes several seconds. So it is not the blood flow that one feels, but the wave of increased pressure. Even if all blood flow is stopped due to an obstruction to the artery beyond the point of measurement or by tying the vessel, the pulse is still present. This may be easily confirmed by occluding the radial artery by pressure with one finger. The pulse may still be felt by the other fingers proximally but not distally to the occlusion point.

A second and smaller pressure rise may occur shortly after the first [FIG. 67]. This is known as the *dicrotic notch* and is due to closure of the aortic valve and reflections of pressure waves in the arterial system.

On occasions the radial pulse may not be available for the determination of heart rate. The radial pulse may be absent due to an obstruction to the radial artery, or it may be impalpable (as in the case of shock where the blood pressure is very low) or the limb may be in plaster. In such a case an alternative pulsation must be employed. In addition, the presence of a pulse confirms that the arterial tree is patent between the heart and the point at which the pulse is taken.

Upper Limb Pulsations

The *ulnar pulse* can usually be felt quite easily on the ulnar side of the wrist [FIG. 74].

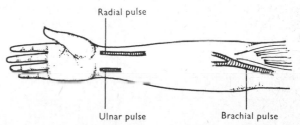

FIG. 74. Upper limb arterial pulses. In addition to the radial pulse, the ulnar and brachial pulses are readily palpable.

The *brachial artery* in the cubital fossa is used for blood pressure determination. It lies on the medial side of the biceps tendon. The artery can often be seen pulsating when the elbow is extended. The brachial artery pulsations can be traced along the medial aspect of the upper arm towards the axilla.

Lower Limb Pulsations

The principal artery supplying the lower limb is the femoral artery. Although it is a large diameter artery with a powerful pulsation, the *femoral pulse* is often a surprisingly difficult pulse to locate. One method of finding it rapidly is to run one's fingers forward round the iliac crest of the pelvis until the anterior superior spine of the ilium is reached. Then bisect the line joining this anterior superior spine to the symphysis pubis in the midline [FIG. 75(A)]. The femoral artery can be palpated as it enters the

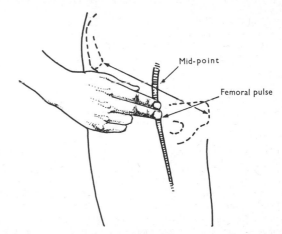

FIG. 75A. Femoral pulse. The femoral artery enters the thigh at the mid-point between the anterior superior spine of the ilium and the symphysis pubis. It is readily palpable just below this point.

thigh at this mid-point. Pulsations can be traced down the thigh for a few inches in the femoral triangle if the leg is laterally rotated.

The popliteal artery lies deep to the surface on the posterior aspect of the femur and tibia. The *popliteal pulse* is difficult to palpate. The subject should lie face downwards with the knee flexed. The observer should take the weight of the leg by holding the foot with one hand and allowing the subject to relax the thigh and leg muscles. The popliteal pulse is then palpated in the centre of the popliteal fossa [FIG. 75(B)]. If it cannot be palpated the presence

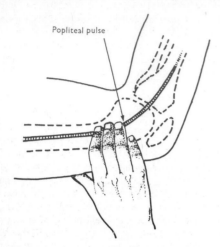

FIG. 75B. Popliteal pulse. The popliteal pulse is palpated in the popliteal fossa with the subject face downwards and the knee flexed.

of a popliteal pulse can usually be confirmed by getting the subject to sit with the legs crossed. The upper leg then acts as a magnifying lever and the foot shows a small forward movement with each popliteal pulsation.

The *posterior tibial pulse* is palpated as the artery passes behind

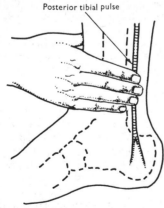

FIG. 75C. Posterior tibial pulse. The posterior tibial artery is palpated as it passes behind the medial malleolus.

the medial malleolus on the inner side of the ankle [FIG. 75(C)]. The *dorsalis pedis artery*, which is a continuation of the anterior tibial artery, is palpated on the dorsum of the foot midway between the two malleoli. The artery continues down the dorsum of the foot between the first and second toes [FIG. 75(D)]. These pulsations may be impalpable if the feet are cold.

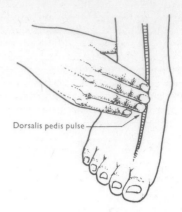

Dorsalis pedis pulse

FIG. 75D. Dorsalis pedis pulse. The anterior tibial artery passes midway between the two malleoli on the dorsum of the foot. It extends down to between the big and adjacent toe as the dorsalis pedis artery.

Head and Neck Pulsations

The *carotid pulse* is the most important pulse in the body, and should be searched for whenever a cardiac arrest is suspected. The common carotid artery is covered by the sternomastoid muscle in the lower part of the neck, and the carotid pulse is therefore felt more easily high in the neck [FIG. 28, p. 29]. The *facial artery pulse* and *temporal artery pulse* (these vessels are branches of the external carotid artery) may be used to determine the heart rate in an anaesthetized patient when the rest of the body is covered by sterile towelling.

(Interstitial n)

TISSUE FLUID

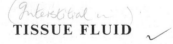

Blood stays in the blood vessels and does not come into direct contact with cells except in the liver and the spleen where the capillary epithelium is absent. The cells are bathed in tissue fluid (interstitial fluid) which acts as an intermediary between the blood and the cells.

Oxygen and food substances diffuse from the blood through this tissue fluid to reach the cells. Carbon dioxide and waste products diffuse in the reverse direction back into the blood.

Tissue fluid formation by the capillaries is modified by any change in either the precapillary or postcapillary resistance.

Formation and Reabsorption

It has been seen that the blood pressure at the arterial end of a capillary is approximately 32 mm. Hg (*4·3 kPa*) [FIG. 63]. The pressure falls progressively along the length of the capillary and has dropped to 12 mm. Hg (*1·6 kPa*) by the time the venous end is reached.

Capillaries consist of endothelial membranes alternating with interstices which are intercellular slits. Lipid-soluble molecules are able to pass through the cells. Water, ions and lipid-insoluble substances of molecular size smaller than plasma albumin are able to pass through the slits.

Capillaries are thus permeable to all the constituents of plasma except the plasma proteins. The blood pressure will tend to force fluid out through the capillary walls into the tissue spaces. Due to the presence of protein in the blood and its absence in the tissue fluid surrounding the capillary, an osmotic pressure gradient of 25 mm. Hg (*3·3 kPa*) is set up which tends to suck fluid back into the capillary. There are thus two opposing forces. At the arterial end of the capillary the blood pressure is in excess of the osmotic pressure and fluid passes from the capillary into the tissue spaces. At the venous end of the capillary, the osmotic pressure exceeds the blood pressure and fluid is returned to the blood.

Tissue fluid is formed at the arterial end of capillaries and is absorbed at the venous end [FIG. 76].

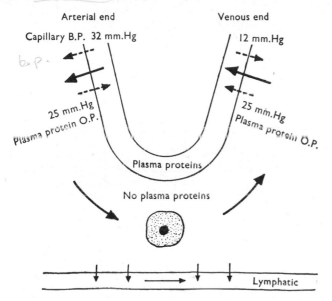

FIG. 76. The formation of tissue fluid. Tissue fluid is formed at the arterial end of the capillary loop. It is reabsorbed at the venous end and via the lymphatics.

Oedema

An excess of tissue fluid formation or a failure of absorption will lead to an accumulation of excess tissue fluid which is known as oedema or dropsy.

It may be due to a variety of causes. Basically it is caused by an increase in the blood pressure in the capillaries which will increase production, or to a decrease in the effective osmotic pressure of the plasma proteins which will decrease reabsorption.

The capillary blood pressure increases:

(a) When there is arteriolar dilatation. When this occurs there is a reduction in the pressure drop across the arterioles and this has the effect of raising the pressure in the capillaries.

(b) When there is obstruction to the veins. The pressure in the veins distal to the obstruction will be raised and this will have the effect of raising the pressure at the venous end of the capillaries. Oedema of the legs may occur in pregnancy due to pressure by the baby on the veins in the pelvis. It is seen when a plaster cast or bandage on a limb is too tight and obstructs the venous return. The part of the limb distal to the bandage may become oedematous. Generalized oedema is seen in patients with a high venous pressure associated with *heart failure* [p. 47].

The effective osmotic pressure is reduced:

(a) When there is a deficiency of the plasma proteins. This will occur when there is insufficient intake of protein (malnutrition) and when there is excessive protein loss in the urine (kidney disease). Both these conditions are associated with oedema.

(b) When the capillaries become permeable to protein. Once protein enters the tissue spaces, the osmotic effect which is drawing fluid back into the blood is reduced and ultimately disappears. One of the results of a generalized oxygen lack in the body is to cause capillary damage with the passage of protein into the tissue spaces. This is an additional cause of oedema associated with heart failure.

The swelling associated with an insect bite or an injury is the result of capillary damage which leads to a localized oedema. The capillary damage is in part the result of the trauma and in part due to the release of histamine and related substances.

Oedema may also be due to the retention of sodium due to disorders of sodium chloride excretion (see the Adrenal Cortex, page 166). Water is retained in addition to the sodium chloride and this leads to generalized oedema.

Lymphatics and Tissue Fluid

Lymphatics arise as blind-ended vessels in the tissue spaces. The lymphatics remove protein from the tissue spaces and by so doing prevent the occurrence of oedema due to the gradual leak of protein from the capillaries. They also remove surplus tissue fluid.

The rise in capillary pressure in exercise due to the dilatation of the arterioles means that very little reabsorption of tissue fluid will take place at the venous end of the capillary. In addition there will be an increased production due to the increase in the mean capillary pressure. Furthermore, the breakdown of the large glycogen molecules in the active muscles into smaller soluble carbohydrate units and the production of metabolites will increase the osmotic pressure of the tissue fluid and this will oppose the osmotic pressure of the plasma proteins. It is not surprising, therefore, that active muscles become water-logged (oedematous), stiff and painful particularly after the first game of the season. Further exercise presumably opens up the lymphatic channels which are the main route for tissue fluid reabsorption in active muscle.

Blockage or damage to the lymphatic system may lead to oedema after infestation with the filaria parasite worm. Oedema may also occur following the surgical removal of lymph nodes for cancer. The oedema occurs in the part of the body that has lost its lymphatic drainage.

A congenital absence or abnormal pattern of lymphatics in the legs gives rise to gross swelling of the affected limbs, a condition known as *lymphoedema*.

Lymphatic System

The lymphatic vessels ramify throughout the tissue spaces. They are difficult to see unless injected with a dye. However, in cases of an infection of, say, the hand, the fine red lines of lymphatic inflammation may be seen running up the arm.

The blind-ended lymphatic capillaries in the tissue spaces join to form lymphatic vessels. The vessels contain a fluid known as lymph which is moved on by the contraction of the surrounding muscles and the presence of valves in the lymphatic channels, in much the same way as the muscle pump aids the venous return [p. 48].

Lymphatic glands or nodes are found at intervals along the path of the lymphatics. These nodes consist of a reticular framework with lymphocytes in various stages of development in the meshwork. The lymph nodes act as a filter. They are also an important site of formation of lymphocytes.

The intestinal lymphatics are the route of absorption of fat from the digestive tract. The lymphatics from the lower limbs join those from the digestive tract to form the thoracic duct. The thoracic duct and the right lymphatic duct join the great veins in the neck.

Tissue Fluid and the Restoration of the Blood Volume Following a Haemorrhage

Haemorrhage is associated with profound vasoconstriction of the arterioles, that is, in the precapillary resistance vessels [p. 52]. The effect of this will be to increase the pressure drop across the arterioles so that by the time the capillaries are reached the pressure will have fallen to a lower value than normal [FIG. 77].

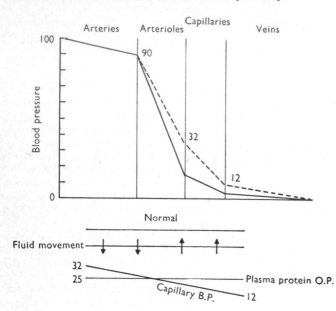

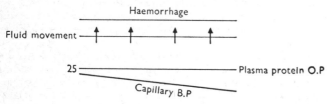

FIG. 77. Fluid movement from the tissue spaces to the blood after a haemorrhage. The profound arteriolar vasoconstriction in haemorrhage causes a fall in the mean capillary blood pressure. Tissue fluid formation ceases and fluid is reabsorbed from the tissue spaces into the blood capillaries along the entire capillary length.

As a result the formation of tissue fluid will be reduced and its reabsorption increased. Fluid will enter the blood from the tissue spaces, thereby restoring the blood volume. Initially, since whole blood (cells and plasma) is lost in a haemorrhage, the haemoglobin content, haematocrit, and red cell count of the subject will be unchanged. These estimations will, at that time, give no indication that a haemorrhage has taken place. As tissue fluid enters the circulation over the course of the next 2–3 hours, the remaining red cells will be diluted, and then a fall in haemoglobin content, haematocrit, and red cell count can be detected.

The lost blood cells are replaced during the course of the next few weeks due to increased bone marrow activity.

THE PULMONARY CIRCULATION

The pulmonary circulation is in series with the systemic circulation. The blood which returns to the right side of the heart from the systemic circulation is pumped by this side of the heart through the lungs to the left side of the heart and back to the systemic circulation again [FIG. 78].

PULMONARY CIRCULATION

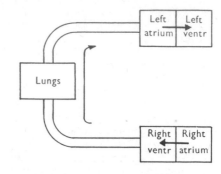

FIG. 78. The pulmonary circulation. The blood flows from the right ventricle through the lungs to the left side of the heart.

The lungs offer a lower resistance to blood flow than the arterioles and capillaries of the systemic circulation. There are no *resistance vessels*, like the systemic arterioles, in the lungs, and what resistance there is, is due to the smaller branches of the pulmonary artery, the lung capillaries and the tributaries of the pulmonary veins.

The absence of arterioles makes the blood flow through the lung capillaries pulsatile. The pulmonary blood flow is high during systole and it falls to a low value at the end of each diastole.

As a result of this low resistance, the right ventricle develops a lower pressure than the left ventricle in pumping the same volume of blood per minute. The pressure in the pulmonary artery is only 25/8 mm. Hg, that is, the systolic pressure is 25 mm. Hg (*3·3 kPa*) and the diastolic pressure is 8 mm. Hg (*1 kPa*), as against 120 and 80 mm. Hg for the aortic pressure [see FIG. 26, p. 28]. The pressure in the pulmonary veins is the same as that in the systematic veins close to the heart (approximately atmospheric pressure = 0 mm. Hg).[5] It will be noted that the pulmonary artery pulse pressure (= 17 mm. Hg) is greater than the mean pressure (= 16·5), whereas the aortic pulse pressure (= 40 mm. Hg) is less than the mean pressure (= 100 mm. Hg).

The lung capillaries are surrounded by the air sacs or *alveoli*. Gas exchange takes place between the air in the alveoli and the blood in the capillaries, so that the blood leaving the capillaries and returning to the left side of the heart, has taken up oxygen and has lost carbon dioxide. This blood is pumped to the tissues.

The pulmonary blood vessels are thin-walled and distensible. They have a large capacity so that the pulmonary circulation, at any one time, contains about one-fifth of the blood (=1 litre).

[5] Direct measurements in man indicate the left atrial pressure is usually about 5 mm. Hg and the right atrial pressure about 0 mm. Hg.

Their capacity is variable, for example, it increases during inspiration as the pressure around the vessels falls. These capacity changes will affect the inflow to the left atrium and thus the beat to beat output of the left ventricle.

Since the volume of blood filling the circulation is constant, any over-all decrease in the capacity of the lung vessels will make more blood available for the systemic circulation, and will increase the cardiac output by increasing the ventricular filling.

Control of Pulmonary Blood Vessels

The pulmonary blood vessels are supplied by sympathetic vasoconstrictor fibres. On the sensory side, pulmonary artery baroreceptors and chemoreceptors have been shown to exist, but their importance has yet to be evaluated as far as the regulation of the resistance of the pulmonary circulation is concerned. The pulmonary resistance does change. In exercise, when the cardiac output increases by six times, there is very little change in pulmonary arterial pressure. This implies that the resistance to blood flow offered by the lungs must have decreased by six times! The mechanism is obscure.

The pulmonary blood vessels are constricted by anoxia and as a consequence there is an increase in the resistance to blood flow through the lungs when the oxygen tension falls. Chronic anoxia leads to pulmonary hypertension. This is seen in people who live at high altitudes. Pulmonary vasoconstriction also occurs when there is a fall in the blood pH (acidaemia).

The pulmonary blood vessels as well as the air passages are constricted by histamine.

Bronchial Arteries

The larger air-passages receive their nutrition from the bronchial arteries which arise from the aorta. Anastomoses exist between the bronchial and pulmonary arteries in the lungs which may be said to have a double blood supply. The shunting of blood from the systemic to the pulmonary circulation that this causes is insignificant.

Pulmonary Oedema

The pressure in the pulmonary capillaries is lower than that in the systemic capillaries. Under normal conditions nowhere along the capillary length does the pressure exceed the osmotic pressure of the plasma proteins (=25 mm. Hg). Thus no tissue fluid is formed. If, however, the capillary pressure rises, due, for example, to the left side of the heart failing to transfer the blood from the pulmonary veins to the aorta ('left heart failure'), tissue fluid will form in the lungs giving rise to pulmonary oedema, and severe breathlessness (cardiac asthma).

Effect of Gravity on Lung Blood Flow

It has been seen that the blood pressure is reduced by 8 mm. Hg for every 10 cm. distance above the heart due to the effect of gravity. This is of importance in the pulmonary circulation owing to the low pressure in the pulmonary artery.

In the upright posture the blood pressure in the branches of the pulmonary artery in the superior parts of the lungs (apices) will be very low, leading to a relatively poor blood flow in this region. The pressure in the inferior parts of the lungs (lung bases) will be high, giving a higher blood flow but with the increasing likelihood of pulmonary oedema in these regions.

These pressure differences between the apices and bases will disappear in the lying position and be replaced by anterior-posterior pressure differences.

The pressure differences may lead to inequality between the blood flow and air flow in different parts of the lungs [see shunting, p. 99].

REFERENCES AND FURTHER READING

Anrep, G. V., Pascual, W., and Rossler, R. (1936) Respiratory variations of the heart rate, *Proc. roy. Soc. B*, **119**, 191 and 218.

Asmussen, E., and Nielsen, M. (1955) Cardiac output during muscular work and its regulation, *Physiol. Rev.*, **35**, 778.

Bainbridge, F. A. (1915) The influence of venous filling upon the rate of the heart, *J. Physiol. (Lond.)*, **5**, 65.

Barcroft, H., and Swan, H. J. C. (1953) *Sympathetic Control of Human Blood Vessels*, London.

Bayliss, W. M. (1902) On the local reaction of the arterial wall to changes in internal pressure, *J. Physiol. (Lond.)*, **28**, 220.

Brecher, G. A. (1956) *Venous Return*, New York.

Bronk, D. W., and Stella, G. (1932) Afferent impulses in the carotid sinus nerve, *J. cell. comp. Physiol.*, **1**, 113.

Daley, R., Goodwin, J. F., and Steiner, R. E. (1960) *Clinical Disorders of the Pulmonary Circulation*, London.

Daly, M. de B., and Scott, M. J. (1958) The effects of stimulation of the carotid body chemoreceptors on heart rate in the dog, *J. Physiol. (Lond.)*, **144**, 148.

De Cyon, E., and Ludwig, C. (1866) Die Reflexe eines der sensiblen Nerven der Herzen auf die motorischen Nerven der Blutgefässe, *Ber. Sachs. ges. Wiss.*, **18**, 307.

Elkington, E. J. (1968) Finger blood flow in Antarctica. (CIVD). *J. Physiol. (Lond.)*, **199**, 1.

Folkow, B. (1955) Nervous control of the blood vessels, *Physiol. Rev.*, **35**, 629.

Folkow, B. (1964) Description of the myogenic hypothesis, *Circulat. Res.*, **14–15**, Suppl. 1, 279.

Folkow, B., Heymans, C., and Neil, E. (1965) *Handbook of Physiology*, Section 2 Circulation, Vol. 3, Washington, D.C.

Folkow, B., and Neil, E. (1971) *Circulation*, New York.

Franklin, K. J. (1937) *A Monograph on Veins*, London.

Grollman, A. (1932) *The Cardiac Output of Man in Health and Disease*, London.

Heymans, C., and Neil, E. (1958) *Reflexogenic Areas of the Cardiovascular System*, London.

Keatinge, W. R. (1959) The effect of increased filling pressure on rhythmicity and atrio-ventricular conduction in isolated hearts, *J. Physiol. (Lond.)*, **149**, 193.

Kinsman, J. M., Moore, J. W., and Hamilton, W. F. (1929) Studies on the circulation, *Amer. J. Physiol.*, **89**, 322.

Kolin, A. (1936) An electromagnetic flowmeter, *Proc. Soc. exp. Biol. (N.Y.)*, **35**, 53.

Landis, E. M. (1934) *Capillary Pressure and Capillary Permeability*, New Haven.

McDowall, R. J. S. (1938 and 1956) *The Control of the Circulation of the Blood*, London.

Moore, J. W., Kinsman, J. M., Hamilton, W. F., and Spurling, R. G. (1929) Studies on the circulation, *Amer. J. Physiol.*, **89**, 331.

Patterson, S. W., Piper, H., and Starling, E. H. (1914) The regulation of the heart beat, *J. Physiol. (Lond.)*, **48**, 465.

Riva-Rocci, S. (1896) Un nuovo sfigmomanometro, *Gazz. Med. di Torino*, **47**, 981.

Rushmer, R. F., and Smith, O. A. (1959) Cardiac control, *Physiol. Rev.*, **39**, 41.

Sarnoff, S. J., and Berglund, E. (1954) Ventricular function, *Circulation*, **9,** 706.

Starling, E. H. (1918) *The Law of the Heart*, Linacre Lecture, London.

Wetterer, B. (1939) Eine Neue Methode zur registriering der Blutstromungeschwindig Kein am Uneroffmeter Gefass, *Z. Biol.*, **98,** 26.

Wood, J. E. (1968) The venous system, *Scientific American*, **218,** No. 1, 86.

5. RESPIRATION

Breathing ensures an adequate intake of oxygen from the room or environmental air for the oxidative processes of the body, and also ensures the removal of sufficient carbon dioxide to maintain a constant hydrogen ion concentration in the blood. At rest, 250 ml. of oxygen are absorbed from the room air per minute and 200 ml. of carbon dioxide are exhaled.

Blood in its passage through the lung capillaries does not come in contact with room air, but with the air in the depth of the lungs, in the air sacs or alveoli. This air is known as **alveolar air.** It contains less oxygen than room air and more carbon dioxide [TABLE 5].

TABLE 5

Inspired air (Room air)		Alveolar air	
O_2	20·96%	O_2	14%
CO_2	0·03%	CO_2	6%
N_2	79·01%	N_2	80%

The blood passing through the lungs removes oxygen from the alveolar air and adds carbon dioxide. The composition of the alveolar air is kept constant by the intermittent exchange of a small part of it for room air. The mechanism whereby this exchange is accomplished is known as *breathing*.

Tidal Volume and Dead Space

The lungs after a normal quiet expiration still contain 3 litres of air in the erect posture. (The volume is slightly less when lying

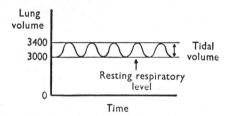

FIG. 79A. Respiration. With a quiet inspiration the lung volume increases from 3,000 ml. (resting respiratory level) to 3,400 ml.

down due to the pressure of the abdominal contents on the diaphragm.) At the next inspiration the volume of the thorax is increased by 400 ml. by contraction of the inspiratory muscles. The lungs increase in volume correspondingly to 3.4 litres (FIG. 79(A)], and 400 ml. room air enter the air passages. This is termed the *tidal volume*. Only the first 250 ml. of this air, however, reaches the alveoli; the last 150 ml. is still in the air passages when inspiration ceases, expiration follows and the air is exhaled as unchanged room

air [FIG. 79(B)]. It is known as *dead space air*. The last part of the exhaled air will be alveolar air. Thus, expired air is a mixture of room air from the dead space and alveolar air and has a composition intermediate between the two.

It follows that the tidal volume of each breath must be in excess of the dead space volume, otherwise the respiration will be ineffective. It is important to remember this when artificial respiration is being employed. Unless the manoeuvre adopted produces a large tidal air flow, the lung ventilation will be inadequate.

The dead space is increased when breathing takes place through any tube that is common to both the inspired and expired gases, by an amount equal to the volume of the tube [FIG. 79(B)].

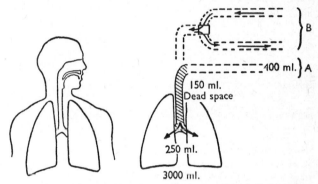

FIG. 79B. Dead space. The tidal volume is thus 400 ml. Only the first 250 ml. of the inspired air reaches the lungs. The last 150 ml. remains in the dead space. If the dead space is increased by a tube common to both the inspired and expired air the tidal volume must be increased. With a long tube efficient ventilation of the lungs may be impossible (A). If a two-way valve is used so that the inspired air passes along one tube and the expired air along a second tube (B) then there is no increase in the dead space volume and no increase in tidal volume is necessary.

It follows, therefore, that when swimming under water, breathing through a 'snorkel' or breathing tube, the tidal volume must be increased considerably. Effective ventilation of the lungs will be impossible if the snorkel tube is too long. If valves are used so that the inspired and expired gases pass along different tubes, then there is no increase in dead space and no limit to the length of tubing that may be employed on this account [FIG. 79(B)]. Underwater swimming would, however, still be restricted to a few feet close to the surface because at a greater depth, with the air in the lungs at atmospheric pressure, the water pressure on the chest wall would make inspiration impossible.

The Resting Respiratory Level

The lungs and contents of the mediastinum, such as the heart and blood vessels, completely fill the thorax. The outer coat of the lungs, the pleura, slides easily over the chest wall pleura when chest movements take place. Inflammation of these layers is termed *pleurisy*. Should these layers become roughened then a sound known as a 'pleural rub' may be heard by the application of a stethoscope to the affected area.

At the resting respiratory level (after a normal quiet expiration) the elastic recoil tending to collapse the lungs is opposed by an equal and opposite traction outwards due to the elasticity of the chest wall.

The equilibrium at the resting respiratory level is maintained unless air is allowed to enter between the two layers of pleura, when the lung will collapse inwards and the chest wall will expand outwards [FIG. 80(*B*)].

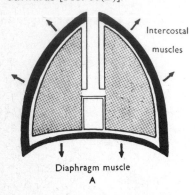
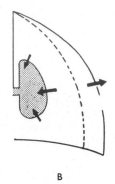

FIG. 80

A. Inspiration is brought about by the descent of the diaphragm and the movement of the chest wall upwards and outwards. Expiration is brought about by the elastic recoil of the lungs and chest wall. At the resting respiratory level the lungs and chest wall are in equilibrium.

B. When air enters the pleural cavity (a pneumothorax) the chest wall moves outwards and the lungs collapse inwards.

The fluid in the theoretical space between the two layers of pleura is at a sub-atmospheric pressure, that is at a mean pressure of 755 mm. Hg when the barometer reads 760 mm. Hg. The mediastinal contents will also be at a similar sub-atmospheric pressure. This pressure varies about the mean with respiration. It becomes −2 mm. Hg during expiration and −8 mm. Hg during inspiration.

Respiratory Muscles

The chief muscles of respiration are the diaphragm, supplied by the phrenic nerves which arise in the neck from the cervical roots C.3, 4, 5, and the intercostal muscles supplied by the thoracic intercostal nerves. These are inspiratory muscles and when they contract the volume of the thoracic cavity and therefore of the lungs is increased. The dome of the diaphragm moves downwards. The intercostal muscles move the chest wall upwards and outwards. Inspiration is thus an *active* process involving muscular activity.

Expiration is brought about by the elastic recoil of the lungs and chest wall following the relaxation of the inspiratory muscles. It is a *passive* process and the lungs and chest wall return to the *resting respiratory level*.

Expiratory muscles, such as the muscles of the abdominal wall, are employed when a forced voluntary expiration is made, such as by blowing hard. They are also employed when the chest volume is taken below the resting respiratory level in a prolonged expiration.

Pulmonary Ventilation

Pulmonary ventilation = Respiratory rate × Tidal volume
(minute volume)

The respiratory rate at rest is 15–20 breaths per minute. Since the tidal volume is 400 ml. the minute volume or pulmonary ventilation will be 15 × 400 to 20 × 400 ml. = 6–8 litres per minute.

Pulmonary Ventilation in Exercise

In exercise the oxygen uptake increases from 250 ml. per minute at rest to 1,500 ml. per minute in moderate exercise, and 3,000 ml. per minute in severe exercise. In moderate exercise, the respiration is increased in both rate and depth, but expiration is still a passive process. In severe exercise both inspiration and expiration are active processes involving muscular activity, and the pulmonary ventilation is 50–100 litres per minute. Although this is not the maximum possible pulmonary ventilation, any further increase in ventilation would not result in more oxygen being available for the body since all the additional oxygen would be needed for the respiratory muscles.

Alveolar Ventilation

Owing to the dead space *the volume of room air that reaches the lungs per minute* is always less than the pulmonary ventilation. This volume is termed the *alveolar ventilation*.

Alveolar ventilation = Respiratory rate × (Tidal volume
— Dead space volume)

With a tidal volume of 400 ml., a dead space volume of 150 ml. and a respiratory rate of 15

Alveolar ventilation = 15(400 − 150) ml. per minute
= 3·75 litres per minute.

Pulmonary ventilation = 15 × 400 ml. per minute
= 6 litres per minute.

With rapid shallow breathing the alveolar ventilation may be very low, even though the pulmonary ventilation is normal. Thus if the respiratory rate rises to 30 and the tidal volume falls to 200 ml., the pulmonary ventilation will still be 6 litres per minute but:

Alveolar ventilation = 30 (200 − 150) ml. per minute
= 1·5 litres per minute.

The work done by the respiratory muscles is related to the pulmonary ventilation whilst the supply of oxygen to the body and the elimination of carbon dioxide depends on the alveolar ventilation. It follows that rapid shallow breathing is very inefficient. Should the tidal volume fall to 150 ml. or less, no pulmonary ventilation, however large, will produce any effective alveolar ventilation. The resultant oxygen lack and retention of carbon dioxide (carbon dioxide excess) is termed *asphyxia*.

RECORDING RESPIRATORY MOVEMENTS
Chest Movements

The movements may be recorded in humans by using any device which responds to the increase in chest circumference which occurs in inspiration. Such methods include the use of strain gauges, and flexible tubes filled with mercury which increase their electrical resistance when the tube is elongated. One of the simplest methods is the stethograph attached to a tambour which writes on a kymograph drum [FIG. 81(A)]. The stethograph, or pneumograph, as it is

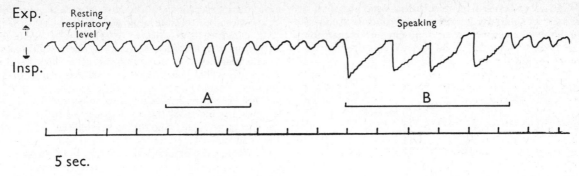

FIG. 81B. A stethograph record of quiet breathing, voluntary over-ventilation (*A*) and talking (*B*). Note that the respiratory pattern is modified so that speaking takes place during expiration.

sometimes called, consists of flexible corrugated tubing ('elephant' tubing) filled with air. On inspiration the tubing is stretched and its volume increases; air is displaced from the rubber tambour and the writing point descends. On expiration the tambour pointer rises.

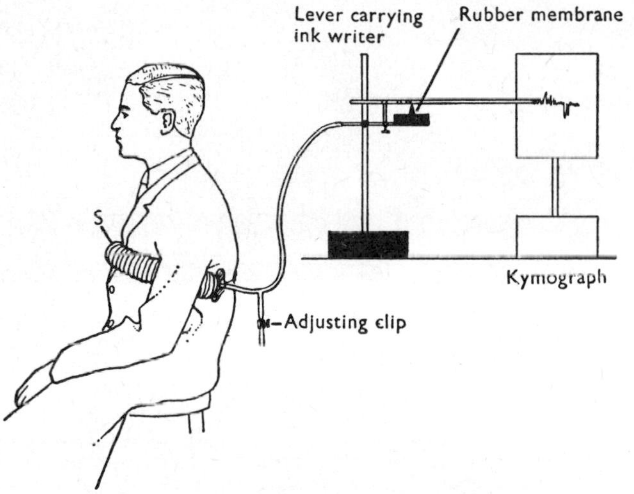

FIG. 81A. Recording respiratory movements in man using the stethograph (S). Expansion of the chest increases the volume of the stethograph and the lever attached to the tambour descends. Thus inspiration corresponds to a downward movement on the record.

Such a method gives the respiratory rate and qualitative changes in depth. It is unsuitable for quantitative measurements of tidal volume and pulmonary ventilation.

FIGURE 81(B) shows a stethograph record of a subject breathing quietly, breathing deeply and talking.

FIGURE 82 shows an electronic form of the stethograph which enables a similar record to be made using an electronic recorder. Such a stethograph was used to record respiration in the sinus arrhythmia record [FIG. 62, p. 51]. The stethograph consists of a rubber or plastic tube filled with a solution of copper sulphate. It is strapped round the chest. As the chest expands with inspiration, the electrical resistance between the two copper electrodes increases. A Wheatstone bridge circuit converts this change in

electrical resistance into an electrical voltage which after suitable amplification operates the recorder.

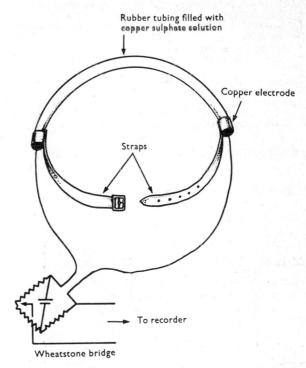

FIG. 82. An electronic stethograph consisting of a flexible tube filled with a copper sulphate solution. When stretched the electrical resistance between the two copper electrodes increases. A Wheatstone bridge circuit converts the electrical resistance change into a voltage change to operate a recorder.

Intratracheal Pressure

The pressure in the trachea increases with expiration and decreases with inspiration. A pressure manometer connected to the trachea will, therefore, indicate the phases of respiration. An electronic manometer similar to that used for recording blood pressures may be employed provided that it has sufficient sensitivity, i.e. will respond to a few mm. of mercury pressure change.

Alternatively, a rubber tambour similar to that used above with a stethograph may be employed. Such a method is frequently employed in animal experimentation following a tracheostomy [FIG. 83].

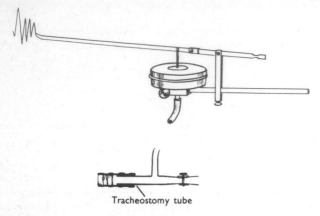

Tracheostomy tube

FIG. 83. Recording intratracheal pressure changes. A tambour is connected to the side arm of a tracheostomy tube. The intra-tracheal pressure falls during inspiration and rises during expiration. As a result the lever attached to the tambour descends with inspiration, and rises with expiration. An increase in respiration is associated with greater pressure changes and an increased excursion on the record.

Recording Pulmonary Ventilation

Since each inspiration is followed by the expiration of practically the same volume of gas in the opposite direction, the total air flow per minute, measured by breathing back and forth through a recording gas meter, would be zero. To determine pulmonary ventilation the flow in each direction must, therefore, be measured separately.

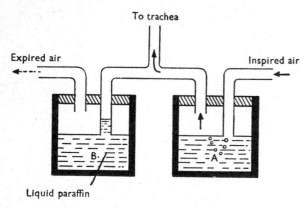

To trachea

Expired air

Inspired air

B·

·A

Liquid paraffin

FIG. 84. Inspiratory and expiratory valves (Muller valves). The fluid used is water or liquid paraffin. During inspiration air bubbles through A to the trachea, but the fluid seal in B prevents the inspiration of the previously expired air. During expiration the air bubbles out through B, but the fluid seal in A prevents expiration via this route.

The flow may be obtained in a roundabout way by measuring the velocity of the air movement using a *pneumotachygraph* and integrating the velocity in each direction with respect to time. One form of pneumotachygraph introduces a resistance into the air flow in the form of a fine metallic mesh. The pressure across this

mesh is measured by a sensitive electronic manometer, and is proportional to the rate of air flow.

If, however, the inspired gases can be separated from the expired gases, then their volumes can be readily measured.

This is achieved by using two valves, an inspiratory valve and an expiratory valve in such a way that inspiration takes place only through the first valve and expiration only through the second. As has been seen, the connecting tubing between these valves and the air passages increases the dead space and for this reason it is kept as short as possible. The tubing beyond the valves has no effect on the dead space.

A simple valve arrangement is shown in FIGURE 84. Glass tubes dip just below the surface of the water, or liquid paraffin. During inspiration the inspired air bubbles through the fluid in *A*. The fluid in *B* rises up the tube for a short distance but the air-seal is maintained.

In expiration, the expired air bubbles out through *B*. Such an arrangement is frequently used in animal experimentation.

FIGURE 85 shows a respiratory recording made using a tambour

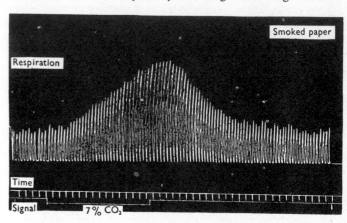

Smoked paper

Respiration

Time

Signal 7% CO_2

FIG. 85. The recording of respiration using the valves as in FIG. 84 with a tambour and leak connected to the expiratory valve. The record shows the stimulation of breathing resulting from the addition of 7 per cent. CO_2 to the inspired air at signal (cat). Time = 5 seconds.

and such a valve system. The tambour was connected to the expiratory side of the valve, and the expired air allowed to escape through a constriction in a side-arm. The tambour thus measures the pressure developed by the flow of expiratory gas through this constriction.

The record shows that there is a marked increase in respiration when the carbon dioxide level in the inspired air is raised.

Collection of Expired Air

A two-valve system complete with mouth-piece for use in man is shown in FIGURE 86. A nose clip is applied and the subject breathes through the mouth-piece of the valve. Inspired air enters at *X*. Expired air leaves via *Y*.

The volume of expired air produced per minute may be measured by collecting the gas for a known time in a suitable container such as a Douglas bag [FIG. 86].

The expired air leaves the body saturated with water vapour at 37 °C. The volume of the gas in the Douglas bag is reduced slightly as it cools to room temperature. In addition water condenses out leaving the gas saturated with water vapour at room temperature.

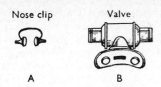

Nose clip   Valve

A   B

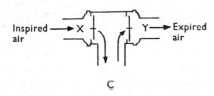

Inspired → X   Y → Expired
air   air

C

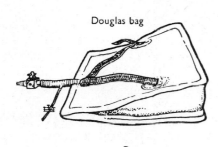

Douglas bag

D

Rubber tubing

FIG. 86. The collection of expired air in man using a mouthpiece and Siebe-Gorman valve (*B*) and a Douglas bag (*D*). A nose-clip (*A*) is applied and breathing takes place through the mouth. The valve (*C*) allows inspired air to enter at *X* and expired air passes from *Y* through the tubing to the Douglas bag.

This water condensation sometimes presents problems in the design of respiratory apparatus (and brass musical instruments where facilities must be provided so that the condensed water can be removed). In apparatus where it is essential to prevent water condensation the tubing is heated electrically to above body temperature.

The volume of the expired air collected in the Douglas bag is measured by passing the contents of the bag through a gas meter. For accurate work this volume is converted to N.T.P. (dry) that is, 760 mm. Hg (*101·3 kPa*) and 0 °C., using the formula:

$$V_{\text{NTP}} = V \times \frac{273}{(273 + t)} \times \frac{(B - p)}{760}$$

where *V* is the volume of gas at room temperature, *t* °C. is the room temperature, *B* mm. Hg is the barometric pressure, and *p* mm. Hg is the saturated water vapour pressure at *t* °C. [see Tables on p. 222].

Alternatively the volume may be converted to the volume it occupied in the body. This is referred to as B.T.P.S. (which stands for body temperature and pressure saturated). It is the volume the gas would occupy in the lungs at 37 °C., at the barometric pressure and saturated with water vapour. It is given by the formula:

$$V_{\text{BTPS}} = V \times \frac{(273 + 37)}{(273 + t)} \times \frac{(B - p)}{(B - 47)}$$

By dividing the pulmonary ventilation, determined in this manner, by the respiratory rate, the tidal volume can be determined.

Recording-spirometer

For a more accurate study of the pulmonary ventilation, the tidal volume and the volume of the respiratory gases absorbed and evolved, the recording-spirometer is used. This consists of a counterbalanced bell free to move in the vertical plane [FIG. 87].

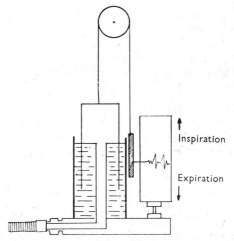

Inspiration

Expiration

FIG. 87. Recording-spirometer. Inspiration causes the bell to descend and the writing lever to move upwards on the chart. The chart is calibrated so that measurements of lung volume changes may be made.

It has a water-seal so that it forms an airtight chamber of variable volume. The changes in volume of this chamber are recorded on a chart.

When breathing through the mouth-piece into the spirometer with a nose clip applied, the air in the lungs, air passages and the spirometer is in a closed system. Thus when the lung volume increases in inspiration the volume of gas in the spirometer is reduced, and when the lung volume decreases in expiration, the spirometer volume increases. Thus the volume changes in the spirometer reflect equal and opposite changes in lung volume. In this manner changes in lung volume may be accurately followed.

From the spirometer tracing, the tidal volume, the respiratory rate and hence the pulmonary ventilation may be determined.

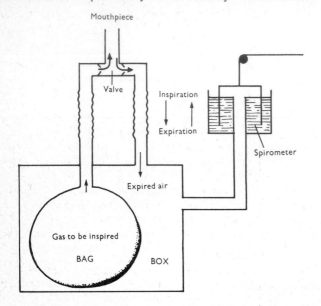

FIG. 88. The bag-box arrangement which allows a spirometer to record tidal volume and respiratory rate without the expired gas being rebreathed. Gas is inspired from the bag and expired into the box. The volume of gas inspired and expired is recorded by the movements of the spirometer bell. Several bags of gas may be enclosed in the same box and change-over taps used to select any desired gas. The recordings can continue until the bags of gas are empty.

The simple spirometer arrangement is unsuitable for recording more than a few breaths. This is because the expired air (which contains more carbon dioxide and less oxygen than the inspired air) is breathed back into the spirometer. As a result the composition of the inspired gas will be continually changing, and the subject will be subjected to a progressive asphyxia as the carbon dioxide level in the spirometer rises and the oxygen level falls.

To overcome this disadvantage, the carbon dioxide can be eliminated by including a CO_2-absorber in circuit [see FIG. 129, p. 121]. Although the carbon dioxide content of the inspired gas is now constant (at zero), the oxygen content of the spirometer will still progressively fall and the spirometer bell will gradually descend. Such an arrangement is potentially dangerous to the subject, particularly if the spirometer is filled with room air. He will be subjected to a progressive oxygen lack (anoxia) as the oxygen is used up and the gas in the spirometer changes towards pure nitrogen. As will be seen in Chapter 8 [p. 98], the body has very little protection against anoxia and the subject will 'black out' and faint, usually without warning, when the oxygen level reaches a low figure. Hence this type of experiment should never be carried out without supervision.

The oxygen content of the spirometer gas must be kept at an approximately constant level by running pure oxygen into the spirometer at the same rate as it is used up.

For a limited period of time (10 minutes) the descent of the spirometer bell can be used as a measure of the oxygen uptake [p. 122]. In this case the spirometer is initially filled with 100 per cent. oxygen and the experiment is stopped well before all the oxygen has been used up.

Bag-box Apparatus

To overcome these disadvantages and to allow the subject (or an experimental animal) to breath an inspired gas of constant composition whilst still recording the tidal volume and respiratory rate using a spirometer, the bag-box technique has been devised.

A thin flexible bag containing the gas to be inspired is enclosed in an air-tight box [FIG. 88]. The box is connected to the spirometer. During inspiration the gas is breathed in from the bag via inspiratory valves. The spirometer bell falls as air leaves the bag. During expiration, the expired air is returned via expiratory valves to the box outside the bag and the spirometer bell rises. Since the volumes inspired and expired are very similar the bell will fluctuate about a mean, and the recording can continue until the bag is empty without the spirometer tracing reaching the edge of the paper.

Lung Volumes and Capacities[6]

FIGURE 89 shows the lung volumes and lung capacities of a subject recorded on a spirometer. The volume of air in the lungs

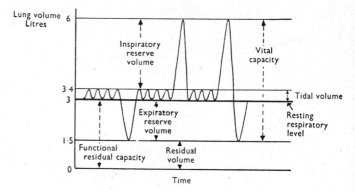

FIG. 89. Lung volumes.

at the end of a quiet expiration (i.e. at the resting respiratory level) is termed the *functional residual capacity*. This volume is equal to 3 litres. By bringing the expiratory muscles into play it is possible to reduce the volume of air in the lungs down to a minimum of 1·5 litres. This minimal volume is termed the *residual volume*.

The residual volume represents the volume of air trapped in the alveoli of the lungs that cannot be removed by making a maximal expiratory effort. This volume has to be measured in some other way since it cannot be measured using a spirometer. To measure it, the standard method is to introduce a gas into the lungs, to allow it to mix with the air there and then to measure its dilution. Helium or nitrogen may be used.

[6] Volume changes which are brought about by inspiratory and expiratory efforts are termed 'volumes'. Volumes which are determined by the size of the lungs and thorax are termed 'capacities'.

To reach the residual volume by making an expiratory effort, the volume of air in the lungs must be decreased from 3 litres to 1·5 litres. Thus the lung volume must be decreased by $3 - 1·5 = 1·5$ litres by expiring this volume of air. This volume change is termed the *expiratory reserve volume*.

As has been seen, with quiet breathing the lung volume increases from the resting respiratory level of 3 litres to 3·4 litres, a tidal volume of 400 ml. It is, however, possible to make a greater inspiratory effort and to increase the volume of the lungs to a maximum of 6 litres. The additional volume of air taken in is termed the *inspiratory reserve volume*.

The maximum possible breath is determined by the *vital capacity*. To measure it a maximum possible inspiration is followed by a maximum possible expiration. The volume of air expired will be the vital capacity. It is the sum of the inspiratory reserve volume, the normal tidal volume and the expiratory reserve volume [FIG. 89]. It is related to the size and development of the subject. It is increased when standing due to the decrease in blood in the lungs. Typical values are 4·5 litres in the male and 3·2 litres in the female.

Timed Vital Capacity

In a normal subject at least 83 per cent. of the vital capacity can be expired in the first second. The volume of air expired in this time is denoted by FEV_1 (forced expiratory volume) [FIG. 90].

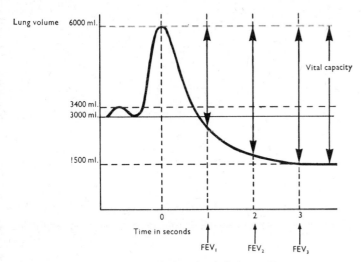

FIG. 90. Spirometer tracing during timed vital capacity manœuvre. A light-weight spirometer bell must be used in conjunction with a fast paper speed. The subject makes a maximal inspiration and then exhales as rapidly and as fully as possible. The volume of gas expired after 1, 2 and 3 seconds is measured and related to the total volume of gas expired ('forced' vital capacity). The result is expressed as a percentage of this volume. A normal subject will expire over 83 per cent. in 1 second, 91 per cent. in 2 seconds and 97 per cent. in 3 seconds.

It is greatly reduced in some forms of lung disease and such a measurement gives a better indication of the severity of the disease than the vital capacity itself.

Peak Flow Meter

As an alternative to the timed vital capacity test, the maximum velocity of air-flow that can be produced during a forced expiration is measured using a peak flow meter [FIG. 91]. Low values are obtained when the expiratory respiratory muscle activity is weak and when there is an obstruction to expiratory air-flow as in asthma and other forms of bronchoconstriction.

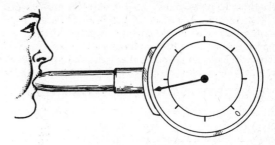

FIG. 91. Peak flow meter.

Maximum Ventilation Volume

Another test of lung function is to determine the maximum breathing capacity of the subject. Since rapid deep breathing cannot be kept up for a whole minute, the subject is asked to breathe as rapidly and as deeply as possible for 15 seconds and the results are multiplied by 4. The pulmonary ventilation per minute calculated from the tracing on a recording spirometer is termed the maximum ventilation volume. Alternatively the expired air may be collected in a Douglas bag and its volume measured. Normal subjects can exceed 100 litres per minute.

Elasticity and Resistance

The lungs and chest wall are elastic structures. The combined elasticity may be measured, by relaxing the respiratory muscles, and then seeing by how much the volume increases, when the pressure in the lungs is raised. Elasticity is $\frac{\Delta P}{\Delta V}$ where ΔP is the change of pressure and ΔV is the change of volume. In respiratory mechanics it is usual to consider not elasticity, but its reciprocal *compliance*, i.e. $\frac{\Delta V}{\Delta P}$.

The compliance is reduced in lung disease, such as pulmonary fibrosis.

In quiet breathing, work is done in overcoming the elasticity during inspiration, but this work is available for the elastic recoil in expiration. A small quantity of energy is lost due to friction.

In more rapid breathing the air passages offer a resistance to the air movement, and high pressure differences must be created between the lungs and external air to enable rapid air-flow to occur. This resistance is increased when narrowing of the airway occurs, as in asthma.

If the volume of the lungs and the pressure in the thorax are recorded simultaneously, then in quiet breathing where the elastic resistance predominates, the volume and pressure will change together.

In rapid breathing, where the resistance to flow in the air passages predominates, it is the air-flow (i.e. rate of change of volume) that will change with the pressure. Thus, the peaks of volume and pressure will occur at different times. As the rate increases, the volume flow associated with the same pressure changes becomes less and less until for very high respiratory rates there is scarcely time for the air to move at all, and the breathing becomes very inefficient.

Surfactant

The compliance of the lungs is made greater by the presence of **surfactant** in the alveoli. This substance acts like a detergent and reduces the surface tension of the water in the alveoli. It allows the alveoli to expand more readily with air during inspiration. Without surfactant, any alveolus which loses its air and contains only water during expiration would not fill with air during the next inspiration because the respiratory effort would be insufficient to overcome the surface tension forces [see also Laplace's Law p. 223].

ANALYSING RESPIRATORY GASES

The percentage of carbon dioxide in a respiratory gas may be measured by taking a known volume of gas x, absorbing the carbon dioxide with a solution of potassium hydroxide (potash) and measuring the new volume y.

The percentage of $CO_2 = \dfrac{x - y}{x} \times 100$ per cent.

Similarly, the percentage of oxygen may be determined by next absorbing the oxygen with a solution of chromous chloride or pyrogallol. If the new volume is z then the percentage of

$$O_2 = \dfrac{y - z}{x} \times 100 \text{ per cent.}$$

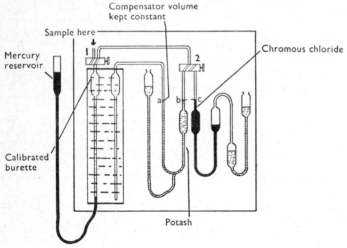

FIG. 92. The Haldane gas analysis apparatus. All readings are taken with the calibrated burette connected to potash and with fluid levels at a, b and c.

In the Haldane apparatus [FIG. 92] the gas is drawn into a calibrated burette by lowering the mercury reservoir. Taps (1) and (2) are turned to connect the burette with the solution of potassium hydroxide. By raising and lowering the mercury reservoir the gas is brought into contact with the potash. When absorption of CO_2 is complete, tap (2) is turned to bring the gas in contact with the solution of chromous chloride and the process of raising and lowering the reservoir is continued until all the O_2 has been absorbed.

Alterations in atmospheric temperature and pressure are a source of error in this type of analysis. The Haldane apparatus compensates for any such change by having a second burette alongside the calibrated burette. The volume of air in this second burette is maintained constant throughout by raising or lowering the potash reservoir. This reservoir applies a correcting pressure and exactly

the same pressure is transmitted to the calibrated burette when this burette is in communication with the potash.

Thus, provided that all three readings, x, y and z, are taken with the calibrated burette connected to potash, no allowance need be made for atmospheric temperature or pressure changes occurring during the analysis.

The gases in the Haldane apparatus are saturated with water vapour throughout the determinations, but the percentage results obtained are equal to those of the dry gas.

Other methods available for the analysis of respiratory gases include the Lloyd modification of the Haldane apparatus, infrared CO_2 analyser, paramagnetic O_2 analyser, and the mass spectrometer.

Infra-red CO₂ Analyser

The infra-red CO_2 analyser is based on the principle that carbon dioxide absorbs radiant heat. Thus if infra-red radiation from a source is passed through a sample chamber to a detector [FIG. 93],

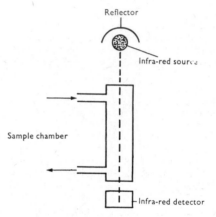

FIG. 93. Infra-red carbon dioxide analyser. Carbon dioxide absorbs infra-red radiation. The higher the percentage of carbon dioxide in the sample chamber the lower the amount of infra-red reaching the detector.

the amount of radiation reaching the infra-red detector will depend on the carbon dioxide content of the gas in the sampling chamber. It will be less when carbon dioxide is present in the sample chamber than when it is absent.

In a more refined version of the apparatus the heat transmitted through the sample chamber is compared with that passing through the control tube which contains the same gases without carbon dioxide.

This apparatus has a reasonably rapid response and can be used for breath-to-breath analysis.

Paramagnetic O₂ Analyser

Oxygen is paramagnetic. When liquid oxygen is poured over the pole-piece of a magnet it adheres to the magnet instead of falling to the ground under the influence of gravity. In this respect oxygen resembles iron filings. The other respiratory gases (nitrogen, argon, carbon dioxide and water vapour) do not have this property.

Oxygen will displace an evacuated sphere from the magnetic field which exists between the two poles of a permanent magnet. This fact forms the basis of the paramagnetic oxygen meter. The displacing force depends on the number of oxygen molecules present which will depend on the percentage of oxygen at a given barometric pressure.

Two versions of this type of analyser are in use [FIG. 94]. In the first, the displacement of the sphere by the oxygen is opposed by a

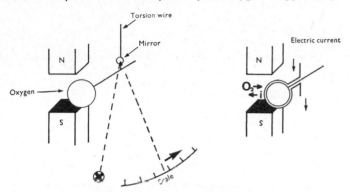

FIG. 94. Paramagnetic oxygen meter. Oxygen is paramagnetic and exerts a force which tends to displace an evacuated sphere from the magnetic field between the pole-pieces of a permanent magnet. *Left:* The displacement is limited by a torsion wire. A mirror reflects a beam of light on to the scale to give a reading of the displacement which is proportional to the oxygen present. *Right:* The displacement of the sphere by the oxygen is opposed by passing an electric current (i) through a coil surrounding the sphere. The current is adjusted manually or electronically until there is no displacement of the sphere. The current flowing is then proportional to the oxygen present.

torsion wire to which a small mirror is attached. The resultant displacement is measured by the movement of a light beam across a scale in a similar manner to a mirror galvanometer. In the second, the displacement of the sphere is opposed by passing an electric current through a single coil of wire around the sphere. This coil in a magnetic field acts like a moving-coil galvanometer and the direction of current flow is arranged so that the direction of movement due to the current is the opposite of that due to the oxygen. A mirror and light beam are again used to detect displacement of the sphere and to enable the current to be adjusted for a *nul-deflection.*

The adjustment of the current is carried out manually using a multi-turn variable resistor, the dial of which is directly calibrated in 'percentage oxygen'. Alternatively the current may be set by means of a servo-mechanism which uses two photo-transistors one on each side of the light beam at the nul-point. Should the light beam fall on the 'too low' photo-transistor, the current is increased. Should the light beam fall on the 'too high' photo-transistor the current is reduced. In this way the current is automatically set to the correct value for a nul-deflection. The higher the oxygen content the higher the current required to return the sphere to the zero position. The resultant current is passed through a milliammeter directly calibrated in 'percentage oxygen'.

Paramagnetic oxygen meters require a long time (30 seconds or longer) to reach their equilibrium stage, and are more suited to the analysis of static gas samples than the continuous analysis of respiratory gases.

Respiratory Mass Spectrometer

The sample of respiratory gas, which is introduced into the mass spectrometer, is first reduced to a very low pressure (about 10^{-6} mm. Hg). It is then ionized by an electron beam. Positively

changed ions of nitrogen, oxygen, argon and carbon dioxide are formed. These are deflected in a semi-circle by a magnetic field.

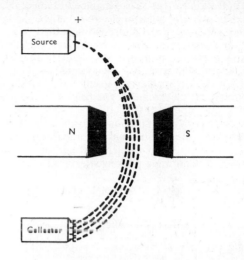

MASS SPECTROMETER

FIG. 95. The gas to be analysed is reduced to a very low pressure $(10^{-6}$ mm. Hg) and ionized by bombarding it with electrons. The positively charged ions of nitrogen, oxygen, carbon dioxide and argon take paths through the magnetic field to the collector electrodes. By using one electrode for each gas and measuring the electrode current the percentage composition of the mixture can be determined.

In another form of the respiratory mass spectrometer, only one electrode is used and the gases are measured sequentially by changing the voltage on the collector electrode.

The path taken depends upon the mass of the ions which is given by their 'molecular' weights: $N_2 = 28$, $O_2 = 32$, $A = 40$ and $CO_2 = 44$. The light ions take a short path whilst the heavy ions take a longer path.

Each ion may be collected on a separate electrode [FIG. 95] and

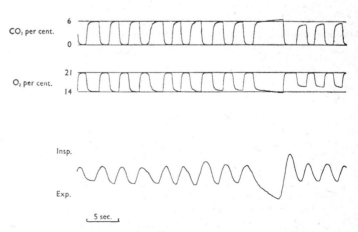

FIG. 96. Recordings of the CO_2 and O_2 levels in the inspired and expired air using the mass spectrometer (upper traces) and respiratory movements using a stethograph (lower trace) during quiet breathing and following a prolonged expiration. The end-expiratory level of CO_2 and O_2 gives an indication of the alveolar air composition. (Duke, Helen N. and Green, J. H.)

the electrical charge will be a measure of the amount of the gas in the mixture. After suitable amplification a continuous recording of the concentration of these gases may be obtained.

Alternatively only one collector electrode may be used. The path taken by the ions also depends on the magnitude of the negative voltage on the collector electrode. Thus by varying this negative voltage between two limits repetitively twenty-five times a second, the ions of the different gases may be collected sequentially on the collector electrode. Such an arrangement will allow a spectrum of molecular weights to be scanned. The heights of the peaks which correspond to the respiratory gases may be extracted electronically and may be used to give a continuous recording. FIGURE 96 shows a typical recording made using such an apparatus. It shows the breath-to-breath changes in oxygen and carbon dioxide during quiet breathing and during a prolonged expiration.

The Collection and Analysis of a Sample of Alveolar Air

	Inspired air (per cent.)	Alveolar air (per cent.)
Oxygen	21	14
CO_2	0	6
N_2	79	80

FIGURE 97 shows in diagrammatic form the change in composition of the respiratory gases at the nose or mouth during one respiratory cycle. It will be seen that when expiration starts the first air breathed out is dead space air (unchanged room air containing 21 per cent. O_2 and 0 per cent. CO_2). This is followed by a

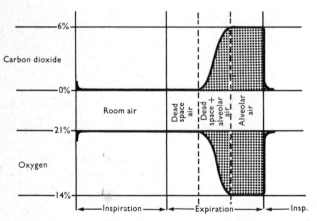

FIG. 97. Diagram showing the changes in composition of the respiratory gases recorded at the mouth (or nose) during a single respiratory cycle. During inspiration the composition is that of room air. During expiration the composition changes from that of room air (dead space air) to that of alveolar air at the end of expiration. An analysis of an end-expiratory sample will thus enable the composition of alveolar air to be determined.

mixture of dead space air and alveolar air. Finally, provided that the tidal volume is great enough, alveolar air is exhaled (14 per cent. O_2 and 6 per cent. CO_2).

Thus, to sample alveolar air, an end-expiratory sample of expired air is required. Using an instantaneous gas recorder, such

as the mass spectrometer, the level of CO_2 and O_2 at the end of expiration may be noted [FIG. 97].

A simpler method involves the use of the Haldane alveolar air tube [FIG. 98]. At the end of a normal quiet inspiration or expiration the mouth is applied to the tube and a maximal expiration

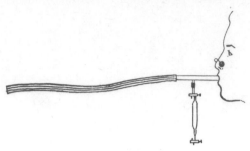

FIG. 98. The Haldane alveolar tube with sampling tube attached to the side-tube.

made down the tube. The tongue is placed over the end of the tube until a sample has been drawn off via the side tube.

An alternative method is to use a Y-tube connected to two rubber bags. The bags are initially empty and the necks are closed with the fingers [FIG. 99]. At the end of a normal quiet expiration,

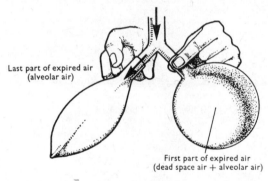

Last part of expired air (alveolar air)

First part of expired air (dead space air + alveolar air)

FIG. 99. The two-bag technique for obtaining a sample of alveolar air. At the end of a quiet inspiration the Y-tube is placed in the mouth and a maximal expiratory effort made into the first bag. Just before the end of expiration a change is made with the fingers to the second bag and the end-expiratory sample is collected in this bag. It may then be analysed using an infra-red carbon dioxide analyser and paramagnetic oxygen analyser. Alternatively the Haldane gas analysis apparatus may be used.

the tube is placed in the mouth and a maximal expiration is made. The fingers are removed from the first bag and the first part of the expired air is collected in this bag. Towards the end of expiration, when all the dead space air has been expired, a change is made with the fingers to the second bag. At the end of the expiration when the alveolar air sample has been collected in the second bag, the fingers close the neck of this second bag which can then be taken to an infra-red CO_2 analyser and paramagnetic O_2 analyser for analysis. The content of the first bag is discarded.

An automatic end-tidal sampler may be employed triggered by respiration. A device which employs the next inspiration to collect the previous end-tidal sample is the Otis–Rahn sampler. This is shown combined with a respiratory valve in FIGURE 100. During inspiration the balloon expands and draws in end-expiratory air from beyond the expiratory valve. During expirations the balloon is compressed and prevents any dead space air from entering the analyser. An end-tidal sampler only gives an accurate sample of alveolar air when the tidal volume is high enough to ensure that no dead space air is mixed with the expired alveolar air at the end of expiration. The normal tidal volume of 400 ml. is barely sufficient for this purpose.

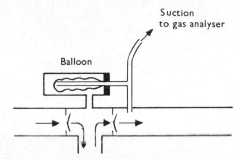

FIG. 100. Combined respiratory valve and Otis–Rahn alveolar air sampler.

Analysis of Expired Air

Expired air, collected in a Douglas bag over a known period of time, may be analysed by taking a sample from the bag via the side tube. When the analysis has been completed, the rest of the bag's content may be passed through a gas meter to estimate the total volume and the minute volume of expired air.

The volume of inspired air has not been measured, but it may be calculated as follows from the analysis of the two gases:

	Inspired air (per cent.)	Expired air (per cent.)
Oxygen	21	16
CO_2	0	4
N_2	79	80

The percentage of nitrogen is obtained by subtracting the combined percentages of oxygen and CO_2 from 100 per cent.

Nitrogen in the lungs is in equilibrium with that in the blood. It is not metabolized by the body and as a consequence nitrogen is neither absorbed nor excreted in the lungs. Yet the percentage of nitrogen appears to be different in the inspired air and expired air.

Percentages can only be compared if they are percentages of the same thing! The percentage of nitrogen in expired air is different from that in the inspired air because although the nitrogen is the same, the volumes are different. The expired volume is less than that of the inspired volume and, therefore, the same quantity of nitrogen gives an apparently higher percentage.

The expired volume is less than the inspired volume because not all the oxygen absorbed is converted to expired carbon dioxide. Some of the oxygen is converted to water which is excreted via the kidneys.

The nitrogen percentages enable the volume of inspired air to be calculated from that of expired air.

If V_{INSP} is the inspired volume and V_{EXP} the expired volume:

$$79\% \; V_{INSP} = 80\% V_{EXP} \text{ using the above figures.}$$

$$\therefore \quad V_{INSP} = \frac{80}{79} V_{EXP}$$

The oxygen used and the carbon dioxide evolved per minute may be obtained approximately by multiplying the pulmonary ventilation by the percentage differences.

Thus,

	Inspired Air %	Expired Air %	Difference %
O_2	21	16	5
CO_2	0	4	4

If $V_{EXP} = 6$ litres per minute
O_2 used is approximately 5 per cent. of 6 litres = 300 ml.
CO_2 given off is 4 per cent. of 6 litres = 240 ml.

The oxygen used is, however, not exactly 5 per cent. of the pulmonary ventilation because the inspired and expired volumes are different. Assuming that the nitrogen percentages are as above, then *with reference to the expired volume* V_{EXP} the corrected figures are:

$$O_2 \text{ uptake} = \left[\frac{80}{79} \times 21\% - 16\%\right] V_{EXP}$$

$$= 5{\cdot}27\% \; V_{EXP}$$

$$= \frac{5{\cdot}27}{100} \times 6000$$

$$= 316 \text{ ml. per minute}$$

$$CO_2 \text{ evolved} = \left[-\frac{80}{79} \times 0\% + 4\%\right] V_{EXP}$$

$$= 4\% \; V_{EXP}$$

$$= \frac{4}{100} \times 6000$$

$$= 240 \text{ ml. per minute}$$

Calculation of Dead Space Volume

The size of the dead space may be calculated from the tidal volume and an analysis of the inspired, expired and alveolar airs.

At the end of each inspiration the dead space is full of room air. The next expiration will consist of this dead space air and the remainder will be alveolar air.

Let the dead space volume be x ml.
Then considering carbon dioxide:

Volume of CO_2 expired with each breath

$$= \frac{\text{Tidal volume}}{100} \times \%CO_2 \text{ in expired air (by definition)}$$

It also equals

$$\frac{x}{100} \times \%CO_2 \text{ in room air}$$

$$+ \frac{(\text{Tidal volume} - x)}{100} \times \%CO_2 \text{ in alveolar air.}$$

Since $\%CO_2$ in room air is zero:

$$x = \text{Tidal volume} \ \frac{\%CO_2 \text{ in alveolar air} - \%CO_2 \text{ in expired air}}{\%CO_2 \text{ in alveolar air}}$$

EXAMPLE: If tidal volume = 400 ml.

Percentage of CO_2 in alveolar air = 6%
Percentage of CO_2 in expired air = 4%

Then,

$$\text{Dead space} = 400 \times \frac{6-4}{6}$$

$$= 133 \text{ ml.}$$

A similar calculation may be made using oxygen.

Relationship Between Ventilation and Alveolar CO_2

In the equilibrium state, the amount of CO_2 lost per minute in the expired air is equal to the CO_2 production per minute by the body. The percentage composition of the alveolar air is set by the fact that:

CO_2 lost = $\%CO_2$ in alveolar air $\times$ alveolar ventilation
  = CO_2 produced

Thus:

$$\%CO_2 \text{ in alveolar air} = \frac{CO_2 \text{ production}}{\text{alveolar ventilation}}$$

Example:

CO_2 production = 200 ml. per minute
Alveolar ventilation = 3,300 ml. per minute

Hence,

$$\%CO_2 \text{ in alveolar air} = \frac{200}{3,300} = 6\%$$

BREATH SOUNDS

The movement of air into and out of the thorax produces audible sounds. When a stethoscope is applied to the chest two varieties of *breath sound* are heard.

Vesicular Breathing

This is a soft rustling sound caused by the passage of air into the alveoli. The inspiratory sound lasts for at least twice as long as the expiratory sound and there is no pause between the two sounds. This sound is heard over healthy lung tissue. Typical vesicular breathing may be heard in the axilla and in the infrascapular region.

Bronchial Breathing

This is a loud harsh sound (like a sharp whispered 'hah') that is heard over the larger air passages, the larynx, trachea and bronchi. There is a distinct pause between the inspiratory and expiratory components and the latter is prolonged. It may be heard by placing the stethoscope over the trachea and at the back of the chest between the scapulae at the level of the 4th thoracic vertebra where the trachea bifurcates. In other parts of the chest propagation of this sound is prevented by the healthy lung. If the lung has become solid, as for example, in pneumonia, the sounds will be conducted easily from the large airways to the chest wall and bronchial breathing will be heard over parts of the chest remote from the large air passages.

Absence of Breath Sounds

An absence of breath sounds indicates either that air is not entering that part of the lung, or that the lung is separated from the chest wall by, for example, fluid in a pleural effusion.

Rhonchi and Râles

When fluid is present in the air passages, additional sounds may be heard. If the sound has a musical quality (a wheeze) it is referred to as a *rhonchus*. If it is non-musical in quality it is referred to as a *râle*. Fine râles are also known as *crepitations*. These sounds are similar to those heard when rubbing the hair just above the ear.

ARTIFICIAL RESPIRATION

If the respiratory muscles are no longer able to provide an adequate ventilation of the lungs, artificial respiration must be employed.

The first essential is to establish and maintain a clear airway from the nose or mouth to the lungs. When a person becomes unconscious the tongue may fall back across the pharynx and obstruct the entrance to the trachea [FIG. 101]. (A partial blockage

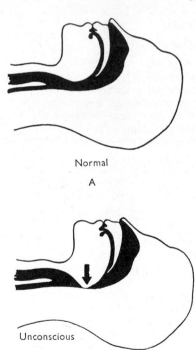

Normal
A

Unconscious

Tongue blocking throat
B

FIG. 101. Airway blockage. Air passages are shown in black. In the unconscious subject the tongue may fall back and block the throat. To clear the airway the head is extended, and the lower jaw forced forwards.

of the airways by the tongue during sleep leads to turbulence in the air flow called *snoring*.) To remove this obstruction the tongue must be pulled forward. If this is not possible, the same result may be achieved by pulling the jaw forwards and extending the head backwards. Alternatively, an 'airway' may be inserted round the back of the tongue to hold the tongue forward.

Artificial respiration machines or respirators may be divided into two classes—cabinet respirators and positive pressure respirators. Theoretically, both work on the same principle, which is that the lungs may be expanded by making the internal pressure in the lungs higher than the external pressure acting on the chest wall.

Cabinet Respirator

A cabinet respirator (iron lung) consists of a box in which the patient is placed with his head outside and an air-tight seal around his neck. Respiratory movements are brought about by the cyclical lowering and raising of the pressure in the box. The pressure in the lungs remains at atmospheric pressure. When the box pressure is lowered, the pressure inside the lungs exceeds the external pressure on the thorax and the chest expands causing inspiration. When the pressure in the box changes back to atmospheric pressure or above, air is expelled. Such an apparatus is used when the patient's respiratory muscles have been paralysed by injury or an infection such as anterior poliomyelitis.

Intermittent Positive Pressure Lung Inflation

If the paralysis affects the medulla (bulbar paralysis) the cough reflex may be lost. To prevent secretions and vomit entering the lungs, a tracheostomy is often performed and a cuffed tube inserted into the trachea. Such a tube may be connected to a positive pressure pump which inflates the lungs by cyclically blowing air into them (intermittent positive pressure ventilation, I.P.P.V.). Expiration may be brought about by the passive recoil of the lung and chest wall. Alternatively a negative pressure phase may be incorporated in the pump so that the expired air is sucked out. Nursing procedures are easier with this type of machine than with the cabinet respirator. The patient may even move about in a wheeled-chair with the positive pressure machine attached. A similar arrangement using a face-piece or a tube inserted into the trachea via the mouth, in place of the tracheostomy tube, is used to maintain respiration in patients under anaesthesia, particularly when a muscle relaxant drug has been given.

EMERGENCY RESUSCITATION

Mouth-to-mouth Respiration (Expired Air Resuscitation)

If no mechanical pump is available, in an emergency one's own lungs may be used as a positive pressure pump. Air may be blown into the lungs of the victim by way of his mouth or nose. This is known as mouth-to-mouth or mouth-to-nose artificial respiration. The subject's head is tilted backwards as far as possible to prevent blockage of the airway by the tongue [FIG. 101]. In addition the lower jaw may be pulled forward so that the lower teeth are in front of the upper teeth. The tongue is attached to the lower jaw and this pulls the tongue away from the posterior pharyngeal wall. The extension of the head should be at the atlanto-occipital joint (between C.1 and skull). It may be difficult to do this without first flexing the neck in the lower cervical region. The final position of flexion and extension has been likened to that employed when craning the head forwards and upwards to 'sniff the morning breeze'.

The mouth is sealed around the victim's mouth to prevent air leakages.

Either the cheek should be pressed against the victim's nose or the nose should be closed with the other hand [FIG. 102 (3) and (1)].

Alternatively, the mouth is kept closed and the air is blown in through the nose [FIG. 102 (4)]. This mouth-to-nose method is just as efficient as the mouth-to-mouth method and is often easier to carry out.

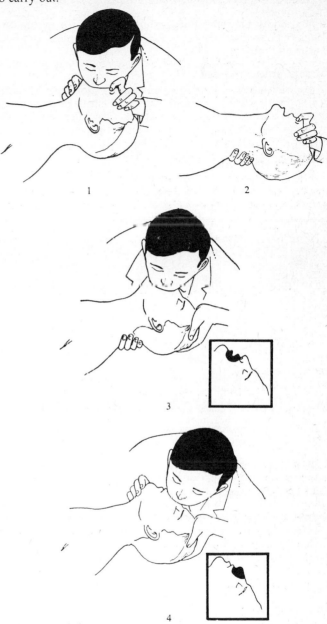

FIG. 102. Expired air resuscitation (mouth-to-mouth artificial respiration).

The alternative methods are shown. To ensure a clear airway the head is extended (2). In addition the lower jaw may be pulled forward (1 and 4).

When using the mouth-to-mouth method the nose must be closed. This may be carried out by using the fingers (1) or using the side of the cheek (3). If the mouth-to-mouth artificial method is not possible, the mouth-to-nose technique (4) may be employed. In children the mouth may be applied to both the nose and mouth simultaneously.

After inflation of the victim's lungs the operator's head is turned to one side and expiration allowed to occur by the elastic recoil of the lungs and chest wall (2).

Air is blown into the lungs until the chest is seen to expand. The mouth is then removed and the chest watched whilst expiration takes place. The recommended rate of inflating the lungs is 12–15 times a minute.

If the mouth is firmly shut the lungs are inflated by blowing through the nose. In infants the mouth may be applied over both the nose and mouth.

External Cardiac Massage

If respiration ceases, then after a relatively short time the heart will stop beating (cardiac arrest) due to the effects of oxygen lack and carbon dioxide excess on the heart muscle. Alternatively if the heart stops beating, respiration will cease. In such cases artificial respiration is ineffective unless accompanied by external cardiac massage to restore the circulation of blood [FIG. 103].

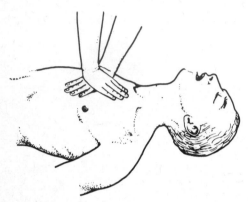

FIG. 103. External Cardiac Massage.
Intermittent pressure is applied with the heel of the hands to depress the sternum 1–1½ in. towards the vertebral column, at the rate of once a second. The pressure is applied in the midline one third of the way up the sternum. The procedure should be interrupted briefly after every 6–8 compressions whilst the lungs are inflated by mouth-to-mouth artificial respiration.

Cardiac arrest should be suspected when sudden deep unconsciousness occurs in a person who was previously conscious. It should also be suspected in a person with respiratory failure if there is no improvement in the victim's colour, that is, if the cyanosis does not disappear, after six lung inflations. It may be confirmed by the absence of any pulse and the absence of heart sounds. The pupils will be dilated.

External cardiac massage should not be employed if the heart is still beating. It is only used in conjunction with expired air resuscitation and never on its own.

The victim should be placed on a hard surface in the supine position and external cardiac compression applied at a rate of 60–90 times per minute. The site of application of the pressure is important. It should be applied to the sternum exactly in the midline at the level of the 4th intercostal space (level of the nipples in the male). If applied too high it will be ineffective; if too low, the liver may be damaged. Pressure is applied to depress the sternum 1 to 1½ inches and to compress the heart in its pericardium against the underlying vertebral column. This manœuvre forces blood from the ventricles into the aorta and pulmonary arteries. When the pressure is removed, the elastic recoil of the costal cartilages and ribs allows the ventricles to fill from the veins. The pressure is applied with the hands placed on top of one another, but only the heel of the palm of the lower hand should touch the chest.

Since it is not possible to inflate the lungs by expired air resuscitation whilst cardiac massage is being applied, these procedures must be employed alternatively. After every 5–13 compressions, the external cardiac massage is interrupted briefly whilst the operator (or another operator) inflates the lungs by mouth-to-mouth artificial respiration.

If these manœuvres are successful the colour of the subject will improve and the pupils will constrict.

When the subject starts to recover he should be rolled over into the semi-prone position, preferably with his pelvis raised and head turned to one side to prevent the inhalation of any vomit [FIG. 104].

FIG. 104. When the resuscitated subject starts to recover he should be rolled over into the 'coma' position to prevent the inhalation of any vomit.

Mouth-to-mouth respiration and external cardiac massage are essentially 'first aid' measures which will maintain a subject for up to one hour or more. Subsequent action will depend on the facilities available.

If the respiration is not restored, the trachea can be intubated and artificial respiration maintained using a positive pressure pump. An external cardiac defibrillator [p. 35] may be used to bring about defibrillation if the ventricles are fibrillating. It may be necessary to correct the developing acidaemia by hyperventilation or an injection of sodium bicarbonate before this is successful. It may be necessary to insert an external cardiac pacemaker electrode into the right ventricle via a superficial neck vein under X-ray control [p. 34], and to employ an external pacemaker to 'pace' the heart.

Since further cardiac arrests may occur within the subsequent 24 hours, the subject is usually attached to an E.C.G. monitor for this period.

Pulmonary Obstruction

If a victim chokes and inhales a bolus which lodges in the larynx, a complete pulmonary obstruction may result. The victim becomes greatly distressed, is unable to speak and rapidly becomes cyanosed. Mouth-to-mouth resuscitation is of no use in such a situation.

An immediate tracheostomy below the obstruction would allow air to reach the lungs, but, as a first-aid measure, an attempt should be made to dislodge the bolus. A certain amount of air will still be present in the lungs and the principle is to squeeze some of this residual air out of the lungs so that the bolus pops out like a cork from a bottle. Pushing sharply upwards below the diaphragm towards the head is one method. Hemlich (1974) has described a technique whereby the operator stands behind the victims with both arms around the waist at belt level. A sharp hug is given and at the same time the victim is allowed to slump forward with his head and arms dangling in an attempt to dislodge the bolus.

REFERENCES AND FURTHER READING

Barcroft, J. (1928) *The Respiratory Function of the Blood*, 2nd ed., Cambridge.

Campbell, E. J. M. (1958) *The Respiratory Muscles and the Mechanics of Breathing*, London.

Dittmer, D. S., and Grebe, R. M. (Eds) (1958) *Handbook of Respiration*, Philadelphia.

Douglas, C. G., and Priestley, J. G. (1948) *Human Physiology*, 3rd ed., Oxford.

Hugh-Jones, P., and Campbell, E. J. M. (Eds) (1963) Respiratory physiology, *Brit. med. Bull.*, **19**, No. 1.

Miller, W. S. (1947) *The Lung*, 2nd ed., Springfield.

Rahn, H., and Otis, A. B. (1949) Continuous analysis of alveolar gas composition during work, hyperpnoea, hypercapnia and anoxia, *J. appl. Physiol.*, **1**, 717.

Severinghaus, J. W. (1962) Respiration, *Ann. Rev. Physiol.*, **24**, 421.

ARTIFICIAL RESPIRATION

Cox, J., Woolmer, R., and Thomas, V. (1960) Expired-air resuscitation, *Lancet*, i, 727.

Drinker, P., and McKhann, C. F. (1929) The use of a new apparatus for the prolonged administration of artificial respiration, *J. Amer. med. Ass.*, **92**, 1658.

Eve, F. C. (1932) Actuation of the inert diaphragm by a gravity method, *Lancet*, ii, 995.

Hemlich, H. (1974) Hemlich's hug, *J. Amer. med. Ass.*, **229** No. 7, 746.

Paul, R. W. (1935) The Bragg–Paul pulsator, *Proc. roy. Soc. Med.*, **28**, 436.

Whittenberger, J. L. (1955) Artificial respiration, *Physiol. Rev.*, **35**, 611.

FILM

Principles of Respiratory Mechanics, Parts I and II, British Film Institute.

6. THE CARRIAGE OF GASES

CARRIAGE OF GASES

Partial Pressures

The inspired air, expired air and alveolar air are composed of the three gases, oxygen, carbon dioxide, and nitrogen plus one vapour, water-vapour, in varying proportions.

Dalton's Law of Partial Pressures states that:

(1) The pressure exerted by a mixture of gases is equal to the sum of the pressures which each would exert if it alone occupied the space.

(2) The pressure exerted by a saturated vapour depends only upon the temperature and the particular liquid considered.

It follows from the first part of Dalton's Law that the partial pressure that each gas exerts is the same whether or not the other gases are present and depends only on the concentration of that gas. From Boyle's Law it follows that the partial pressure of any gas is proportional to its percentage by volume in the mixture. Thus when three gases oxygen, carbon dioxide and nitrogen occupy the same space each will contribute to the total pressure in proportion to its concentration.

Consider first dry room air which contains 21 per cent. oxygen and 79 per cent. nitrogen. The carbon dioxide percentage is only 0·03 and it may be ignored. If the barometric pressure is 760 mm. Hg then the oxygen and nitrogen will contribute to this in proportion to their percentages. The partial pressure of oxygen will be 21 per cent. of 760 mm. Hg = 160 mm. Hg. The partial pressure of nitrogen will be 79 per cent. of 760 = 600 mm. Hg.

The barometric pressure at sea-level fluctuates slightly about 760 mm. Hg (*101·3 kPa*) from day to day with the prevailing meteorological conditions. There will, correspondingly, be small fluctuations in the partial pressures of oxygen and nitrogen.

Above sea-level there is a progressive decrease in the barometric pressure with altitude [FIG. 105]. Thus at 18,000 feet the barometric pressure has fallen to 380 mm. Hg or half that at sea-level. At the high altitudes reached by manned space-craft (over 100 miles) the external barometric pressure is very close to 0 mm. Hg (or a vacuum).

The barometric pressure is represented by P_B (or more simply by **B**). It is expressed in *mm. Hg* (or alternatively in kilopascals).

At an altitude of 16,000 ft. where the barometric pressure is only 400 mm. Hg the percentage composition of the air is the same as at sea-level, but the partial pressure of oxygen will now be only 21 per cent. of 400 = 82 mm. Hg and that of nitrogen only 318 mm. Hg [TABLE 6].

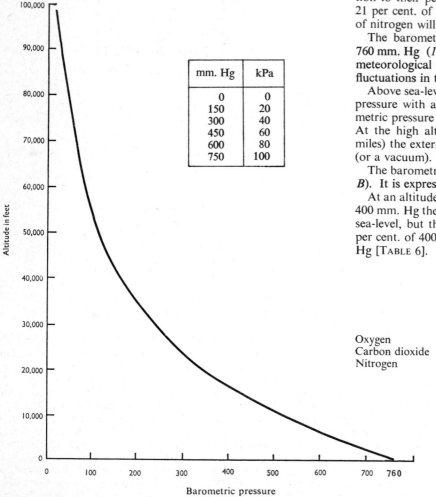

mm. Hg	kPa
0	0
150	20
300	40
450	60
600	80
750	100

Altitude in feet

Barometric pressure

TABLE 6

INSPIRED AIR

	Per cent. dry gas	Partial pressure at sea level (mm. Hg)	Partial pressure at 16,000 ft. (mm. Hg)
Oxygen	21	160	82
Carbon dioxide	0	0	0
Nitrogen	79	600	318
		760	400

FIG. 105. The barometric pressure decreases with altitude in approximately a logarithmic fashion. The pressure has halved (=380 mm. Hg) by the time 18,000 feet is reached.

If the air is moist, water vapour pressure must be taken into account and the partial pressures will be reduced correspondingly.

Saturated Water Vapour Pressure [FIG. 106]

Alveolar air is saturated with water vapour at body temperature. At 37 °C., water has a vapour pressure of 47 mm. Hg (*6·3 kPa*) and this is independent of the other gases and the total barometric pressure. It is the same at sea-level as at 16,000 ft.

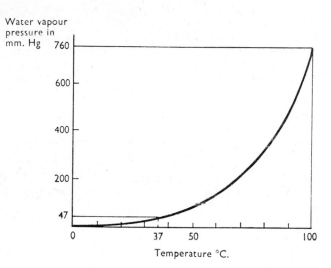

FIG. 106. The saturated water vapour pressure at different temperatures. At boiling point (= 100 °C.) the water vapour pressure equals the atmospheric pressure (= 760 mm. Hg). The water vapour pressure at body temperature (37 °C.) is 47 mm. Hg.

When water vapour is present, the partial pressures of the other gases will total the barometric pressure less the water vapour pressure.

Thus, the partial pressure of oxygen in the alveolar air will equal 14 per cent. of (760 − 47) mm. Hg ≃ 100 mm. Hg. The partial pressure of carbon dioxide will equal 6 per cent. of (760 − 47) mm. Hg ≃ 40 mm. Hg. [See TABLE 7 for mm. Hg and kPa values.]

TABLE 7

ALVEOLAR AIR

	Per cent. dry gas†	Partial pressure (mm. Hg)	(kPa)
Oxygen	14	100	13·3
Carbon dioxide	6	40	5·3
Nitrogen	80	573	76·4
Water vapour	—	47	6·3
Total = barometric pressure		760	101·3

† As determined, for example, by the Haldane apparatus, p. 70.

Water vapour complicates the problem of supplying oxygen at high altitudes because no matter how low the barometric pressure, the water vapour pressure is still 47 mm. Hg. Thus, at 70,000 ft. where the barometric pressure is only 80 mm. Hg, only 33 mm. Hg total is available for the oxygen, carbon dioxide and nitrogen in the alveolar air. Even on pure oxygen, consciousness would be lost.

Non-saturated Water Vapour Pressure

Although alveolar air and expired air are saturated with water vapour, inspired air is not. When the gas considered is not saturated with water vapour, the pressure exerted by the water vapour will lie between zero and the saturated water vapour pressure for the temperature. It will depend on the humidity. The water vapour pressure of inspired air may be determined from tables of temperature and humidity.

Henry's Law of Solution—Concept of Tension

Henry's Law states that the quantity of a gas going into simple solution at constant temperature is proportional to the pressure. Dalton's extension to this states: 'In a mixture of gases, the solubility of each gas varies proportionally with its partial pressure'.

Thus, if water is exposed to air at a pressure of one atmosphere, the nitrogen, oxygen and carbon dioxide will dissolve in amounts proportional to their solubilities (absorption coefficients) and their partial pressures [FIG. 107].

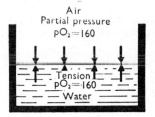

FIG. 107. When a liquid is exposed to a gas having a partial pressure *P* mm. Hg, the gas enters the fluid until the tension of the gas in the fluid has built up to *P* mm. Hg. Thus, when water is exposed to room air having an oxygen partial pressure of 160 mm. Hg, oxygen enters the water until the oxygen tension in the water has reached 160 mm. Hg. The gas and liquid are then in equilibrium.

The solubilities of oxygen, carbon dioxide and nitrogen are in the ratio of 2:50:1.

When dissolved the gas molecules will try to leave at the surface with a **tension** equal to the partial pressure.

The direction of movement of a gas across a liquid/gas interface or a liquid/liquid interface is determined by the tension gradient. The movement is always from the region of high tension to a region of low tension.

In the lungs, oxygen passes from the alveolar air into the blood because the oxygen 'tension' (more accurately *partial pressure*) in the alveolar air is higher than that in the blood arriving at the lungs. When the same blood reaches the tissues, oxygen will pass out to the tissues because the blood oxygen tension is higher than that in the tissue fluid.

In the case of carbon dioxide the blood arriving at the lungs has a higher tension than that in the alveolar air. Carbon dioxide, therefore, passes in the opposite direction to oxygen and is given off. In the tissues, the tissue carbon dioxide tension exceeds that of the blood and carbon dioxide, therefore, passes into the blood.

By the time the blood has passed through the lung capillaries, it has come into equilibrium with the alveolar air and the tensions of oxygen and carbon dioxide in the pulmonary veins will be equal to the partial pressure or 'tension' of the gases in the alveoli of the lungs.

The tension of oxygen in the blood is written as pO$_2$ and that of carbon dioxide as pCO$_2$. Both are measured in mm. Hg or kPa.

If the partial pressure or tension of a gas in a specific part of the body is under consideration, it is written as $P_{suffix\ O_2}$ for oxygen and $P_{suffix\ CO_2}$ for carbon dioxide. The following suffixes are employed:

PA_{O_2} (or PA_{CO_2}) partial pressure of oxygen (or CO_2) in alveolar air

PI_{O_2} (or PI_{CO_2}) partial pressure of oxygen (or CO_2) in inspired air

PE_{O_2} (or PE_{CO_2}) partial pressure of oxygen (or CO_2) in expired air

Pa_{O_2} (or Pa_{CO_2}) tension of oxygen (or CO_2) in arterial blood

Pv_{O_2} (or Pv_{CO_2}) tension of oxygen (or CO_2) in venous blood

Pc_{O_2} (or Pc_{CO_2}) tension of oxygen (or CO_2) in capillary blood

A bar over the **a** or **v** means the average value. Thus

$P\bar{v}_{O_2}$ tension of oxygen in mixed venous blood.

Alveolar Air Partial Pressures

It has been seen that if the barometric pressure is 760 mm. Hg and the alveolar air contains 14 per cent. oxygen, the partial pressure of oxygen is given by 14 per cent. of (760 − 47) mm. Hg.

TABLE 8

PARTIAL PRESSURES OF O_2 AND CO_2 IN ALVEOLAR AIR
(mm. Hg. Divide by 7·5 for kPa)

%	BAROMETRIC PRESSURE (mm. Hg)					
	730 97·3 kPa	740 98·7 kPa	750 100·0 kPa	760 101·3 kPa	770 102·7 kPa	780 104·0 kPa
0·1	0·7	0·7	0·7	0·7	0·7	0·7
0·2	1·4	1·4	1·4	1·4	1·4	1·5
0·3	2·1	2·1	2·1	2·1	2·2	2·2
0·4	2·7	2·8	2·8	2·9	2·9	2·9
0·5	3·4	3·5	3·5	3·6	3·6	3·7
0·6	4·1	4·2	4·2	4·3	4·3	4·4
0·7	4·8	4·9	4·9	5·0	5·1	5·1
0·8	5·5	5·5	5·6	5·7	5·8	5·9
0·9	6·2	6·2	6·3	6·4	6·5	6·6
1·0	6·8	6·9	·7·0	7·1	7·2	7·3
2·0	13·7	13·9	14·1	14·3	14·5	14·7
3·0	20·5	20·8	21·1	21·4	21·7	22·0
4·0	27·3	27·7	28·1	28·5	28·9	29·3
5·0	34·2	34·7	35·2	35·7	36·2	36·7
6·0	41·0	41·6	42·2	42·8	43·4	44·0
7·0	47·8	48·5	49·2	49·9	50·6	51·3
8·0	54·6	55·4	56·2	57·0	57·8	58·6
9·0	61·5	62·4	63·3	64·2	65·1	66·0
10·0	68·3	69·3	70·3	71·3	72·3	73·3
11·0	75·1	76·2	77·3	78·4	79·5	80·6
12·0	82·0	83·2	84·4	85·6	86·8	88·0
13·0	88·8	90·1	91·4	92·7	94·0	95·3
14·0	95·6	97·0	98·4	99·8	101·2	102·6
15·0	102·4	104·0	105·4	107·0	108·5	110·0
16·0	109·3	110·9	112·4	114·1	115·7	117·3
17·0	116·1	117·8	119·5	121·2	122·9	124·6
18·0	122·9	124·7	126·5	128·3	130·1	131·9
19·0	129·8	131·7	133·6	135·5	137·4	139·3
20·0	136·6	138·6	140·6	142·6	144·6	146·6
21·0	143·4	145·5	147·6	149·7	151·8	153·9

In general terms the partial pressure of any gas in the alveolar air at any barometric pressure is given by:

$$P = X \text{ per cent. of } (B - P_w) \text{ mm. Hg} \qquad (1)$$

where X is the percentage of the gas in the alveolar air, B is the barometric pressure in mm. Hg and P_w is the water vapour pressure in the lungs. Since $P_w = 47$ mm. Hg at 37 °C. equation (1) becomes:

$$P = \frac{X}{100} \times (B - 47) \text{ mm. Hg} \qquad (2)$$

TABLE 8 has been compiled from equation (2) and enables the partial pressure to be read off from the composition of alveolar air (determined as on page 72) and the barometric pressure. It is applicable to both oxygen and carbon dioxide.

EXAMPLE:

The composition of alveolar air as determined using a Haldane alveolar air tube [p. 72] is:

$$CO_2 = 5·6\%$$
$$O_2 = 14·2\%$$

Barometric Pressure = 770 mm. Hg.

Since the table does not list 5·6%, look up 5% and add 0·6% to it.

Thus $5\% + 0·6\% \equiv 36·2 + 4·3 = 40·5$ mm. Hg

Similarly $14\% + 0·2\% \equiv 101·2 + 1·4 = 102·6$ mm. Hg

Hence

and $PA_{CO_2} = 40·5$ mm. Hg
             $PA_{O_2} = 102·6$ mm. Hg

mm. Hg can be converted to kilopascals by dividing by 7·5:

Thus $PA_{CO_2} = 40·5 \div 7·5 = 5·4$ kPa
             $PA_{O_2} = 102·6 \div 7·5 = 13·7$ kPa

It will be noted that the answers in kPa are approximately the same as the original percentages. The correction factors for barometric pressure to convert alveolar percentages to kPa partial pressures are:

98	100	102	104	kPa
×0·92	×0·94	×0·96	×0·98	

HAEMOGLOBIN

The haemoglobin molecule plays an important part in the carriage of both oxygen and carbon dioxide in the blood. Recent work using the techniques of X-ray crystallography and amino acid sequence analysis has thrown light on the structure of this unique molecule. Haemoglobin is found in the red blood cell. Each red cell contains about 30 picograms of haemoglobin (mean corpuscular haemoglobin = 30 pg, p. 13) or about 300,000,000 molecules of haemoglobin.

The haemoglobin molecule is built up of four sub-units. Each sub-unit consists of a ferrous iron containing haem group attached near the centre of a protein chain made up of just over 140 amino acid units. In normal haemoglobin (haemoglobin-A) these protein chains are two sets of pairs, two 141 amino acid alpha (α) chains and two 146 amino acid beta (β) chains.

The sub-units with the beta chains resemble the molecule of myohaemoglobin (the oxygen-carrying protein in muscle) whose structure has been worked out by Kendrew using X-ray crystallography. Myohaemoglobin, having only a single chain, has a molecular weight of 17,000, one-quarter that of haemoglobin.

Each protein chain in haemoglobin is in the form of a straight helix folded five or seven times to give a three dimensional structure enfolding the haem group. The four sub-units interlock with one another to form a compact molecule. Although there do not appear to be chemical bonds between the sub-units, the protein chains inter-act with one another and the two beta chains shift relative to the alpha chains making the molecule smaller when it is carrying oxygen.

The amino acid sequence in haemoglobin is different in different animals. In man abnormal haemoglobins have been demonstrated. The replacement of the amino acid, glutamic acid, with valine in the sixth position (from the amino group end) in the beta chain leads to *sickle-cell anaemia*. This haemoglobin is termed Haemoglobin-S. It is an inherited mutation. Foetal blood contains a different haemoglobin from that in adult blood in that the beta chains are replaced by a different protein chain known as a gamma chain. The foetal haemoglobin (Haemoglobin F) disappears from the circulation a few months after birth. Other abnormal haemoglobins include Haemoglobin-H where all four chains are beta chains. This haemoglobin does not change shape when taking up oxygen (Perutz, 1964).

CH—CH
CH CH   Pyrrole ring
NH

The haem group is a large flat heterocyclic structure consisting of four pyrrole rings, lying at the corners of a square, and joined by —CH= groups. At the centre is an atom of ferrous iron Fe^{2+}.

HAEM

Haem is a pigment and gives blood its characteristic colour. In the oxyhaemoglobin state a molecule of oxygen is bound to each ferrous iron atom so that each haemoglobin molecule will bind a maximum of four molecules of oxygen. The state of oxygenation of the haem group in some complex way affects the protein chain and modifies the ionization of particularly the histidine amino acid units. This plays an important part in the carriage of carbon dioxide.

When blood cells break down, the iron is stored for use in the formation of new red cells whilst the rest of the haem molecule gives rise to the bile pigment bilirubin [p. 10].

BILIRUBIN

CARRIAGE OF OXYGEN

Oxygen is carried in two ways by the blood.

(a) Dissolved in the Plasma

The amount of oxygen in solution in the plasma is proportional to the tension. At a tension of 100 mm. Hg, 0·3 ml. oxygen is dissolved in every 100 ml. blood. This may be abbreviated to 0·3 vol. per cent. At a tension of 40 mm. Hg only 0·13 ml. oxygen is dissolved in every 100 ml. blood.

These small volumes of oxygen may be neglected as far as the oxygen supply to the tissues is concerned, but they are very important in determining the oxygen tension gradients from the plasma to the tissues.

(b) Combined with Haemoglobin in the Red Cell

The amount of oxygen combined with haemoglobin in the red cell (the oxygen content) also depends upon the tension, but in this case the relationship is not a linear one but an S-shaped curve known as the **oxygen dissociation curve** [FIG. 108].

At tensions above 100 mm. Hg the haemoglobin in the red cells is fully saturated with oxygen and the dissociation curve is usually plotted as percentage saturation against tension.

One possible explanation for the S-shaped nature of the curve is that the four haem groups in the haemoglobin molecule differ in their ability to take up oxygen molecules as the oxygen tension increases. Let us assume they do so in sequence forming haemoglobin-O_2 (25 per cent. saturation), haemoglobin-O_4 (50 per cent. saturation), haemoglobin-O_6 (75 per cent. saturation) and finally haemoglobin-O_8 (100 per cent. saturation).[7] The first haem group accepts oxygen with moderate difficulty as the oxygen tension increases from 0 mm. Hg towards 20 mm. Hg. The dissociation curve shows that the saturation rises slowly at first and then more rapidly to 25 per cent. saturation. The second and third haem groups accept their oxygen readily as the tension is raised to 45 mm. Hg and the saturation rises to 75 per cent. The uptake of oxygen by the fourth haem group appears to be the most difficult and full saturation is not reached until the oxygen tension is in excess of 100 mm. Hg. However, when the oxygen tension changes the **rate** at which oxygen is taken up or given off by this fourth group is faster than with the first three.

The quantity of oxygen carried by the saturated blood will depend upon the haemoglobin content of the red cells. With a

---

[7] The oxyhaemoglobin form of haemoglobin (100 per cent. saturation) is conveniently denoted by 'HbO₂'. HbO₈ would be more accurate.

normal haemoglobin content of 14·5 g./100 ml. blood, 20 ml. oxygen will combine with the haemoglobin in every 100 ml. blood (20 vol. per cent.). This amount carried when fully saturated is termed the **oxygen capacity.**

Thus

1. *The quantity of oxygen carried by 100 ml. blood at a given tension is termed the* **OXYGEN CONTENT** *of the blood at that tension.*
2. *The quantity of oxygen carried by 100 ml. blood when fully saturated is termed the* **OXYGEN CAPACITY** *of the blood. It is independent of tension. It depends on the blood haemoglobin content.*

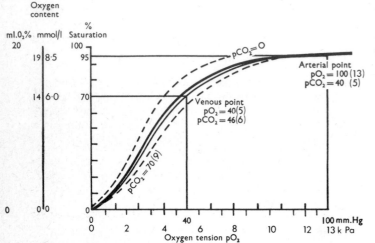

FIG. 108. The oxygen dissociation curve. The curve is usually plotted as 'percentage saturation of oxygen' against 'oxygen tension'. The corresponding oxygen content in 'ml. O₂ per cent.' and mmol/l is also given assuming an *oxygen capacity* of 20 ml. per cent. (8·8 mmol/l) for the blood. kPa in brackets.
Heavy curve pCO₂ = 40 mm. Hg (arterial blood carbon dioxide tension). Light curve pCO₂ = 46 mm. Hg (venous blood carbon dioxide tension). Dotted curves pCO₂ = 0 and pCO₂ = 70 mm. Hg.

By plotting the oxygen dissociation curve as percentage saturation instead of vols. per cent. it becomes more universal as it is still applicable in anaemia where the oxygen capacity is low.

For example, if a subject is anaemic and has a blood haemoglobin content of only 50 per cent. (=7·25 g. haemoglobin/100 ml. blood) the same dissociation curve will be applicable provided that the 'percentage saturation' ordinate scale is used. A new oxygen content scale must, however, be prepared since 100 per cent. saturation now corresponds to only 10 ml. oxygen/100 ml. blood and all oxygen content values will be correspondingly reduced.

In addition to tension and haemoglobin content, the oxygen content of the blood depends on the carbon dioxide being carried simultaneously. FIGURE 108 shows how an increase in pCO₂ from the normal value of 40 mm. Hg shifts the oxygen dissociation curve to the right, that is, less oxygen is carried at a given tension. This is known as the Bohr effect.

Oxygen Transport

We have seen that the blood in its passage through the lungs becomes in equilibrium with the alveolar air. Since the oxygen tension (partial pressure) in the lungs is 100 mm. Hg, the tension of oxygen in the blood leaving by the pulmonary veins will equal 100 mm. Hg (*13·3 kPa*). At this tension 0·3 ml. oxygen will be dissolved in each 100 ml. plasma and the haemoglobin will be 95 per cent. saturated with oxygen. Thus, the *oxygen content* of the blood will be 95 per cent. of the *oxygen capacity*, that is, 19 ml. O₂ per 100 ml. blood (assuming a normal haemoglobin content and an oxygen capacity of 20 ml. O₂ per 100 ml. blood).

The oxygen tension and content remains unchanged in the passage through the left side of the heart and the arteries. When the tissues are reached the blood comes in contact with tissue fluid having an oxygen tension which is very much less, say 40 mm. Hg (*5·3 kPa*). Oxygen now passes from the blood through the wall of the capillaries until the tension has fallen to 40 mm. Hg. At this tension the oxygen in solution will have fallen to 0·13 ml. per cent. and the oxygen combined with haemoglobin to 14 ml. per cent. (= 70 per cent. saturation [FIG. 108]). Under these conditions every 100 ml. blood will have given up 5 ml. oxygen.

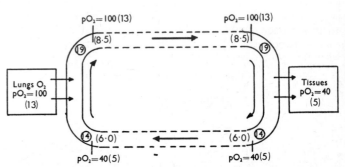

FIG. 109. Oxygen transport. Blood leaves the lungs at an oxygen tension of 100 mm. Hg (13 kPa) with 19 ml. O₂ per cent. combined with haemoglobin (8·5 mmol/l). S.I. units in brackets.

The blood returns to the lungs with an oxygen tension of 40 mm. Hg and during the build-up of tension to 100 mm. Hg, the oxygen in solution increases to 0·3 ml. per cent. and that combined with haemoglobin to 19 ml. per cent. Every 100 ml. blood thus absorbs 5 ml. oxygen from the alveolar air.

Fick Principle for Determining Cardiac Output

It is now possible to discuss an important method for the determination of cardiac output or, more accurately, the lung blood flow, which is the same. From a knowledge of the oxygen absorbed by every 100 ml. blood during its passage through the lungs, and the total oxygen uptake per minute, the lung blood flow may be calculated. Since we are dealing with a circulation, this will be equal to the output from the left ventricle and the venous return to the right side of the heart.

If the oxygen uptake per minute is 250 ml. and only 5 ml. oxygen are taken up by each 100 ml. blood, then clearly 50 lots of 100 ml. blood must pass through the lungs to carry the oxygen. The lung blood flow is thus 50 × 100 ml. = 5 litres per minute.

This principle was first proposed by the German physiologist Adolf Fick and when applied to the determination of cardiac output by a consideration of oxygen becomes:

$$\text{Cardiac output} = \frac{O_2 \text{ uptake in ml. per minute}}{A - V \text{ oxygen difference}} \times 100 \text{ ml.}$$

where '$A - V$ oxygen difference' is the difference in the oxygen content between the blood leaving the lungs and blood arriving at the lungs expressed as ml. oxygen per 100 ml. blood.

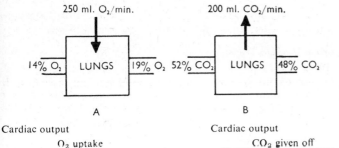

$$\text{Cardiac output}$$

$$= \frac{O_2 \text{ uptake}}{A - V \text{ oxygen difference}}$$

$$= \frac{250}{19 - 14} \times 100$$

$$= 5,000 \text{ ml. per minute}$$

$$\text{Cardiac output}$$

$$= \frac{CO_2 \text{ given off}}{V - A \ CO_2 \text{ difference}}$$

$$= \frac{200}{52 - 48} \times 100$$

$$= 5,000 \text{ ml. per minute}$$

FIG. 110. Application of the Fick principle to determine the cardiac output from the uptake of O_2 or from the CO_2 given off. The $A - V$ difference is the difference between the gas content in the 'arterial' and 'mixed venous' blood expressed as ml. gas per 100 ml. blood (% O_2, % CO_2 in this figure). The mixed venous sample is obtained by cardiac catheterization. The catheter is passed from a superficial vein through the right side of the heart into the pulmonary artery.

The oxygen uptake may be determined using a spirometer. An arterial sample will give the oxygen content of blood leaving the lungs. A sample of blood arriving at the lungs is more difficult to obtain and a catheter is usually passed from a superficial vein to the great veins and through the right atrium, right ventricle and into the pulmonary artery to obtain such a sample.

Muscle Oxygen Requirements of Exercise

We are now in a position to consider the increased oxygen supply to a muscle during activity. A muscle requires approximately fifty times more oxygen per minute when active than when at rest. This is achieved by:

(a) An increase in lung blood flow and cardiac output from 5 litres per minute to 30 litres per minute, an increase of six times.

(b) Redistribution of the blood so that a greater percentage of the cardiac output goes to the active muscles. This gives an increase of three times.

The muscle blood flow has thus been increased by eighteen times.

(c) More oxygen is extracted from every 100 ml. blood passing through the muscle. Whereas at rest every 100 ml. blood arrives carrying 19 ml. oxygen and leaves with 14 ml. oxygen, thus giving up 5 ml., in exercise the blood leaves with only 4 ml., thus giving up 15 ml. oxygen per 100 ml. blood, an increase of three times. This is the result of a reduction in the oxygen tension in the tissues surrounding the blood capillaries in the muscle.

Thus, $3 \times 18 = 54$ times more oxygen is made available for the active muscles. Respiration will be increased so that this additional oxygen may be supplied to the body.

The ability to use and obtain information from a dissociation curve is of prime importance for the understanding of the carriage of gases in the blood. The reader should plot the point referred to

above (4 ml. O_2 per cent. = 20 per cent. saturation) on the oxygen dissociation curve [FIG. 108] and determine for himself the oxygen tension (pO_2) of the blood leaving the muscle, remembering that the pCO_2 will be higher than normal. It will be found that the pO_2 is less than 20 mm. Hg.

The importance of the 'steep part' of the oxygen dissociated now becomes apparent. A comparatively small fall in tissue oxygen tension below the normal of 40 mm. Hg allows a relatively large amount of additional oxygen to be given off from the blood to supply the tissues.

This oxygen dissociation curve will be considered in detail again when breath-holding is discussed on page 100.

Anoxia (Hypoxia)

Oxygen lack in the body is termed *anoxia*. Alternatively the term *hypoxia* may be used. Four types of anoxia are recognized:

1. Anoxic Anoxia

This is anoxia due to a low tension of oxygen in the arterial blood. As a result the oxygen content of the blood is too low.

It may be caused by a low partial pressure of oxygen in inspired air due to breathing a low oxygen mixture or a low atmospheric pressure (high altitudes). Alternatively, it may be caused by disease whereby either the alveolar epithelium is modified and oxygen diffusion is impaired, or some of the blood is by-passing the aerated alveoli—such as in lobar pneumonia and some forms of congenital heart disease. In severe cases the patient will be cyanosed [p. 7].

An alternative name for this type of anoxia is *hypoxic hypoxia*.

2. Anaemic Anoxia

This is anoxia due to a reduction in the haemoglobin available for oxygen carriage, but the blood oxygen tension is normal.

It occurs in anaemia, methaemoglobinaemia, and carbon monoxide (CO) poisoning. Carbon monoxide has affinity for haemoglobin 300 times that of oxygen, and a carbon monoxide tension of only 0·3 mm. Hg will result in half the haemoglobin no longer being available for oxygen carriage. This corresponds to a concentration of less than 0 1 per cent. CO in the inspired air.

The compound carbon monoxide forms with haemoglobin is termed **carboxyhaemoglobin** [p. 13]. This should not be confused with the neutral compound carbon dioxide forms with haemoglobin which is called **carbaminohaemoglobin** [p. 84].

3. Stagnant Anoxia

In this form of anoxia, the oxygen tension in the blood leaving the lungs is normal, and so is the haemoglobin content of the blood, but the circulation is so sluggish that insufficient oxygen may be available for the tissues' needs.

It may occur in heart failure, haemorrhage and shock.

4. Histotoxic Anoxia

The cells are poisoned in this type of anoxia and are unable to utilize the oxygen supplied to them.

Cyanide inhibits the enzyme *cytochrome oxidase* and interferes with tissue oxidation.

Anoxaemia (Hypoxaemia)

An oxygen deficiency in the blood is termed *anoxaemia* or *hypoxaemia*.

CARRIAGE OF CARBON DIOXIDE

Carbon dioxide is carried in the blood in three ways:

(a) In Solution

Carbon dioxide is a more soluble gas than oxygen. At physiological pressures it obeys Henry's Law and the amount in solution is proportional to the tension. At a tension of 40 mm. Hg 3 ml. carbon dioxide dissolve in every 100 ml. blood.

Carbon dioxide in solution forms carbonic acid which ionizes at blood pH into hydrogen ions and bicarbonate ions (see Chapter 7):

$$CO_2 + H_2O \rightleftharpoons H_2CO_3 \rightleftharpoons H^+ + HCO_3^-$$

The equilibrium lies to the left. Only 1 molecule in 1,000 of carbon dioxide is in the form of carbonic acid, the rest are in simple solution. Carbonic acid is a weak acid and relatively few

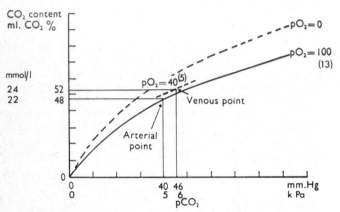

FIG. 111. The CO_2 dissociation curve. Unlike the O_2 dissociation curve this curve is always plotted as 'CO_2 content' against the tension of carbon dioxide. A 'percentage saturation' has no meaning as far as carbon dioxide is concerned. S.I. units in brackets.

of its molecules are dissociated into hydrogen ions and bicarbonate ions. The dissociation into carbonate ions does not occur at body pH.

(b) Combined with Protein

Carbon dioxide forms a neutral *carbamino* compound with haemoglobin and to a lesser extent ,with the plasma proteins.

It combines with the haemoglobin in the red cell at a different part of the molecules from that at which the oxygen combines. The oxygen combines with the iron haem radical whereas the carbon dioxide combines with the amine ($-NH_2$) groups of the protein.

At tensions in excess of 10 mm. Hg saturation occurs. The amount combined then depends not so much on the partial pressures as on the state of oxygenation of the haemoglobin molecule. With fully reduced haemoglobin 8 ml. carbon dioxide are carried as carbamino by 100 ml. blood, whereas with oxyhaemoglobin only 3 ml. carbon dioxide are carried.

(c) As Bicarbonate

The greatest proportion of carbon dioxide in the blood is in the form of bicarbonate—sodium bicarbonate in the plasma and potassium bicarbonate in the red cells.

The relationship between the total carbon dioxide content and tension is given by the carbon dioxide dissociation (or association) curve [FIG. 111]. Since saturation does not occur, percentage saturation is not employed and the ordinate is ml. CO_2/100 ml. blood.

The carbon dioxide dissociation curve is affected by the oxygen content and is moved to the right by an increase in oxygen tension.

Carbon Dioxide Transport

Blood leaves the lungs and arrives at the tissues carrying 48 ml. CO_2 per 100 ml. blood at a tension of 40 mm. Hg (*5·3 kPa*). It leaves the tissues and arrives at the lungs carrying 52 ml. CO_2 per 100 ml. blood at a tension of 46 mm. Hg (*6·1 kPa*) [FIG. 112].

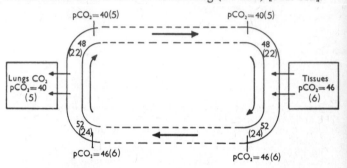

FIG. 112. Carbon dioxide transport by the blood. Very little carbon dioxide is given off in the passage through the lungs compared with the amount in the blood, but the amount given off equals the amount taken up from the tissues. S.I. units in brackets.

It will be noted that the tension gradients are very much less for carbon dioxide than they are for oxygen. Thus:

Tensions	Arterial Blood (mm. Hg)	Venous Blood (mm. Hg)	Difference
Oxygen	100	40	60
Carbon dioxide	40	46	6

In arterial blood 3 ml. CO_2 per cent. are in solution, 3 ml. as carbamino and 42 ml. as bicarbonate, making a total 48 ml. CO_2 per 100 ml. blood.

The additional 4 ml. CO_2 per cent. picked up from the tissues is carried as follows:

In solution . . .	$3 + 0·5 =$	3·5 ml.
Carbamino . . .	$3 + 0·7 =$	3·7 ml.
Bicarbonate . . .	$42 + 2·8 =$	44·8 ml.
		————
		52·0 ml.

Formation of Bicarbonate

Carbon dioxide enters the blood in solution from the tissue fluid. It passes through the plasma into the red cell which contains the enzyme **carbonic anhydrase.** It is a protein having a molecular weight of 30,000 and contains a zinc atom in the molecule. This enzyme accelerates the reaction:

$$CO_2 + H_2O \rightarrow H_2CO_3$$

which would otherwise have insufficient time to occur during the short time the blood is in the capillaries—less than one second.

At the same time oxygen is being given off by the red cell.

Reduced haemoglobin is a weaker acid than oxyhaemoglobin. The liberation of oxygen from the haemoglobin releases base in the form of potassium hydroxide which immediately reacts with the carbonic acid to form potassium bicarbonate.

The red cell thus acts as a factory manufacturing bicarbonate. Owing to the 'sodium pump', any positively charged sodium ions entering the red cell are rapidly ejected. Sodium therefore remains in the plasma and the positively charged potassium ions remain in the red cell. The bicarbonate ions are negatively charged and some of these leave the cell in exchange for a corresponding number of chloride ions, which enter the cell from the plasma. The bicarbonate ions entering the plasma form sodium bicarbonate. The end result is an increase in potassium bicarbonate in the red cell and an increase in sodium bicarbonate in the plasma, with the migration of chloride in the cell. This is known as the *chloride shift*.

$$CO_2 \longrightarrow CO_2 + H_2O \xrightarrow{\text{Carbonic anhydrase}} H \cdot HCO_3$$

$$KHbO_2 + H \cdot HCO_3 \rightarrow H \cdot Hb + KHCO_3$$

Na$^+$ K$^+$ Cl$^-$ K$^+$ HCO$_3^-$ Cl$^-$ K$^+$ HCO$_3^-$ Cl$^-$ Na$^+$

The presence of carbonic anhydrase in the red cell is necessary for the formation of the bicarbonate in the short time available (less than one second in tissue capillaries) because considerable molecular rearrangement is necessary as will be seen from the spatial configuration of the reactants:

O=C=O + H—O—H ⇌ O=C(O—H)(O—H)

Carbon dioxide   Water   Carbonic acid

Presumably carbonic anhydrase acts as a template to twist the molecules of carbon dioxide and water into such a shape that the reaction occurs readily.

The formation of bicarbonate will increase the osmotic concentration in the red cell and as a result water will be sucked into the red cells from the plasma. This accounts for the 3 per cent. increase in haematocrit of venous blood [p. 5].

The whole process is reversed when the blood reaches the lungs. Chloride leaves the cells. Bicarbonate enters and is converted by the carbonic anhydrase to carbon dioxide which enters the plasma and diffuses through the alveolar membrane into the alveolar air. The liberated potassium ions associate with the haemoglobin which has simultaneously been converted into the oxyhaemoglobin form [see also p. 93 for haemoglobin-histidine and CO_2 carriage].

It will be seen how closely interconnected is the carriage of these two gases.

Since respiration is concerned not only with the transport of oxygen and carbon dioxide, but also with the regulation of blood pH, these matters will be considered further [Chapter 8] after the role of respiration in the regulation of the hydrogen ion concentration has been discussed [Chapter 7].

CARRIAGE OF GASES IN S.I. UNITS

1 mmol. of oxygen = 22·4 ml.
1 mmol. of carbon dioxide = 22·26 ml.

In S.I. units the normal blood haemoglobin level is 2·2 mmol/l. Since each molecule of haemoglobin will carry 4 molecules of oxygen, the *oxygen capacity* will be 8·8 mmol of oxygen. The gas tensions (and contents) in *arterial* blood can be rounded off to $P_{O_2} = 13$ kPa (8·5 mmol/l), $P_{CO_2} = 5$ kPa (22 mmol/l) and in *venous* blood $P_{O_2} = 5$ kPa (6·0 mmol/l), $P_{CO_2} = 6$ kPa (24 mmol/l)

LUNGS

$$O_2 \text{ content} = 8·5 \quad P_{O_2} = 13$$
$$CO_2 \text{ content} = 22 \quad P_{CO_2} = 5$$

$$O_2 \text{ content} = 6·0 \quad P_{O_2} = 5$$
$$CO_2 \text{ content} = 24 \quad P_{CO_2} = 6$$

TISSUES

In S.I. units content is expressed in mmol/l and tension in kPa.

The oxygen difference at rest is thus 2·5 mmol/l and the carbon dioxide difference is 2·0 mmol/l. An oxygen uptake of 12·5 mmol (=5 × 2·5 mmol) will require a lung blood flow of 5 1/minute.

mm. Hg	0	15	30	45	60	75	90	105
kPa	0	2	4	6	8	10	12	14

PLOTTING A DISSOCIATION CURVE

Dissociation curves are plotted by exposing blood to different partial pressures of oxygen and carbon dioxide in a tonometer [FIG. 113]. The partial pressure of oxygen and carbon dioxide is

FIG. 113. A tonometer. A small quantity of blood (3 ml.) is exposed over a large surface to a gas mixture by rotation of the tonometer in a water bath at the appropriate temperature. The blood gas content is determined and the partial pressure of the gas in the mixture calculated from the composition of the gas and its total pressure.

calculated from the total pressure, and the percentage composition of the gas. A small quantity of blood (3 ml.) is used and it is spread to give a large surface area by rotation in a water bath at 37 °C. so that it comes into equilibrium with the gas. The tensions of oxygen and carbon dioxide in the blood will then equal the partial pressures of oxygen and carbon dioxide in the gas. The oxygen and carbon dioxide content of the blood is then determined.

Carbon Dioxide Content

The carbon dioxide content of *whole* blood may be determined by exposing the blood to a vacuum. All three gases are evolved, oxygen, nitrogen and carbon dioxide and their volume measured. The carbon dioxide is absorbed by a solution of sodium hydroxide and the reduction in volume gives the carbon dioxide in the blood. The answer is expressed in ml. CO_2 per 100 ml. blood or converted to mmol. CO_2 per litre. Care has to be taken to prevent the blood sample from coming in contact with room air before analysis otherwise the gas content will not remain constant.

To Determine the Carbon Dioxide Content of Plasma

Plasma contains no carbonic anhydrase and only evolves some of its carbon dioxide when exposed to a vacuum. The addition of an acid such as lactic acid, however, causes all the carbon dioxide to be evolved.

This forms the basis of the Van Slyke blood gas analysis apparatus. The volumetric form of this apparatus is shown in FIGURE 114.

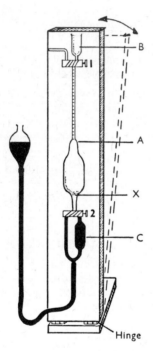

FIG. 114. The Van Slyke volumetric blood gas apparatus (see text).

Briefly the mode of operation is as follows. The calibrated burette A is filled with mercury to exclude all the air by raising the mercury reservoir. A solution of lactic acid is placed in the cup B, and 1 ml. blood (or plasma) run under the surface of the lactic acid into the bottom of the cup using an Ostwald pipette. The blood and most of the lactic acid is run into the burette.[8] Tap (1) is closed and the blood is exposed to a vacuum by lowering the mercury reservoir by a distance exceeding 760 mm. until the aqueous solutions have reached the bottom of the pipette—mark X.

Tap (2) is shut and the apparatus is shaken backwards and forwards for 3 minutes. To facilitate shaking, it is hinged at its base.

[8] For an accurate analysis the blood is run directly into the burette from the Ostwald pipette.

The aqueous fluid is then trapped in the right-hand trap C, leaving the gas in A. The volume of gas evolved is now measured in the calibrated stem of A after the mercury reservoir has been raised to restore the pressure to atmospheric. NaOH solution is carefully run in and the reduction in volume, after certain correction factors have been applied [TABLE 9], gives the CO_2 evolved from the blood. The result is expressed as ml. CO_2 per 100 ml. blood or converted to mmol. per litre by dividing by 2·2.

TABLE 9

Temperature °C.	Factor by which the volume of CO_2 (in ml.) extracted from 1 ml. blood or plasma is multiplied to give ml. CO_2 per 100 ml. blood or plasma									
	BAROMETRIC PRESSURE (mm. Hg)									
	730	735	740	745	750	755	760	765	770	775
15°	96·2	96·9	97·6	98·2	98·9	99·5	100·2	100·9	101·5	102·2
16°	95·6	96·2	96·9	97·5	98·2	98·8	99·5	100·2	100·8	101·5
17°	95·0	95·6	96·3	96·9	97·6	98·2	98·9	99·6	100·2	100·9
18°	94·4	95·1	95·7	96·4	97·0	97·7	98·3	98·9	99·6	100·2
19°	93·9	94·6	95·2	95·9	96·5	97·2	97·8	98·4	99·1	99·7
20°	93·4	94·0	94·6	95·3	95·9	96·6	97·2	97·8	98·5	99·1
21°	92·8	93·4	94·1	94·7	95·3	96·0	96·6	97·2	97·9	98·5
22°	92·2	92·8	93·5	94·1	94·7	95·4	96·0	96·6	97·3	97·9
23°	91·6	92·3	92·9	93·5	94·1	94·8	95·4	96·0	96·7	97·3
24°	91·1	91·7	92·3	92·9	93·6	94·2	94·8	95·4	96·0	96·7
25°	90·5	91·1	91·7	92·3	93·0	93·6	94·2	94·8	95·4	96·1
26°	89·9	90·5	91·1	91·8	92·4	93·0	93·6	94·2	94·8	95·4
27°	89·4	90·0	90·7	91·3	91·9	92·5	93·1	93·7	94·3	94·9

Oxygen Content

The oxygen content of the blood is estimated in the same apparatus in a similar way, but using different reagents. The red

TABLE 10

Temperature °C.	Factor by which the volume of O_2 (in ml.) extracted from 2 ml. blood is multiplied to give ml. O_2 per 100 ml. blood									
	BAROMETRIC PRESSURE (mm. Hg)									
	730	735	740	745	750	755	760	765	770	775
15°	45·0	45·3	45·6	45·9	46·2	46·5	46·8	47·1	47·4	47·7
16°	44·8	45·1	45·4	45·7	46·0	46·3	46·6	46·9	47·2	47·5
17°	44·6	44·9	45·2	45·5	45·8	46·1	46·4	46·7	47·0	47·3
18°	44·4	44·7	45·0	45·3	45·6	45·9	46·2	46·5	46·8	47·1
19°	44·2	44·5	44·8	45·1	45·4	45·7	46·0	46·3	46·6	46·9
20°	43·9	44·2	44·5	44·8	45·1	45·4	45·7	46·0	46·3	46·6
21°	43·7	44·0	44·3	44·6	44·9	45·2	45·5	45·8	46·1	46·4
22°	43·5	43·8	44·1	44·4	44·7	45·0	45·3	45·6	45·9	46·2
23°	43·2	43·5	43·8	44·1	44·4	44·7	45·0	45·3	45·6	45·9
24°	43·0	43·3	43·6	43·9	44·2	44·5	44·8	45·1	45·4	45·7
25°	42·8	43·1	43·4	43·7	44·0	44·3	44·6	44·9	45·2	45·5
26°	42·5	42·7	43·0	43·3	43·6	43·9	44·2	44·5	44·8	45·1
27°	42·3	42·6	42·8	43·1	43·4	43·7	44·0	44·3	44·6	44·9

cells are first haemolysed by saponin forming a solution of haemoglobin. The haemoglobin then has its oxygen carrying power destroyed by oxidation of its ferrous iron to ferric iron using potassium ferricyanide, thus forming methaemoglobin. Then it is exposed to a vacuum to expel the oxygen. Since carbon dioxide and nitrogen will also be given off, the carbon dioxide is absorbed first by sodium hydroxide and then the oxygen is absorbed with sodium dithionite (hydrosulphite) $Na_2S_2O_4$. This latter reduction in volume gives the oxygen content of the blood. The result is given in ml. O_2 per 100 ml. blood, after the correction factors have been applied [TABLE 10].

Measurement of Oxygen Saturation Using Light Absorption

Since oxyhaemoglobin and reduced haemoglobin differ in colour and have different light absorption spectra, the oxygen saturation of a sample of blood may be deduced from its light transmission properties at different wavelengths. This principle is frequently used in oxygen saturation recorders, where a comparison is made between transmission at two wavelengths.

Kramer (1951) has shown that the absorption of whole blood using monochromatic red light having a wavelength of 6300 Å (630×10^{-9} metres wavelength) is linearly related to the oxygen saturation.

DETERMINATION OF BLOOD GAS TENSIONS

Although the tensions of oxygen and carbon dioxide in the arterial blood can be deduced from the partial pressure of these gases in the alveolar air, techniques have been developed to enable these tensions to be determined more directly in any blood sample.

One method of determining tension is to determine the oxygen and carbon dioxide contents of the blood and to read off the corresponding tensions from the dissociation curves. Such a method is very inaccurate for oxygen tensions above 70 mm. Hg since large changes in tension are associated with quite small changes in oxygen content [see FIG. 108].

In the bubble technique a small quantity of gas is introduced into the blood and allowed to come into equilibrium with the blood. Provided that the bubble is small enough compared with the volume of blood, this manœuvre will not alter the gas tensions in the blood. The gas bubble is analysed for oxygen and carbon dioxide using micro-gas analysis apparatus (such as the Roughton–Scholander apparatus) and the partial pressures of these gases determined. Since equilibrium has been established these partial pressures will be equal to the blood gas tensions.

These methods of determining blood gas tensions are tending to be replaced by oxygen and carbon dioxide electrodes which measure the gas tensions directly.

Oxygen Electrodes

If a platinum electrode is maintained at a voltage of minus 0·6 volts with reference to a neutral silver electrode in an aqueous solution, the current flowing is linearly related to the oxygen tension in the solution. The voltage used is too low for electrolysis of the water into hydrogen and oxygen. Instead the oxygen in solution is reduced at the platinum electrode (forming H_2O_2) and this reaction causes an electric current to flow. Since oxygen is used up, the steady current flowing will also depend on the rate at which oxygen can diffuse through the solution to the platinum cathode. Controlled stirring is sometimes employed to increase the current flowing.

The current flowing (of the order of a few microamps) has to be amplified before it can be displayed on a meter calibrated in pO_2 units or used to operate a recorder. Alternatively a resistance is included in the circuit in series with the platinum electrode and the voltage across the resistance, after suitable amplification, used as an indication of the current flowing. Precautions have to be taken to see that the voltage drop across this resistance does not alter the 0·6 polarizing voltage.

Such a system satisfactorily measures the oxygen tension in water, but when applied to blood fails because the plasma proteins 'poison' the platinum electrode and cause a fall in sensitivity. To enable the oxygen electrode to be used with blood, the platinum electrode is separated from the blood by a plastic membrane which will allow oxygen to diffuse through but will prevent the passage of the plasma proteins [FIG. 115]. A thin polythene film has this property.

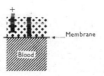

FIG. 115. Oxygen electrode enables the oxygen tension in blood to be measured directly. The current which flows through water when a negative voltage of 0·6 volts is maintained between a platinum electrode ($-$) and a neutral silver electrode ($+$) depends on the tension of oxygen. Since proteins in blood 'poison' the platinum electrode, the blood is separated from the electrodes by a membrane permeable to oxygen but impermeable to protein.

Oxygen electrodes have been built into the tips of catheters so that the oxygen tension can be determined as the blood flows through the circulation.

If a blood sample is withdrawn for the determination, great care must be taken to ensure that it is not exposed to air while it is being transferred to the oxygen electrode. Any such exposure to room air would immediately cause a change in gas tension. The determinations must be carried out at the correct temperature, usually 37 °C.

Carbon Dioxide Electrodes

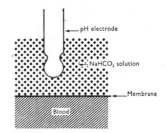

FIG. 116. Carbon dioxide electrode enables the carbon dioxide tension in blood to be measured directly. The blood is separated from the dilute bicarbonate solution by a membrane permeable to carbon dioxide. Once the bicarbonate solution has reached the same carbon dioxide tension as the blood, the reduction in pH due to the formation of carbonic acid will be related to the pCO_2 in the blood. The pH change is measured using a glass-electrode pH meter. The volume of the bicarbonate solution is kept as low as possible to obtain a rapid response.

When carbon dioxide enters a very dilute bicarbonate solution it forms carbonic acid which neutralizes the alkaline bicarbonate. The lowering of the pH is a measure of the carbon dioxide tension. This forms the basis of the carbon dioxide electrode. As in the case of the oxygen electrode, the blood is separated from the bicarbonate solution by a plastic membrane [FIG. 116]. This membrane is permeable to carbon dioxide but impermeable to blood constituents which would cause a pH change. A PTFE membrane has this property.

The pH change can be measured using a glass-electrode pH meter specially calibrated in pCO_2 units to take account of the fact that there is a logarithmic relationship between the pH change and the pCO_2.

In order to obtain a rapid response the volume of the bicarbonate solution must be kept as small as possible. In one form of carbon dioxide electrode the bicarbonate solution is replaced by cellophane soaked in bicarbonate between the membrane and the glass-electrode of the pH meter.

Gas Tensions in Body Fluids other than Blood

It has been seen that the tensions of oxygen and carbon dioxide in tissue fluid are the same as those in the *venous* blood. The gas tensions in urine are the same as those in *arterial blood*. This is because oxygen is not used up by the urine.

The gas tensions in the cerebrospinal fluid (C.S.F.) approximate more closely to those of venous blood than to those of arterial blood. The *bicarbonate* concentration in the cerebrospinal fluid, on the other hand, is regulated by its secretion into the cerebrospinal fluid at the choroid plexuses [p. 190]. The importance of this fact will be discussed in Chapter 8.

Hyperbaric Oxygen

If pure oxygen is breathed under a pressure of several atmospheres giving a partial pressure of several thousand mm. Hg, so much oxygen may dissolve in the plasma that the body's requirement may be satisfied from this source. Under these conditions there may be no need for the oxyhaemoglobin to dissociate (it will not do so if the tissue oxygen tension builds up to 100 mm. Hg), and no base will be released to assist the carriage of carbon dioxide. Oxygen under such pressure is an irritant gas and it can poison enzyme systems if breathed for long periods. For short periods, however, **hyperbaric** oxygen (as it is called) is used in the treatment of anoxia, carbon monoxide poisoning, and to facilitate heart surgery particularly in babies. It is possible that a small amount of oxygen may enter via the skin.

In pressure chambers, such as diving bells, and aqualungs where the inspired gas is under several atmospheres pressure, air is a safer gas to use.

Decompression

At high atmospheric pressures, such as are necessary in diving bells and caissons, the dissolved nitrogen must be taken into account. If the pressure is released too rapidly this dissolved nitrogen will come out of solution and form bubbles throughout the body. The pain associated with this condition is known as the 'bends'. More seriously, bubbles may form in the spinal cord and brain leading to neurological trauma.

When the 'bends' occur the victim is placed in a pressure chamber and recompressed to drive the nitrogen back into solution. He is then decompressed slowly.

At great depths an oxygen and helium mixture is used [p. 101].

REFERENCES AND FURTHER READING

Barcroft, J. (1928) *The Respiratory Function of the Blood*, 2nd. ed., Cambridge.

Bohr, C., Hasselbalch, K., and Krogh, A. (1904) Ueber einin in biologischer Beziehung wichtigen Einfluss, den die Kohlensäurespannung des Blutes auf dessen Sauerstoffbindung übt, *Skand. Arch. Physiol.*, **16**, 402.

Christiansen, J., Douglas, C. G., and Haldane, J. S. (1914) The absorption of carbon dioxide by human blood, *J. Physiol. (Lond.)*, **48**, 244.

Douglas, C. G., Haldane, J. S., and Haldane, J. B. S. (1912) The laws of combination of haemoglobin with carbon monoxide and oxygen, *J. Physiol. (Lond.)*, **44**, 275.

Gibson, Q. H. (1959) The kinetics of reactions between haemoglobin and gases, *Progr. Biophys.*, **9**, 1.

Kramer, K., Elam, J. O., Saxton, C. A., and Elam, W. N. (1951) Influence of oxygen saturation, erythrocyte concentration, and optical depth upon the red and near infra-red light transmittance of whole blood, *Amer. J. Physiol.*, **165**, 229.

Perutz, M. F. (1964) The haemoglobin molecule, *Scientific American*, **211**, No. 5, 64.

Roughton, F. J. W. (1943) Some recent work on the chemistry of carbon dioxide transport by the blood, *Harvey Lect.*, **39**, 96.

Torrance, R. W. (1968) *Arterial Chemoreceptors*, Oxford.

FILM

Cardiac Output in Man, I.C.I. Film Library, London.

7. RESPIRATION AND HYDROGEN ION CONCENTRATION

Respiration plays an important part in the regulation of the hydrogen ion concentration of the body. Before studying its role, the principal factors which determine the hydrogen ion concentrations will be considered.

HYDROGEN ION CONCENTRATION

The fluids of the body are aqueous solutions. The water exists in two forms, as molecules of H_2O and as molecules dissociated into hydrogen ions H^+ and hydroxyl ions OH^-. Only a very small number of the water molecules are dissociated, the vast majority stay as undissociated H_2O molecules.

$$H_2O \rightleftarrows H^+ + OH^-$$

The hydrogen atom consists of one positively-charged proton with one negatively-charged orbital electron. Hydrogen is ionized by the removal of the electron, but the resultant hydrogen nucleus or proton is too reactive to exist in the free state and is attached to another water molecule making the hydronium ion H_3O^+. It is this ion that will be referred to as the hydrogen ion and denoted by H^+.

The number of hydrogen ions present multiplied by the number of hydroxyl ions present is a constant in all aqueous solutions and depends only on the temperature. It is independent of the source of the ions and of the other ions present in the solution. It is usual to measure the concentration of ions present in the units *moles per litre* (mol/l).

Since the molar weight of the hydrogen ion is 1 and that of the hydroxyl ion is $16 + 1 = 17$ (see atomic weights, p. 221), then both:

	1 g. of hydrogen ions per litre
and	17 g. of hydroxyl ions per litre

are counted as 1 mole per litre.

If the hydrogen and hydroxyl concentrations are measured in this manner, they are written as $[H^+]$ and $[OH^-]$ respectively, and the product of their concentrations will be a constant. An alternative to $[H^+]$ is cH.

At 23 °C.

$$[H^+] \times [OH^-] = 10^{-14}$$

This is the figure that is generally used.

Should the product ever exceed this value, some of the hydrogen and hydroxyl ions will recombine to form undissociated H_2O molecules until the number of ions has been reduced and the product once again restored to 10^{-14}.

At body temperature, 37 °C., the product of the hydrogen and hydroxyl ions is slightly greater and is equal to $10^{-13 \cdot 6}$.

$$[H^+] \times [OH^-] = 10^{-13 \cdot 6}$$

Neutral Solutions

Water is said to be neutral when the number of hydrogen ions present equals the number of hydroxyl ions present. Thus at 23 °C. the concentration of both the hydrogen ions and the hydroxyl ions in pure water will be 10^{-7} since:

$$[H^+] \times [OH^-] = 10^{-7} \times 10^{-7} = 10^{-14}$$

Thus in pure water at this temperature there are 1×10^{-7} g. of hydrogen ions per litre, and 17×10^{-7} g. of hydroxyl ions. There will be 1 g. of hydrogen ions per 10,000,000 litres of water! Incidentally this volume of water contains 1,100,000,000 g. of hydrogen of which only 1 g. is in the form of hydrogen ions.

Many substances when dissolved in water do not alter the equality between the hydrogen and hydroxyl ions. Such substances are said to be neutral. They include sodium chloride, glucose and urea.

Acids

Certain substances, when dissolved in water, upset the balance between the hydrogen ions and hydroxyl ions and cause the hydrogen ions to exceed the hydroxyl ions. Such substances are called acids. This change may be due to the substance adding hydrogen ions to the solution, that is, it acts as a *hydrogen ion donor*, or to the substance removing hydroxyl ions from the solution. The net result will be the same in the two cases since the product of these two ions is a constant.

Hydrochloric acid donates hydrogen ions to the solution:

$$HCl \rightarrow H^+ + Cl^-$$

The product of the $[H^+] \times [OH^-]$ is now too high and some of the OH^- will disappear from the solution (together with an equal, but relatively small, number of hydrogen ions) to form water.

A molar solution of hydrochloric acid (1 mol per litre) will supply 1 mol per litre of hydrogen ions, and the concentrations of hydroxyl ions will fall to 10^{-14} since:

$$[H^+] \times [OH^-] = 1 \times 10^{-14} = 10^{-14}$$

An $\dfrac{M}{10}$ solution of HCl, such as is produced by the gastric glands in the stomach will contain $\frac{1}{10}$ mol per litre of hydrogen ions (cH = 10^{-1}) and the hydroxyl ion concentration will fall to 10^{-13}.

Ammonium chloride is an example of a substance that makes a solution acid by the removal of hydroxyl ions. Some of the ammonium ions combine with the hydroxyl ions to form undissociated ammonia and water leaving a surplus of hydrogen ions:

$$NH_4Cl + H^+ + OH^- \rightarrow NH_4OH + H^+ + Cl^-$$
$$\searrow$$
$$NH_3 + H_2O$$

Bases

Other substances when dissolved in water upset the hydrogen/hydroxyl ion balance in the reverse direction and cause the hydroxyl ions to exceed the hydrogen ions. Such substances are called alkalis or more generally bases. The change this time may be due to the addition of hydroxyl ions to the solution or to the removal of hydrogen ions due to the substance acting as a *hydrogen ion acceptor*. Once again the net result will be the same in the two cases since the hydrogen ion concentration falls as the hydroxyl ion concentration rises.

Thus sodium hydroxide donates hydroxyl ions to the solution:

$$NaOH \rightarrow Na^+ + OH^-$$

whereas ammonia removes hydrogen ions:

$$NH_3 + H^+ + OH^- \rightarrow NH_4^+ + OH^-$$

Both make the solution alkaline.

A molar solution of NaOH will supply 1 mol per litre of hydroxyl ions and the concentration of the hydrogen ions will fall to 10^{-14} since:

$$[H^+] \times [OH^-] = 10^{-14} \times 1 = 10^{-14}$$

pH NOTATION

The degree of acidity or alkalinity of a solution is given by the hydrogen ion concentration. Once the hydrogen ion concentration is known, the hydroxyl ion concentration is immediately fixed.

We have seen that in a highly acid solution produced by molar hydrochloric acid the hydrogen ion concentration $= 1 = 10^0$. In neutral water the hydrogen ion concentration is 10^{-7} and in a highly alkaline solution such as molar sodium hydroxide the hydrogen ion concentration is 10^{-14}.

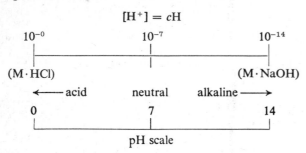

A scale has been devised to express the degree of acidity and alkalinity which is based, not on the actual hydrogen ion concentration, but on the indices of the concentration (the 0, 7 and 14 above). This is known as the pH scale.

The pH of a solution is defined as the *negative logarithm of the hydrogen ion concentration*, or, in other words, minus the number to which 10 has to be raised to give the hydrogen ion concentration.

The neutral point on this scale is pH 7 at 23 °C. (pH 6·8 at 37 °C.). Any pH lower than this will be acid. Any pH higher than this will be alkaline.

Blood pH

The pH of blood is 7·4. It is therefore an alkaline fluid. The range of blood pH compatible with life is only 7·0 to 7·8. This corresponds to a range of six times in the hydrogen ion concentration. Over such a limited range there is no advantage in using a

logarithmic scale such as pH, and the direct consideration of the hydrogen ion concentration (cH) often gives a better indication of any change that has taken place. The normal blood cH is 40×10^{-9} which equals 40 *nanomoles* of H^+ per litre.

$[H^+] = cH$	pH	
100×10^{-9}	7·0	
64×10^{-9}	7·2	
40×10^{-9}	7·4	NORMAL
25×10^{-9}	7·6	
16×10^{-9}	7·8	

It will be noted that blood at a pH of 7·4 has more hydroxyl ions than hydrogen ions, and that it has fewer hydrogen ions than pure water. This is due to the presence of hydrogen ion acceptors (bases) in the blood which have removed some of the hydrogen ions.

Strong and Weak Acids and Bases

A strong acid such as hydrochloric acid has such a loose bond between the hydrogen and the rest of the molecule that it is completely dissociated in solution and provides the concentration of hydrogen ions expected from a consideration of the concentration of acid in the solution. Thus:

M HCl provides 1 g. H^+ per litre and pH = 0   cH = 1

$\dfrac{M}{10}$ HCl provides 10^{-1} g. H^+ per litre and pH = 1   cH = 10^{-1}

$\dfrac{M}{100}$ HCl provides 10^{-2} g. H^+ per litre and pH = 2   cH = 10^{-2}

$\dfrac{M}{1,000}$ HCl provides 10^{-3} g. H^+ per litre and pH = 3   cH = 10^{-3}

On the other hand $\dfrac{M}{10}$ carbonic acid provides only 2×10^{-4} g. hydrogen ions per litre (and not 10^{-1}). This is because many of the carbonic acid molecules remain unionized as H_2CO_3 and do not dissociate into hydrogen ions and bicarbonate ions.

A sodium bicarbonate solution is slightly alkaline (pH = 8) because some of the bicarbonate ions act as hydrogen ion acceptors and combine with hydrogen ions from the water to form undissociated carbonic acid. This leaves a surplus of hydroxyl ions and a deficiency of hydrogen ions.

$$NaHCO_3 + H_2O \rightarrow Na^+ + \underbrace{HCO_3^- + H^+}_{H_2CO_3} + OH^-$$

Salts of strong acids and strong bases tend to be completely ionized in solution, and on this ionic theory, the reaction between a strong base and a strong acid involves only the formation of undissociated water from the hydrogen and hydroxyl ions. Thus when sodium hydroxide is neutralized by hydrochloric acid the reaction may be expressed as:

$$\underbrace{Na^+ + OH^-} + \underbrace{H^+ + Cl^-} \rightarrow Na^+ + H_2O + Cl^-$$

The sodium and chloride ions are present before and after the reaction.

BUFFER SYSTEMS

A buffer system is a solution which will accept hydrogen or hydroxyl ions with very little change in over-all pH or cH. Such a system may consist of a weak acid and its salt with a strong base, or a weak base and its salt with a strong acid. Consider a buffer system composed of a weak acid HA, and its sodium salt NaA which will exist in solution as Na^+ and A^-. Such a system is able to resist any change in the hydrogen ion concentration in the following manner.

If additional hydroxyl ions are added, some of the undissociated acid will dissociate and supply hydrogen ions which will remove the surplus OH^- ions as water:

$$HA \rightarrow H^+ + A^-$$
$$OH^- + H^+ \rightarrow H_2O$$

If additional hydrogen ions are added the negatively charged anions will combine with them forming undissociated acid. These anions will have been provided by the completely dissociated salt.

$$H^+ + A^- \rightarrow HA$$

Such a system is thus able to mop up both H^+ and OH^-. The relationship between the pH and the concentrations of HA and A^- may be deduced as follows: by the Law of Mass Action, the rate at which the weak acid HA dissociates into H^+ and A^- is proportional to the concentration of HA. The rate at which H^+ and A^- recombine to form HA depends on the product of the concentrations of H^+ and A^-. In the equilibrium state these two rates must be equal, thus:

$$k_1[HA] = k_2[H^+] \times [A^-]$$

or

$$[H^+] = \frac{k_1}{k_2} \frac{[HA]}{[A^-]}$$

Since A^- is being supplied almost entirely by the salt NaA (as the acid itself is only feebly dissociated into A^-) we may approximate this equation to:

$$[H^+] = \text{constant} \times \frac{[HA]}{[NaA]}$$

If we wish to use the pH nomenclature we must invert both sides of this equation and then take logarithms.

Thus

$$\frac{1}{[H^+]} = \text{constant} \times \frac{[NaA]}{[HA]}$$

$$\log_{10} \frac{1}{[H^+]} = \log_{10} \text{constant} + \log_{10} \frac{[NaA]}{[HA]}$$

$$\log_{10} \frac{1}{[H^+]} = pH \text{ (by definition)}$$

Let $\log_{10}$ constant be pK.

Then

$$pH = pK + \log_{10} \frac{[NaA]}{[HA]}$$

or, more generally,

$$pH = pK + \log_{10} \frac{[salt]}{[acid]}$$

pK is the value of pH at which the concentrations of the salt and acid will be equal since when the ratio is equal to unity:

$$pH = pK + \log_{10} 1$$
$$= pK + 0$$
$$= pK$$

A buffer functions most efficiently at a pH which is close to its pK value. It will then neutralize equally well both acids and alkalis.

Since A^- accepts hydrogen ions, it may be referred to as a **conjugate base.**

$$\underset{\text{Acid}}{HA} \rightleftharpoons \underset{\substack{\text{Hydrogen} \\ \text{ion}}}{H^+} + \underset{\substack{\text{Conjugate} \\ \text{base}}}{A^-}$$

Using this terminology, a weak acid HA and its conjugate base A^- form a conjugate acid-base pair, the pH of which is given by:

$$pH = pK + \log_{10} \frac{[\text{conjugate base}]}{[\text{weak acid}]}$$

MAINTENANCE OF BODY pH

Because of their ability to minimize changes in pH and cH, it is the **weak acids and bases** together with their ions that are of great importance in enabling the body to maintain its hydrogen ion concentration. Although hydrogen bicarbonate (carbonic acid) is a weak acid, the bicarbonate ions, derived from this acid or from its sodium or potassium salt, have a very strong affinity for hydrogen ions and will remove them from solution as carbonic acid which then splits into CO_2 and water. It is the bicarbonate in the blood that 'buffers' any additional hydrogen ions entering the blood. Any carbon dioxide formed is excreted in the expired air [see p. 94]. Thus if the acid HX enters the blood the following reaction takes place:

$$\underset{\text{Acid}}{HX} + NaHCO_3 \rightarrow \underset{\text{Neutral}}{NaX} + CO_2\uparrow + H_2O$$

or more accurately:

$$H^+ + HCO_3^- \rightarrow CO_2\uparrow + H_2O$$

Hydrochloric acid on the other hand is a **strong acid,** and the chloride ions have such a weak affinity for hydrogen ions that the sodium chloride in the blood plays no part in the removal of the surplus hydrogen ions.

Phosphoric acid occupies a unique position in the buffering systems of the body because it forms two salts that are always present in the blood and urine. The disodium hydrogen salt (Na_2HPO_4) is completely dissociated and the HPO_4^- ions have a strong affinity for H^+ and will remove them from the solution. It acts as a hydrogen ion acceptor in the same way as a base. The other salt, monosodium dihydrogen phosphate (NaH_2PO_4) dissociates into $H_2PO_4^-$ ions which act as H^+ donors and will remove OH^- ions from the solution by converting them to water. It acts as a hydrogen ion donor in the same way as an acid.

Thus, $\qquad HPO_4^{--} + H^+ \rightarrow H_2PO_4^-$

and $\qquad H_2PO_4^- \rightarrow HPO_4^{--} + H^+$

When both ions are present simultaneously the solution will 'mop up' both H^+ and OH^- ions. The pH of the solution will be given by:

$$pH = 6{\cdot}8 + \log_{10} \frac{[HPO_4^{--}]}{[H_2PO_4^-]}$$

PROTEINS AS BUFFERS

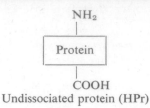

NH$_2$

Protein

COOH

Undissociated protein (HPr)

The proteins of the body behave as weak acids and weak bases and therefore play an important part in buffering. It has been seen [p. 21] that the way in which they ionize depends upon the hydrogen ion concentration of their immediate environment. The enzymes of the body are proteins, and it is because their activity depends on their ionization that the body pH must be kept at exactly the correct level.

Proteins and the amino acids of which they are formed, possess both acidic groups, usually the —COOH radical, and basic groups, usually the —NH$_2$ radical.

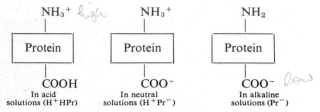

NH$_3^+$ *high*   NH$_3^+$ NH$_2$

Protein   Protein   Protein

COOH   COO$^-$ COO$^-$ *low*

In acid   In neutral   In alkaline
solutions (H$^+$HPr)   solutions (H$^+$Pr$^-$)   solutions (Pr$^-$)

Such a molecule can ionize in two different ways. The molecule can act as a hydrogen ion acceptor, the —NH$_2$ group becomes —NH$_3^+$. This form of ionization predominates when the hydrogen ion concentration is high, that is, in acid solution. If we consider the molecule to be a substituted ammonia compound, the changes may be seen to be similar to the conversion of ammonia NH$_3$ to the ammonium ion NH$_4^+$ which occurs in acid solutions.

In alkaline solutions, and with the exception of gastric juice, the body fluids which contain protein are alkaline, the —COOH group acts as a hydrogen ion donor and ionizes as —COO$^-$, that is, as an anion giving the molecule a predominant negative charge.

The plasma proteins have a strong affinity for hydrogen ions which unite with the anions to form undissociated protein molecules HPr, and with the —NH$_2$ groups to form doubly ionized H$^+$Pr$^-$ molecules.

Since the main cation in plasma is sodium (Na$^+$) we may consider the plasma proteins as existing as hydrogen proteinate (mainly H$^+$Pr$^-$) and completely dissociated sodium proteinate (Na$^+$ + Pr$^-$). These plasma proteins are able to act as buffers to neutralize, for example, any hydrochloric acid entering the blood.

Thus: HCl + NaPr → NaCl + HPr
                   Acid   Neutral

The proteins also buffer hydroxyl ions. Thus:

HPr + NaOH → NaPr + H$_2$O

The proteins account for about one-sixth of the buffering power of the blood.

HAEMOGLOBIN

The majority of the —NH$_2$ and —COOH groups of the amino acids forming a protein play no part in the buffering because they are forming the peptide links —CO·NH— between the amino acid units. The only intact —NH$_2$ and —COOH groups are those at the ends of the peptide chains.

H$_2$N ☐ CO·NH ☐ CO·NH ☐ CO ... NH ☐ COOH

However, certain amino acids contain additional —NH$_2$ and —COOH groups in their molecules [see p. 114], and these groups are available for buffering as they do not form the peptide links. Histidine is such as amino acid. There are 38 histidine units in the haemoglobin molecule and, as a result, haemoglobin is a particularly good buffer.

The histidine molecule contains the equivalent of a second —NH$_2$ group in its iminazole ring. The ≡N in this ring is a hydrogen acceptor and becomes ≡NH$^+$. It is thus basic and will neutralize acids

. . . . . NH·CH·CO . . .   . . . NH·CH·CO . . .
            | |
            CH$_2$ CH$_2$
        HC═══C   HC═══C
       | | | |
      HN   N   HN   NH$^+$
          ╲   ╱   ╲   ╱
            C   C
            | |
            H   H
Histidine unit   Histidine unit
(unionized)   (ionized)

Considering the histidine units, haemoglobin will exist in two forms which may be represented as:

Hb   and   HbH$^+$ OH$^-$

Using this concept, the carriage of CO$_2$ in the blood and the formation of bicarbonate in the red cell [cf. p. 84] may be considered as:

Hb + H$_2$CO$_3$ → HbH$^+$ + HCO$_3^-$

with the CO$_2$ carried as 'haemoglobin-histidine bicarbonate'.

The ionization of the histidine depends upon the degree of oxygenation of the haemoglobin molecule. When oxygen is given off in the tissues, more HbH$^+$ and OH$^-$ ions are formed which will be available to neutralize the carbonic acid formed from the CO$_2$ entering the red blood cell.

A Further Consideration of Haemoglobin as a Buffer

The majority of intact —COOH and —NH$_2$ groups in haemoglobin play little part in buffering because the pK values of these groups are far removed from the pH of blood (=7·4). The relationship between the unionized and ionized forms, and pH is given by:

$$pH = pK_1 + \log_{10} \frac{[-COO^-]}{[-COOH]}$$

and $$pH = pK_2 + \log_{10} \frac{[-NH_2]}{[-NH_3^+]}$$

When the pH is equal to the pK value (pK$_1$), the concentrations of the —COOH form and the —COO$^-$ form will be equal and the system will act as an efficient buffer. At higher pH's (that is, on the alkaline side of pK_1) the —COO^- form will predominate. At lower pH's (on the acid side of pK_1), the —COOH form will predominate.

Haemoglobin consists of four peptide chains, and thus has four terminal —COOH groups. It has also 56 side-chain —COOH groups associated with the dicarboxylic acids, glutamic and aspartic, in these peptide chains. All these —COOH groups have pK values around 4 [TABLE 11]. Thus at pH 7·4 they will exist almost entirely in the ionized —COO^- form and will play a negligible part in buffering.

TABLE 11

HAEMOGLOBIN

Principal ionizing groups

Group	No. of groups	Approxi-mate pK	Principal form at pH 7·4
Terminal —COOH	4	4	—COO^-
Side-chain —COOH	56	4	—COO^-
Terminal —NH_2	4	7·7	—NH_2 and —NH_3^+
Side-chain —NH_2	44	11	—NH_3^+
Histidine ≡N	38	7	≡N and ≡NH^+

At a different pH (equal to pK_2) the concentrations of the —NH_2 form and the —NH_3^+ form will be equal. The terminal —NH_2 groups have a pK value close to the pH of blood (pK ⌒ 7·7). These groups will therefore be playing a part in buffering, but there are only four of them in the haemoglobin molecule (one for each peptide chain). The 44 side-chain —NH_2 groups, associated with the lysine in haemoglobin will not be efficient buffers since their pK value is approximately 11; these groups will exist almost entirely in the ionized —NH_3^+ form at the pH of blood.

The 38 imidazole groups associated with the histidine in haemoglobin are so very important because their pK values are around seven and thus close to the pH of blood. They form the principal buffering system by changing between the ≡N and ≡NH^+ forms with very small changes in pH.

$$pH = pK_3 + \log_{10} \frac{[≡N]}{[≡NH^+]}$$

Buffer Action

Haemoglobin at pH 7·4 may thus be considered as having a large number of relatively permanent —COO^- groups, and a slightly smaller number of relatively permanent —NH_3^+ groups; hence the over-all net negative charge on the molecule. In addition there are approximately equal numbers of ≡N and ≡NH^+ groups. The ≡NH^+ groups change to the ≡N form, with the release of H^+, when the hydrogen ion concentration starts to fall (pH rises). The ≡N groups change to the ≡NH^+ form, with the acceptance of H^+, when the hydrogen ion concentration starts to rise (pH falls). In this way, haemoglobin minimizes changes in hydrogen ion concentration, that is, acts as a buffer.

Reduced and Oxy-haemoglobin

The pK values are higher for reduced haemoglobin than oxy-haemoglobin. Thus when oxygen is given off, some of the haemo-globin ≡N groups combine with hydrogen ions and change to ≡NH^+. As a result the blood hydrogen ion concentration falls and the blood becomes more alkaline.

In tissue capillaries such a fall in the hydrogen ion concentration is opposed by the simultaneous entry of carbon dioxide into the blood from the surrounding tissue fluid. This carbon dioxide combines with water in the presence of the enzyme carbonic anhydrase in the red cells and forms hydrogen ions and bicarbonate ions.

A Further Consideration of the Formation of Potassium Bicarbonate in the Red Cell

In the last chapter it was seen [p. 84] that the carbon dioxide entering the red cell forms *potassium bicarbonate* whereas in this chapter we have seen that it forms *haemoglobin-histidine bicarbonate*. These are two alternative ways of looking at the same chemical changes.

It has been seen that at pH 7·4 the haemoglobin molecule has more negatively charged —COO^- groups than positively charged —NH_3^+ and ≡NH^+ groups. The molecule therefore has an over-all negative charge. For electrical neutrality in the red cell there must be an additional number of positively charged ions and these are the potassium ions (K^+).

Although the important change in the haemoglobin molecule, when oxygen is given off, is the formation of more positively charged histidine ≡NH^+ groups, the positively charged groups will still be outnumbered by the negatively charged —COO^- groups. The net result is that when oxygen is given off, it is more accurate to consider that the *haemoglobin molecule becomes less negatively charged* than to consider that it has become more positively charged.

As the negative charge on the haemoglobin is reduced, not so many of the potassium ions are needed to match the haemoglobin charge for electrical neutrality. If no carbon dioxide enters the red cell, the blood will become progressively more alkaline as the oxygen is given off. This is because histidine ≡N groups react with water and form ≡NH^+ and OH^- ions. There will then be a surplus of potassium ions (K^+) and a surplus of hydroxyl ions (OH^-), a combination which for simplicity can be considered as equivalent to the formation of potassium hydroxide [p. 85].

$$≡N + H_2O \rightleftharpoons ≡NH^+ + OH^-$$

When carbon dioxide enters the red cell, the carbonic acid, formed by the enzyme carbonic anhydrase, supplies the hydrogen ions for the formation of the ≡NH^+ groups in place of the water, leaving bicarbonate ions (HCO_3^-). This may be considered to be the formation of *haemoglobin-histidine bicarbonate*.

$$≡N + H_2CO_3 \rightleftharpoons ≡NH^+ + HCO_3^-$$

However, there is now a surplus of potassium ions and a surplus of bicarbonate ions. This combination can be considered to be equivalent to the formation of *potassium bicarbonate*.

It should be remembered that the potassium ions play no part in the chemical changes and are constant throughout. Further-more, the total number of negatively charged ions, Hb^- + HCO_3^- + OH^-, is also a constant throughout, and the OH^- ions make the total equal to that of the positively charged K^+.

If one wishes to combine both approaches the compound formed may be referred to as *potassium haemoglobin-histidine bicarbonate*.

MAINTENANCE OF BLOOD pH BY RESPIRATION

The maintenance of the blood pH at 7·4 is of prime importance to the body. Life is only possible if the pH is within the range 7·0 to 7·8.

The pH of the blood (plasma) depends upon the ratio of bicarbonate/carbonic acid in the plasma.

Consider the equations:

$$CO_2 + H_2O \rightleftharpoons H_2CO_3 \quad . \quad . \quad . \quad (1)$$
$$H_2CO_3 \rightleftharpoons H^+ + HCO_3^- \quad . \quad . \quad . \quad (2)$$

By the Law of Mass Action applied to equation (2) the speed of the forward reaction is given by:

$$v_1 = K_1 \times \text{concentration of carbonic acid}$$
$$= K_1[H_2CO_3]$$

The speed of the reverse reaction is given by:

$$v_2 = K_2 \times \text{concentration of hydrogen ions} \times \text{concentration of bicarbonate ions}$$
$$= K_2[H^+] . [HCO_3^-]$$

When the system is in equilibrium:

$$v_1 = v_2$$
$$\therefore \quad K_1[H_2CO_3] = K_2[H^+] . [HCO_3^-]$$
$$\therefore \quad \frac{[H^+] . [HCO_3^-]}{[H_2CO_3]} = \frac{K_1}{K_2} = K \quad . \quad . \quad (3)$$

K is known as the equilibrium constant.

When using the pH scale, equation (3) may be rewritten:

$$\frac{[H_2CO_3]}{[H^+][HCO_3^-]} = \frac{1}{K}$$
$$\therefore \quad \frac{1}{[H^+]} = \frac{1}{K} \times \frac{[HCO_3^-]}{[H_2CO_3]}$$

Take $\log_{10}$ of both sides:

$$\log_{10} \frac{1}{[H^+]} = \log_{10} \frac{1}{[K]} + \log_{10} \frac{[HCO_3^-]}{[H_2CO_3]} \quad . \quad (4)$$

By definition:

$$\log_{10} \frac{1}{[H^+]} = pH$$

Let

$$\log_{10} \frac{1}{[K]} = pK$$

The equation (4) may be written:

$$pH = pK + \log_{10} \frac{[HCO_3^-]}{[H_2CO_3]} \quad . \quad . \quad (5)$$

pK in this equation is approximately 3·1.

(Note that K in the above equations is a constant and does not refer to potassium!)

At a pH of 3·1 (pK) the concentrations of bicarbonate and carbonic acid would be equal, and the system acts as a 'buffer' which can absorb both hydrogen and hydroxyl ions with a minimum change in pH.

The pH of plasma, however, is far removed from this pH.

Additional bicarbonate ions are provided by the sodium bicarbonate and very little of the carbon dioxide in solution is in the form of carbonic acid. The ratio of bicarbonate to carbonic acid is of the order of 20,000 to 1, giving a pH of 7·4. At this pH, the bicarbonate–carbonic acid system would not make efficient buffer *in vitro*. In the body, however, the situation is different, as will be seen later.

Henderson–Hasselbalch Equation

In practice it is more convenient to consider all the CO_2 in solution instead of only that which has become carbonic acid. This is permissible provided that the pK value is changed to take into account equation (1). The equation now becomes:

$$pH = 6·1 + \log_{10} \frac{[HCO_3^-]}{[CO_2 \text{ in solution}]} \quad . \quad . \quad (6)$$

This is known as the Henderson–Hasselbalch equation.

The bicarbonate ion concentration $[HCO_3^-]$ is determined almost entirely by the sodium bicarbonate content of the plasma. Sodium bicarbonate may be considered to be completely dissociated into sodium ions and bicarbonate ions. The carbonic acid contribution to the bicarbonate ion concentration is negligible.

The CO_2 in solution, as has been seen, is directly proportional to the carbon dioxide tension in the blood.

The Henderson–Hasselbalch equation may thus be written as:

$$pH = 6·1 + \log_{10} \frac{[HCO_3^-]}{\alpha . pCO_2} = 6·1 + \log_{10} \frac{[HCO_3^-]}{0·03 . pCO_2} \quad . \quad (7)$$

when $[HCO_3^-]$ is expressed in mEq. per litre and pCO_2 in mm. Hg. If pCO_2 is measured in kPa, $\alpha = 0·24$.

pH and Respiration

The CO_2 in solution in the arterial blood will be determined by the partial pressure of carbon dioxide in the alveolar air which is related to the percentage of carbon dioxide in the alveolar air. This, in turn, depends on the pulmonary ventilation. It is inversely proportional to the *alveolar ventilation* (see page 74). Thus:

$$[CO_2 \text{ in solution}] \propto \text{arterial } pCO_2 \propto \frac{1}{\text{alveolar ventilation}}$$

Over-ventilation will wash out carbon dioxide from the lungs and lower the CO_2 in solution in the blood. Under-ventilation will have the opposite effect and the blood CO_2 in solution will rise. As a rough approximation, at sea-level, the quantity of CO_2 in solution equals the percentage of carbon dioxide in the alveolar air divided by 2. Thus when the CO_2 concentration in the alveolar air is 6 per cent. there will be 3 ml. CO_2 dissolved in each 100 ml. plasma.

The *in vivo* buffering action of the bicarbonate–carbonic acid system is due to the respiratory centre. This centre is stimulated by any fall in pH and, as a result, there is an increase in the pulmonary ventilation. The alveolar pCO_2 falls and the carbon dioxide in solution is reduced. From equation 6 it will be seen that the blood pH change is resisted. Conversely, the respiratory centre is inhibited by a rise in pH, pulmonary ventilation is reduced, CO_2 in solution in the blood rises and the pH change is minimized.

Other Buffer Systems in the Blood

The plasma contains the two sodium salts of phosphoric acid, disodium hydrogen phosphate (Na_2HPO_4) and monosodium dihydrogen phosphate (NaH_2PO_4) in the ratio of 4:1. These form a buffer system since they will remove both hydrogen ions and hydroxyl ions from the solution [see p. 91]. Thus,

$$HCl + Na_2HPO_4 \rightarrow NaCl + NaH_2PO_4$$

and

$$NaOH + NaH_2PO_4 \rightarrow Na_2HPO_4 + H_2O$$

The pK value is 6·8. Thus,

$$pH = 6·8 + \log_{10} \frac{[Na_2HPO_4]}{[NaH_2PO_4]}$$

This buffering system is more important in urine than in blood [p. 138]

As has been seen, the haemoglobin in the red cell and, to a lesser extent, the proteins in the plasma play an important role in buffering the blood.

Body's buffers = bicarbonate carbonic acid

REFERENCES AND FURTHER READING

Davenport, H. W. (1958) *The ABC of Acid-Base Chemistry*, 4th ed., Chicago.
Roughton, F. J. W. (1965) Transport of oxygen and carbon dioxide, *Handbook of Physiology*, Section 3, Vol. 1, 767.

[handwritten notes]

alter it by 0·2 = drastic change.
pH of blood = 7·4.

Acid is produced from $CO_2 + H_2O$ by C. Anhydrase. ① $CO_2 + H_2O \rightleftharpoons H_2CO_3$. Oxyntic cells of stomach. ↘ HCl.

Kidney tubule cells — esp distal tubule — secrete H^+ ion exchange for Na^+.

Small int. cells secrete HCO_3^- ↗ so have to produce H^+ ions.

Carbonic anhy — in all body cells (resp. for synthesis of aqueous humour). ⟹ can ↓ glaucoma by X c.A. (acetazolamide). When lot of carbonic acid in blood xs CO_2 thru' lungs

$\{ H^+ + HCO_3^- \rightleftharpoons H_2CO_3 \rightleftharpoons H_2O + CO_2 ↑$

$NaHCO_3$ — neutralises acid in stomach ②

$NaHCO_3 + HCl \rightleftharpoons NaCl + H_2CO_3 \rightleftharpoons H_2O + CO_2$ causes belching.

↳ weak acid.

Amt Sod Bicarb far exceeds amt. req. to neutralize the acid in the body @ pH. 7.1.

$NaHCO_3 \rightarrow Na^+ + HCO_3^-$. This xs $HCO_3^- \rightarrow$ systemic Alkalosis — kidney presented c̄ xs bicarbonate — some of which will be excreted making alk. urine.

For overdose of acidic drug — make urine alk $NaHCO_3$ I/U or Na citrate,

$NH_4Cl \longrightarrow NH_3 + (H^+ + Cl^-) \longrightarrow$ system acidosis.

RESP ACIDOSIS - (eg in asthma) — alkaline infusion indicated.
↳ c̄ drugs that suppress resp. centre - barbs, antiepileptine.

Resp ALKALOSIS — by toxic doses of Na salicylate — due to effect on resp centre

convulsions result

Diabetes ⟶ v. severe system acidosis.

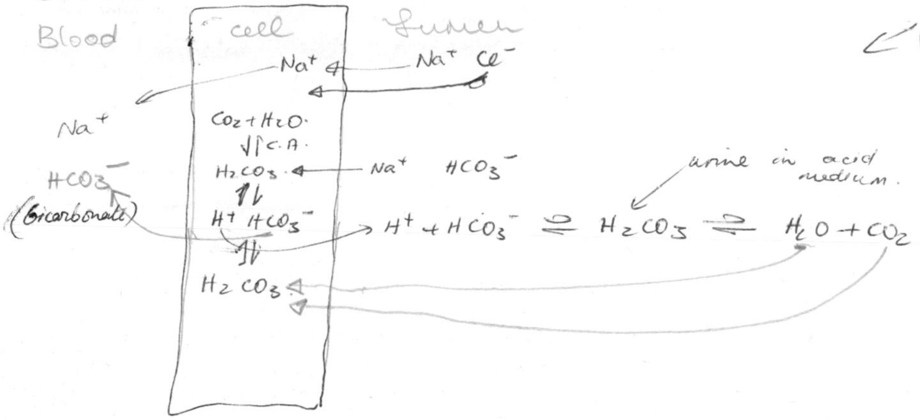

Blood | cell | lumen

urine in acid medium.

Acetazolamide = weak diuretic cos it doesn't interfere c̄ reabsorption of $Na Cl^-$ [used to treat glaucoma].

Body wants to maintain alkali reserves. ∴ preserves HCO_3^-. This molec is too large to pass thru' cell membrane ∴ the body cause the reabsorption of HCO_3^- by c. Anhydrase syn

Areas in body where pH alters greatly stomach 1.5-7: urine 5.5 — 6.5: S int. ≠ 8.

pH - maintained by alkali reserve which consists mainly of HCO_3^-, HPO_4^-

8. THE REGULATION OF RESPIRATION

Now that the roles of haemoglobin and of the respiratory centre in the regulation of the *hydrogen ion concentration* of the blood have been considered, we may return to a consideration of the factors concerned in the regulation of respiration.

Respiration is brought about by the activity of a group of nerve cells in the medulla termed the Respiratory Centre. These cells bring about rhythmical activity in the *striated* respiratory muscles via the somatic nervous system. Both the alpha and gamma control systems are employed [p. 199].

It should be noted that although breathing is to a large extent, an 'involuntary' act, the 'voluntary' nervous system (and not the autonomic nervous system) is the efferent system used to control the respiratory muscles.

The respiratory centre is under the influence of many different stimuli, some excitatory some inhibitory. Probably the most

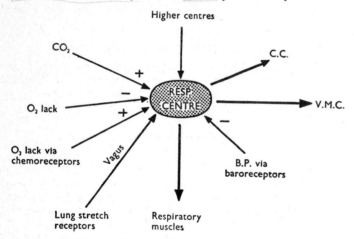

FIG. 117. Factors affecting the activity of the respiratory centre.

important factor is the **hydrogen ion concentration** in the blood supplying the centre or more accurately in the respiratory cells themselves. We have seen that the blood pH is determined by the ratio of bicarbonate to carbon dioxide in solution according to the Henderson–Hasselbalch equation which we will restate as:

$$pH = 6 \cdot 1 + \log_{10} \frac{[\text{Bicarbonate}]}{[\text{CO}_2 \text{ in solution}]} \qquad (1)$$

The normal bicarbonate to CO_2 ratio is 20:1 (60 vol. per cent. for the bicarbonate; 3 vol. per cent. for the CO_2 in solution).

Expressed as millimoles per litre the figures are: 26 mmol/litre for the bicarbonate and $1 \cdot 3$ mmol/litre for the CO_2 in solution.

This gives a pH of $7 \cdot 4$ since

$$6 \cdot 1 + \log_{10} 20 = 6 \cdot 1 + 1 \cdot 3 = 7 \cdot 4$$

The bicarbonate blood level is determined by the diet (see ingestion of acids and alkalis, p. 102) but it is usually constant since the diet is usually constant. The bicarbonate level is regulated by the kidneys [p. 138].

With the bicarbonate constant, the blood hydrogen ion concentration depends on the CO_2 in solution which, as has been seen, is determined by the carbon dioxide tension in the blood. We can now rephrase the above sentence and say that the most important factor acting on the respiratory centre is the level of **carbon dioxide** in the blood or more accurately the carbon dioxide tension in the respiratory cells themselves. Any increase in the blood carbon dioxide will stimulate the respiration, whereas any decrease will reduce respiration.

Other factors acting on the respiratory centre are shown in FIGURE 117.

Oxygen lack depresses the respiratory centre by direct action, and when severe enough respiration ceases. However, acting via the **chemoreceptors,** moderate degrees of anoxic anoxia and stagnant anoxia will stimulate respiration reflexly. Chemoreceptors, which would be better known as 'oxygen lack receptors', are found in collections of glomus tissue supplied by small arteries from the aorta and carotid arteries. The two main aggregates are termed the carotid and aortic bodies [p. 42]. As has been seen, these bodies share the same nerve trunks with the neighbouring baroreceptors so that adjacent neurones may be conveying either baroreceptor or chemoreceptor traffic [see FIG. 50, p. 41]. The activity in the chemoreceptor nerves is slight under normal conditions but it increases markedly when the oxygen tension falls. FIGURE 118

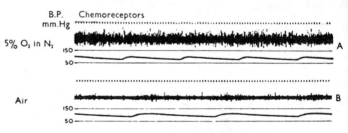

FIG. 118. The increase in activity of the chemoreceptor fibres in the sinus nerve in response to anoxia (cat).

A. Activity when breathing a low oxygen mixture.
B. Activity in the same nerve preparation when breathing air.

shows the difference in impulse traffic from a few fibre preparation from the chemoreceptors in the carotid body of a cat, *A* when the animal was anoxic due to the inspired gas mixture containing only 5 per cent. oxygen and *B* when breathing room air (21 per cent. oxygen). It will be seen that the impulse traffic is very much greater in *A* than *B*.

Chemoreceptors are little affected by a carbon dioxide excess (hypercapnia) when this occurs in the presence of high oxygen tensions, but, when the oxygen tension is low a concurrent hyper-capnia augments the chemoreceptor activity. Such a state with an oxygen lack and a carbon dioxide excess is termed *asphyxia*.

The principal effects of an increase in chemoreceptor activity may now be summarized as follows:
1. *Stimulation of the respiratory centre and thus an increase in breathing.*
2. *Stimulation of the vasomotor centre and thus vasoconstriction and a rise in blood pressure [p. 42].*
3. *A complex effect on heart rate with cardiac slowing by a direct action on the cardiac centre which is opposed by cardiac acceleration reflexes from the lungs as a result of the stimulation of breathing [p. 50].*

An increase in the blood pressure causes a slight reduction in respiration due to **baroreceptor** stimulation.

An increase in baroreceptor activity has thus been shown to have three principal effects:
1. *Inhibition of the vasomotor centre and hence vasodilatation [p. 40].*
2. *Stimulation of the cardio-inhibitory centre (and inhibition of the cardio-accelerator centre) and hence cardiac slowing [p. 50].*
3. *Inhibition of the respiratory centre and hence depression of respiration.*

The respiratory centre is under the influence of **higher centres.** Respiration is increased in emotional excitement. The respiratory centre is depressed during **sleep** leading to an increase in the alveolar carbon dioxide.

E.M.G.

Insp. M

Exp M

FIG. 119. Activity from the inspiratory and expiratory parts of the respiratory centre (cat).

Insp. M = electromyogram of the diaphragm,
Exp. M. = electromyogram of adductor muscle of the larynx.

The rhythmicity of respiration involves the vagi and the pneumotaxic centre in the upper pons. The stretch receptors in the lungs and bronchi are stimulated by lung inflation in inspiration and their nerve impulses up the vagi (**Hering-Breuer fibres**) inhibit the inspiratory part of the respiratory centre and stimulate the expiratory part.

Head demonstrated that distension of the lungs produces a prolonged expiratory effort, whereas collapse causes a strong inspiratory effort. He used a slip of the diaphragm in the rabbit which has an independent nerve supply and could be detached from the xiphoid and attached to a lever. Distension of the lungs caused the slip to relax. Collapse of the lungs caused contraction. This does not occur after vagotomy.

If the vagi are cut or damaged rhythmical respiration continues, but with increased depth provided that the pneumotaxic centre is intact. If this centre is also destroyed, rhythmical respiration ceases and a state of sustained inspiration known as *apneusis* occurs with only the occasional brief expiratory phase. Apneusis is due to the unopposed activity of an *apneustic centre* in the middle pons which stimulates the inspiratory part of the medullary respiratory centre. If this apneustic centre is destroyed leaving only the medullary respiratory centre, rhythmical respiration returns in the form of a series of gasps. If the medullary centre is destroyed, respiration ceases.

The intercostal muscles, and to a lesser extent the diaphragm, contain **muscle spindles.** Recent work suggests that the rhythmical resetting of the muscles spindles from expiration to inspiration (by increasing the gamma efferent activity to the intrafusal fibres of the muscle spindles) may be more important in determining the depth of breathing in man than the Hering–Breuer reflex.

Respiration is temporarily arrested during **swallowing.** Impulses from the pharynx via the IXth cranial nerve inhibit the respiratory centre.

Whilst talking, singing, etc., the respiratory pattern is completely changed and inspiration occurs at the end of each phrase [see FIG. 81, p. 65].

Normal breathing is termed *eupnoea*. A transient cessation of breathing is termed *apnoea*. An increase in the depth of breathing is termed *hyperpnoea*. Difficult or laboured breathing is termed *dyspnoea*.

Much of our knowledge concerning the regulation of respiration in man has been derived from a study of the alterations to breathing (and other systems of the body) which occur when the oxygen and carbon dioxide tensions are changed. The effects of such changes will now be considered.

The knowledge gained has enabled man to take the precautions necessary for survival in extremes of environmental pressure ranging from the near vacuum of outer space to the enormous pressures 600 ft. down on the sea-bed.

BREATHING PURE OXYGEN

The effect of breathing pure oxygen at sea-level will be to increase the arterial oxygen tension from 100 mm. Hg to the region of 700 mm. Hg. Once all the nitrogen has been washed out the oxygen tension will be:

760 — vapour pressure of water at 37 °C. — tension of CO_2
$$= 760 - 47 - 40 \text{ mm. Hg}$$
$$= 673 \text{ mm. Hg}$$

This will increase the oxygen in solution from 0·3 ml. O_2 per 100 ml. blood to 2·0 ml. O_2 per 100 ml. blood. The oxygen combined with haemoglobin will only increase very slightly from 19 ml. O_2 per 100 ml. blood at a tension of 100 mm. Hg to 20 ml. O_2 per 100 ml. blood at a tension of 673 mm. Hg.

The chemoreceptor drive will be reduced leading to a slight reduction in the pulmonary ventilation.

There will be a slight reduction in heart rate, but no change in blood pressure and no change in the mental state of the subject.

Oxygen has toxic properties when breathed at tensions in excess of 400 mm. Hg for long periods. It has an irritant effect on the lungs leading to a type of pneumonia. In premature babies breathing pure oxygen leads to the formation of fibrous tissue behind the lens of the eye (retrolenticular fibroplasia) and blindness.

CARBON DIOXIDE LACK

Since inspired room air contains zero carbon dioxide, it is not possible to lower the carbon dioxide in the body by changing to an inspired gas with less carbon dioxide. It is the presence of the dead space that maintains the alveolar air at a higher CO_2 level than room air. This CO_2 level can be reduced by either increasing the alveolar ventilation or by reducing the dead space volume. The former is used to study the effects of carbon dioxide lack under experimental conditions. The latter, in the form of a tracheostomy, is used to lower the carbon dioxide level in patients with respiratory distress.

Voluntary Over-ventilation

If a subject breathes deeply and rapidly for two minutes and then allows the respiratory centre to take over, a period of depressed breathing follows. Respiration may cease altogether for a time (apnoea) and frequently a period of periodic breathing sets in.

The over-ventilation changes the composition of the alveolar air towards that of the inspired room air. The carbon dioxide tension falls from 40 mm. Hg towards zero (a typical figure of 15 mm. is reached). The oxygen tension rises from 100 mm. Hg to 140 mm. Hg. The respiratory depression is entirely due to the carbon dioxide lack.

Prolonged over-ventilation tends to increased alkalinity of the blood, a feeling of dizziness, faintness and light-headedness, **tetany** develops with carpopedal spasm and stiffness of the face and lips [see p. 165]. The blood flow to the brain is impaired and with it the cerebral activity [p. 203].

OXYGEN LACK AND CARBON DIOXIDE EXCESS

Oxygen lack associated with carbon dioxide excess is termed **asphyxia.** Carbon dioxide excess is termed **hypercapnia.** Oxygen lack occurring alone is termed **anoxia** (or if less severe, **hypoxia**). The body responds powerfully to asphyxia and hypercapnia, but poorly to anoxia.

Asphyxia

Asphyxia may be brought about by breathing in a confined unventilated space, when there is airway obstruction and in respiratory failure.

Experimentally asphyxia may be induced by rebreathing expired air from a spirometer or a small bag. The level of oxygen in the lungs and spirometer progressively falls whilst the level of carbon dioxide progressively rises. The excess CO_2 stimulates breathing by its direct action on the respiratory centre. (The low O_2 in the presence of high CO_2 acts as a powerful stimulus to increased chemoreceptor activity which, in turn, further stimulates the respiratory centre by its direct action.) Pulmonary ventilation is markedly increased and as the asphyxia progresses accessory muscles of respiration are brought into play. These include sterno-mastoid and the scalenes which elevate the rib cage, and the pectoral muscles which increase the size of the thoracic cavity provided that the shoulder girdles are fixed by grasping a firm object. The activity of these muscles in asphyxia can be seen by inspection and confirmed electromyographically.

Asphyxia is associated with a feeling of apprehension which may lead to panic. If severe, the subject 'fights for breath' using most of the body musculature. When attempting to rescue a drowning man who is suffering from asphyxia, it may be necessary to render him unconscious or the rescuer may be drowned in the ensuing struggle.

Hypercapnia

Carbon dioxide excess (**hypercapnia, hypercarbia**) may be induced by raising the level of CO_2 in the inspired air (normally only 0.03 per cent.). With low percentages of CO_2, the respiratory stimulation keeps the alveolar CO_2 at approximately normal levels (6 per cent.), but when the percentage of CO_2 in the inspired air approaches and exceeds this value, the alveolar CO_2 tension rises markedly leading to acidaemia.

Five per cent. CO_2 in oxygen is sometimes used to stimulate breathing. The addition of the CO_2 to the oxygen is of no value if the CO_2 level is already high due to respiratory insufficiency.

High levels of CO_2 (over 12 per cent.) depress the respiratory centre, and other cerebral activity.

Anoxia (Hypoxia)

The types of anoxia and their causes have been discussed on page 83. All forms of anoxia, but particularly anoxic anoxia and stagnant anoxia stimulate breathing via the chemoreceptors. But anoxia also depresses the respiratory centre by a direct action, and if the anoxia is severe, the chemoreceptor drive may be unable to stimulate the breathing.

Anoxia may be induced experimentally by breathing gas mixtures containing low percentages of oxygen or by rebreathing expired air with a CO_2-absorber (such as soda-lime) in circuit to prevent the accumulation of carbon dioxide as the oxygen is used up. Such experiments must be carefully supervised by an observer who is not himself exposed to the anoxia, since acute anoxia may produce unconsciousness without prior warning.

Acute anoxia, such as is produced by entering an atmosphere of nitrogen or helium, leads to unconsciousness in a very short time often without prior warning or marked respiratory stimulation. For this reason care must be taken when descending wells or manholes to ensure that one is not entering a pocket of nitrogen due to the oxygen having been used up.

Carbon monoxide, present in the exhaust fumes of the internal combustion engine and in coal gas, is a particularly dangerous gas since it produces a severe anaemic anoxia which depresses the brain and removes all sense of danger and the will to survive.

A person who is subjected to progressive anoxia becomes confused and disorientated. Unconsciousness follows, and ultimately all nervous activity, including that of the respiratory centre, ceases. Anoxia of the cardiac muscle leads to cardiac arrest. Breathing may be stimulated slightly at first but later is depressed.

Anoxia Due to High Altitudes
Acclimatization

The slow ascent to high altitudes by mountaineers exposes them to the gradual onset of anoxia as the partial pressure of oxygen in the inspired air falls with altitude [p. 78]. Respiration is stimulated by the chemoreceptors, but the increased ventilation washes out carbon dioxide from the lungs. This causes alkalaemia. Referring to the Henderson–Hasselbalch equation [p. 96], the blood pH is raised because the [CO_2 in solution] is now reduced. This pH change acts on the respiratory centre and opposes the anoxic chemoreceptor drive. As a result, the increase in pulmonary ventilation is, at this stage, inadequate. The subject suffers from discomfort and fatigue, and breathing may be periodic.

It should also be remembered that haemoglobin without oxygen is more alkaline than haemoglobin with oxygen [p. 93] and this fact further adds to the *alkalosis*.

After a few days the kidneys excrete an alkaline urine containing sodium bicarbonate, thereby reducing the [Bicarbonate] in the Henderson–Hasselbalch equation. With the disappearance of the profound alkalaemia, the respiration increases and CO_2 once again becomes of importance in the regulation of breathing. The bone marrow is stimulated by the anoxia and the red cell count increases [p. 10]. At the same time the well-being of the subject improves. These changes are referred to as **acclimatization.**

When respiratory volumes are converted to 0 °C. and 760 mm. Hg, there is little change in the oxygen uptake per minute, the CO_2 production, the RQ, and the pulmonary ventilation at altitudes up to 22,000 ft. However, at an altitude of 18,000 ft., where the barometric pressure is half that at sea-level, this means that the pulmonary ventilation at the ambient pressure will be double that at sea-level. The same number of oxygen molecules that would occupy 250 ml. at sea-level, will now occupy 500 ml., so that the oxygen uptake at rest will be 500 ml. per minute at this altitude. Similarly the CO_2 production will be 400 ml. per minute (double the sea-level value).

The percentage compositions of the gases in the inspired, expired and alveolar air will be close to those at sea-level, but the corresponding partial pressures will be reduced at high altitudes (due to the low barometric pressure) and, in particular, the arterial pO_2 and pCO_2 will be low.

The high pulmonary ventilation at high altitudes, where the inspired air is very dry due to the cold, leads to a high water loss in the expired air and the subject becomes very thirsty.

VENTILATION–PERFUSION INEQUALITIES

Anoxia due to Shunting

The normal exchange of gases between the alveolar air and the blood in the lungs will only take place if the alveolar ventilation in different regions of the lungs matches the blood flow. If, for example, a large part of one lung has a blood flow but is unventilated, the blood flowing through it will not take up any oxygen [FIG. 120]. This blood, containing reduced haemoglobin, will

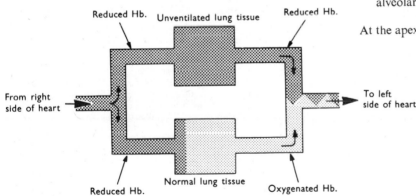

mix in the pulmonary veins and left side of the heart with the oxyhaemoglobin blood from the rest of the lungs. Thus the blood passing to the arteries will be desaturated with a low oxygen tension and content. The subject will be subjected to anoxic anoxia and may be cyanosed.

Since the blood from the normal part of the lungs is already fully saturated, the hyperventilation brought about by the anoxic chemoreceptor drive will not improve the oxygen uptake. This is because it is not possible for the blood passing through the normal lung tissue to take up any more oxygen in order to make up for that lost by some of the blood passing through unaerated lung tissue. As a result the anoxia persists.

The carbon dioxide retention does not occur since the hyperventilation causes the healthy lung tissue to eliminate additional carbon dioxide. This is essentially an oxygen problem due to the shape of the oxygen dissociation curve.

Similar shunts arise in congenital heart defects where not all the blood passes through the lungs with each circulation. If the shunt is large these infants are cyanosed ('blue babies').

Increased Dead Space

If a part of a lung has insufficient blood flow, much of the ventilation to this part of the lung will be wasted. This part of the lung will be acting as additional dead space. The **'physiological' dead-space** is thus greater than the **'anatomical' dead space** (nasal cavity, pharynx, larynx, trachea and bronchi) since it includes parts of the lungs where excessive ventilation occurs.

Physiological dead space = anatomical dead space
+ alveolar dead space

It has already been seen [p. 61] that even under normal conditions the pulmonary blood flow to the apices of the lungs in the erect posture is poor compared with that to the rest of the lungs because of the effect of gravity. This part of the lungs will be over-ventilated relative to the perfusion and will be contributing to the physiological dead space. At the bases of the lungs the perfusion is luxurious and here the lungs are relatively under-ventilated.

Ventilation–Perfusion Ratio

The ratio of alveolar ventilation to the blood flow is termed the ventilation/perfusion ratio $= \dfrac{\dot{V}_A}{\dot{Q}}$. Considering the lung as a whole, the alveolar ventilation is 3·75 litres per minute [p. 64] and the pulmonary blood flow (=cardiac output) is 5 litres per minute.

Thus:

$$\text{alveolar ventilation/perfusion ratio} = \frac{3\cdot75}{5} \simeq 0\cdot8 \text{ (over-all)}$$

At the apex of the lung the ratio is 3. At the base it is 0·6.

FIG. 120. Anoxic anoxia due to shunting.
If part of the lung tissue is unventilated, the blood flowing through it will not take up oxygen. This blood containing reduced haemoglobin will desaturate the blood which has passed through normal lung tissue. The oxygen tension and content of the blood leaving the lungs falls and the subject may become cyanosed. It is not possible to compensate for this type of anoxic anoxia by hyperventilation since the blood passing through the normal lung tissue is already saturated with oxygen and is unable to take up any more.
If the blood flow to the unaerated part of the lung ceases, the anoxia disappears.

The gas tensions in areas of the lung where the ratio is high will approach those of room air with a relatively high pO_2 and a low pCO_2. Where the ratio is low, the gas tensions will approach those of mixed venous blood with a low pO_2 and a high pCO_2.

Four generalizations may be made:

1. A low ventilation–perfusion ratio indicates **shunting.** It will be associated with a low arterial pO_2, anoxic anoxia, and possibly cyanosis.
2. A high ventilation–perfusion ratio indicates a **large physiological dead space.** It will be associated with a high arterial pCO_2 and CO_2 retention.
3. **Dead space** = wasted ventilation.
4. **Shunting** = wasted perfusion.

ASPHYXIA DUE TO BREATH-HOLDING

It is possible, for a time, to inhibit the normal respiratory rhythm with inhibitory impulses from higher centres. The time for which the breath can be held depends on many factors. Nevertheless, simple breath-holding experiments demonstrate the relative importance of CO_2 and O_2 levels in stimulating respiration, by equating their drive with the higher centre inhibition.

When the breath is held, the CO_2 production (200 ml. per minute) continues, but as no CO_2 is being expired the CO_2 level in the body rises. At the same time the oxygen level falls. The lungs at the end of a quiet respiration contain only 420 ml. oxygen (14 per cent. of 3000 ml.), and with an oxygen utilization of 250 ml. per minute this represents a reserve supply of less than two

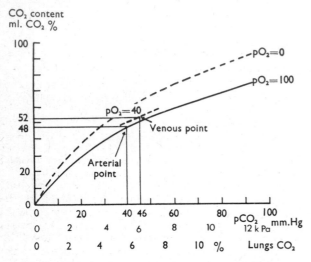

FIG. 121. Carbon dioxide dissociation curve. This shows the relationship between the CO_2 content of blood (measured in ml. CO_2 per 100 ml. blood $\equiv$ ml. $CO_2\%$) and the tension of carbon dioxide (measured in mm. Hg and kPa). A decreased oxygen tension shifts the CO_2 dissociation curve to the left.

Since arterial blood is in equilibrium with the alveolar air, the arterial pCO_2 will equal the alveolar pCO_2. The alveolar pCO_2 is proportional to the percentage of carbon dioxide in the alveolar air [p. 80]. [Alveolar $pCO_2 = \%CO_2 \times$ (Barometric pressure $-$ Vapour pressure of water at $37\,°C.$) $= \%CO_2 \times (760 - 47)$ mm. Hg $= \%CO_2 \times 713$ mm. Hg]. A scale showing the percentage of carbon dioxide in the alveolar air for different *arterial* tensions has been included. (This scale is not applicable to *venous* tensions.)

As the breath-holding proceeds, the pCO_2 in the arterial blood rises, and with it, the percentage of CO_2 in the alveolar air. A rise in pCO_2 from 40 mm. Hg to 50 mm. Hg will be associated with a rise from 6% (more accurately 5·6%) to 7% CO_2 in the alveolar air.

in the alveolar air) to 0 mm. Hg (0 per cent. O_2 in the alveolar air), whereas if the blood gains an equal amount of CO_2, so that its content increases from 48 ml. to 67 ml./100 ml. blood, the tension will only increase from 40 mm. Hg (6 per cent. in the alveolar air) to 56 mm. Hg (8 per cent. CO_2 in the alveolar air).

FIGURES 121 and 122 have identical 'tension' axes, but the 'content' scales are different. If the two dissociation curves were plotted on the same scale, the oxygen curve would appear smaller and more compressed since the content only extends up to 20 ml. per 100 ml. blood (see this point on the CO_2 dissociation curve, FIG. 121).

If the subject hyperventilates before breath-holding, the starting alveolar pCO_2 will be lowered. The breath-holding time is now

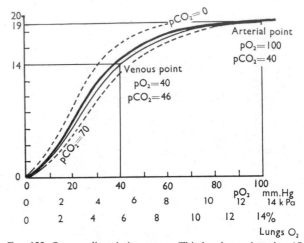

FIG. 122. Oxygen dissociation curve. This has been plotted as 'O_2 content of blood' instead of the more usual 'percentage saturation' [FIG. 108]. The curve shows the oxygen content of blood having a normal oxygen capacity of 20 ml. O_2 per 100 ml. blood at different oxygen tensions in mm. Hg and KPa. An increased carbon dioxide tension shifts the O_2 dissociation curve to the right.

A scale showing the percentage of oxygen in the alveolar air for different arterial tensions has been included [see FIG. 108].

As the breath-holding proceeds, the pO_2 in the arterial blood falls, and so does the percentage of oxygen in the alveolar air. The $\%O_2$ falls more rapidly than the $\%CO_2$ rises [see text]. A fall in arterial pO_2 from 100 mm. Hg to 57 mm. Hg will be associated with a fall in oxygen concentration from 14% to 8% in the alveolar air.

Note that FIGS. 121 and 122 have identical 'tension' axes, but that the 'content' scales are different. If the two dissociation curves were plotted on the same scale, the oxygen curve would appear smaller and more compressed since the content only extends up to 20 ml. per 100 ml. blood (see this point on the CO_2 dissociation curve, FIG. 121).

minutes. This volume could be doubled by breath-holding at full inspiration.

The rise in the percentage of CO_2 in the alveolar air is less than the corresponding fall in oxygen percentage because of the differences in dissociation curves and because CO_2 is a soluble gas and any excess will dissolve in the body fluids. At breaking point typical figures are that the alveolar oxygen has fallen from 14 to 8 per cent. (alveolar $pO_2 = 57$ mm. Hg) whilst the alveolar CO_2 has only risen from 6 to 7 per cent. (alveolar $pCO_2 = 50$ mm. Hg).

If the dissociation curves are consulted [FIGS. 121 and 122], it will be seen that if 100 ml. of arterial blood loses all its 19 ml. of oxygen, the oxygen tension falls from 100 mm. Hg (14 per cent. O_2

greatly increased and the subject may become cyanosed due to the low pO_2 before the CO_2 has built up to a sufficient value for breaking point to be reached. Such an experiment demonstrates that the CO_2 drive is of more importance than the anoxic drive.

Amongst the other factors which determine the breath-holding time are the mental state of the subject and his determination to continue the breath-holding, the discomfort as the chest volume progressively decreases with more oxygen being removed than carbon dioxide replaced, and an apparent inherent desire for rhythmical movements of the chest. The degree of asphyxia reached is less than that found when 'signs of distress' occur rebreathing expired air from a small bag or spirometer (typical alveolar air figures being 7·8 per cent. CO_2 and 6 per cent. O_2) or when re-breathing 5 per cent. CO_2 in oxygen (hypercapnia only) when the subject may continue until 9 per cent. CO_2 (pCO_2 = 64 mm. Hg) is reached.

If gas of a similar composition to that in the lungs is rebreathed at breaking-point, the breath can be held for a further time, although this rebreathing will not have altered the CO_2 and O_2 levels.

BREATH-HOLDING & BREATHING UNDER WATER

In recent years under-water swimming has increased in popularity. When simple apparatus such as a face mask is used, the breath is held while swimming under water. With more elaborate equipment, a cylinder or tank of compressed air enables the swimmer to continue to breath whilst submerged.

The pressure exerted on the body by water increases rapidly as the swimmer descends to a greater depth. It has already been seen that at only a few feet down, the breathing of air at atmospheric pressure from a tube extending to the surface becomes impossible because of the water pressure on the chest and abdomen [p. 63]. At a depth of only five feet the increase in pressure will exceed 100 mm. Hg, and it is not possible to make an inspiratory effort against even this small pressure. It follows, therefore, that respiratory movements will only be possible if the gas mixture breathed is at the same pressure as that applied externally to the body.

At a depth of 33 feet the pressure on the body is twice that at the surface (2 atmospheres or 2 × 760 mm. Hg = 1,520 mm. Hg). When diving *rapidly* to this depth, holding one's breath, the gases in the lungs are compressed to half their volume. It is, therefore, essential to take a deep breath before making such a dive to prevent excessive caving in of the chest.

The fact that the environmental pressure at a depth of 33 feet is double that at the surface, means that the partial pressures of oxygen and carbon dioxide in the lungs will be doubled even though the percentage composition remains unchanged. At greater depths the partial pressures will be increased correspondingly. The blood gas tensions, however, will initially be unchanged because the blood is incompressible. As a result the tension of the carbon dioxide in the lungs will be higher than that in the venous blood returning to the lungs. Carbon dioxide will, therefore, pass in the reverse direction to normal, that is, it will pass from the lungs to the blood. (It will be remembered that gases always pass from a region of high tension to a region of low tension.) This may be minimized by lowering the percentage of carbon dioxide in the lungs by overventilation before diving.

Even though oxygen is being used up, it will continue to enter the blood from the lungs because of the increased partial pressure of oxygen in the lungs. Oxygen will continue to enter the blood and maintain an adequate arterial oxygen tension until the percentage of oxygen in the lungs has fallen to a very low figure. This fact constitutes a hazard during the return to the surface. During the ascent the environmental pressure falls and with it the partial pressure of oxygen in the lungs. If the ascent is rapid, the partial pressure of oxygen in the lungs during the ascent will fall to a lower value than the tension of oxygen in the pulmonary arterial blood. As a result oxygen will leave the blood and pass into the lungs. This will lead to acute anoxia and may lead to unconsciousness.

A similar hazard arises during an escape from a sunken submarine. Air must be breathed out during ascent as the gases in the chest expand, but at the same time the partial pressure of oxygen in the lungs will fall because of the reduction in environmental pressure. If the percentage of oxygen in the lungs was initially low, consciousness may be lost during the ascent.

Cardiovascular Changes Following Total Immersion

Man shows a 'diving response' following total immersion including the face. These cardiovascular changes are shown best if the breath-holding commences at the moment of face immersion. The changes include an increased forearm vascular resistance, a fall in forearm blood flow, and increased blood pressure and slowing of the heart. It is not shown when breath-holding in air, immersion of torso only, or whilst snorkel-breathing.

Cardiovascular Changes at Depth

At depth the water pressure is transmitted to the cardiovascular system and the blood pressure will be correspondingly increased in terms of *absolute units*, that is with reference to a vacuum. Thus, at 33 feet the systolic blood pressure will be 1,520 + 120 = 1,640 mm. Hg absolute (instead of 880 mm. Hg, p. 28). Relative to the environmental pressure, however, the blood pressure will be unchanged, the systolic pressure will still be 120 mm. Hg.

In order to see clearly in the water it is necessary to wear a face mask. If goggles only are worn, the increased blood pressure in the vessels of the eye will be greater than that of the air trapped in the goggles and conjunctival haemorrhages will occur. To prevent this, the face mask also covers nose, and air is breathed out from the lungs into the mask so as to maintain an air pressure in the mask equal to that of the environmental pressure outside.

In cold climates the duration of an under-water swim is limited by the fall in body temperature which results from the excessive heat loss by the body since the water temperature is lower than the skin temperature [see p. 149]. The duration can be increased by wearing heat-insulating clothing such as a rubber wet suit. Even so, in water at 0 °C. the duration of immersion is limited to about 15 minutes. Excessive cold leads to coma.

Respiration at Great Depths

When using under-water breathing apparatus (aqua-lung or scuba) the air is supplied to the mouth at the environmental pressure so that the normal pressure difference between the lungs and the outside of the chest wall can be maintained.

At very great depths, 400–1,000 feet, helium and oxygen mixtures (such as 98 per cent. helium, 2 per cent. oxygen) are used in place of air. This is because nitrogen in the air is poisonous at these pressures. Only 1–2 per cent. oxygen is needed because the pressure is 12–30 atmospheres. Experiments with specially made under-water craft with hatches to allow aquanauts to swim out and work on the ocean floor have shown that man can live for extended periods at very high environmental pressures. The helium atmosphere present in such craft presents problems. It changes the quality of the voice, and makes it very high in pitch and squeaky so that speech communication becomes difficult. Neon and oxygen, or helium and oxygen with a little nitrogen added, produces a more normal voice, and a temporary change to these gases can be made whilst talking by telephone to the surface. Helium under pressure is also a very good conductor of heat, and so much heat is lost from the body surface, that a very high environmental temperature (90 °F., 32 °C.) is required for comfort.

A very slow decompression, lasting several days, is needed after such a dive, to prevent 'bends' [p. 85] which in this case would be due to the formation of helium bubbles.

RESPIRATION DURING SPACE FLIGHTS

The space-suit (when worn) and the space-craft (when the suit is removed) have to provide an environment in which the astronaut's respiratory system can function as normally as possible. An alveolar oxygen tension of 100–120 mm. Hg can be achieved in two ways. First, the space-craft can be maintained at 1–1·2 atmospheres pressure and an environment of 21 per cent oxygen and 79 per cent. nitrogen used to maintain sea-level conditions. Alternatively 100 per cent. oxygen can be used and the pressure in the space-craft reduced until a similar oxygen partial pressure is achieved.

The first method has the advantage that the conditions approximate very closely to normal, but the nitrogen carried adds to the weight of the space-craft. The second method means that only oxygen need be carried. In addition there will be a smaller pressure difference between the inside of the space-craft and the vacuum of outer space outside, so that small leaks are not so serious and a thin walled craft can be used for, say, a moon landing; 100 per cent. oxygen however constitutes a fire hazard, especially before take-off.

The Russian space programme appears to favour the first method whilst the American space programme uses the second. For a link up in space an interchange zone arrangement has to be used.

In both systems carbon dioxide absorbers are required to remove the expired CO_2. Lithium hydroxide is used in place of soda-lime since it is lighter. The excess water is absorbed to maintain the humidity at an acceptable level. In addition the cabin temperature must be carefully controlled.

When astronauts re-enter the lunar module after a time on the surface of the moon they continue to wear their space-suits since the pressure in the module is zero (a vacuum). Having sealed the hatches, the oxygen cylinders are turned on to pressurize the module. When the pressure reaches 2 lb/in² (103 mm. Hg) they would be barely conscious without their space suits. Once the cabin pressure reaches 3·5 lb/in² (181 mm. Hg) the suits may be removed. The pressure is increased to 4·6 lb/in² (238 mm. Hg) to allow for an adequate safety margin. In the space craft itself a pressure of 6·1 lb/in² (315 mm. Hg) is employed. The system is designed so that if a small hole (1 inch diameter) develops in the space craft sufficient cabin pressure can be maintained for a long enough period of time to enable the astronauts to put on their space suits.

It should be remembered that in the lungs the sum of the partial pressures of the gases present will equal the pressure in the capsule itself but the water vapour in the lungs will continue to have a partial pressure of 47 mm. Hg and with the carbon dioxide which is also present in the lungs, the alveolar oxygen tension will always be very much less than the cabin pressure.

A FURTHER CONSIDERATION OF ACIDAEMIA AND ALKALAEMIA

Plasma Bicarbonate

The bicarbonate content of the plasma is available to neutralize fixed acids entering the blood. The bicarbonate content at a CO_2 tension of 40 mm. Hg, expressed as the volume of carbon dioxide which can be evolved from 100 ml. plasma, is known as the *standard bicarbonate* or *alkali reserve*. Its value is 60 ml. CO_2 per cent. or 26 mmol/l. More CO_2 is carried as bicarbonate per 100 ml. plasma than per 100 ml. red cells; hence this figure is higher

than that given on page 84 for the bicarbonate in whole blood (= 42 ml. CO_2 per cent).

Acidaemia (Acidosis)

The term *acidaemia* is used when there is an increase in the hydrogen ion concentration in the blood. A tendency to such a change is termed *acidosis*.

Acids may enter the circulation and produce a fall in the blood pH as the result of production in the following ways.

① Exercise, leading to the formation of lactic acid.
② High protein diet.
　　The protein sulphur is oxidized to sulphuric acid.
　　The protein phosphorus is oxidized to phosphoric acid.
③ Ingested acids.
　　Vinegar (= 4 per cent. acetic acid).
　　Acids in soft drinks (phosphoric and, sometimes, sulphuric acid).
　　Ammonium chloride. This is converted to urea (neutral) and hydrochloric acid.
④ In untreated diabetes mellitus, keto acids resulting from the imperfect oxidation of fats (acetoacetic acid and β-hydroxybutyric acid).

As soon as they enter the blood these acids are neutralized by the plasma sodium bicarbonate. Thus,

$$HCl + NaHCO_3 \rightarrow NaCl + H_2CO_3$$

or in general terms

$$HX + NaHCO_3 \rightarrow NaX + H_2CO_3$$

where X = the acid radical.

There is thus a reduction in the *plasma bicarbonate* and from equation 1, page 96, a fall in blood pH. This fall is limited by the respiratory stimulation which reduces the alveolar CO_2 and thus the CO_2 in solution. A new equilibrium is set up with both the bicarbonate and CO_2 in solution reduced. The pH has changed only slightly, say from 7·4 to 7·35, but the plasma bicarbonate (alkali reserve) has been reduced and respiration stimulated. This state is termed *compensated metabolic acidosis*.

It has been seen from the Henderson-Hasselbalch equation [p. 96] that provided that the ratio of [bicarbonate] to [CO_2 in solution] remains at 20:1 the blood pH will remain unaltered. The blood pH will remain unchanged if the proportional decrease in the pCO_2 of the blood (due to the respiratory stimulation) is one-twentieth the decrease in the bicarbonate.

Acidaemia may also result from a loss of alkaline intestinal fluids in diarrhoea, biliary and pancreatic fistulae.

In severe acidaemia the respiratory stimulation may be so marked as to give rise to the clinical condition of 'air hunger'. This is seen in severe ketosis.

The final correction takes place in the kidney. The non-volatile acid is excreted in the urine, and both the plasma bicarbonate and the respiration return to normal.

The rise in the hydrogen ion concentration of the blood which occurs as a result of a failure to excrete sufficient carbon dioxide is termed *respiratory acidaemia*. It occurs in asphyxia and hypercapnia which result from respiratory obstruction, respiratory insufficiency, an excessive dead space and when breathing a gas mixture containing a high concentration of carbon dioxide (7 per cent.).

It should be noted that both respiratory and metabolic acidaemia are associated with an increase in pulmonary ventilation (hyperventilation). However, in *respiratory acidaemia* there is an increase in the blood pCO_2 and an increase in the bicarbonate, whilst in *metabolic acidaemia* there is a decrease in the blood pCO_2 and a decrease in the bicarbonate.

The slight fall in blood pH is due to the fact that in *respiratory acidaemia* the increase in blood pCO_2 is relatively greater than the increase in bicarbonate, whilst in *metabolic acidaemia* the decrease in bicarbonate is relatively greater than the decrease in pCO_2.

Alkalaemia (Alkalosis)

The term *alkalaemia* is used when there is a reduction in the hydrogen ion concentration in the blood (an increase in the blood pH). A tendency to such a change is termed *alkalosis*.

Alkalaemia may result from the ingestion of alkalis such as sodium bicarbonate. A vegetarian diet may tend to a high intake of such salts as sodium citrate and sodium tartrate which are metabolized to carbon dioxide and sodium bicarbonate. An increase in plasma bicarbonate occurs, the blood pH rises, respiration is depressed and there is an increase in the alveolar carbon dioxide and carbon dioxide in solution which compensates for the bicarbonate increase. The rise in pH is very small. This state is termed *compensated metabolic alkalosis*.

Alkalaemia may result from the loss of acid gastric juice in vomiting.

In severe alkalaemia the rise in blood pH may lead to tetany [see p. 164].

As in the case of acidaemia the final correction takes place in the kidneys. The excess sodium bicarbonate is excreted in the urine, and the plasma bicarbonate and respiration return to normal.

The excretion of an excessive amount of carbon dioxide due to voluntary or mechanical overventilation will lead to a fall in the hydrogen ion concentration of the blood and to *respiratory alkalaemia*.

It should be noted that both respiratory and metabolic alkalaemia reduce the activity of the respiratory centre and lead to a decrease in the pulmonary ventilation (hypoventilation). In *respiratory alkalaemia* there is a decrease in the blood pCO_2 and bicarbonate. In *metabolic alkalaemia* there is an increase in blood pCO_2 and bicarbonate.

A FURTHER CONSIDERATION OF pH AND RESPIRATION

Cerebrospinal Fluid

It has been seen [p. 88] that the gas tensions in the cerebrospinal fluid (C.S.F.) approximate closely to those in the venous blood, but that the bicarbonate concentration in the cerebrospinal fluid is regulated independently by its secretion into the C.S.F. by the choroid plexuses.

Carbon dioxide in solution diffuses rapidly and it readily crosses the *blood-brain barrier*. As a result changes in blood pCO_2 are followed by similar changes in the C.S.F. pCO_2. Changes in blood bicarbonate are, however, not immediately followed by changes in C.S.F. bicarbonate.

In metabolic acidaemia, although the blood becomes more acid, the C.S.F. may become more alkaline. This is because the metabolic acidaemia lowers the pCO_2 in the blood (see above) in addition to lowering the bicarbonate in the blood. The resultant lowering of the C.S.F. pCO_2 (without a concurrent lowering of the C.S.F. bicarbonate) will raise the C.S.F. pH, i.e. make it more alkaline.

Respiratory acidaemia, on the other hand, is associated with an increase in the blood and C.S.F. pCO_2 and will thus lower the C.S.F. pH, i.e. make it more acid.

	Blood & C.S.F. pCO_2	C.S.F. pH	Blood pH	Respiration
Metabolic acidaemia	falls	rises	falls	stimulation
Respiratory acidaemia	rises	falls	falls	marked stimulation
Metabolic alkalaemia	rises	falls	rises	depression
Respiratory alkalaemia	falls	rises	rises	marked depression

Since the pH of the cells of the respiratory centre in the brain depends, in part, on the pH of the C.S.F. (in addition to the pH of the blood), *respiratory acidaemia* will produce a greater pH reduction in these cells, and hence a greater respiratory stimulation, than *metabolic acidaemia* for the same blood pH change.

By a similar argument, *respiratory alkalaemia* produces a greater reduction in respiratory drive than *metabolic alkalaemia* of the same blood pH magnitude.

Cerebrospinal fluid is a poor buffer compared with blood. Its pH appears to be normally regulated by the secretion of bicarbonate into it by the choroid plexuses. Thus part of the process of acclimatization to high altitudes [p. 98] is probably the reduction in the secretion of bicarbonate into the C.S.F. The pH rise associated with the increased pulmonary ventilation disappears. It no longer opposes the anoxic chemoreceptor drive. Breathing increases and carbon dioxide once again becomes important in the regulation of the respiration.

REFERENCES AND FURTHER READING

Adrian, E. D. (1933) Afferent impulses in the vagus and their effect on respiration, *J. Physiol. (Lond.)*, **79**, 332.

Campbell, J. M. H., Douglas, C. G., Haldane, J. S., and Hobson, F. G. (1913) The response of the respiratory centre to carbonic acid, oxygen and hydrogen ion concentration, *J. Physiol. (Lond.)*, **46**, 301.

Douglas, C. G., Haldane. J. S., Henderson, Y., and Schneider, E. C. (1913) Physiological observations made on Pike's Peak, Colorado, with special reference to adaptation to low barometric pressures, *Phil. Trans. B*, **203**, 185.

Gray, J. S. (1949) *Pulmonary Ventilation and its Physiological Regulation*, Springfield.

Haldane, J. S., and Priestley, J. G. (1935) *Respiration*, 2nd ed., Oxford.

Heymans, C., and Neil, E. (1958) *Reflexogenic Areas of the Cardiovascular System*, London.

Wang, S. C., Ngai, S. H., and Frumin, M. J. (1957) Organisation of central respiratory mechanisms in the brain stem of the cat: genesis of normal respiratory rhythmicity, *Amer. J. Physiol.*, **190**, 333.

Winterstein, H. (1911) Die Regulierung der Atmung durch das Blut, *Arch. ges. Physiol.*, **138**, 167.

9. DIGESTION

THE ALIMENTARY TRACT

Mouth and Oesophagus

Food is needed as the source of heat and energy, and for growth and the repair of tissues. The food, after entering the mouth, is chewed, mixed with saliva, and formed into a *bolus*. The bolus is forced backwards into the pharynx by the voluntary act of swallowing. From then onwards the passage of food along the digestive tract ceases to be under voluntary control. The stimulation of the sensory nerve endings in the pharynx causes the involuntary phase of swallowing by reflex action. The larynx is raised and the epiglottis prevents the food from entering the larynx. At the same time respiration ceases as the bolus passes down the pharynx to the **oesophagus;** breathing in, at this time, would cause choking

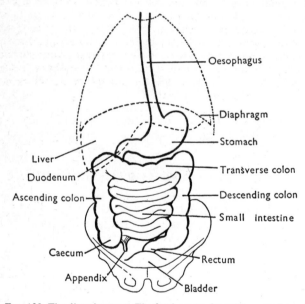

FIG. 123. The digestive tract. The food passes from the oesophagus through the stomach and duodenum, to the small intestine. The undigested residue passes through the ileocaecal valve to the caecum, ascending colon, transverse colon, descending colon, rectum and anal canal. The pancreatic duct and the bile duct join the duodenum.

as a result of the food entering the larynx. Peristaltic waves then convey the bolus down the oesophagus to the **stomach.**

Peristalsis is the mechanism by which the food is moved along the digestive tract. It consists of a wave of contraction of the muscle coat which is usually preceded by a wave of relaxation. The bolus is forced onwards by the contraction into the segment ahead. Peristalsis in the oesophagus only occurs when the vagal nerve fibres are intact.

Stomach and Intestines

Gentle contractions of the stomach wall mix the food with the *gastric juice.* The muscular wall of the stomach is stronger in the *pyloric* region, and the peristaltic waves in this part of the stomach propel a small part of the stomach's contents at intervals through the *pyloric sphincter* into the **duodenum** and **small intestine** [FIG. 123]. By now it has become *chyme.*

In the duodenum the chyme comes in contact with the *pancreatic juice* and *bile* from the liver.

In the small intestine the chyme is propelled along by a series of short peristaltic waves which carry it on several centimetres at a time. In addition the chyme is mixed with the intestinal juice by pendulum and segmental movements. Pendulum movements are alternate shortening and lengthening of the intestine. Segmental movements cause the intestine to be divided into sections by a series of contractions and relaxations of the muscle wall along its length. The regions alternately change their state. The contracted region relaxes, and the relaxed region contracts, thereby mixing the contents, but without causing any forward movement.

The unabsorbed residue of the food passes through the *ileo-caecal valve* from the small intestine to the **large intestine** which consists of the *caecum, ascending, transverse* and *descending colons, pelvic (sigmoid) colon, rectum* and the *anal canal.*

General Principles of Digestion

As the food passes along the digestive tract, it is acted upon by biological catalysts known as **enzymes** which break down the constituents of the food into small molecules suitable for absorption into the blood. This process is known as digestion.

Not all the items of the diet are modified in this way. Some are already in a suitable form for absorption, and for these no change occurs. Such substances include water, mineral salts, vitamins and glucose. They are absorbed unchanged from the small intestine.

Other substances are not digested because the appropriate enzymes are absent from the digestive tract. These substances have no food value, and are excreted in the faeces as roughage.

The foods of greatest value for nutrition have been chosen, by trial and error over the centuries, because enzymes exist which will convert these foods into absorbable and utilizable form. These foods may be classified into three main classes; carbohydrates, proteins and fats.

Carbohydrates

This class of food substances includes the sugars and starches. They are composed of the three elements, carbon, hydrogen and oxygen, and have a general chemical formula $C_x(H_2O)_y$. It is because this formula is equivalent to 'hydrated carbon' that the group has received the name 'carbohydrate'. The group includes

104

monosaccharides $C_6H_{12}O_6$, disaccharides $C_{12}H_{22}O_{11}$, and poly-saccharides (starches) which are built up of a large number of monosaccharide units.

Only the monosaccharides are absorbed into the blood. More complex sugars and starches need to be broken down before absorption. The enzymes available are *amylase*, which will act on starch, *sucrase* which will split sucrose (cane or beet sugar) into glucose and fructose, *maltase* which will convert maltose into glucose, and *lactase* which will split lactose (milk sugar) into glucose and galactose. Glucose, fructose and galactose are mono-saccharides. Sucrose, maltose and lactose are disaccharides.

Many other carbohydrates exist in nature, but without the appropriate enzymes they are of no value in human nutrition. We do not have, for example, any enzyme for splitting cellulose although it is built up of glucose units.

$$\text{Starch} \xrightarrow{\text{amylase}} \text{Maltose} \xrightarrow{\text{maltase}} \text{Glucose}$$

$$\text{Sucrose} \xrightarrow{\text{sucrase}} \text{Glucose + Fructose}$$

$$\text{Lactose} \xrightarrow{\text{lactase}} \text{Glucose + Galactose}$$

The sugar in the blood is **glucose**. Galactose, after absorption, is converted to glucose by the liver. Fructose is changed into glucose, not only by the liver cells, but also by many other cells of the body. The absorption of glucose, galactose and fructose from the small intestine is an *active process* which involves the expenditure of energy by the intestinal cells.

The end-product of carbohydrate in the diet is thus glucose in the blood and it is immaterial whether the source is starch such as bread and potatoes, or sucrose or even glucose itself. Sucrose, as such, is not found in the blood or urine.

Protein

Protein in the diet is required for the formation of body protein which is needed for the growth and repair of tissues. The surplus is used as a source of heat and energy.

About twenty different amino acids are found in body proteins. These proteins are extremely complex molecules. They are composed of thousands of amino acid units in various combinations. Not all the amino acids are found in any one molecule.

The food proteins are the source of amino acids for the protein synthesis. Eight of the amino acids are 'essential'. This means that they must be constituents of the dietary protein otherwise growth and repair of tissues will cease. The remaining amino acids may be ingested and used; if not present in the diet they will be made by the body when required. The body is unable to make the essential amino acids in sufficient quantities to satisfy its needs.

An amino acid contains at least one amino group NH_2 (or its equivalent) and at least one carboxyl group —COOH. The amino acids are joined together in the protein molecule by the *peptide linkage* (amide linkage) between these two groups.

$$-CO \boxed{OH + H} HN— \rightarrow —CO . NH— + H_2O$$
Peptide linkage

Dietary protein is broken down into the constituent amino acids by enzymic activity in the digestive tract. The chief enzymes are *pepsin* in the stomach, *trypsin* and *chymotrypsin* in the pancreatic juice and a collection of peptidase enzymes known as *erepsin* in the mucosal cells and juice of the small intestine.

The stages in the breakdown are:

$$\text{Food protein} \xrightarrow{\text{pepsin}} \text{Long chain peptides}$$
$$\xrightarrow{\text{trypsin}} \text{Short chain peptides}$$
$$\xrightarrow{\text{erepsin}} \text{Amino acids}$$

Each stage represents the breaking of more and more of the peptide linkages existing in the protein molecule so that smaller and smaller molecules are produced. Finally, the amino acid stage is reached.

The amino acids are absorbed from the small intestine into the blood. Each growing cell removes the appropriate amino acids from the blood for the synthesis of its own cell protein. The surplus amino acids are broken down by the liver [p. 155].

All the essential amino acids must be circulating at the same time and should be present in the diet eaten at one meal. All protein from animal sources (including fish) contains the essential amino acids. However, some forms of vegetable protein may be relatively deficient in one or more of the essential amino acids. In this case the nutritional value will be increased greatly by having a mixed diet. For example, bread contains all the essential amino acids, but is relatively deficient in the essential amino acid *lysine*. Milk, on the other hand, not only contains all the essential amino acids, but has a surplus of lysine. Thus bread and milk taken together at the same meal have a higher protein nutritional value than the foods taken separately on different occasions.

Maize (sweet corn) is relatively deficient in both tryptophan and lysine. Children living in parts of the world where maize is the principal item of diet show a marked incidence of the protein deficiency syndrome known as *kwashiorkor* even when their Calorie intake is adequate. Work is being carried on to develop strains of corn which have improved amino acid content.

Fats

Fats are compounds of glycerol (glycerin) and three fatty acids. They are triglycerides.

CH$_2$OH	CH$_2$O·fatty acid	CH$_2$O·fatty acid
CHOH	CHO·fatty acid	CHO·fatty acid
CH$_2$OH	CH$_2$O·fatty acid	CH$_2$OH
glycerol	triglyceride (fat)	diglyceride

a fatty acid $C_{17}H_{35}COOH$

The digestion and absorption of fat is a complex physico-chemical process because the ingested fats and oils are inherently insoluble in water whilst both the intestinal juice and the blood are watery fluids.

The bile salts from the liver have a 'detergent' action and emulsify the fat into small droplets which form a suspension with the water. The action of the bile salts is aided by the action of the fat-splitting enzyme lipase. In addition, lipase splits some of the fat into mono- and diglycerides and free fatty acids.

Soap is the sodium or potassium salt of fatty acids, but at the pH of the intestine (6·5–7·5) no soaps are formed.

The minute droplets of fat (microchylons) are absorbed into the intestinal lymphatics. These are known as *lacteals* because of the milk-like appearance of their contents due to the fat droplets. The fat droplets enter the venous blood at the root of the great veins via the thoracic duct. These droplets of neutral fat pass to the fat depots of the body which are situated under the skin and in the abdomen.

Some of the fat split by the lipase is absorbed into the blood and reaches the liver via the portal vein where it is reconstituted into neutral fat.

When required, the fat is withdrawn from the fat depots and transported back to the liver as soluble phospholipids.

DIGESTIVE SECRETIONS AND THEIR CONTROL

The main stimulus for the secretion of the digestive juices is the presence of food in the digestive tract. The physiological mechanism by which this secretion occurs changes as the food proceeds along the digestive tract. In the mouth the secretion of saliva is under nervous control. In the stomach and duodenum the control is both nervous and humoral whilst in the intestines it is the mechanical presence of food that stimulates the production of the appropriate juices.

Saliva

Saliva is produced in the mouth by the parotid, submandibular and sublingual salivary glands. The secretomotor nerves are parasympathetic fibres associated with the IXth and VIIth cranial nerves. The IXth nerve supplies the parotid gland. The VIIth nerve supplies the other two. Details of the nerve pathways will be found on page 206. Stimulation of these parasympathetic nerves not only produces a flow of saliva but also causes an increase in the blood flow to the glands (vasodilatation).

The production of saliva is studied by cannulating the salivary duct with a fine catheter. To facilitate the cannulation in an experimental animal, Pavlov's method was to free the terminal part of the duct and mucous membrane and to transfer it to the external skin surface.

The stimulus for saliva production is:

(1) the presence of food in the mouth;
(2) the sight, smell and thought of food.

The first is an inborn or unconditioned reflex. The second is a *conditioned reflex*. Pavlov (1910) noted that salivation occurred in his experimental dogs when the bells of St. Petersburg, as it then was, now Leningrad, chimed at midday. This was the usual feeding time and an association between the presentation of food and the bells had developed into a conditioned reflex. Once this reflex was established the chiming of the bells, without the food, produced salivation.

The parasympathetic activity which brings about the salivary secretion originates in the superior and inferior salivary nuclei which lie in the reticular formation in the lower pons and medulla.

Functions of Saliva

Saliva aids the formation of the food bolus and acts as a lubricant to facilitate swallowing. In addition, it contains salivary amylase *ptyalin*. Ptyalin acts in a neutral or slightly acid medium and commences the breakdown of cooked starch to maltose. Its action ceases soon after the food has been swallowed. The hydrochloric acid of the stomach juice penetrates to the centre of the bolus in approximately 20 minutes, after which time the pH is too low for the continued action of the ptyalin. In the dog, which has frequently been used as the experimental animal for the study of digestive secretions, the saliva contains no ptyalin.

In addition to acting as a lubricant and containing salivary amylase, the saliva has many other functions. It moistens the tongue and mouth and makes speech possible. It has a cleaning action on the tongue; in fevers the salivary secretion is reduced and the tongue becomes 'dirty'. The sensations of taste (sweet, salt, acid and bitter) which arise from the taste buds on the tongue depend on solutions being formed. The sensation of thirst and the associated dry mouth give an indication of a water deficiency in the body [p. 129]. In dogs the evaporation of saliva from the moist tongue and the resultant cooling this produces, play an important part in the temperature regulation in this animal in hot environments [p. 151].

Gastric Juice (Stomach)

The stomach acts as a storage organ so that food may be eaten at intervals as a meal instead of continuously. The food enters the stomach during a meal and is passed on, a small quantity at a time, through the pyloric sphincter into the duodenum during the next 3–4 hours.

The stomach produces the protein-splitting enzyme *pepsin*. It is released as the precursor *pepsinogen* in the form of granules which dissolve in the hydrochloric acid produced by the *oxyntic cells*. The pepsin converts the proteins to smaller molecules. The strength of the hydrochloric acid released is one-tenth molar (3·65 g. HCl per litre of water) which has a pH = 1 (cH = 10^{-1}). The resultant dilution by the food raises the pH of the stomach contents to between 1·5–2·5. The corresponding hydrogen ion concentration (cH) will be $10^{-1 \cdot 5}$–$10^{-2 \cdot 5}$ or (30–3) × 10^{-3} g. of hydrogen ions per litre.

The oxyntic cells in the stomach are unique in that they can make and secrete hydrochloric acid. They do so by means of a H^+ pump and a Cl^- pump. The cells contain *carbonic anhydrase* and the hydrogen ions are obtained indirectly[9] by the reaction of CO_2 with water. The energy required comes from the oxidation of carbohydrate. The chloride ions are obtained from NaCl.

The over-all reaction is thus:

$$CO_2 + NaCl + H_2O \rightarrow \underset{\text{(stomach)}}{HCl} + \underset{\text{(venous blood)}}{NaHCO_3}$$

It follows that whilst HCl is being secreted the venous blood leaving the stomach will contain more sodium bicarbonate than the arterial blood arriving at the stomach. This production of sodium bicarbonate is termed the **alkaline tide.** The blood pH will rise slightly, respiration will be depressed and the arterial pCO_2 will rise [see alkalaemia, p. 103].

Rennin is found in the stomach of young animals. It solidifies milk by converting the *caseinogen* (casein in U.S.A.) into *casein* (paracasein in U.S.A.) which is precipitated as calcium caseinate. Under the name of 'rennet', rennin extracted from the stomach of a calf is sold to make junket. Rennin is absent in man, but the pepsin present has a similar action.

Gastric Activity

Gastric activity has been studied in man by direct observation of patients who, as a result of an injury or operation, have an opening from the stomach to the exterior which is known as a gastric fistula.

The first account of such a fistula was by Beaumont who in 1833 published his account of his researches into gastric secretion using

---

[9] Indirectly because the water is first split into H^+ and OH^- and then the CO_2 is used to remove the OH^- formed as HCO_3^-.

Alexis St Martin as his subject. Alexis St Martin had a gastric fistula following a gun-shot wound and he was studied by Beaumont over the course of a number of years. Beaumont found that in the absence of food the stomach did not secrete any gastric juice and was contracted. More recently Wolf and Wolff (1947) reported using their laboratory assistant Tom, who also had a gastric fistula, that the stomach responds to emotional states. Fear is associated with a reduction in secretion and vasoconstriction whilst anger is associated with an increased acid secretion, increased motility and vasodilatation.

The movements of the stomach are studied by X-rays. A meal of barium sulphate, which is insoluble and opaque to X-rays is given. This outlines the cavity of the stomach on the X-ray screen.

The interior of the stomach may be viewed without a surgical incision by means of an optical instrument known as a gastroscope. This is passed down the oesophagus into the interior of the stomach.

The original instruments were rigid tubes containing the optical lenses. More recently fibre optics have enabled flexible gastroscopes to be employed which can be attached to a film camera or television unit incorporating a video-tape recorder so that a permanent record can be obtained.

A sample of gastric juice may be obtained by passing a thin rubber tube, such as a Ryle's tube, down into the stomach. Such a tube is passed through the nose or mouth and swallowed.

Gastric activity may be analysed in greater detail in animals in which an isolated stomach pouch has been constructed [FIG. 124].

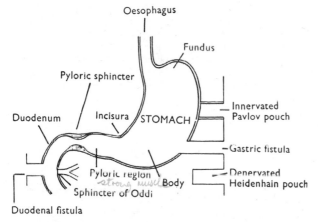

FIG. 124. The pouches and fistulae associated with the study of digestive secretions of the stomach and duodenum. The food enters the stomach from the oesophagus. It is passed on, a small quantity at a time, through the pyloric sphincter into the duodenum, by peristaltic waves which pass over the pyloric region. Stomach pouches enable gastric juice to be collected which is uncontaminated with food. A gastric fistula enables the gastric mucosa to be inspected. A duodenal fistula enables the pancreatic duct to be cannulated.

This has the advantage that gastric juice may be collected from this pouch in response to food in the main stomach without being contaminated with the food. Such a pouch may be denervated (Heidenhain pouch) or have its nerve supply intact (Pavlov pouch).

Sham feeding is a method whereby the effect of swallowing on the secretion of gastric juice may be studied without the food reaching the stomach. The oesophagus is brought to the surface and the swallowed food leaves via this oesophagostomy.

Sham feeding associated with a stomach pouch demonstrates that the presence of food in the mouth and the act of swallowing bring about gastric secretion. This flow of gastric juice is seen only in the innervated pouch and is mediated by the vagus nerve. It is known as the primary or nervous phase of gastric secretion.

The simple experiment of studying the effect of electrical stimulation of the vagus nerve on gastric secretion in an animal experiment is complicated by the fact that vagal stimulation slows the heart and the blood pressure falls. However, if the vagus nerve is cut, the inhibitory fibres to the heart degenerate before those running to the stomach. After 48 hours the cardio-inhibitor fibres have degenerated and vagal stimulation then produces a flow of gastric juice rich in pepsin.

Another way of producing stimulation of the vagal fibres to the stomach, which is applicable to man, is to lower the blood glucose level. This may be achieved by an injection of insulin [p. 168]. The lowered blood sugar level stimulates the vagal nucleus in the medulla. It produces no response after the vagus nerve has been cut.

If the rate of production of gastric juice is studied after sham feeding, it is seen that the flow starts shortly after the food enters the mouth, reaches a maximum after half an hour and then declines [see FIG. 125, vagus curve] whereas after a normal meal the secretion continues for 3–4 hours. Furthermore the denervated pouch, which does not secrete in sham feeding, secretes when a normal meal enters the stomach. Thus some other mechanism, other than a nervous one, exists for the production of gastric juice.

This secondary phase of gastric juice secretion is brought about by the release of the humoral agent *gastrin* in response to both the presence of food in the stomach and duodenum and the vagal activity. Gastrin is absorbed into the blood and stimulates the glands of the stomach (gastric glands) via the circulation. This humoral phase reaches a maximum about 1½ hours after the food is taken and lasts for 3–4 hours [FIG. 125, humoral curve]. In addition the mechanical

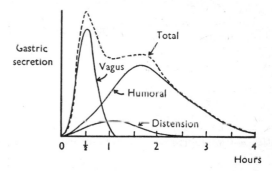

FIG. 125. The phases of gastric secretion. The nervous phase (vagus) reaches a maximum in ½ hour and then declines. The humoral phase (gastrin) reaches a maximum after 1½ hours and persists for 3–4 hours. There is also a small production of juice due to the distension of the stomach by the food.

presence of food in the stomach further stimulates the production of gastric juice. The summation of the secretions produced by these three phases gives the total gastric response to a meal [FIG. 125, dotted curve].

Recent work by Gregory and his co-workers has shown that gastrin is a peptide containing 17 amino acids. Gastrin I has the following sequence:

pyr—gly—pro—try—leu—glu—glu—glu—glu—glu—ala—
               tyr—gly—try—met—asp—phe NH_2

Gastrin II has the same sequence but with a sulphate group on the tyrosine.

It appears that the terminal tetra-peptide amide sequence . . . try—met—asp—phe NH_2 is the important part of the molecule.

A synthetic penta-peptide known as *pentagastrin* has a gastrin-like activity, and is used to stimulate gastric secretion.

Enterogastrone is released by the stomach and small intestine when fat is present in the food. This humoral transmitter circulates in the blood and inhibits the activity of the stomach. It decreases the movements of the stomach and delays emptying.

The stomach produces the intrinsic factor which facilitates the absorption of vitamin B_{12} and thereby prevents pernicious anaemia [p. 10]. The intrinsic factor is a protein released by the main gastric glands. An absence of the intrinsic factor from the gastric juice is associated with a concurrent absence of hydrochloric acid (achlorhydria).

The stomach mucosa is protected from the digestive action of pepsin by the secretion of mucus. Should erosion of the stomach or first part of the duodenum take place a *peptic ulcer* is formed. If the peptic ulcer is in the stomach it is termed a *gastric ulcer*. If it is in the duodenum it is termed a *duodenal ulcer*. Such ulcers are treated by making sure that the stomach always contains food (feeding every 2 hours), by neutralizing the acid with weak alkalis such as aluminium hydroxide and magnesium trisilicate, and by vagotomy. The operation of vagotomy consists of cutting the vagal nerve fibres as they pass through the diaphragm. This has the effect of reducing the pepsin content of the digestive juice.

With the possible exception of some water and alcohol, very little absorption takes place in the stomach.

Fractional Test Meal

In order to test the production of gastric juice (including the hydrochloric acid) and the rate of emptying of the stomach, a fractional test meal is given. The procedure is briefly as follows.

A stomach tube is passed and the stomach emptied. A stimulus is then given to cause the secretion of gastric juice. Samples of gastric juice (usually 10 ml.) are withdrawn at 15 minute intervals over the next 2 hours and the juice is analysed.

The various stimuli used are:

(a) 1 pint of thin gruel
(b) 50 ml. 7 per cent. alcohol
(c) a subcutaneous injection of histamine (0·04 mg./kg.)
(d) a deep intramuscular injection of pentagastrin (0·006 mg./kg.)
(e) an injection of insulin (0·1 Units/kg.)

The histamine acts on the oxyntic cells and causes the secretion of a highly acid juice, but this juice is poor in enzymes. To minimize the side-effects of histamine [p. 44] an antihistamine such as *mepyramine maleate* is injected intramuscularly 30 minutes before the histamine.

Pentagastrin has a quicker action than histamine and gastric juice is collected every 10 minutes for 40 minutes.

The insulin produces a profound fall in blood glucose (hypoglycaemia) which stimulates the vagus and produces a juice rich in enzymes.

It has no effect after vagotomy, and this fact may be used to test the completeness of such a vagotomy operation.

For the maximum stimulation of the gastric glands histamine and insulin are used simultaneously.

The hydrochloric acid in the samples is partly in the form of free acid and partly combined with the proteins which will be acting as bases in this acid medium [p. 92]. The hydrochloric acid is estimated by neutralization with sodium hydroxide using two indicators in succession. Töpfer's reagent (Methyl Yellow) changes colour between pH 3 and 4. The amount of sodium hydroxide added when this colour change takes place is used to calculate the amount of free hydrochloric acid present. Phenolphthalein is added and further sodium hydroxide is added until the solution becomes pink (pH 8·5). This gives the hydrochloric acid combined with protein.

In achlorhydria no titratable acid is produced in response to histamine or pentagastrin. It occurs in association with pernicious anaemia.

Pancreatic Juice (Pancreas)

The pancreas has a microscopical structure similar to that of a salivary gland. Its duct enters the duodenum with the common bile-duct at the sphincter of Oddi.

Pancreatic juice contains trypsinogen, chymotrypsinogen, amylase, maltase and lipase. The trypsinogen is activated by enterokinase which is released by the duodenum and small intestine to form trypsin. The chymotrypsinogen is converted to chymotrypsin by trypsin. These powerful proteolytic enzymes continue the breakdown of the ingested protein to the polypeptide stage (short amino acid chains).

The pancreatic juice contains sodium bicarbonate which neutralizes the hydrochloric acid present in the chyme when it enters the duodenum from the stomach. As a result the contents of the small intestine are approximately neutral.

Pancreatic juice in an animal is studied by the direct cannulation of the pancreatic duct through a tube inserted into the duodenum which leads to the surface of the abdominal wall [FIG. 124].

In man the duodenum around the entry of the pancreatic duct may be isolated by two inflatable balloons and the juice aspirated from this segment. A three lumen catheter is swallowed and manœuvred through the pyloric sphincter into the duodenum [FIG. 126]. Such pancreatic juice will be contaminated with bile and duodenal juice.

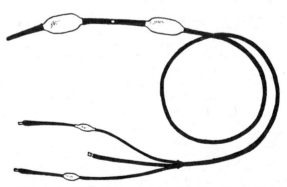

FIG. 126. A double balloon tube for collecting duodenal juice. The tube has three lumens, two for inflating the balloons and the third for aspirating the juice. The two balloons isolate a region of the duodenum around the entrance of the pancreatic duct.

Such studies have shown that there is a small continuous secretion under fasting conditions, but that the flow increases markedly a few minutes after taking food. This is a vagal reflex. Stimulation of the vagus nerve produces a small quantity of viscid pancreatic juice, rich in enzymes, and is associated with the disappearance of granules from the secreting cells. But the vagal control of the pancreas is relatively unimportant compared with the humoral control.

The presence of food in the small intestine releases the humoral agents *secretin* and *pancreozymin* which stimulate pancreatic secretion via the blood stream.

Bayliss and Starling (1902) demonstrated that an extract of the intestinal mucosa stimulated pancreatic secretion when injected into the circulation. They introduced the name *secretin*. The pancreatic juice produced by secretin is poor in enzymes. A second humoral agent *pancreozymin* was discovered by Harper and Raper in 1943. This substance produces a flow of pancreatic juice which is rich in enzymes similar to that produced by vagal stimulation.

Secretin and pancreozymin also play a part in the release of glucagon and insulin from the pancreas [p. 168].

Tests of Pancreatic Function

Starch Tolerance Test

Ingested starch is broken down mainly by pancreatic amylase. A deficiency of this enzyme may be detected by comparing the increase in blood sugar obtained by the ingestion of 50 grams of glucose and 50 grams of starch. Blood samples are taken every 20 minutes for 2 hours as in a glucose tolerance test.

If the pancreatic amylase production is normal, the maximum blood glucose rise obtained with the starch should be at least half that obtained with the glucose.

Duodenal Juice Collection

Using a double balloon tube, the response of the pancreas is tested to intravenous injections of firstly secretin (1·5 units/kg.) and secondly pancreozymin (1·5 units/kg.). The duodenal juice produced is collected for three 10 minute periods.

After the secretin injection, the volume, pH and bicarbonate content are determined. The volume over the 30 minutes should exceed 1·8 ml./kg. body weight. The maximum bicarbonate level should exceed 80 mEq./litre. After the pancreozymin injection, the trypsin content of the juice should rise.

Bile (Liver)

Bile is a viscous greenish fluid secreted continuously by the liver cells at the rate of 500–1,000 ml. per day. Complete denervation of the liver leaves the bile secretion unaltered.

The bile is stored and concentrated in the gall-bladder. When food, and particularly fat, enters the duodenum, the humoral agent *cholecystokinin* is released into the circulation. On reaching the gall-bladder, it causes a contraction and bile passes down the common bile-duct to enter the duodenum with the pancreatic juice. The gall-bladder itself is not essential to life. The horse and the rat do not have a gall-bladder. It may be removed in man without ill effect.

Cholecystokinin (CCK) and pancreozymin (PZ) have been shown to be identical as a 33 amino acid peptide. They are now often referred to collectively as: *cholecystokinin-pancreozymin* (CCK-PZ).

Bile Salts

The bile salts are the sodium salts of glycocholic and taurocholic acids. They are formed in the liver by the conjugation of glycine and taurine with cholic acid and related acids (lithocholic and desoxycholic acids).

These bile salts react with substances like fatty acids, cholesterol and the fat-soluble vitamins to form choleic acid complexes which are soluble in water. In addition, the bile salts reduce the surface tension and with the aid of lipase form a stable emulsion with the ingested fat.

Very little of the bile salts is lost in the faeces. They are re-absorbed in the ileum by a special transport mechanism, and returned via the portal vein to the liver for re-secretion about six times a day. The small loss that occurs is made good by synthesis in the liver cells.

$$CH_2 \cdot NH_2 \cdot COOH \quad \text{(Glycine)}$$

$$\begin{array}{c} CH_2 \cdot SO_3H \\ | \\ CH_2 \cdot NH_2 \end{array} \quad \text{(Taurine)}$$

Cholic acid

Substances which increase the secretion of bile are termed *choleretics*. Bile salts taken by mouth act in this way. Substances which empty the gall-bladder are termed *cholagogues*.

Bile is the route by which cholesterol, bile pigments, and certain drugs such as phenolphthalein, are excreted from the body.

Small Intestine (Succus Entericus)

In the small intestine a series of enzymes both in the lumen and in the mucosal cells complete the digestion of the food. These include *carboxypeptidases* which split off amino acids from the —COOH (carboxyl) end, *aminopeptidases* which split them off from the other —NH_2 (amino) end of the peptide chain, and *dipeptidases* which split dipeptides (two amino acid units). Collectively these enzymes are termed *erepsin*.

Erepsin converts the polypeptides to amino acids; sucrase, maltase and lactase convert the disaccharides to monosaccharides.

The flow of intestinal juice (succus entericus) is slight during the first 2 hours following a meal, but it increases markedly during the 3rd hour. Mechanical stimulation of the intestinal wall by food is probably the main stimulus for the secretory activity.

A large amount of water is absorbed in the small intestine. In addition to ingested fluids, five litres of digestive juice are produced each day by the salivary glands, stomach, pancreas and liver. Only 500 ml. of ileal content enters the caecum, the remaining 4,500 ml. of water has therefore been reabsorbed.

Local Nerve Plexuses

The wall of the small intestine consists of an inner coat or mucosa surrounded by the submucosa, a muscle coat and on the outermost surface the peritoneum. The muscle has an inner circular layer and an outer longitudinal layer. Meissner's plexus is found in the submucosa. Auerbach's plexus is found between the circular and longitudinal muscle layers. These are nerve plexuses of the parasympathetic (vagus and sacral parasympathetic) and sympathetic nervous systems.

Although nervous mechanisms play little part in the secretion of the succus entericus, they are important in regulating the speed of movement of the food along the intestine.

Stimulation of the vagus (parasympathetic) increases the peristaltic movement of the small intestine and increases the tone of the intestinal wall, whereas stimulation of the splanchnic nerves (sympathetic) reduces peristalsis and tone. Peristalsis continues if all the nerves are cut, but ceases if the nerve plexuses are paralysed. Furthermore if a length of intestine is resected and replaced in reverse, there is a complete blockage at this point.

Movements of the intestine may be studied by direct inspection during abdominal operations, and by X-ray examinations following a barium meal. More recently the progress of food along the small intestine has been studied by radio pills. This consists of a small radio-transmitter with enclosed batteries which is small enough to be swallowed. By incorporating sensing elements into the transmitter, information with regard to pressure changes, temperature changes etc., may be recorded from the pill as it passes along the digestive tract.

Large Intestine

The undigested food residue passes through the ileocaecal valve into the large intestine where there is a further absorption of water.

The gastro-colic reflex, due to the entry of food into the stomach causes peristalsis throughout the large intestine which moves the contents on to the rectum. It is the distension of the rectum that gives the desire for defaecation. At other times the rectum is empty, as retrograde peristalsis returns any unexcreted faeces back from the rectum to the descending colon.

The faeces contain the undigested food residues. Substances such as cellulose (a carbohydrate found in the cell membranes of plants; also the main constituent of paper and cotton wool) and keratin and other substances are not digested because the appropriate enzymes are absent from the digestive tract. Fruits and green vegetables contain a high proportion of cellulose. Certain inorganic radicals such as sulphates and certain salts such as those of magnesium are poorly absorbed and are excreted in the faeces. Hence ingested magnesium sulphate (Epsom salts) will appear in the faeces; it will retain water and increase the bulk of the faeces. This causes it to act as a laxative.

The faeces will contain the insoluble compounds taken in by mouth such as barium sulphate for an X-ray examination. The formation of insoluble calcium salts (calcium phosphate, calcium phytate) and insoluble iron salts (iron phosphate) may lead to poor absorption of these metallic ions and their loss in the faeces.

About half the weight of the faeces consists of bacteria (mostly dead) which have grown in the digestive tract. The faeces also contain desquamated epithelial cells from the intestinal mucosa, mucus, cholesterol and the bile-pigment derivatives (stercobilinogen and stercobilin). Faeces are still produced during starvation when no food is eaten.

Defaecation

The digestive tract is once again under voluntary nervous control at its caudal end. The anal canal is closed by two sphincters. The internal sphincter is composed of smooth muscle which is supplied by the autonomic nervous system. The sacral parasympathetic fibres (S.2, 3, 4) cause relaxation of the sphincter whilst the sympathetic fibres (L.1, 2) cause constriction of the sphincter. This sphincter is normally closed.

The external sphincter is composed of striated muscle and is controlled by the somatic (voluntary) nervous system. The nerve supply is the pudendal nerve which has the same nerve roots as the sacral parasympathetic (S.2, 3, 4).

In the new-born baby, distension of the rectum by faeces brings about a spinal reflex which results in the relaxation of the internal sphincter, contraction of the colon and rectum and defaecation. During the first few years of life tone develops in the external sphincter and the reflex is brought under the control of the higher centres. Defaecation then only takes place when the external voluntary sphincter is relaxed. Following spinal cord or brain injury the voluntary nature of defaecation may be lost and it reverts to being a spinal reflex again.

The expulsion of faeces from the rectum is aided by an increased abdominal pressure. To achieve this the diaphragm and abdominal muscles are contracted. In addition, a forced expiration may be made against a closed glottis (cf. Valsalva manœuvre, page 48) to increase the thoracic pressure and force the diaphragm downwards. This is termed straining at stool.

Vomiting

Vomiting is usually preceded by a feeling of nausea, excessive salivation, pallor and sweating. The stomach is divided into two by a contraction band at the region of the incisura [FIG. 124]. A deep inspiration is taken, and then with a strong contraction of the diaphragm and abdominal muscles the stomach contents are regurgitated up the oesophagus.

Vomiting is brought about as the result of stimulation of the vomiting centre in the medulla. This stimulation may be brought about in many different ways. An injection of the drug apomorphine will produce vomiting by its action on the vomiting centre. Other ways are: tickling the back of the throat, irritating the stomach with, for example, mustard and water, and stimulation of the vestibular mechanism and the eyes by a rocking motion (sea-sickness). Vomiting may also be produced by disorders of the cerebral cortex and the cerebellum.

REFERENCES AND FURTHER READING

Bayliss, W. B., and Starling, E. H. (1902) *J. Physiol.* (*Lond.*), **28,** 325.

Beaumont, W. (1833) *Experiments and Observations on the Gastric Juice, and the Physiology of Digestion*, Plattsburgh.

Borrison, H. L., and Wang, S. C. (1953) Physiology and pharmacology of vomiting, *Pharmacol. Rev.*, **5,** 193.

Davenport, H. W. (1959) Digestive system, *Ann. Rev. Physiol.*, **21,** 183.

Dixon, M., and Webb, E. C. (1958) *Enzymes*, London.

Grace, W. J., Wolf, S., and Wolff, H. G. (1951) *The Human Colon*, London.

Gregory, R. A. (1974) The gastrointestinal hormones. A review of recent advances, *J. Physiol.* (*Lond.*), **241,** 1.

Harper, A. A., and Raper, H. S. (1943) Pancreozymin, *J. Physiol.* (*Lond.*), **102,** 115.

Haslewood, G. A. D. (1955) Bile salts, *Physiol. Rev.*, **35,** 178.

Hunt, J. N. (1959) Gastric emptying and secretion in man, *Physiol. Rev.*, **39,** 491.

Ivy, A. C., Grossman, M. I., and Bachrach, W. H. (1951) *Peptic Ulcer*, New York.

James, A. H. (1957) *The Physiology of Gastric Digestion*, London.

Kahlson, G. (1948) The nervous and humoral control of gastric secretion, *Brit. med. J.*, **2,** 1091.

Wolf, S., and Wolff, H. G. (1947) *Human Gastric Function*, New York.

FILMS

Gastric Secretion, I.C.I., Film Library, London.
Pancreatic Secretion, I.C.I., Film Library, London.

10. METABOLIC PATHWAYS

Production of Heat and Energy

Heat and energy are produced in the body by the oxidation of the carbon and hydrogen in the food to carbon dioxide and water.

Carbon + oxygen → heat and energy + carbon dioxide
Hydrogen + oxygen → heat and energy + water

This oxidation is brought about in a series of steps each involving a specific enzyme aided in many cases by a co-enzyme.

When the energy is not required immediately as heat, it is stored in high energy molecules. Two such molecules are adenosine triphosphate (ATP) and creatine phosphate.[10]

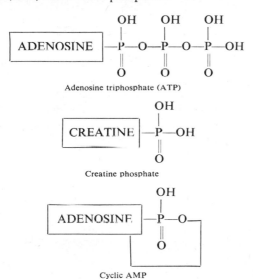

Adenosine triphosphate (ATP)

Creatine phosphate

Cyclic AMP

Although a very large number of different substances may be eaten as food, the process of digestion reduces the number of metabolic pathways needed for their oxidation. By the process of digestion all the starches and sugars are reduced to glucose or closely related monosaccharides. The thousands of different protein molecules eaten each day are reduced to relatively few amino acids. The fat in the diet is not changed chemically by digestion, but it only consists of glycerol combined with long-chain fatty acids. There are about ten different but closely related fatty acids in the ingested fat.

A further convergence of the metabolic pathways occurs in the cells of the body. Most carbohydrates, fats and amino acids not needed for the synthesis of body protein are converted into an active form of acetic acid known as *acetyl co-A* before being completely oxidized to carbon dioxide and water.

[10] Substances such as dinitrophenol allow the energy to be lost as heat instead of being converted to the high energy molecules. They are known as *uncoupling agents*.

Acetyl co-A = acetyl co-enzyme A
= $CH_3 \cdot CO—S \cdot co-A$
= $CH_3 \cdot CO—S \cdot CH_2CH_2NH$-pantothenic acid diphosphate-ribose phosphate-adenine

It is readily converted to acetic acid.

Hydrolysis and Phosphorolysis

The enzymes acting in the digestive tract break down the ingested food by the process of *hydrolysis*. Hydrolysis is the splitting of large molecules with the incorporation of water in the form of —H attached to one fragment and —OH to the other.

For example, the peptide links of the dietary proteins are split thus:

$$R_1—CO—NH—R_2 \xrightarrow{+H_2O} R_1—COOH + R_2—NH_2$$

In the cells the breakdown is often brought about by the process of *phosphorolysis*. The molecule is split with the incorporation of phosphoric acid (H_3PO_4) in the form of —H_2PO_3 attached to one fragment and —OH to the other.

These processes of breakdown are termed *dissimilation* or *catabolism*.

Molecules may be built up (*assimilation* or *anabolism*) by the removal of water or by the removal of phosphoric acid. Glucose is converted first into glucose-6-phosphate and then to glucose-1-phosphate. Glucose-1-phosphate is converted into glycogen or animal starch by the removal of the phosphoric acid molecules. Glycogen is reconverted to the glucose-1-phosphate again by phosphorolysis.

Glycogen (store)
⇅
Glucose-1-phosphate
⇅
Glucose → Glucose-6-phosphate
↓
CO_2 and water,
heat and energy

Conversion of Carbohydrate to Fat

Surplus carbohydrate is converted to fat and laid down in the fat depots. As long ago as 1866 Lawes and Gilbert found that the carcasses of pigs contained more fat than the animals had eaten during their lifetime, and this fat must have been synthesized from non-fat sources.

The conversion of carbohydrate to the fatty acids for the formation of fat is probably along these lines:

Carbohydrate → pyruvic acid
→ acetyl co-A → long chain fatty acid

CARBOHYDRATE METABOLISM

The sugar and starches in the food appear in the blood after digestion and absorption as glucose ($C_6H_{12}O_6$). The normal blood glucose level is 60–100 mg. glucose per 100 ml. of blood (3·3–5·5 mmol/l). With a blood volume of 5 litres, there will thus be 5 g. of glucose in the blood. Glucose diffuses readily into the tissue fluid and into the cells and if we assume a uniform concentration throughout the whole body water of 45 litres, there will be 45 g. of glucose in the intracellular body fluids. *polysacch*

Monos Glucose is stored in the liver and muscles as *glycogen*. Glycogen has a high molecular weight and is insoluble in water. The conversion of glucose into glycogen is facilitated by insulin from the islet cells of the pancreas and is inhibited by pituitary and adrenal cortex hormones.

Glucose insulin → glycogen
– pit. ,
, adrenal cortex

Liver Glycogen

Liver glycogen is readily converted back to glucose. This process is known as *glycogenolysis*. This reconversion of liver glycogen back to blood glucose is stimulated by adrenaline from the adrenal medulla and glucagon from the pancreas.

The liver contains 100 g. of glycogen and this acts as a carbohydrate store to maintain the blood glucose level. It will satisfy the body's requirements for 24 hours.

Nervous tissue has no reserve of carbohydrate (no glycogen store) and depends upon an adequate level of blood glucose. Low blood glucose levels lead to convulsions and coma.

Liver glycogen is also formed by the process of *neoglucogenesis* (gluconeogenesis) from deaminated amino acids, glycerol and other non-carbohydrate sources.

Muscle Glycogen

Skeletal muscle contains 400 g. of glycogen. It is used up during muscular activity and replenished after the exercise from the blood glucose. Muscle glycogen, unlike liver glycogen, is not readily converted to glucose, and it is therefore not readily available as a reserve to maintain the blood glucose level.

Muscle glycogen is, however, broken down to glucose phosphate and then to two molecules of pyruvic acid[11] ($CH_3CO \cdot COOH$) with the release of energy which is stored as high energy ATP molecules (*Embden-Meyerhof glycolytic pathway*).

Under anaerobic conditions (insufficient oxygen) no further breakdown towards carbon dioxide and water occurs. The co-enzyme associated with the formation of the pyruvic acid becomes blocked in the reduced state by the addition of two hydrogen atoms. Under these circumstances pyruvic acid is converted to lactic acid with the removal of the hydrogen from the co-enzyme. This formation of lactic acid is a 'side turning' along the main metabolic pathway. The formation of lactic acid allows more pyruvic acid to be made with the further production of energy.

[11] Pyruvic acid, lactic acid, citric acid and the other acids in these metabolic pathways will be buffered in the cells and blood, and will exist in the form of pyruvate, lactate, citrate, etc. It is important, however, not to forget the hydrogen ions associated with these acids since they are tending to lower the pH of the body (metabolic acidosis, p. 102). For this reason they will be referred to as acids. It should be noted also that when lactate or citrate is given to a patient as, for example, in an intravenous infusion the sodium (or potassium) salt will be used. This will have the opposite effect of pH (will raise it) since this sodium lactate or citrate will be metabolized to sodium bicarbonate [p. 103].

$$CH_3CO \cdot COOH + \text{Co-enzyme 2H}$$
$$\text{Pyruvic acid}$$
$$\rightarrow CH_3CH(OH)COOH + \text{Co-enzyme}$$
$$\text{Lactic acid}$$

The resting blood lactic acid level is 10 mg. per 100 ml. of blood (1·1 mmol/l). In severe exercise it may rise to 200 mg. lactic acid per cent. (22·2 mmol/l). When oxygen becomes available after the exercise, about one-fifth of the lactic acid is oxidized to carbon dioxide and water; the energy resulting is used to convert the remaining lactic acid back to glycogen.

Anaerobic muscular activity leads to an 'oxygen debt' which has to be paid off when the exercise has been completed. Respiratory stimulation maintains an increased pulmonary ventilation until the excess lactic acid has disappeared from the blood.

When adequate oxygen is available (**aerobic metabolism**) the pyruvic acid is converted to carbon dioxide and water with a large release of energy which is stored in the muscle as ATP and creatine phosphate.

The aerobic oxidation of pyruvic acid is in itself complex and involves many enzymes and co-enzymes in a cycle of events known as the Krebs cycle (citric acid, tri-carboxylic acid cycle).

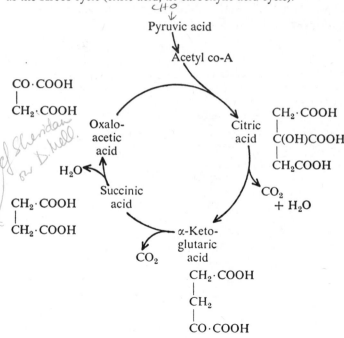

Krebs cycle (simplified)

The pyruvic acid is converted first to acetyl co-A which then combines with the oxaloacetic acid to form citric acid. The conversion of citric acid to α-ketoglutaric acid to succinic acid and back to oxaloacetic acid (via numerous intermediaries) results in the formation of two molecules of carbon dioxide and two molecules of water from each molecule of pyruvic acid. Altogether about forty molecules of high energy ATP are synthesized from the breakdown of every glucose molecule by aerobic dissimilation.

1 g. carbohydrate + 820 ml. O_2 → 17·2 kilojoules or 4·1 Calories

Thus,

1 litre of oxygen used to metabolize carbohydrate → 21 kJ or 5 Cal.

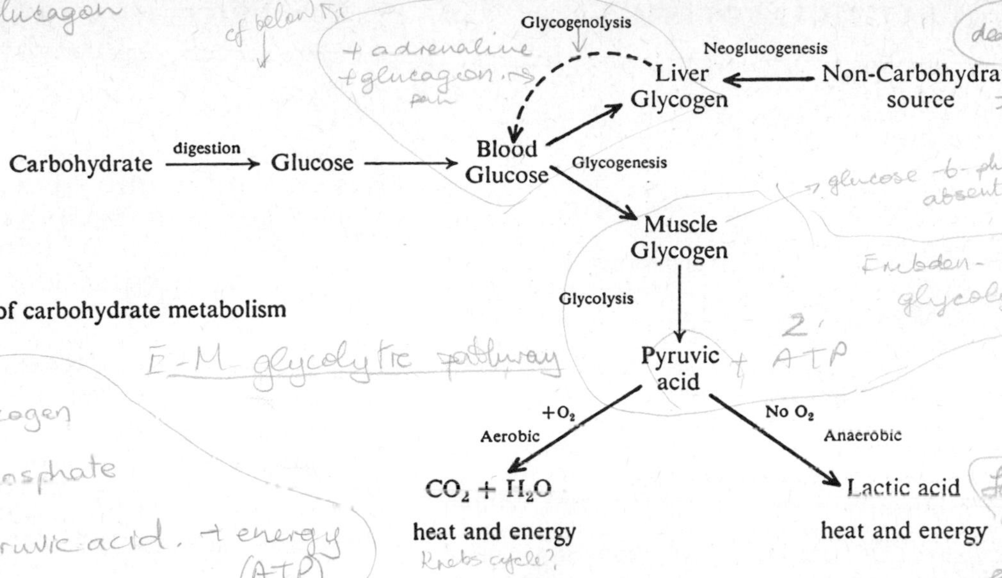

[handwritten: Pancreas — Insulin, Glucagon]

[handwritten annotations around diagram: "of below ?", "+ adrenaline + glucagon → path", "deaminated A.A. Glycerol etc", "glucose-6-phosphatase absent ∴ not ⇌", "Embden–Meyerhof glycolytic pathway", "2 ATP", "Krebs cycle?", "lactic acidosis → lethal caused by phenformin"]

General plan of carbohydrate metabolism (simplified)

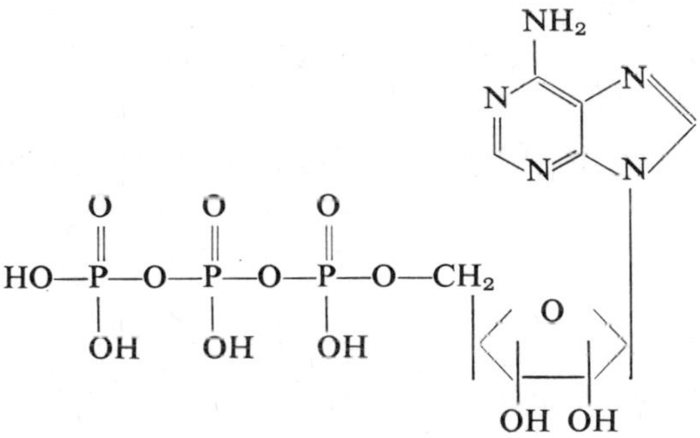

[handwritten: E–M glycolytic pathway]

*[handwritten:
Muscle glycogen
↓
glucose phosphate
↓
2 moles pyruvic acid. + energy (ATP)]*

ATP

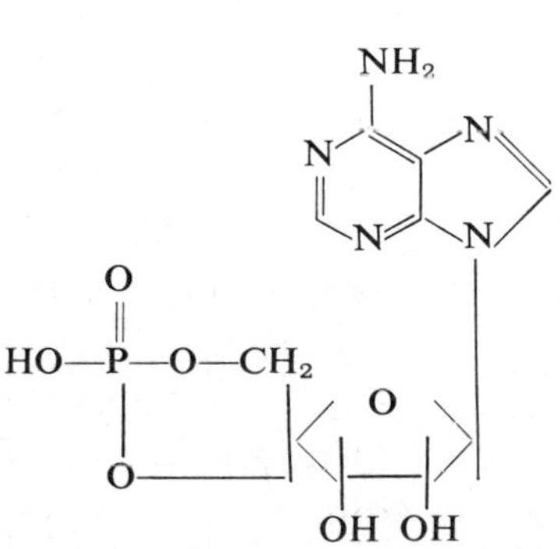

CYCLIC AMP (Cyclic adenylic acid, cyclic 3'-5' adenosine monophosphate)

A Further Consideration of the Breakdown of Liver Glycogen to Blood Glucose

The conversion of the glycogen in liver cells to glucose by adrenaline may be demonstrated using a suspension of liver cells *in vitro*. However, if the liver cells are first broken down, the formation of glucose only occurs if **cyclic-AMP** is added to the suspension. This fact has led to the discovery of the importance of cyclic-AMP as an *enzyme activator* in cells and its possible function as a 'second messenger' or 'intra-cellular hormone' acting as a go-between in the cell between the blood-based hormone and the enzyme system involved.

Cyclic-AMP (cyclic adenosine monophosphate) is a derivative of ATP in which two phosphates are removed, and the remaining one forms a cyclic derivative with the adenosine. The enzyme which brings about this change is termed **adenyl cyclase.**

[handwritten: → glucagon]

$$\text{ATP} \xrightarrow[\text{cyclase}]{\text{adenyl}} \text{CYCLIC AMP}$$ *[handwritten: = enz activator.]*

[handwritten: hyperglycaemia → Pancreas (B cells) (Cyclic AMP) insulin]

The steps in the breakdown of glycogen to glucose by adrenaline and glucagon may be summarized as follows:

1. Adrenaline activates the enzyme adenyl cyclase
2. $\text{ATP} \xrightarrow{\text{adenyl cyclase}}$ cyclic AMP
3. Inactive phosphorylase (b) $\xrightarrow{\text{cyclic AMP}}$ Active phosphorylase (a)
4. Glycogen $\xrightarrow{\text{active phosphorylase}}$ glucose-1-phosphate
5. Glucose-1-phosphate $\xrightarrow[\text{phosphoglucomutase}]{\text{LIVER}}$ glucose-6-phosphate
6. Glucose-6-phosphate $\xrightarrow[\text{glucose-6-phosphatase}]{\text{LIVER}}$ blood glucose and inorganic phosphate *[handwritten: (enz)]*

In skeletal muscle, the enzyme glucose-6-phosphatase is absent, hence muscle glycogen cannot be converted to blood glucose. It can, however, be utilized after the formation of lactic acid.

FAT METABOLISM

fatty acids → acetyl co-a. → Krebs cycle

$$CH_2O \cdot CO \cdot C_{17}H_{35}$$
→ in pairs →
$$CHO \cdot CO \cdot C_{17}H_{35}$$
acetoacetic Acid,
(a Keto acid)
$$CH_2O \cdot CO \cdot C_{17}H_{35}$$

Glycerol tristearate (saturated fat)

After absorption the fat which is not immediately required for heat and energy is deposited as neutral fat in the fat depots. It may be stored in large amounts in the cells of the adipose tissue. The fat depots are situated under the skin and at the back of the abdomen. Some of the fat becomes incorporated into the structure of cells and this fat is not available as a future source of heat and energy.

When required, fat is withdrawn from the fat depots and passes to the liver. Here it is split by the liver lipase to glycerol and fatty acids. The glycerol is oxidized via the carbohydrate pathways. The fatty acids are oxidized to acetyl co-A and either join the Krebs cycle to be completely oxidized, or else join in pairs to form the keto-acid acetoacetic acid (CH_3COCH_2COOH). Acetoacetic acid and other ketone bodies, β-hydroxybutyric acid and acetone, formed by the liver, circulate in the blood and are oxidized by other tissues to carbon dioxide and water plus heat and energy.

Liver

KETONE BODIES

ACETOACETIC ACID	$CH_3CO \cdot CH_2 \cdot COOH$
β-HYDROXYBUTYRIC ACID	$CH_3 \cdot CHOH \cdot CH_2COOH$
ACETONE	$CH_3 \cdot CO \cdot CH_3$

from Krebs cycle

On a high fat, low carbohydrate diet insufficient oxaloacetic acid may be available for the oxidation of the acetyl co-A. The accumulating acetyl co-A forms a surplus of ketone bodies and **ketosis** develops.

This is the modern biochemical explanation of the phrase that fat 'can only be burnt in the fire of carbohydrate'. Ketosis may be induced in a normal person by reducing the carbohydrate intake and increasing the fat considerably. After a few days ketone bodies may be detected in the urine. Severe ketosis is a feature of untreated diabetes mellitus. The carbohydrate metabolism is deranged in this condition by a deficiency of insulin from the pancreas or by an excess of anterior pituitary and adrenal cortical hormones. Blood glucose is not converted to muscle glycogen in sufficient amounts to oxidize the acetyl co-A from the fat. The accumulation of ketone bodies in the blood may lead to coma (a diabetic or ketotic coma).

Ketosis causes an acidaemia. The acetoacetic acid and the β-hydroxybutyric acid are buffered by the bicarbonate/CO_2 buffer system of the blood [p. 102] with a decrease in the plasma bicarbonate and an increase in the pulmonary ventilation. Patients with severe ketoacidosis breathe more rapidly and more deeply and give the impression of having an 'air hunger'.

The acids are excreted in the urine as the ammonium salts of acetoacetic acid and β-hydroxybutyric acid.

1 g. of fat oxidized by 1·98 litres of oxygen

→ 38 kilojoules or 9·3 Calories

Thus,

1 litre of oxygen used to oxidize fat → 20 kJ or 4·7 Cal.

PROTEIN METABOLISM

The amino acids formed by the digestion of dietary protein are:

Glycine	$CH_2(NH_2)COOH$
Alanine	$CH_3 \cdot CH(NH_2)COOH$
Valine	$(CH_3)_2CH \cdot CH(NH_2)COOH$
Leucine	$(CH_3)_2CH \cdot CH_2CH(NH_2)COOH$
Isoleucine	$(CH_3)(C_2H_5)CH(NH_2)COOH$
Cysteine	$HS \cdot CH_2CH(NH_2)COOH$
Cystine	$S \cdot CH_2CH(NH_2)COOH$
	$S \cdot CH_2CH(NH_2)COOH$
Methionine	$CH_3SCH_2CH_2CH(NH_2)COOH$
Serine	$CH_2(OH)CH(NH_2)COOH$
Threonine	$CH_3CH(OH)CH(NH_2)COOH$
Asparagine	$CH_2CO \cdot NH_2$
	$CH(NH_2)COOH$
Aspartic acid	CH_2COOH
	$CH(NH_2)COOH$
Glutamic acid	CH_2CH_2COOH
	$CH(NH_2)COOH$
and its amide glutamine	$CH_2CH_2CONH_2$
	$CH(NH_2)COOH$

Arginine

$$NH$$
$$\|$$
$$NH_2-C-NH \cdot CH_2CH_2CH_2CH(NH_2)COOH$$

Lysine $NH_2CH_2CH_2CH_2CH_2CH(NH_2)COOH$

Histidine

$$CH=\!=\!=C \cdot CH_2CH(NH_2)COOH$$
$$HN \qquad N$$
$$CH$$

Phenylalanine

Tyrosine $HO-$⟨benzene ring⟩$-CH_2CH(NH_2)COOH$

Tryptophan

Proline

$$CH_2- \quad CH_2$$
$$CH_2 \quad CH \cdot COOH$$
$$NH$$

and Hydroxy-proline

$$HO-CH---CH_2$$
$$CH_2 \quad CH \cdot COOH$$
$$NH$$

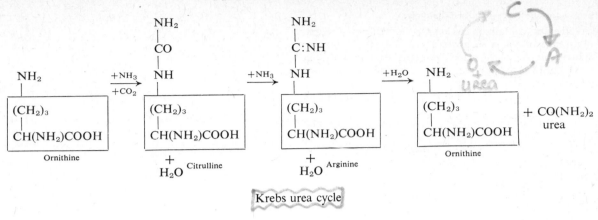

Krebs urea cycle

Aspartic acid and glutamic acid have two COOH groups and only one —NH₂ group. They are acidic substances. Arginine, lysine and histidine are basic substances. All the other amino acids have an equal number of —COOH groups and —NH₂ groups (or their equivalent) and are inherently neutral substances.

These are all α-amino acids, that is, the —NH₂ and —COOH groups are both attached to the same C atom. The peptide chains forming the dietary and body proteins have a helical (spiral) configuration:

$$\cdots\text{NH}\cdot\text{C}\cdot\boxed{\text{CO}\cdot\text{NH}}\cdot\text{C}\cdot\boxed{\text{CO}\cdot\text{NH}}\cdot\text{C}\cdot\boxed{\text{CO}\cdot\text{NH}}\cdot\text{C}\cdot\text{CO}\cdots$$

Peptide link   Peptide link   Peptide link

Amino acids can be interconverted and new amino acids can be formed by the amination of carbohydrate and fat residues. As a result only eight of the above amino acids are 'essential', that is they must be present in the ingested protein. They are indispensable and cannot be synthesized in sufficient quantities to satisfy the body's requirements for growth and repair of tissues.

These are:

Lysine	Isoleucine	Valine	Phenylalanine
Leucine	Threonine	Methionine	Tryptophan

Rose demonstrated in 1950 that it is possible for man to live on a diet in which the only nitrogen compounds were pure synthetic amino acids.

The remaining amino acids are 'unessential' but this is not to say that they are undesirable. Their intake relieves the body of the need to synthesize them.

Amino Acid Pool

Protein is needed for growth and repair of tissues. In children it is needed for the increase in body tissue associated with skeletal growth. In adults, although growth in stature has ceased, there is a continued need for protein for the formation of skin, hair, finger and toe nails as well as for the repair of injured tissues.

The ingested and synthesized amino acids form a pool on which the cells of the body may draw in order to make body protein. In addition, there is a continuous exchange of amino acids in the already formed body protein with the amino acids of the metabolic pool. About 100 g. of body protein are broken down and resynthesized each day.

If the protein intake is insufficient, as in starvation, some of the broken down body protein will not be resynthesized because some of the amino acids released will have been incorporated in more essential molecules such as for the formation of creatine in muscles. Creatine is constantly being lost from the body as creatinine in the urine and has to be continually replaced (see below).

Starvation is thus associated not only with a loss of the carbohydrate and fat stores, the subject becoming thin, but also with a loss of body protein and wasting of the muscles.

Deamination — removing N.

The diet usually contains a surplus of protein and, therefore, of amino acids. These amino acids are deaminated in the liver and elsewhere, thus removing nitrogen from the molecule. The remainder of the molecule consisting only of carbon, hydrogen and oxygen is utilized as a source of heat and energy. For example, the deamination of alanine is associated with oxidation of the molecule and the formation of ammonia:

$$2\,\text{CH}_3\text{CH(NH}_2)\text{COOH} + \text{O}_2 \rightarrow 2\,\text{CH}_3\text{CO}\cdot\text{COOH} + 2\,\text{NH}_3$$

Alanine   Pyruvic acid   Ammonia

The ammonia formed may be used for the synthesis of new amino acids. The surplus is converted into urea and excreted in the urine by the kidneys.

$$\text{carbon dioxide} + \text{ammonia} \rightarrow \text{urea} + \text{water}$$
$$\text{CO}_2 + 2\,\text{NH}_3 \rightarrow \text{CO(NH}_2)_2 + \text{H}_2\text{O}$$

Urea formation occurs only in the liver. The ammonia and carbon dioxide are converted into urea in the ornithine, arginine cycle (Krebs urea cycle):

$$\text{Ornithine} \xrightarrow[-\text{H}_2\text{O}]{\substack{+\text{NH}_3 \\ +\text{CO}_2}} \text{Citrulline} \xrightarrow[-\text{H}_2\text{O}]{+\text{NH}_3} \text{Arginine} \xrightarrow{+\text{H}_2\text{O}} \text{Urea} + \text{Ornithine}$$

Neither the amino acid ornithine nor the amino acid citrulline (see top of page) are normal constituents of the dietary protein. They are synthesized in the body from glutamic and aspartic acids.

After deamination

1 g. of protein → 17·2 kJ or 4·1 Calories

Other Pathways for the Excretion of Nitrogen

The dietary proteins contain on the average 16 per cent. of nitrogen. A daily intake of 100 g. of protein will yield 16 g. of waste nitrogen which is excreted in the urine. The daily excretion of 30 g. of urea accounts for 14 g. of this nitrogen. The remainder is excreted as uric acid, creatinine and ammonium ions.

It will be noted that urea contains just slightly less than half its weight of nitrogen.

If the daily excretion of nitrogen exactly equals the daily nitrogen intake the subject is said to be in a state of **nitrogen equilibrium.** If the excretion exceeds the intake so that protein is being lost by the body, the subject is said to be in a **negative nitrogen balance.** If the intake exceeds the loss and protein is being laid down the subject is in a **positive nitrogen balance.**

Creatine is a compound of guanidine and acetic acid. It is methyl-guanidino-acetic acid.

Creatine is formed in the liver from amino acids, arginine, methionine and glycine.

The normal blood level is 0·5 mg. creatine per 100 ml. of blood (38 μmol/l). Creatine passes via the blood to the muscles where it is phosphorylated to give creatine phosphate. This creatine phosphate is continuously changing into creatinine (the anhydride of creatine)

$\sim$ = high energy bond

and phosphoric acid. The creatinine level in the blood is 1 mg. per 100 ml. of blood (90 μmol/l). It is excreted in the urine, and this represents a loss of amino acids and nitrogen to the body. The amount excreted depends on the size of the creatine phosphate store which is related to the muscle mass.

About 2 per cent. of the body creatine is lost in this way each day.

Purine Compounds and Nucleic Acids

In addition to the amino acids in the proteins, the diet contains another source of nitrogen in the purine and pyrimidine compounds. Purine consists of a six-membered pyrimidine ring fused to a five-membered iminazole ring. Each ring is made up of carbon atoms and two nitrogen atoms.

Purine

6:-NH$_2$ is adenine (6 aminopurine)
2:-NH$_2$; 6: = 0 is guanine (2 amino 6 oxypurine)

Pyrimidine

2: 0; 6: = 0 is uracil (2:6 dioxy-pyrimidine)
2: = 0; 6: —NH$_2$ is cytosine (6 amino 2 oxypyrimidine)
2: = 0; 5: —CH$_3$; 6: = 0 is thymine (5 methyl 2,6 dioxypyrimidine)

A purine derivative combined with the 5-carbon sugar ribose ($C_5H_{10}O_5$) forms a *nucleoside*. The phosphate of this compound is termed a *nucleotide*.

Adenine is the 6-amino derivative of purine and when combined with ribose forms the compound *adenosine* which is a nucleoside.

Adenosine with two phosphate radicals forms adenosine diphosphate (ADP), and with three phosphate radicals forms the high energy compound adenosine triphosphate (ATP). These are both nucleotides.

Adenine
(6-amino purine)

Polymers of ribose nucleotides with *adenine* (A) and related bases *uracil* (U), *guanine* (G) and *cytosine* (C) form ribonucleic acid (RNA). Polymers of a slightly different sugar, deoxyribose, and the same bases, except that *thymine* (T) replaces uracil, form deoxyribonucleic acid (DNA). The DNA in the nucleus of cells forms the **genetic code** for the transmission of hereditary factors. The DNA is believed to exist as two helical chains linked by hydrogen bonds between bases in the adjacent chains (Watson-Crick hypothesis). Since links occur only between adenine and thymine, and between guanine and cytosine, the sequence of bases in one chain will determine the sequence of bases in the other. Such an arrangement will also allow one chain to act as a template for the reduplication of the other after splitting.

It is the arrangement of the four bases in sequence in this extremely long molecule that enables characteristics to be transmitted to the daughter cells. The code for the synthesis of protein appears to be a triplet code, that is, there is a sequence of three bases (codon) for each of the twenty amino acids used in the formation of body protein. Some amino acids are coded by more than one triplet. The appropriate part of the DNA code is thought to be transferred first to messenger ribonucleic acid (mRNA). The messenger RNA passes out of the nucleus and through the cytoplasm to the ribosomes where it acts as a template for the linking up of amino acids to form the protein molecule. Smaller transfer RNA (tRNA) molecules, specific to each amino acid, are thought to 'read' off the code in sequence and to attach amino acids in their correct order.

It should be remembered that the long helical chains of DNA and RNA are made up of **pentose sugar** molecules connected together by phosphate groups. They should not be confused with proteins which consist of long helical chains of **amino acids** [p. 115].

Phosphate   Phosphate   Phosphate

Pentose Sugar   **Pentose Sugar**   **Pentose Sugar**   **Pentose Sugar**

Base   Base   Base   Base

Nucleic acid basic structure (DNA and RNA)

The pentose sugars, ribose and deoxyribose, are made from glucose by the pentose phosphate (Warburg-Lipman-Dickens) pathway. In this 'pentose shunt' pathway glucose 6 phosphate is converted to 6-phosphoglucuronic acid and then to pentose sugars, instead of to pyruvic acid, acetyl co-A and the Krebs cycle as in the glycolytic (Embden-Meyerhof) pathway on page 112. The pentose shunt pathway is important for the formation of DNA and RNA. The degradation may proceed further with the formation of heat and energy. It provides the cells with a subsidiary route for the utilization of glucose.

Purine Metabolism

Cells in the food which are rich in nuclei, such as yeast, liver and kidney, will yield nucleotides and nucleosides which will be broken down to purine compounds. The caffeine and theobromine found in tea, coffee and cocoa are purine derivatives. All these purine compounds are oxidized to hydroxy-purine. The 6-hydroxy compound is called *hypoxanthine*. The 2:6 dihydroxy compound is *xanthine* and the final oxidation product of all these compounds in man is the 2:6:8 trihydroxy-purine which is better known as *uric acid*.

Uric Acid

Uric acid was discovered as a chemical substance in urine by Scheele in 1776. It will be noted that although it is an organic acid the molecule contains no —COOH group. It exists in two tautomeric forms due to the presence of:

$$—NH \cdot CO— \rightleftharpoons —N:C(OH)—$$

The formula of uric acid is:

Uric acid (2:6:8 trihydroxy-purine)

Uric acid not only arises from the breakdown products of nucleic acids and ingested purine derivatives, some is synthesized from glycine.

Ammonium urate is the chief mode of excretion of nitrogen in birds, reptiles and insects. In most animals, including the dog, uric acid is oxidized to *allantoin* and, therefore, uric acid is not present in the urine. The exception is the Dalmatian coach dog which excretes the uric acid formed. In man most of the uric acid formed is excreted as such.

The normal blood uric acid level is 3 mg. per 100 ml. of blood (180 μmol/l). The daily output in the urine is 1 g. per day.

An excess of uric acid in the blood leads to its deposition in the joints (as sodium mono-urate) with recurrent attacks of acute joint pain which is commonly known as gout. In addition, urates are deposited under the skin forming swellings which are known as gouty *tophi*. It is a metabolic disorder associated with an increased production of uric acid from the amino acids.

Metabolism of Sulphur Compounds

The amino acids methionine, cysteine and cystine contain sulphur. This sulphur is oxidized to sulphate and is excreted as sodium, potassium and ammonium sulphate in the urine.

A high protein diet will give not only a high urea excretion per day but also a high sulphate excretion.

Phenols and similar organic compounds are detoxicated in the liver by the formation of organic sulphates (ethereal sulphates) which are excreted in the urine.

$$R \cdot OH \xrightarrow{+H_2SO_4} R \cdot O—SO_3 \cdot H + H_2O$$
Phenol   Ethereal sulphate

Aspirin is excreted in this form.

REFERENCES AND FURTHER READING

Baldwin, E. (1967) *Dynamic Aspects of Biochemistry*, 5th ed., Cambridge.

Baron, D. N. (1969) *A Short Textbook of Chemical Pathology*, 2nd ed., London.

Bayliss, L. E. (1959) *Principles of General Physiology*, Vol. I, 5th ed., London.

Carter, C. W., Coxon, R. V., Parsons, D. S., and Thompson, R. H. S. (1959) *Biochemistry in Relation to Medicine*, 3rd ed., London.

Clark, B. F. C., and Marcker, K. A. (1968) How proteins start, *Scientific American*, **218**, No. 1, 36.

Davidson, J. N. (1969) *The Biochemistry of Nucleic Acids*, 6th ed., London.

Dixon, M., and Webb, E. C. (1964) *Enzymes*, 2nd ed., London.

Duncan, G. G. (Ed.) (1969) *Diseases of Metabolism*, 6th ed., Philadelphia.

Fruton, J. S., and Simmonds, S. (1958) *General Biochemistry*, 2nd ed., New York.

Greenberg, D. M. (Ed.) (1969) *Metabolic Pathways*, 3rd ed., New York.

Krebs, H. A. (1943) The intermediary stages in the biological oxidation of carbohydrate, *Advanc. Enzymol.*, **3**, 191.

Lawes, J. B., and Gilbert, J. H. (1866) On the sources of the fat of the animal body, *Phil. Mag.*, (4) **32**, 439.

Lovern, J. A. (1957) *The Chemistry of Lipids of Biochemical Significance*, 2nd ed., London.

Rose, W. C. (1950) The amino acid requirements of man, *J. biol. Chem.*, **182,** 541.

Thorpe, W. V. (1970) *Biochemistry for Medical Students*, 9th ed., London.

Watson, J. D., and Crick, F. H. C. (1953) A structure for deoxyribose nucleic acid, *Nature (Lond.)*, **171,** 737.

Whipple, G. H. (1956) *The Dynamic Equilibrium of the Body Proteins*, Springfield.

11. NUTRITIONAL REQUIREMENTS

Having dealt with the metabolic pathways of the individual food constituents, we now have to consider the body as a whole.

Each day the diet must supply the body with:

1. Adequate Calories.
2. Vitamins.
3. Mineral salts.
4. Water.

These will be considered in turn.

It will be noticed that no special mention has been made of protein for growth and repair of tissues. Under normal circumstances, if the total Calories provided by the diet are adequate so also will be the protein intake.

CALORIES

The Fuel Requirements of the Human Machine

The total heat and energy requirements of man have been studied in a respiration chamber. The Atwater–Benedict *respiration calorimeter* is shown in FIGURE 127. In such a chamber a man can live

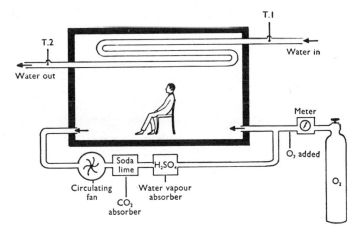

FIG. 127. The Atwater–Benedict chamber for the determination of metabolism in man by direct calorimetry. The heat given off by the subject raises the temperature of the water in the cooling tubes and the heat evolved is calculated from this temperature rise (T.2–T.1) and the rate of water flow.

for an extended period of time and his total energy production measured as the heat gained by the circulating water. As the oxygen is used up, fresh oxygen is supplied from the cylinder. The water vapour and carbon dioxide expired are removed by sulphuric acid and soda lime. The chamber is maintained at a constant temperature by passing water through cooling coils. The heat produced by the subject is calculated from the difference between the inflow and outflow temperatures of the water and the rate of water flow.

Heat and Work Done

Even the mechanical work done such as riding a stationary bicycle [FIG. 128] results in heat.

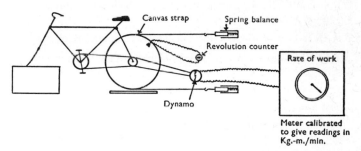

FIG. 128. The bicycle ergometer. This enables standardized exercise to be carried out without moving. A brake-band is applied to the rear wheel and the work done calculated from the tension difference between the two sides of the bands and the linear velocity of the circumference of the wheel. With a standardized tension difference, the work done depends on the rate of pedalling which is determined by the revolution counter attached to the rear wheel. A modification consists of a dynamo in place of the brake-band. The work done is calculated from the electrical power produced. A meter, calibrated in kg. m per minute, is placed above the front of the bicycle so that it may be seen by the subject who maintains a constant work output.

Heat and mechanical work are both forms of energy, and for all practical purposes we can consider that all external work done ultimately results in the production of heat. It is convenient to express all forms of energy either in heat units or work units (kilojoules).

The heat unit employed is the large Calorie (spelt with a capital C) and is the amount of heat required to raise the temperature of 1 kg. of water by 1 °C. It is equal to 1,000 calories (spelt with a small c) which is the amount of heat required to raise only 1 g. of water by 1 °C.

$$1 \text{ Calorie} = 1 \text{ kcal.} = 1{,}000 \text{ calories}$$

As long ago as 1798 Count Rumford noted that heat was developed during the boring of a cannon in Munich. Davy in 1799 showed that ice could be melted by rubbing two pieces together.

Mayer in 1842 suspected that there was a connexion between heat and work in humans when he noticed venous blood was brighter in the tropics than elsewhere. He argued that the heat gain was greater in the tropics than in colder regions and that less oxygen was used up as the blood flowed through the tissues.

It was Joule (1848) who established conclusively that mechanical work and heat were interchangeable and determined the mechanical equivalent of heat, thus giving us the formula:

$$W = K.H$$

where W is the work done, and H is the heat produced and K is the constant, the mechanical equivalent of heat.

If W is in joules and H is in Calories

$$K = 4 \cdot 186 \times 10^3 \text{ joules } Cal^{-1}$$

Thus, 4·186 kilojoules of work are equivalent to 1 Calorie or **4,186 joules of work are equivalent to 1 Calorie.**

Other equivalents are:

1 Calorie = 0·239 kilojoules (kJ)
1 kilowatt hour = 860 Calories = 3600 kilojoules
1 horse-power hour = 641 Calories

The energy of combustion of a food substance may be determined by burning it in oxygen in a vessel known as a bomb calorimeter. The heat produced is calculated from the rise in temperature of the bomb and the known mass of water in which the bomb is placed using the formula:

Heat gain = mass × specific heat × temperature change.

The respiration calorimeter demonstrates that living animals obey the law of conservation of energy.

Respiratory Quotient

At rest it has been seen that 250 ml. oxygen are absorbed from the room air per minute and 200 ml. carbon dioxide are given off.

The ratio $\dfrac{CO_2 \text{ given off}}{O_2 \text{ uptake}}$ is known as the Respiratory Quotient (R.Q.) and is an indication of the type of food being metabolized.

In this case the R.Q. $= \dfrac{200}{250} = 0 \cdot 8.$

If only carbohydrate is being metabolized the over-all equation for complete oxidation to CO_2 and water may be written as:

$$C_6H_{12}O_6 + 6\,O_2 \rightarrow 6\,CO_2 + 6\,H_2O$$
Glucose   Oxygen   Carbon   Water
                 dioxide

Thus 6 vol. O_2 gives 6 vol. CO_2

$$\therefore R.Q. = \frac{6}{6} = 1$$

Six vol. of O_2 give 6 vol. of CO_2 and if this was the only food being metabolized the R.Q. would be equal to 1. Nervous tissue such as the brain has an R.Q. = 1 and this suggests that brain cells are metabolizing only carbohydrate.

With a fat, such as glycerol tristearate, the equation for complete oxidation may be written as:

$$CH_2O \cdot CO \cdot C_{17}H_{35}$$
$$2 \; | \atop CHO \cdot CO \cdot C_{17}H_{35} + 163\,O_2 \rightarrow 114\,CO_2 + 110\,H_2O$$
$$CH_2O \cdot CO \cdot C_{17}H_{35}$$

Thus, 163 vol. O_2 will give 114 vol. CO_2.

Therefore, the R.Q. $= \dfrac{114}{163} = 0 \cdot 7.$ fat

On a mixed diet and including protein the R.Q. is approximately 0·80.

We have seen that:

1 g. carbohydrate + 820 ml. $O_2 \rightarrow 17$ kJ or 4·1 Cal.

Thus, with an R.Q. = 1

1 litre O_2 used to metabolize CHO $\rightarrow 21$ kJ or 5 Cal.

     *   *   *   *

1 g. fat + 1,980 ml. $O_2 \rightarrow 38$ kJ or 9·3 Cal.

With an R.Q. = 0·7

1 litre O_2 used to metabolize fat $\rightarrow 20$ kJ or 4·7 Cal.

TABLE 12

R.Q.	Calories per 1 litre of O_2 absorbed	kJ per 1 litre of O_2 absorbed
0·70	4·686	19·62
0·71	4·690	19·63
0·72	4·702	19·68
0·73	4·714	19·73
0·74	4·727	19·79
0·75	4·739	19·84
0·76	4·752	19·89
0·77	4·764	19·94
0·78	4·776	19·99
0·79	4·789	20·05
0·80	4·801	20·10
0·81	4·813	20·15
0·82	4·825	20·20
0·83	4·838	20·25
0·84	4·850	20·30
0·85	4·863	20·36
0·86	4·875	20·41
0·87	4·887	20·46
0·88	4·900	20·51
0·89	4·912	20·56
0·90	4·924	20·61
0·91	4·936	20·66
0·92	4·948	20·71
0·93	4·960	20·76
0·94	4·973	20·82
0·95	4·985	20·87
0·96	4·997	20·92
0·97	5·010	20·97
0·98	5·022	21·02
0·99	5·034	21·07
1·00	5·047	21·13

TABLE 12 gives the Calorie and kilojoule equivalent of 1 litre of oxygen at 0°C. and 760 mm. Hg for different Respiratory Quotients.

There is thus a correlation between the oxygen uptake and the Calories used. This enables the rate of metabolism to be determined by an indirect method whereby the heat produced is deduced from the consumption of oxygen.

It has been seen that the oxygen uptake may be calculated from the difference between the amount of oxygen in the inspired air and the amount in the expired air [p. 73].

It may more readily be measured by modifying the recording spirometer [Fig. 129]. A two-way valve is fitted and a soda-lime

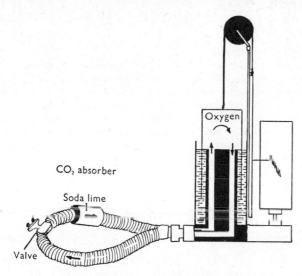

FIG. 129. Indirect calorimetry. The oxygen uptake is determined by the rate of descent of the bell which is filled with oxygen. The expired CO_2 is absorbed by means of soda lime. Movements of the bell are recorded on a chart. The tracing often reads from right to left [see Fig. 131].

canister is incorporated in the circuit to absorb the carbon dioxide formed. This spirometer is filled with pure oxygen and the subject breathes to and from the spirometer through the mouth-piece with a nose clip on. The tidal volume excursions on the tracing rise as the oxygen is used up. The slope of this rise gives the oxygen usage per minute.

Basal Requirements (Basal Metabolic Rate: B.M.R.)

The basal metabolic rate is the energy requirements under basal conditions. For comparison between subjects a standardized rest condition is employed. This is the state of complete bodily and physical rest and 12–18 hours after a meal so that digestion and absorption have been completed.

The basal requirements depend on the size of the person and correlate closely to the body surface area.

For comparison purposes it is usual to express the basal metabolic rate in kilojoules or Calories per square metre surface area per hour. The surface area may be determined by covering the subject all over with silver paper of standard thickness. This silver paper is then weighed and its area calculated. More simply, the surface area may be calculated from the height and weight using Du Bois' formula:

$$A = W^{0.425} H^{0.725} \times 0.00718$$

where A is the area in m.^{2}, W is the weight in kg. and H the height in cm. FIGURE 130 is a nomogram based on this formula. If a ruler or straight edge is placed between the weight on the left and the height on the right, the intersection with the central line will give the body surface area in m.^{2}.

The basal requirements are 170 kilojoules (40 Calories) per m.^{2} per hour for a man and 155 kilojoules (37 Calories) per m.^{2} per hour for a woman.

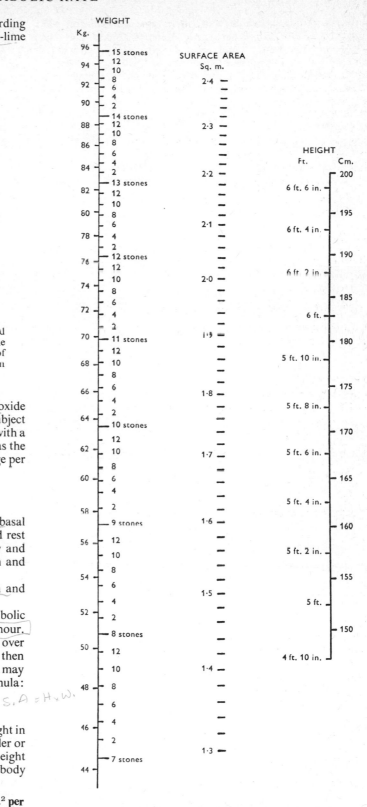

FIG. 130. Nomogram for the determination of the surface area of the body from the height and weight.

In a man with a surface area of 1·8 m.² this works out at 300 kilojoules per hour (72 Calories per hour) = 7200 kJ per day (1,728 Calories per day).

Metabolism per m.² surface area decreases with age. It drops from 210 kilojoules (50 Calories) per m.² per hour at the age of 2 to 125 kilojoules (30 Calories) per m.² per hour at the age of 60.

Example using Calories: O_2 uptake as measured is 250 ml. per minute and R.Q. = 0·8

1 litre $O_2 \equiv 4\cdot8$ Calories

∴ 250 ml. O_2/min. = 1·2 Calories/min. = 72 Calories/hour

The subject's height is 5 ft. 9 in. = 175 cm. and weight is 10 stone 7 lbs. = 147 lbs. = 67 kg.

$$\text{The surface area} = 0\cdot007184 \times \text{Weight}^{0\cdot425} \times \text{Height}^{0\cdot725}$$
$$= 0\cdot007184 \times 67^{0\cdot425} \times 175^{0\cdot725}$$
$$= 1\cdot82 \text{ m.}^2$$

Basal metabolic rate per m.² per hour $= \dfrac{72}{1\cdot8} = 40$ Calories per m.² per hour.

Metabolism in Exercise

Exercise may be carried out under controlled conditions on a bicycle ergometer. FIGURE 131 shows the oxygen uptake recorded using a spirometer at rest and during a bout of moderate to severe exercise of a young female subject. The oxygen uptake increased from 300 ml. per minute to 1·5 litres per minute. Assuming an R.Q. = 0·8 at rest and an R.Q. = 1·0 during the exercise, the metabolic rate is 1·4 Calories per minute at rest, and 7·6 Calories per minute during the exercise. The resting metabolic rate of 1·4 Calories per minute is equivalent to approximately 2,000 Calories per day [see FIG. 131].

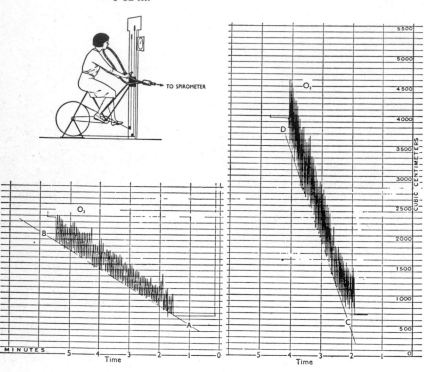

FIG. 131. The determination of respiratory rate, pulmonary ventilation, tidal volume and oxygen uptake using the bicycle ergometer and recording spirometer. This Figure shows the results of such a determination in a young adult (female aged 23) both at rest and during a short bout of moderate to severe exercise.

The subject on the bicycle ergometer breathed oxygen from the recording spirometer. The expired carbon dioxide was absorbed using soda lime. The spirometer tracings read from right to left.

Results

(a) At rest

The subject was sitting on the bicycle but was not pedalling.

Respiratory rate = 17 breaths per minute

Tidal volume = 450 ml. (average)

Thus,   Pulmonary ventilation = 7·65 litres per minute

The oxygen consumption is given by the slope of the line *AB.*

Thus   Oxygen consumption = 300 ml. per minute
= 0·300 litres per minute

Assuming an R.Q. of 0·8 this is equivalent to:

0·300 × 4·801 = 1·44 Calories per minute
= 2,074 Calories per day.

(b) Exercise

The subject was pedalling the bicycle

Respiratory rate = 25 breaths per minute

Tidal volume = 1,000 ml. (average)

Pulmonary ventilation = 25 litres per minute

The oxygen consumption is given by the slope of the line *CD.*

Thus,   Oxygen consumption = 1,500 ml. per minute
= 1·5 litres per minute

Assuming an R.Q. of 1·0, this is equivalent to:

1·500 × 5·047 = 7·57 Calories per minute

This corresponds to 10,900 Calories per day but such a rate of work could only be kept up by the subject for a relatively short time.

Subject: Height, 160 cm.; Weight, 47 kg.

Energy Requirements per Day in Calories and joules

We have seen that the basal requirement of a man, surface area 1·8 m.² is 1,728 Calories per day. This is equivalent to 7,300 kJ.

The rate of metabolism will be approximately equal to the basal level during the 8 hours sleep. During the 16 hours awake, additional Calories will be required for exercise and work. The energy expenditures per hour in some typical pursuits are given below:

	Calories	kJ
Walking at 2½ m.p.h.	140	600
Walking at 4 m.p.h.	240	1000
Sawing wood	450	1900
Coal mining	120	500
Sweeping and cleaning	100	400
Dish washing	70	300
Sewing	40	170
Concentrated brain work	Nil	Nil

It will be seen that concentrated brain activity requires no additional energy!

Allowing an additional 1,000 Calories (4200 kJ) for the waking hours we reach a total of 2,800 Calories or 11,500 kilojoules for the day.

The quantity of food required to give this number of Calories may be calculated from the fact that:

Carbohydrates give 4 Calories (17 kJ) per g.
Fats	„	9	„	(38 kJ) „
Proteins	„	4	„	(17 kJ) „ „

when used for heat and energy.

A typical diet might consist of:

375 g. carbohydrate	= 1,500 Calories	= 6,200 kJ
100 g. protein	= 400 Calories	= 1,600 kJ
100 g. fat	= 900 Calories	= 3,700 kJ
Total	2,800 Calories	11,500 kJ

See TABLE 21 [p. 220] for Calorie–Kilojoule conversion.

Calorie Requirements for Growth

Although protein incorporated in the skeleton gives no heat or energy and therefore no Calories, such protein is an essential part of the diet to provide amino acids for growth and repair of tissues. Let us consider how many additional Calories should be allowed for in the diet for this purpose.

During the time of most rapid growth between the ages of 11 and 16, the body protein increases by 4 kg. per year. Allowing 4 Calories per gram of protein this will be equivalent to an additional 16,000 Calories per year. This works out at only another 44 Calories per day.

Source of Calories

Every day the world has to find food for 3,700 million human beings and this number is steadily growing. It may well reach 7,000 million by the end of the century.

The main sources of carbohydrates are: (1) cereals—wheat, maize, rye, oats, barley, rice, sorghum and millet; (2) roots and tubers—potatoes, sweet potatoes, yams and cassava; and (3) sugar.

The main sources of protein are: (1) meat and fish; (2) milk and eggs; (3) pulses—beans, peas and lentils; (4) edible seeds, nuts and oil-containing fruit.

Well-developed countries are able to produce or purchase enough foods to ensure a well-balanced diet of 2,000–3,000 Calories per day for all their people. In many under-developed countries, where only primitive methods of cultivation exist, less than 2,000 Calories per day are available and this is mainly in the form of carbohydrates with little of the important but expensive protein. It is estimated that 1,500 million of the world's population are suffering from serious undernourishment.

Complete Starvation

When no food is being eaten, the component tissues of the body are used for the production of energy. The glycogen store in the liver is depleted in the first 24 hours, but the blood sugar is maintained by the conversion of amino acids, and glycerol from fats, to glucose. Fat from the fat deposits is mobilized and broken down in the liver to ketone bodies which pass to the tissues via the blood for oxidation. With the fall in carbohydrate metabolism, ketosis develops. Tissue protein is broken down and the amino acids utilized to form glucose, creatine for the muscles, growth of hair, etc. The protein loss is greatest in the liver and spleen and least in the brain and heart. Death occurs in about 4 weeks when the body weight has fallen to one half. If no water is available, death occurs very much earlier from dehydration.

Appetite

Two hypothalamic centres have been demonstrated in animals, an appetite centre and a satiety centre. Electrodes implanted into these centres bring about increased eating (or the opposite for the satiety centre) when the centre is electrically stimulated.

In a physically active person, the appetite equates the energy value of the food eaten with the external work performed. But this does not apply to sedentary people nor to the sick when the appetite is often poorest when need is greatest. An excessive intake leads to obesity for which the only certain remedy is to restrict the food intake. This procedure may be more distasteful and more dangerous than the condition.

VITAMINS

In addition to proteins, carbohydrates and fats, the diet must include a series of organic compounds in order to maintain health. The compounds, which are absorbed from the small intestine are known as vitamins. They are effective in small amounts and do not act as a source of energy. However, they are essential for the transformation of energy and the regulation of the metabolism.

The significant vitamins in human nutrition are:

A
B Complex   Thiamine (aneurine, B_1)
            Nicotinic acid (niacin)
            Riboflavin (B_2)
            Pantothenic acid
            Pyridoxine (B_6)
            Cyanocobalamin (B_{12})
            Folic acid
C
D
K

The bacteria in the intestines synthesize folic acid and vitamin K.

The vitamins came to be allocated letters in the following way. In 1910 Hopkins showed the inadequacy of a diet consisting only of carbohydrates, proteins and fats in feeding experiments on young rats. He fed one batch of rats on starch and cane sugar as the carbohydrate, purified casein as the protein and lard as the fat. These rats failed to put on weight. To the second batch he added 3 ml fresh milk per animal per day. These grew normally. Hopkins postulated that an 'accessory food factor' was essential for growth. This substance was insoluble in water but soluble in fat (fat soluble) and was present in the milk fat. Later in 1915 McCallum and Davis showed that milk also contained a water-soluble growth factor.

Eighteen years previously Eijkman (1897) in Indonesia (Java), studying the disease *beriberi* common among the natives whose diet was rice, noticed that the fowls developed beriberi when fed with polished rice, but not when they ate the crudely milled rice. Funk in 1911, obtained an alcoholic extract of the outer husks of rice which cured beriberi. This was thought to be a vital amine and the term 'vitamine' was introduced. To prevent confusion with chemical amines the e has now been dropped and such substances are referred to as *vitamins*.

Hopkins' fat-soluble growth factor was given the letter A and the anti-beriberi factor the letter B.

Holst's (1912) studies on the withholding of green vegetables from guinea-pigs which led to symptoms of scurvy led to the discovery and later the isolation of a substance which was designated vitamin C. It could probably have been discovered earlier, but the commonly employed laboratory animal, the rat, can synthesize vitamin C and therefore does not develop scurvy.

Shortly afterwards it was discovered that animal fats had the power of preventing rickets. This anti-rachitic vitamin was different from Hopkins' vitamin A; letters B and C having already been allocated this vitamin was lettered vitamin D.

Vitamin B (water soluble) was found to be not a single vitamin, but a complex including the anti-beriberi factor, anti-pellagra factor, and growth promoting factors. These were initially given numbers, but the system became so confused that names are now frequently employed.

Vitamin A

$$CH_3 \quad CH_3$$

$$-CH=CH-C=CH-CH=CH-C=CH-CH_2OH$$

$$CH_3 \qquad CH_3 \qquad CH_3$$

Vitamin A (retinene) is a long-chained alcohol found in the fat of milk, and therefore in butter and cream. It is also found in eggs and liver fat, but mainly in fish-liver oils such as cod-liver oil and halibut-liver oil. Vitamin A is absent from vegetable fats and oils, e.g. linseed oil, ground-nut oil, olive oil. Margarine is made from these oils and vitamin A is added during the manufacture. It is stored in the liver.

Two vitamin A molecules joined together with the elimination of the —OH groups give β-carotene. Carotenes are found in green vegetables and carrots. They are converted to vitamin A in the body.

Both vitamin A and carotene need fat in the diet as a carrier and lipase-bile salt activity for their absorption.

A deficiency of vitamin A leads to epithelial changes. The mucous membranes change into stratified epithelium. The secreting glands, such as the lacrimal glands, salivary glands, and sweat glands degenerate. The conjunctiva of the eyes become cornified. The condition is known as xerophthalmia (dry eye) and leads to blindness.

Vitamin A is essential for the regeneration of visual purple in the eye, which is bleached by bright light. A failure or delay in dark adaptation (night blindness) is one of the earliest signs of vitamin A deficiency.

It is the aldehyde (—CHO in place of the terminal —CH_2OH) derivative of vitamin A, known as vitamin A aldehyde (retinaldehyde), which combines with the protein opsin and forms visual purple (rhodopsin). Light splits this molecule into vitamin A aldehyde and opsin.

$$\text{Vit A aldehyde} + \text{OPSIN} \rightarrow \text{RHODOPSIN} \longrightarrow$$
$$\text{(protein)} \qquad \text{(visual purple)}$$
$$\text{Vit A aldehyde} + \text{OPSIN}$$

In the teeth a vitamin A deficiency leads to degeneration of the enamel organ, which is an epithelial organ, and to hypoplasia of the enamel.

VITAMIN B COMPLEX

All the members of the vitamin B group are water soluble.

Thiamine (Aneurine) (Vitamin B₁)

$$NH_2 \cdot HCl$$

Thiamine hydrochloride

This is the anti-beriberi factor. It is found in yeast, the germ and bran of cereals and pulses. The features of beriberi are polyneuritis with disorders of both sensory and motor nerves, and cardiac failure leading to a raised venous pressure and oedema.

Thiamine is the precursor of a co-enzyme necessary for the oxidation of pyruvic acid. Its deficiency brings about a disorder of carbohydrate metabolism.

Riboflavin (Vitamin B₂)

$$CH_2 \cdot (CHOH)_3 \cdot CH_2OH$$

Riboflavin

Riboflavin is found in meat, milk and wholemeal flour. It is destroyed by light and disappears from milk standing in sunlight.

A deficiency in man leads to inflammation of the mouth and tongue, and seborrhoeic dermatitis of the skin. The cornea of the eye becomes vascularized and retrobulbar neuritis may develop leading to defective vision.

Nicotinic Acid (Niacin) (anti-Pellagra Factor)

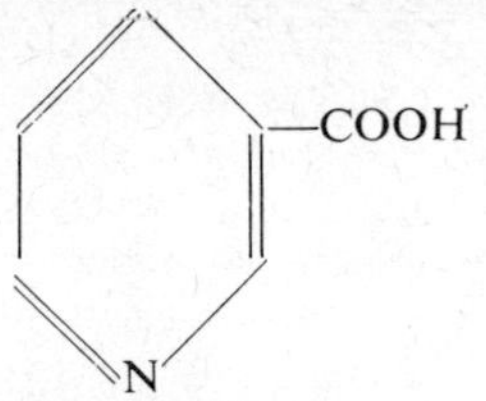

Nicotinic acid
(pyridine-β-carboxylic acid)

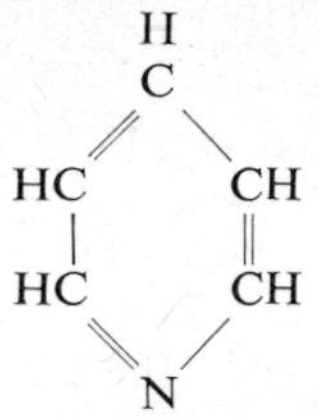

Pyridine

This is the anti-pellagra factor. It is found in liver, kidney and yeast.

Pellagra is a disease occurring among poorly nourished populations living mainly on maize. There are changes in the skin which becomes inflamed and blistered, diarrhoea is common and there are changes in the brain, spinal cord and peripheral nerves. The classical signs of pellagra are therefore dermatitis, diarrhoea and dementia.

Pantothenic Acid

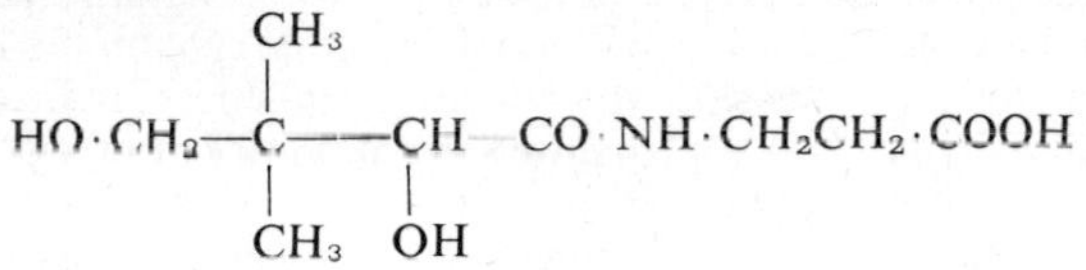

Pantothenic acid

This vitamin plays a part in the formation of co-enzymes, which are essential to the enzymic reactions.

Nicotinic acid, riboflavin and pantothenic acid co-enzymes are concerned in the citric acid cycle.

Pyridoxine (Vitamin B_6)

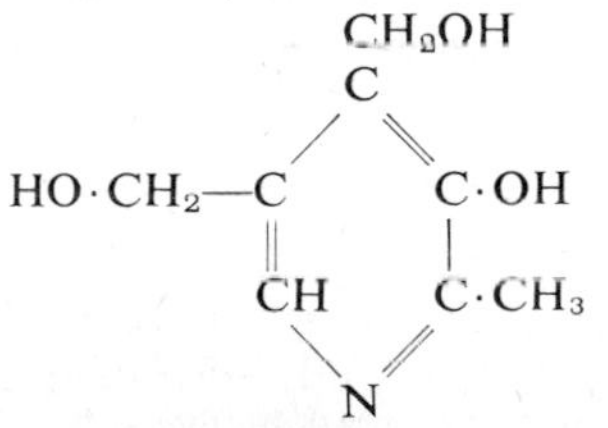

Pyridoxine

This vitamin is widely distributed in all plant and animal tissue. The aldehyde compound pyridoxal is a co-enzyme important in amino acid and fat metabolism.

Cyanocobalamin (Vitamin B_{12})

Vitamin B_{12} consists of four substituted pyrrole rings attached to an atom of cobalt. The cobalt is also joined to a benzimidazole ring which is connected via ribose phosphate to one of the pyrrole side chains. This vitamin is needed for *methylation* in protein synthesis.

Vitamin B_{12}, which was first obtained from liver, is now made as a product of fermentation by the mould *Streptomyces griseus*.

Replacement of the cyanide group (—CN) attached to the cobalt by the hydroxyl group (—OH) gives *hydroxycobalamin* which has a similar physiological action.

This vitamin has already been discussed in connexion with red cell formation [p. 10]. Vitamin deficiency is usually due to a failure of intestinal absorption rather than to a failure of intake. A deficiency leads to a macrocytic anaemia of the pernicious anaemia type, and

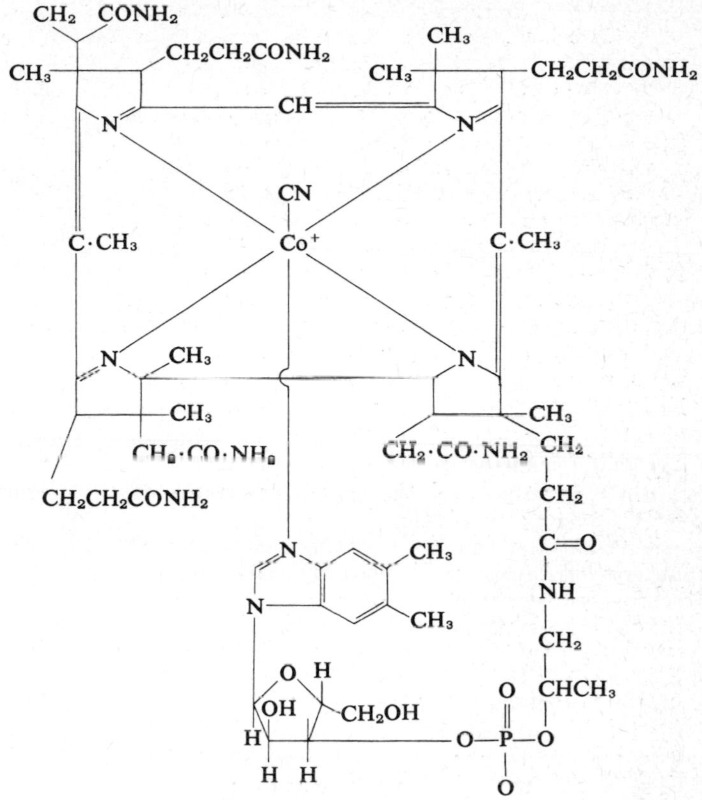

VITAMIN B_{12} (cyanocobalamin)

subacute combined degeneration of the spinal cord. The anaemia may often respond to **folic acid,** but the neurological manifestations do not.

Folic Acid

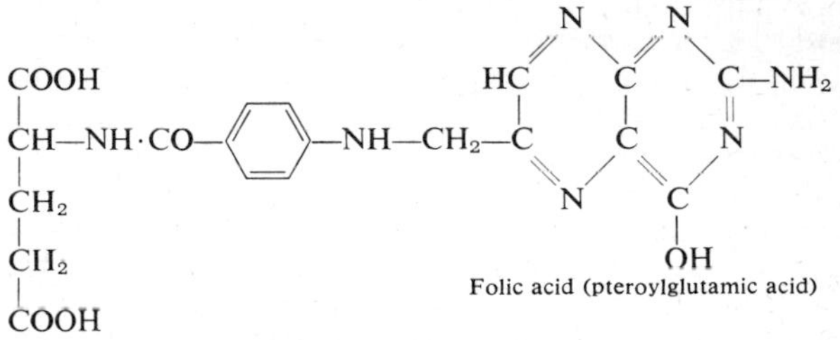

Folic acid (pteroylglutamic acid)

Folic acid (pteroylglutamic acid) was first isolated from green leaves (L. *folium* = leaf) in 1943. It is found in yeast, kidney, liver, leafy vegetables and cereals.

Folic acid is needed for the synthesis of purines and pyramidines and the formation of nucleic acids, as well as acting as a co-enzyme for the interconversion of amino acids. It is particularly required during cell division and the highest requirements are therefore in the red bone marrow, the digestive tract and the skin. It is important for growth in the foetus. The normal plasma folic acid level is

$$10 \text{ ng./100 ml. } (1 \text{ ng.} = 10^{-9} \text{ g.}).$$

Vitamin C (Ascorbic Acid)

Ascorbic acid

Ascorbic acid is water soluble. It is a reducing agent. It is destroyed by cooking, particularly in alkaline solution, i.e. when soda has been added.

Vitamin C is found in fresh fruit, blackcurrant, strawberry, orange, lemon and grapefruit, but there is very little in blackberry, lime, plum, pear and apple. It is present in fresh vegetables, sprouts, cauliflower, cabbage, tomatoes and potatoes.

It is absent from meat, fish, eggs, fats and oils and is not present in milk after pasteurization.

A deficiency of vitamin C leads to scurvy. Historically scurvy was the first disease shown to be due to a dietary deficiency. It occurred amongst sailors and armies without fresh food, and in towns during the winter months.

Jacques Cartier published an account in 1545 of a mysterious illness which broke out amongst the crew of his ship anchored in the St. Lawrence River at Stadacona, Quebec, in 1535–6. He stated that the disease advanced daily so that on his three ships there were not three men in good health. One day whilst walking on the ice he came across a native who ten days before had been suffering from the same disease. The native was now cured. Cartier asked the Indian how he had been cured. He replied that it was the juice and dregs of the leaves of a certain tree which had healed him. A concoction was made of the leaves of the tree as a potion for the sick men.

Only one or two were prepared to test it. They improved miraculously. Then the others took the potion and were cured.

Cartier's description of the disease, swollen legs blotched with drops of purplish blood, with decayed gums, and teeth falling out, was clearly a description of a vitamin C deficiency, i.e. scurvy.

In 1753 Captain Lind published a treatise on scurvy showing that it could be cured in about a week by supplementing the diet with oranges and lemons. Lime juice was thought to be as effective and was issued to the British Navy.

Holst's work on the guinea-pig proved conclusively that scurvy was a deficiency disease. Vitamin C was finally isolated in 1932 by Szent-Györgyi and termed *ascorbic acid*.

Lack of vitamin C leads to a weakening of the intracellular cement substance with the result that wounds heal slowly. The capillaries become more fragile and haemorrhages occur all over the body. Bleeding into the joint capsules gives tender swollen joints (haemarthroses). Bleeding occurs under the skin with comparatively minor injuries. Haemorrhage from the tooth sockets leads to bleeding gums. As a result of the haemorrhages, anaemia may develop.

Scurvy occurs in infants fed on sterilized food, and in old people living alone who have a deficiency of fresh fruit in their diet. In such old people one of the characteristic signs of scurvy, namely the bleeding gums, may be absent since scurvy does not give rise to bleeding gums in the toothless.

Vitamin D

Vitamin D₃ (cholecalciferol)

Vitamin D₂ (ergocalciferol)

Mellanby in 1918 showed that rickets in dogs could be cured with cod-liver oil. A year later Huldschinsky showed that exposure to ultra-violet light was also effective.

Rickets is due to a deficiency of vitamin D. The natural vitamin is vitamin D_3 which is formed by the action of sunlight on the 7-dehydrocholesterol in the skin [p. 149]. A similar substance *ergocalciferol* (vitamin D_2) is formed by the action of ultra-violet light on *ergosterol* from ergot.

Both vitamin D_3 and D_2 prevent rickets when taken by mouth. They are fat soluble and need bile salts for their absorption.

Vitamin D is found in the fat of milk, egg-yolk and fish-liver oils. It is absent from vegetable fats and oils, and like vitamin A is added to margarine during the manufacture.

Rickets is a disease of childhood due to lack of sunlight. It is a disorder of bone ossification. There is excess preparation of the cartilage for ossification, but a deficiency of calcification. This leads to swelling of the epiphysial junction. The bones are soft leading to bent legs and deformed thorax and pelvis.

Osteomalacia is a similar disease in adults, occurring particularly in pregnant women.

Metabolism of Vitamin D

It has been suspected for some time that vitamin D is converted into a more active form in the body, since there is a delay of 12 hours in a vitamin D-depleted animal before administrated vitamin D produces an effect.

1 : 25 Dihydroxycholecalciferol (active form of vitamin D made by the kidney when plasma calcium is lower than normal).

Recent work suggests that the liver first converts vitamin D to a more active hydroxy form which is more soluble. Thus:

Vitamin D_3 → 25-Hydroxycholecalciferol (25 HCC)
Vitamin D_2 → 25-Hydroxyergocalciferol (25 HEC)

This 25-hydroxy compound is then converted to the 1 : 25-dihydroxy-compound (1 : 25-dihydroxycholecalciferol) by the kidney when the plasma calcium is below the normal level. This dihydroxy compound of vitamin D is much more active than the original vitamin in bringing about absorption of calcium from the gut and its incorporation in bone.

Since the formation of the dihydroxy-compound by the kidneys does not occur when the plasma calcium level is higher than normal, 1 : 25 dihydroxy-vitamin D can be considered to be a hormone produced by the kidney which is regulating the uptake of calcium from the gut and hence the plasma calcium level.

If the dietary intake of calcium is high, the plasma calcium level will tend to be higher than normal and very little 1 : 25-dihydroxy-vitamin D will be produced by the kidney, and the absorption of calcium will be low. If on the other hand the dietary intake of calcium is low, the high level of dihydroxycholecalciferol released by the kidney into the blood will stimulate the absorption of calcium from the gut.

Removal of the kidneys (bilateral nephrectomy) is associated with a low plasma calcium.

Vitamin E (Tocopherol)

This is a fat soluble vitamin which occurs in large amounts in wheat-germ and cottonseed oil. It exists in several forms designated α, β, γ and δ-tocopherol, etc. The principal form is α-tocopherol.

Although vitamin E is present in the tissues of the body, is stored in fatty tissue, and has an average level in circulating blood of 1 mg. per 100 ml., there is no evidence that it is essential to man and a deficiency produces no recognisable symptoms.

Vitamin K

This is a fat soluble vitamin found in green vegetables, cabbage, sprouts and spinach. It is essential for the formation of prothrombin by the liver [p. 18]. Since it is synthesised by the bacterial flora of the intestines, deficiencies may occur at birth (when the gut is sterile) or when an ingested broad-spectrum antibiotic has destroyed this flora.

Avitaminosis

Although the gross vitamin deficiency diseases xerophthalmia (vitamin A), beriberi (thiamine), pellagra (nicotinic acid), scurvy (vitamin C) and rickets (vitamin D) are rare in well-developed countries, moderate degrees of vitamin deficiency may occur. This applies particularly to the B group, especially when the requirement is raised by a high carbohydrate intake.

REFERENCES AND FURTHER READING

CALORIES

Atwater-Benedict Calorimeter (1903) original description in Experiments on the Metabolism of Matter and Energy in the Human Body, *U.S. Dept. Agric. Bull.*, **175.**

Davidson, S., Meiklejohn, A. P., and Passmore, R. (1966) *Human Nutrition and Dietetics*, 3rd ed., Edinburgh.

Du Bois, D., and Du Bois, E. F. (1916) Clinical calorimetry. A formula to estimate the approximate surface area if height and weight be known, *Arch. intern. Med.*, **17,** 863.

Food and Agriculture Organisation of the United Nations (1957) *Nutritional Studies*, No. 15, Calorie requirements.

Rubner, M. (1885) Calorimetrische Untersuchungen, *Z. Biol.*, **21,** 250.

Widdowson, E. (1955) Assessment of the energy value of human foods, *Proc. Nutr. Soc.*, **14,** 142.

Wohl, M. G., and Goodhart, R. S. (1964) *Modern Nutrition in Health and Disease*, 3rd ed., London.

VITAMINS

Cartier, J. (1545) *Brief Recit et Succinte Narration* (reprinted on the occasion of the XIXth International Physiological Congress, Montreal, 1953).

Crandon, J. H., Lund, C. C., and Dill, D. B. (1940) Experimental human scurvy, *New Engl. J. Med.*, **223,** 353.

Hopkins, F. G. (1912) Feeding experiments illustrating the importance of accessory factors in normal dietaries, *J. Physiol. (Lond.)*, **44,** 425.

Lind, J. (1753) *A Treatise of the Scurvy*, Edinburgh.

Platt, B. S. (Ed.) (1956) Recent research on vitamins, *Brit. med. Bull.*, **12,** No. 1.

Szent-Györgyi, A. (1933) Identification of vitamin C, *Nature (Lond.)*, **131,** 225.

12. WATER AND MINERAL SALTS

THE DIETARY REQUIREMENTS OF MINERAL SALTS

General Consideration of Dietary Requirements

When shopping in a foreign country using a little understood currency the easy way out is to say to oneself 'when in doubt give too much and accept the change'.

It is not unknown for a government to proffer just this advice to the general public when converting a country's currency to a new and poorly understood decimal system.

A similar principle is adopted when planning a diet to include all the essential inorganic substances. The requirements of mineral salts by the body on any particular day are not generally known with any degree of certainty. Nor is the inorganic content of the food ingested. Ask yourself how many milligrams of magnesium you ingested at your last meal. Another variable factor is the percentage of the amount ingested that is absorbed. It has already been seen [p. 9] that in the case of iron it may be as low as 10 per cent.

With so many unknowns and imponderables the only way to maintain the correct concentration of these substances in the body **is to ingest an excess and to leave it to the kidney to excrete the surplus in the urine.**

The efficiency of such a system is borne out by the comparatively low incidence of electrolyte imbalance that occurs under normal conditions.

In abnormal states, however, an excess or lack may easily arise. Thus a depletion can rapidly occur as a result of, say, vomiting, diarrhoea, excessive sweating or too great a loss in urine. An excess may be the result of too high an intake by mouth or by intravenous infusion, or due to an inability on the part of the kidney to excrete the surplus.

The kidney's ability to excrete a substance is limited both on a time basis and on a concentration basis. It cannot produce solid urine. Water must always be lost when a substance is excreted. There is a limit to the concentration of a substance in the urine. The kidney requires time to excrete a given quantity of a substance. Even water cannot be excreted in excess of a rate of 15–22 ml. per minute. Any intake in excess of this will lead to over-hydration.

Ingesting large quantities of sea-water overloads the body with sodium and chloride ions, because the concentration of sodium ions in sea-water is in excess (except in very rare cases) of that in the kidney's most concentrated urine.

In general, it appears that it is **the concentration of an electrolyte in the body fluids that must be kept constant** to provide the correct environment for the body's cells rather than the absolute amount in grams present in the body.

Sodium

Sodium chloride (NaCl) is the only mineral salt consumed as such in the diet. The intake is usually far in excess of the needs. 10–15 g. per day is the general intake. The requirement is less than 1 g. per day unless sweating is occurring. The surplus NaCl is excreted in the urine.

Fruit and vegetables contain sodium citrate and tartrate. Other ingested sodium salts include sodium bicarbonate (baking soda), sodium nitrite and sodium nitrate (used as a meat preservative).

Potassium

of angina Pectoris.

Potassium is present in all cells and, therefore, a dietary deficiency is unlikely. Although most of the potassium in the body is in the cells, its level in the plasma of 5 mmol per litre must be maintained and potassium supplements are given if this level is not being maintained.

An excess of potassium ions in the plasma has the opposite effect to calcium ions and stops the heart in diastole.

Both a low and a high plasma potassium level leads to muscle weakness because of the alterations to cell membranes.

In acidaemia when the hydrogen ion concentration inside cells increases, the potassium ion concentration in the cells is reduced.

Calcium

Calcium salts are needed by the body: (1) for the ossification of bone and teeth; (2) to regulate the excitability of nerve fibres and nerve centres; (3) for the continued contraction of the heart; and (4) for the clotting of blood and the curdling of milk in the stomach.

Calcium salts are found in milk, cheese, cereal and green vegetables. Flour, and, therefore, bread, may be fortified by the addition of chalk. The absorption of calcium from the intestines is far from complete and as much as 70 per cent. may appear in the faeces. Vitamin D is essential for an adequate calcium absorption and utilization. The blood calcium level is regulated by the formation by the kidney of 1 : 25 dihydroxycholecalciferol, by the parathyroid gland, and by calcitonin from the thyroid gland.

Plasma Calcium

The level of calcium in the plasma is 10 mg. per cent (2·5 mmol per litre). [There is virtually none in the red cell.] Of this 5 mg. per cent. (1·25 mmol/l) is in simple solution as Ca ions and is physiologically active. 4 mg. per cent. is loosely combined with plasma proteins and is not diffusible. The remaining 1 mg. per cent. is diffusable but unionized; this calcium is in the form of complexes with phosphate, bicarbonate and citrate. [See also plasma calcium level, p. 220.]

An excess of calcium ions in the plasma will stop the heart in systole. Calcium chloride injected into the coronary arteries may be used to stop the heart in cardiac surgery where a by-pass is being employed.

Phosphorus

The element phosphorus is needed by the body as organic phosphate in all tissue cells. It is a constituent of the nucleic acids and phosphatides. It provides the energy for muscles in the form of ATP and creatine phosphate. Inorganic phosphate in the form of the calcium phosphate salts gives strength to bone, whilst the alkaline disodium hydrogen phosphate (Na_2HPO_4) and the acid sodium dihydrogen phosphate (NaH_2PO_4) form buffer systems in the blood and urine.

Phosphorus is found in all cells and therefore a dietary insufficiency is unlikely. It is found as a phosphoprotein in the caseinogen of milk and as phosphatides in egg-yolk, liver and pancreas.

Magnesium

Only about one-quarter of the ingested magnesium is absorbed. Like potassium, magnesium in the body is found in the cells but about one-half of the body magnesium is in bone. Since the magnesium in bone acts as a reserve, a deficiency is rare. The normal blood magnesium level is 1·0 mmol per litre (2 mg. per 100 ml. plasma)

A low blood magnesium leads to tremor, muscle twitching, convulsions and delirium.

A high blood magnesium leads to drowsiness and coma.

Iodine

Minute traces of iodine or iodide are required for the proper functioning of the thyroid gland. It is present in sea food and in crops grown near the sea. It may be deficient in areas remote from the sea, leading to swelling of the thyroid gland, a condition known as a *goitre* [p. 163]. Some table-salt is 'iodized', that is, potassium iodide has been intentionally added.

Iron

Iron is found in most foods except milk. It is present in meat, eggs, cheese, bread, peas, cabbage and potatoes.

It is needed for the formation of the haemoglobin [p. 9] and the iron-containing pigments and enzymes.

Zinc

The importance of an adequate dietary intake of zinc has only recently been realized.

It is needed for the formation of many of the enzymes of the body including carbonic anhydrase [p. 84] and carboxypeptidase [p. 109]. A zinc deficiency delays wound healing.

THE DIETARY REQUIREMENTS OF WATER

Of all the items in the diet the daily intake of water is the most important. Without it we are unable to maintain the water balance of the body and dehydration (anhydraemia) follows. Although a starving man can survive for a month without food, he can only live for a few days without water.

Since water makes up 70 per cent. of the body weight and the weight remains constant from day to day, it follows that the water content of the body remains constant, and this implies that the water intake per day equals the water loss. Every surplus litre of water will increase the body weight by 1 kg.

Water Gained

Water is gained by the body in three ways [FIG. 132]. Fluids taken by mouth are equivalent to an equal volume of water. Thus,

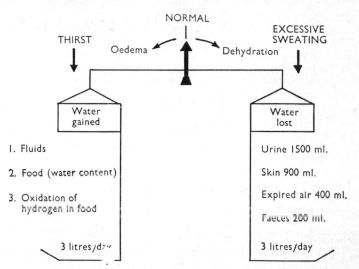

FIG. 132. Water balance of the body.

a glass of milk is equal to a glass of water as far as water intake is concerned. The solid food we eat has a water content. In the case of some fruits, such as the melon, this may be as high as 95 per cent. or even higher. Even the driest biscuit contains some water. Lastly, about 500 ml. water are made in the body each day by the oxidation of the hydrogen in the food. The final stage of the metabolic pathways for the complete oxidation of carbohydrate, fat and de-aminated amino acids is the formation of carbon dioxide and water. Thus, the water gained by the body each day is greater than that present as moisture or fluids in the food taken by mouth.

The average daily intake of water by these three means is 3 litres per day.

Should the intake be insufficient there is a sensation of thirst. The thirst centre is probably in the hypothalamus. Hypertonic saline injected into the hypothalamus or electric stimulation of the same area in the goat has been shown to cause the animal to drink excessively (Andersson, 1957).

Water Lost

Water is lost by the body in the urine, through the skin, in the expired air and in the faeces. In addition, there may be a loss associated with a haemorrhage, vomiting, tears, etc.

The volume of urine is adjusted to maintain the water balance. It is the only true variable on either side. The whole plasma volume of 3 litres passes over fifty times a day into the tubules of the kidney (glomerular filtrate rate 170 litres per day, page 135), but over 99 per cent. of it is reabsorbed further down the tubule. The remainder passes to the bladder and is excreted as urine. The normal volume of urine is 1·5 litres per day, but it can be reduced down to an absolute minimum of 300 ml. per day, or increased to 22·5 litres per day. The minimum obligatory volume is the amount necessary for the removal of the excretory products. The urinary volume is regulated by ADH from the posterior pituitary gland [see p. 136].

Water is lost through the skin both as sweat and as imperceptible (insensible) perspiration. Sweating occurs via the sweat glands which are controlled by the sympathetic nervous system from the heat regulating centre in the hypothalamus. Temperature regulation of the body takes precedence over the water balance and in a hot environment as much as 3 litres of water may be lost in 1 hour as sweat.

The skin is not a completely water-tight covering and some evaporation of water is taking place all the time through the skin. This is known as imperceptible perspiration and the water loss is not from the sweat glands. In temperate climates, the water loss by evaporation amounts to 900 ml. per day.

The expired air is saturated with water vapour and this causes a loss of 400 ml. of water per day.

There is a large water turnover in the digestive tract. Five litres of water are present in the digestive juices produced each day, but 4·8 litres are reabsorbed in the small and large intestines. The remaining 200 ml. are lost in the faeces.

In cases of diarrhoea the loss by this route may be substantial.

The total water loss is on average 3 litres per day, the same as the intake.

FLUID COMPARTMENTS

The 45 litres of water in the body are shared between the cells (30 litres), the tissue or interstitial spaces (12 litres) and the plasma (3 litres). The water in the cells plus its dissolved salts is termed the *intracellular fluid*. The tissue fluid and the plasma together constitute the *extracellular fluid*.

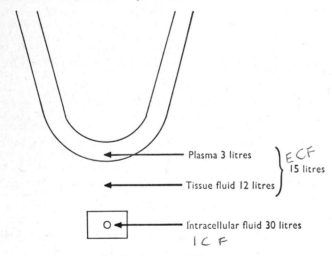

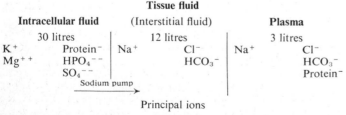

FIG. 133. Fluid compartments of the body.

Most of the sodium, potassium and magnesium salts found in body fluids are completely dissociated, that is, they exist in solution as electrically charged particles known as ions. To become electrically charged they have gained or lost electrons. The negatively charged ions have gained electrons and are termed anions. The number of electrons gained is shown by the suffix $^-$, $^{--}$, or $^{---}$ etc. (or alternatively $^-$, $^{2-}$, $^{3-}$ etc.). Thus, Cl^-, the chloride ion, has one electron more than the chlorine atom, whilst SO_4^{--} has two electrons more than the uncharged sulphate radical.

The cations are ions that have lost electrons and have thereby developed a positive charge. The number of electrons lost is shown by the suffix $^+$, $^{++}$, or $^{+++}$ etc. (or alternatively $^+$, $^{2+}$, $^{3+}$ etc.). Thus Na^+, the sodium ion, has one electron less than the sodium atom. In physiological fluids the number of positively charged ions usually equals the number of negatively charged ions. If they are not equal, then an electrical potential will be generated (see nerve cells, page 181).

In addition, there are many soluble substances in the body fluids that are both neutral and unionized. Two such examples are glucose and urea.

The ions associated with the water in the different fluid compartments are shown in FIGURE 133. The main cation in the cell is potassium. The main cation in the tissue fluid and plasma is sodium. The tissue fluid and plasma have a similar ionic composition, except that plasma contains protein, i.e. the plasma proteins. At the pH of plasma (7·4, slightly alkaline) these proteins are ionized as anions.

The cells of the body actively secrete sodium by a mechanism known as the 'sodium pump'. It is for this reason that the cellular sodium content is low.

Measurement of Fluid Compartments

The volume of the body fluids is determined by dilution techniques in a similar manner to the determination of blood volume [p. 6]. For the total body water, the test substance used is heavy water, either deuterium oxide (D_2O) or tritium oxide (T_2O). The extracellular fluid is measured using sodium thiosulphate, sodium thiocyanate, mannitol and radio-active substances, such as radio-active sulphate, chloride and sodium. The results differ slightly according to the test substance used since some substances penetrate into the cerebrospinal fluid and digestive tract whilst others do not. The tissue fluid volume is obtained by subtracting the plasma volume from the answer. The intracellular volume is calculated by subtracting the extracellular volume from the total body water volume.

Ionic Balance

It is convenient to consider the movements of the cations and anions separately. This is permissible provided that it is remembered that, to remain electrically neutral, a solution cannot gain an anion unless at the same time it gains a cation having an equal and opposite charge.

Prior to the introduction of S.I. units the quantity of each ion present was often measured, not in grams, but in *equivalent weights*. The cations and anions could then be separately totalled and the two totals would be equal.

The equivalent weight (gram-equivalent) is the molecular weight of the ion expressed in grams for a monovalent ion. The molecular weight is the sum of the atomic weights of the atoms making up the ion and no allowance is necessary for the addition of or lack of electrons which differentiate ions from uncharged molecules. For divalent ions (calcium, magnesium, sulphate) the equivalent weight is half the molecular weight.

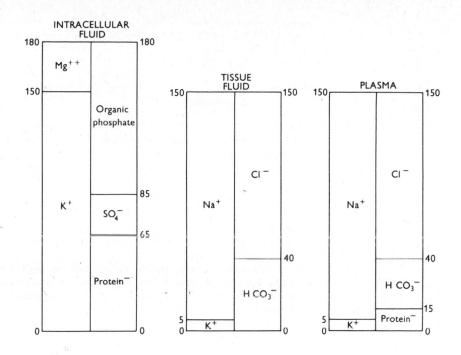

FIG. 134. Positive and negative ions in the water compartments of the body (simplified). In S.I. units the term 'mEq/l' is replaced by the term 'mmol/l' of charges.

For multivalent ions the equivalent weight is the molecular weight in grams divided by the valency.

The equivalent weight of one ion combines with or replaces or is 'equivalent' to the equivalent weight of another.

Thus 23 g. of Na^+ (i.e. 1 Eq. Na^+) are equivalent to 39 g. of K^+ (i.e. 1 Eq. K^+) both combine with 35·5 g. of Cl^- (i.e. 1 Eq. Cl^-).

The equivalent weight was too large a unit for general use in physiology. A more convenient unit was one thousand times smaller, the *milli-equivalent* (mEq.). This was determined by dividing the ionic weight in milligrams by the valency.

Molarity

Ostwald introduced the term *mole* (mol) to represent the quantity of a substance equal to its molecular weight in grams. The *mole* has become the basic unit of quantity in the S.I. Units system, and this definition has been broadened to cover atoms, molecules, ions, etc. [see p. 220]. It replaces the gram-molecule, gram-ion and gram-equivalent which now become obsolescent.

A solution which contains one mole of a substance made up to one litre with water is termed a *molar* solution (1·M or 1 mol/l). One ml. of such a solution will contain one millimole (1 mmol) of the substance. If it is made up to 1 kg. with water, or other solvent, the solution is termed a *molal* solution (1 mol/kg.).

In the case of a substance with a valency of 1 (e.g. HCl) the concentration in milli-equivalents per litre (*normality*) is numerically equal to the *molarity*. For substances with a valency of 2 (e.g. H_2SO_4), the concentration in milli-equivalents per litre will be twice the molarity.

Molarity is of more general application than milli-equivalents per litre since it can be applied to hormones, drugs and other substances of which the molecular weight is known but which have no obvious valency.

Ionic Equality

If the concentrations of ions in a solution are measured in *mmol/l of charges* (rather than in mmol/l of ions) then the addition of cations and anions can be carried out and, even though some ions may be multi-valent, the two totals will be the same. 'Milli-moles per litre of charges' is the expression used in the S.I. units system to take account of the fact that divalent ions such as Mg^{++} carry a double charge. It is numerically equal to the older 'milli-equivalents per litre'. For monovalent ions the concentration in *mmol/l of charges* is numerically the same as that in *mmol/l of ions*.

This summation may be represented graphically by a series of blocks as in FIGURE 134 which illustrate the equality between the positive and negative ions in the body fluids.

Potassium (150 mmol per litre) and magnesium (30 mmol per litre) are the main cations in cells. The anions are protein (65 mmol per litre), organic phosphate (95 mmol per litre) and sulphate (20 mmol per litre).

Sodium (145 mmol per litre) is the main cation outside the cells (extracellular fluid). The main anions are chloride and bicarbonate in the tissue fluid, and chloride, bicarbonate and the plasma proteins in the plasma.

It will be noted that sodium and potassium are present in similar concentrations of 150 mmol per litre in the extracellular and intra-cellular fluids respectively.

The chloride ion is the main anion in the plasma. Its concentration is 110 mmol per litre. It should be remembered that when blood is exposed to the air, carbon dioxide is given off from the red cells (since CO_2 tension in room air is virtually zero) and as a result chloride leaves the plasma and enters the red cell (chloride shift, p. 85). Hence in order to make an accurate determination of the plasma chloride, the separation of plasma from red cells must be carried out without exposure to air.

FLUID REPLACEMENT

15 litres Extracellular fluid Na^+ 145 mmol/litre	Total Na^+ = 15 × 145 = 2,200 mmol = 2·2 moles
30 litres Intracellular fluid K^+ 150 mmol/litre	Total K^+ = 30 × 150 = 4,500 mmol = 4·5 moles

A shortage of water and mineral salts can usually be corrected by giving fluids and electrolytes by mouth. Alternatively an intravenous infusion is used.

All fluids run into veins must be both sterile, and isotonic with blood [p. 12]. The most commonly used fluid is normal 'physiological' saline. Since this fluid contains sodium ions, after absorption it enters only the extracellular fluid compartment of the body, and does not enter the cells.

Normal (Physiological) Saline

This is made by dissolving 0·9 g. sodium chloride in 100 ml. sterile water [p. 12]. The concentration is thus 9 g. NaCl per litre. Since molecular weight of sodium chloride is 23 + 35·5 = 58·5

$$58·5 \text{ g.} = 1 \text{ mole}$$
$$58·5 \text{ mg.} = 1 \text{ millimole}$$
$$\therefore 9 \text{ g. NaCl} = \frac{9 \times 1000}{58·5} = 154 \text{ millimoles}$$

Since this quantity is dissolved in 1 litre of water the concentration of physiological saline is 154 milli-moles/litre ≡ 154 milli-molar solution. Alternatively it may be stated to contain 154 mmol/litre of both sodium ions and chloride ions.

Since it is neutral, its pH will be 7 (6·8 at 37 °C.).

Although this fluid is used in large quantities for intravenous infusions, it will be noted that compared with normal plasma (Na^+ = 145 mmol/l; Cl^- = 110 mmol/l; pH = 7·4) the sodium ion concentration of physiological saline is slightly higher, the chloride ion concentration is very much higher. It will also be noted that the pH of physiological saline is lower (that is, it is relatively more acid) than plasma.

It follows, therefore, that if large quantities of fluid are to be given intravenously, a solution with fewer chloride ions and a higher pH must be used. This is achieved by replacing some of the sodium chloride with sodium bicarbonate (or sodium lactate which will be metabolized in the body to sodium bicarbonate).

Other Intravenous Fluids

Physiological saline gives the subject both water and sodium ions. If the shortage is that of water only, physiological saline is unsuitable. Sterile distilled water cannot be used as an intravenous infusion fluid to correct dehydration because it will haemolyse the red cells [p. 12]. A substance must be added to the water to give the correct osmotic pressure during the infusion but which will later 'disappear' preferably by being converted to carbon dioxide and water. Such a substance is glucose. A 5 per cent. glucose in water solution is isotonic with blood, and is used when the subject is depleted of water only. Other fluids commonly used for intravenous infusion are:

Glucose (dextrose) saline
    0·18 per cent. NaCl   30 mmol/l Na^+ and Cl^-
    4·3 per cent. glucose
Sodium-potassium glucose mixture
    0·225 per cent. NaCl   40 mmol/l Na^+ and Cl^-
    0·3 per cent. KCl   40 mmol/l K^+ and Cl^-
    2·5 per cent. glucose
Saline lactate
    0·69 per cent. NaCl   102 mmol/l Na^+ and Cl^-
    0·62 per cent. Na lactate   56 mmol/l Na^+ and lactate$^-$
Sodium bicarbonate
    1·35 per cent $NaHCO_3$ 160 mmol/l Na^+ and HCO_3^-

Low Plasma Sodium—Hyponatraemia

It is common practice to speak of 'plasma sodium', 'excretion of sodium by the kidneys', etc., when what is meant is concentration of plasma sodium *ions* or the excretion of sodium *ions* by the kidneys and not the metallic element!

A low plasma sodium level is the result of an inadequate intake of sodium salts. It is liable to occur following excessive sweating if only the water, and not the sodium chloride, is replaced, and leads to painful muscular (stoker's) cramps [p. 151].

A more severe hyponatraemia leads to a reduction in blood volume, a low cardiac output, and a low blood pressure with giddiness, fainting, and a cold, clammy, flabby skin.

High Plasma Sodium—Hypernatraemia

A raised plasma sodium level is found when the kidney excretion of sodium is reduced. This occurs when there is an excess of adrenal cortex hormones in the blood (aldosterone and cortisol, p. 167). It may also result from the ingestion of sea-water or from an excess of sodium ions given by intravenous infusion (by the giving of saline when the shortage was only that of water).

Since sodium is principally an extracellular ion, the increased osmotic pressure of the extracellular fluid will withdraw water from the cells. This leads to a reduction in the water in the intracellular fluid compartment and increase in the concentration in this compartment. At the same time there will be an increase in the water in the extracellular fluid compartment. The increase in blood volume will give an increase in cardiac output and blood pressure (see Cushing's syndrome, p. 168).

A progressive hypernatraemia leads ultimately to death due to cellular dehydration.

Low Plasma Potassium—Hypokalaemia

A low plasma potassium may be the result of a decreased intake, an increased loss or the movement of potassium ions into the cells.

A decreased intake occurs in starvation and with the reduced absorption associated with diarrhoea. An increased loss via the kidneys occurs with increased aldosterone (and cortisol) activity, with digitalis and with some diuretics.

It has been seen [p. 37] that the correct balance between potassium and calcium ions is needed for the correct beating of the heart. The effect of hypokalaemia alone is an increased excitability of cardiac muscle with a long P-R interval, inversion of the T-wave of the electrocardiogram and the appearance of a U-wave after the T-wave [FIG. 135(B)]. There is also skeletal muscle weakness and poor conduction in nerves.

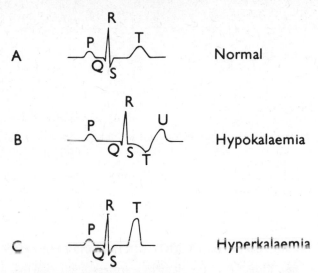

A Normal

B Hypokalaemia

C Hyperkalaemia

Fig. 135. Electrocardiographic changes as a result of
hypokalaemia and hyperkalaemia.

Unlike sodium, potassium ions do not disappear from the urine
when the body is short of potassium and the loss of potassium in the
urine continues.

High Plasma Potassium—Hyperkalaemia

A high plasma potassium may be the result of an excessive intake
by mouth or intravenous infusion (including a transfusion of blood
more than 2 weeks old), a decreased kidney excretion due to kidney
failure or a low aldosterone secretion. It also occurs when there is
an increased breakdown of body protein as in starvation and in
acidaemia where intracellular potassium ions leave the cells and
become extracellular potassium ions.

Hyperkalaemia depresses the heart's activity. It gives a high
peaked T-wave on the electrocardiogram [Fig. 135(C)] and may
lead to ventricular fibrillation. A potassium level in excess of
7 mEq./litre is fatal.

Insulin facilitates the entry of potassium ions into cells. Hence
when treating diabetic ketosis, which is associated with dehydration
and a **high** plasma potassium level, potassium ions should nevertheless be given with the insulin, glucose and water to prevent
hypokalaemia as the acidaemia disappears and the potassium ions
re-enter the cells.

The excretion of potassium ions by the kidney is closely related to
the reabsorption of sodium ions and the excretion of hydrogen ions.
It will be considered further on page 142 after the mechanism for the
formation of urine has been discussed.

REFERENCES AND FURTHER READING

WATER AND MINERAL SALTS

Andersson, B. (1957) in *The Neurohypophysis*, ed. Heller, H., p. 131,
New York.

Elkinton, J. R., and Danowski, T. S. (1955) *The Body Fluids*.
London.

Gamble, J. L. (1954) *Extracellular Fluid*, 6th ed., Cambridge, Mass.

Le Quesne, L. P. (1967) *Fluid Balance in Surgical Practice*, 3rd ed.,
London.

McCance, R. A. (1935) Experimental sodium chloride deficiency in
man, *Proc. roy. Soc. B*, **119**, 245.

Strauss, M. B. (1957) *Body Water in Man; the Acquisition and
Maintenance of the Body Fluids*, London.

Thorn, N. A. (1958) Mammalian antidiuretic hormone, *Physiol.
Rev.*, **38**, 169.

Wolf, A. V. (1958) *Thirst*, Springfield.

13. THE KIDNEY

The two kidneys are situated at the back of the abdomen behind the peritoneum on either side of the vertebral column, between the twelfth thoracic vertebra and the third lumbar vertebra. Each kidney is about 11 cm. long, 6 cm. wide and 3 cm. thick. It weighs about 140 g. The kidney has many functions. It produces erythropoietin [p. 10]. It produces renin [p. 167]. It converts vitamin D into 1:25-dihydroxycholecalciferol [p. 126]. It produces urine which is a yellow fluid due to the presence of the pigment *urochrome*.

The kidneys receive a large blood supply from the renal arteries which arise from the abdominal aorta. The renal veins drain into the inferior vena cava [FIG. 136].

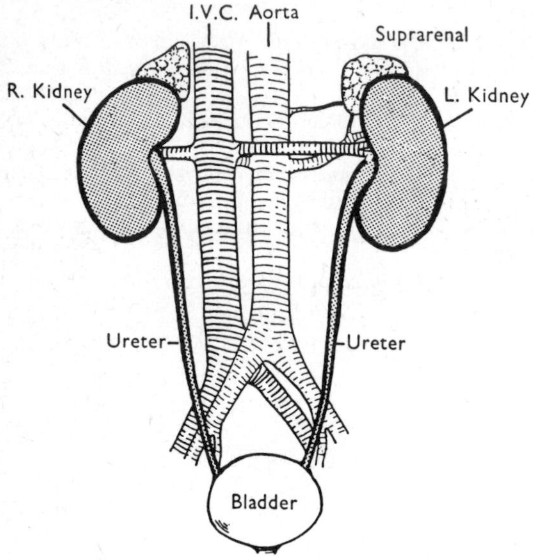

FIG. 136. The kidneys, ureter and bladder in the abdomen.

From each kidney a tube known as the *ureter* conveys the urine to the bladder. Each ureter is 30 cm. long. The urinary bladder acts as a reservoir for the urine which is voided at intervals via the urethra. The urethra in the male is 20 cm. long. In the female it is only 4 cm.

Under normal circumstances about one and a half litres of urine are produced each day containing about 50 g. of solids. The chief constituents are urea (30 g. per day) and sodium chloride (15 g. per day). In addition, the urine contains the surplus of other ingested inorganic ions, and the final stage of the metabolic breakdown of a large number of dietary constituents. It also contains traces of most of the substances circulating in the blood.

The concentration of the constituents in urine increases when the urine production is reduced following copious sweating. If, on the other hand, large quantities of fluid are taken, the concentration decreases and the specific gravity falls as the urinary volume is greatly increased.

The pH of urine varies between 4·5 and 8·5 as the kidney excretes either the surplus ingested hydrogen ions giving an acid urine, or a surplus of ingested hydroxyl ions giving an alkaline urine. In this way the pH of the blood is maintained at a constant level of 7·4.

BASIC PRINCIPLES OF URINE FORMATION

Urine is formed from the blood by a process of filtration followed by selective reabsorption. The basic unit for urine production is the **nephron** [FIG. 137].

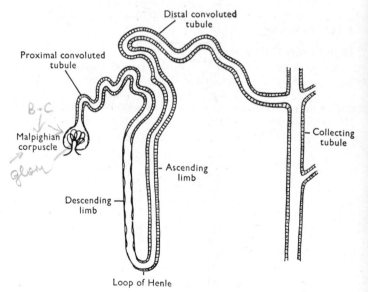

FIG. 137. The nephron.

There are 1,000,000 nephrons in each kidney. Each consists of a blind-ended tube. A tuft of blood capillaries termed a **glomerulus** is invaginated in to the closed end of the tube known as Bowman's capsule. This structure [FIG. 138] was first described by Malpighi who likened it to a blood corpuscle. For this reason it is often referred to as a *Malpighian corpuscle*. We shall refer to the whole structure as the *glomerulus*. Its function is that of a filter as first proposed by the English surgeon Bowman at the beginning of the last century. The remainder of the tube is called the **tubule.** It is divided into the proximal convoluted tubule, the descending and ascending limbs of the loop of Henle, the distal convoluted tubule and the collecting tubule [FIG. 137].

Glomerular Filtration

A filter or sieve is a device which separates large and small particles by retaining the larger particles and allowing the smaller particles to pass through. The glomerulus of each nephron filters the blood. It allows the blood constituents having a molecular weight of less than 68,000 to pass through into the tubule, but retains any larger molecules and particles in the blood capillaries.

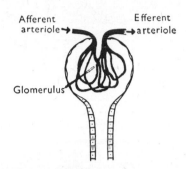

Afferent arteriole →

Efferent arteriole ←

Glomerulus

FIG. 138. Malpighian corpuscle.

Electron microscope studies have revealed pores 0·1 μ in diameter between the capillary and the lumen of the tubule, which would appear to be associated with this filtering action. TABLE 13 lists some of the more important constituents of the blood according to whether or not they exceed this limit.

TABLE 13

Blood constituents having a molecular weight of 68,000 or greater	Blood constituents having a molecular weight of less than 68,000
Red cells	Water
White cells	Food substances (glucose, amino acids, etc.)
Platelets	Inorganic salts
Plasma proteins	Waste products (urea, uric acid, creatinine, etc.)
These substances remain in capillary	These substances enter into tubule

The filtrate, at this stage contains all the blood constituents in the second column of Table 13. The blood cells and the plasma proteins are too large to appear in the filtrate. Haemoglobin (molecular weight 67,000), when present in the plasma following haemolysis of red cells, is just small enough to be filtered.

Glomerular Filtration Rate (GFR)

The kidneys receive a large blood flow. At rest about one quarter of the cardiac output goes to the kidneys. Assuming a cardiac output of 5 litres per minute, the kidney blood flow will be approximately 1,200 ml. per minute. One-tenth of this is filtered as the blood flows through the glomeruli. The volume of filtrate produced by all the glomeruli is thus 120 ml. per minute. This equals 170 litres per day. The volume of filtrate produced per minute is termed the **glomerular filtration rate** (GFR).

The energy for the filtration process is provided by the heart in the form of arterial blood pressure. The filtering force is the blood pressure in the capillaries of the Malpighian corpuscle (*circa* 70 mm. Hg) less the osmotic pressure of the plasma proteins (25 mm. Hg) and the pressure in the tubule (10 mm. Hg). The net filtering pressure is thus 35 mm. Hg.

Auto-regulation

The kidney possesses the property of *auto-regulation*. That is, its blood flow remains constant and independent of blood pressure changes provided that this pressure is within the range of 80–200 mm. Hg. As a result, the glomerular blood flow and glomerular filtration rate are not affected by the small changes in blood pressure which occur in everyday life. Should the blood pressure, however, fall to a very low level, say, following a haemorrhage, the auto-regulation will be lost. The glomerular filtration will be reduced and may cease altogether leading to anuria (= no urine).

Tubular Reabsorption

After the blood has passed through the glomerulus it enters a second arteriole (efferent arteriole) which leads to a second capillary network around the tubule. Only then does the blood return to the renal vein.

The tubule cells selectively reabsorb the constituents of the filtrate as they pass along the tubule. This reabsorption is according to the body's need for the substance. Food substances are completely reabsorbed. The reabsorption of the inorganic salts is variable and depends upon the plasma level of these substances. Waste products are only slightly reabsorbed, the major proportion being excreted.

Tubular reabsorption of many substances is an *active* process. The substance is absorbed at one side of the tubule cell, transported through the cell and secreted in the blood capillary at the other side [FIG. 139]. Such a process requires the utilization of energy by the tubule cells.

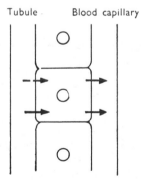

Tubule   Blood capillary

FIG. 139. Active transport of substances by the tubular cells from the lumen of the tubule to the blood capillary. Each substance is taken into the cell on one side, transported across the cell and secreted from the other side.

By the time the filtrate has reached the collecting ducts, the pelvis of the kidney and the ureter, it has become urine. It passes to the bladder and is excreted at intervals via the urethra by the process of *micturition* [p. 207].

FIGURE 140 shows a 'large' hypothetical nephron which represents all the nephrons in the two kidneys. We will use this type of diagram to consider the various constituents of the glomerular filtrate and their reabsorption in more detail.

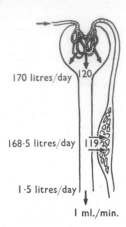

FIG. 140. Filtration and reabsorption of water by the kidneys. This nephron represents all the nephrons in both kidneys. Under the normal conditions shown, over 99 per cent. of the water filtered is reabsorbed leaving only 1 ml. per minute (= 1·5 litres per day) to pass to the bladder for excretion.

(a) Water

Under normal circumstances over 99 per cent. of the water entering the tubule from the glomerulus has been reabsorbed before the ureter is reached. Of the 170 litres per day of water filtered, 168·5 litres are reabsorbed allowing only 1½ litres per day to be passed to the bladder and voided as urine.

The volume of water in the blood (plasma volume) is only 3 litres and the filtration of 170 litres per day represents the filtering and reabsorption of the same water molecules many times during the 24 hours.

Per minute, the volume of glomerular filtrate is 120 ml. [FIG. 140]. 119 ml. of water are reabsorbed, leaving only 1 ml. per minute to pass to the bladder.

The reabsorption of water, as will be seen later, takes place in the proximal tubule, the distal tubule and in the collecting tubule. Seven-eighths of the water is reabsorbed in the proximal tubule. The reabsorption in the distal tubule is under the control of the anti-diuretic hormone (ADH) from the posterior pituitary gland. Failure of production of this hormone leads to a failure of water reabsorption in this part of the nephron. The amount of filtrate passing to the bladder then increases and may amount to 22·5 litres of urine per day. The condition is termed *diabetes insipidus* [see p. 160].

In the distal tubule the reabsorption of water is controlled separately from the reabsorption of sodium which is under the control of a different hormone, **aldosterone.**

The level of ADH in the blood depends on the activity of the posterior pituitary gland. This, in turn, is controlled by the hypothalamus which receives nerve impulses from the osmoreceptor cells, lying in the vascular territory supplied by the internal carotid arteries in the region of the hypothalamus, and possibly also from volume receptors in the thorax [FIG. 141].

The osmoreceptors respond to the osmotic pressure of the blood and extracellular fluid. An increase in tonicity (osmotic pressure) causes an increase in impulse activity to the supra-optic nucleus of the hypothalamus and an increase in the ADH secretion. On the other hand, an increase in blood volume, acting via the volume receptors, inhibits the ADH secretion [see also page 209].

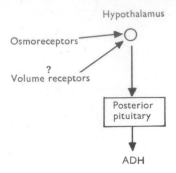

FIG. 141. Factors regulating the release of ADH.

Let us consider the effect on this control mechanism of drinking two glasses of water (500 ml.) when we are not thirsty. We will assume that this water is surplus to the body's need. The water will be absorbed from the small intestine and will pass into the blood, tissue fluid and into the cells. The total body water will increase from 45 litres to 45·5 litres, an increase of nearly 1 per cent. The blood volume will increase by this amount and the volume receptors will be stimulated. The blood's constituents will be diluted thus reducing the osmotic pressure. The osmoreceptor activity will be reduced. Both of these changes will lead to a decrease in the secretion of ADH by the posterior pituitary gland, and less reabsorption of water by the kidney tubules. More water will remain in the tubules and reach the bladder. The volume of urine produced will be increased until the surplus 500 ml. of water has been eliminated. The increased urine flow is termed *diuresis.*

It will be noted that if the reabsorption [FIG. 140] changes from 119 to 118 ml. per minute, the rate of urine formation is doubled. Small changes in the reabsorption make large changes in the urine production [see TABLE 14].

TABLE 14

Glomerular filtration	Tubular reabsorption	Urine formation	
		ml./minute	ml./day
120	119·8	0·2	300
120	119·5	0·5	750
120	119	1	1,500
120	118	2	3,000
120	117	3	4,500
120	115	5	7,500
120	110	10	15,000
120	105	15	22,500

If the body is short of water, the level of circulating ADH increases and the tubular reabsorption rises to a maximum of 119·8 ml. per minute, allowing only 0·2 ml. of water to pass to the bladder. This gives a urine production of 300 ml. per day which is the minimum volume (obligatory volume) necessary to excrete the waste products.

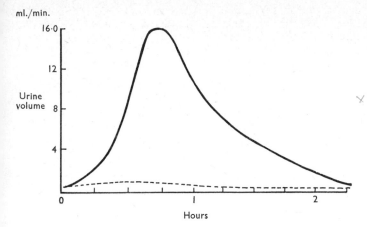

FIG. 142. The effect of drinking 1 litre of water on the rate of urine production per minute.

———, 1 litre of water taken at time zero;
————, 0·5 Units of posterior pituitary extract injected subcutaneously at the same time as taking the water (from an experiment by Helen N. Duke).

The action of ADH on urine production is illustrated in the experiment shown in FIGURE 142. In the upper curve the subject has taken 1 litre of water at time 0 and the urine production has been measured at 15 minute intervals for the next 2¼ hours. It will be seen that there is an increase in urine production which reaches a maximum of 16 ml. per minute after three quarters of an hour.

In the lower curve the subject has taken 1 litre of water at time 0, but the reduction in circulating ADH that this causes has been opposed by a subcutaneous injection of 0·5 Units of posterior pituitary extract. No diuresis occurs and the urine production remains at a low level. The effect of the ADH injection wears off in a few hours and then the diuresis takes place.

(b) Food Substances

Glucose is completely reabsorbed under normal conditions. The tubule cells are limited in the quantity of glucose that they are able to transfer back into the blood in a given time, and if the blood glucose level exceeds 180 mg. per 100 ml. blood (10 mmol/l) glucose appears in the urine. The normal blood glucose level is only 60–100 mg. per 100 ml. blood (3·3–5·5 mmol/l). Patients with untreated diabetes mellitus have a high blood sugar (glucose) level and, therefore, have glucosuria [see page 145].

Amino acids are present only in traces in the urine. Like glucose their reabsorption is virtually complete.

(c) Inorganic Salts

Sodium chloride is the only inorganic compound eaten as such in the diet. As a result there is usually a surplus of intake over requirement which amounts to 10–15 g. of sodium chloride per day. This is the amount excreted per day in the urine.

The 170 litres of plasma filtered per day will have contained something like 1,600 g. of salt. So it will be seen that most of the salt filtered is reabsorbed. Incidentally, the body does not contain 1,600 g. of salt. Like the filtered water, the same NaCl molecules are filtered and reabsorbed many times in the same day.

Sodium chloride is also lost from the body in sweat [p. 151], and should a salt deficiency develop, complete reabsorption occurs in the tubules so that none is lost in the urine.

Seven-eighths of the NaCl is reabsorbed in the proximal tubule, and only one-eighth in the loops of Henle and the distal tubules. The reabsorption in the distal tubules is stimulated by aldosterone produced by the cortex of the suprarenal glands [p. 166].

(d) Waste Products

The waste products found in greatest amounts in the urine are urea, uric acid and creatinine. With the exception of carbon dioxide, practically all the waste products from metabolism are excreted in the urine.

UREA. Thirty grams of urea are excreted per day. After evaporating urine to dryness the urea may be extracted with acetone.

Urea is made in the liver from the ammonia which arises from the de-amination of the surplus amino acids. It is formed by the ornithine-citrulline-arginine cycle (Krebs urea cycle) [p. 115]. The amount produced per day depends upon the protein intake and the nutritional need for amino acids for growth and repair of tissues.

Ammonium salts taken by mouth are converted to urea in the liver. Ammonium chloride taken by mouth thus yields urea and hydrochloric acid. It may therefore be used as an acidifying agent.

The normal level of urea in the blood is 30 mg. per 100 ml. blood (5 mmol/l).

The formation of urea in the liver and its removal by the kidney may be demonstrated experimentally in the following manner. If the liver is removed from an animal the formation of urea ceases and the blood urea falls steadily. However, if the kidneys are also removed the blood urea level remains constant. It follows that the kidneys are the sole route for urea excretion.

In kidney failure in man the blood urea rises and the condition is termed uraemia.

URIC ACID. The level of blood uric acid is 2–3 mg. per 100 ml. blood (0·2 mmol/l). Up to 2 g. of uric acid are excreted per day.

CREATININE. Creatinine comes from the creatine in the muscle. Creatinine is the anhydride of creatine. The amount excreted per day is related to the muscle mass of the body and is greater in a muscular man than in a woman or child.

OTHER METABOLIC PRODUCTS. These include purine compounds such as xanthine, hypoxanthine and adenine; hippuric acid and urinary indican. The indican is derived from the indol which is produced in the intestines by bacterial action. It increases thirtyfold in intestinal obstruction. It disappears if the intestines are sterilized by antibiotics.

Also found in the urine are traces of ketone bodies, and glucuronides which are derivatives of hormones, vitamins, drugs, etc.

FURTHER FUNCTIONS OF THE TUBULE CELLS

Secretion

In addition to being filtered by the glomerulus, certain substances are actively secreted into the lumen by the cells of the tubule. Such substances include para-aminohippuric acid (PAH), diodone and penicillin.

$$O=\!\!\!\!\bigcirc\!\!\!\!-N-CH_2\cdot COOH$$

Diodone (3:5 di-iodo-4-pyridone-N-acetic acid)

PAH and diodone are so ruthlessly excreted both by filtration and tubular secretion that the renal veins are free from these substances.

They may therefore be used to measure renal blood flow by a modification of the Fick principle. The renal blood flow is equal to the amount of PAH removed from the blood per minute divided by the difference in arterial and venous concentrations of this substance [see p. 82]. Since the venous concentration will be zero, the A–V difference will equal the arterial concentration. The amount removed from the blood per minute equals the amount excreted in the urine per minute and may be determined by analysing the urine. Thus,

$$\text{Renal blood flow} = \frac{\text{PAH excreted per minute in the urine in g.}}{\text{Concentration in arterial blood (g. per ml.)}}$$

We shall meet this equation again after a consideration of 'clearance' later in this chapter.

Secretion is an active process on the part of the tubule cells.

$$NH_2\text{---}\langle\bigcirc\rangle\text{---}CO\cdot NH\cdot CH_2\cdot COOH$$

p-aminohippuric acid (PAH)

Maximum Tubular Secretory Capacity

The maximum rate at which the tubules can transport a substance from the blood vessels to the lumen (or vice versa) is limited by the carrier system. The maximum rate is known as the **tubular maximum** and is usually expressed in milligrams per minute.

The secretory tubular maximum (Secretory T_m) for PAH is 80 mg. per minute.

It follows that PAH will only be completely cleared from the blood flowing through the kidneys if its plasma concentration is low. With a high plasma concentration the tubular cells will be unable to secrete all the PAH into the lumen in the time available.

Regulation of pH

The tubule cells are able to secrete either hydrogen ions or hydroxyl ions. In this way they are able to change the pH of the filtrate from 7·4, the pH of the plasma, to anywhere between pH 4·5 to 8·5, the range of urine pH.

Acidaemia (Acidosis)

It has already been seen that a diet rich in protein tends to make the blood more acid (acidaemia). It will be remembered that this is because the sulphur in the protein is oxidized to sulphuric acid, and the phosphorus is oxidized to phosphoric acid. These acids are buffered by the bicarbonate/CO_2 and protein buffer systems of the blood [p. 102]. However, buffering is only a temporary expedient and ultimately the surplus hydrogen ions must be excreted and the alkali reserve (sodium bicarbonate) restored to normal [FIG. 143].

It is the excretion of the hydrogen ions that makes the urine acid, but in the urine the hydrogen ions are buffered. Sulphuric acid[12] is not excreted as such, but is excreted 'indirectly' by the conversion of disodium hydrogen phosphate to monosodium dihydrogen phosphate:

$$2\,Na_2HPO_4 + H_2SO_4 \rightarrow 2\,NaH_2PO_4 + Na_2SO_4$$

[12] A small fraction of the sulphate in the urine (10 per cent.) is in the form of indoxyl sulphate formed in the liver by the union of a phenolic compound with the sulphate ion. Such substances are neutral and are termed ethereal sulphates.

Thus the urine does not contain sulphuric acid and disodium hydrogen phosphate, but acid sodium phosphate and neutral sodium sulphate.

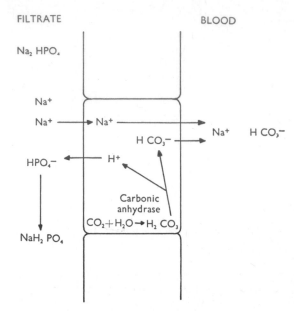

FIG. 143. Secretion of hydrogen ions by the conversion of alkaline sodium phosphate to acid sodium phosphate by the tubule cells with the formation of $NaHCO_3$, that is, an increase in the alkali reserve of the blood.

Similarly, phosphoric acid itself is never present in the urine since:

$$Na_2HPO_4 + H_3PO_4 \rightarrow 2\,NaH_2PO_4$$

Both disodium hydrogen phosphate (Na_2HPO_4) and monosodium dihydrogen phosphate (NaH_2PO_4) are found in the plasma and will therefore be present in the tubular filtrate.

The pH of a solution containing Na_2HPO_4 and NaH_2PO_4 is given by:

$$pH = pK + \log_{10}\frac{[Na_2HPO_4]}{[NaH_2PO_4]}$$

where $pK = 6\cdot8$ [see p. 95]

The pH of the urine is thus determined by the ratio of disodium hydrogen phosphate to the monosodium dihydrogen phosphate. The pH is given by:

$$pH = 6\cdot8 + \log_{10}\frac{[Na_2HPO_4]}{[NaH_2PO_4]}$$

The strongly alkaline salt Na_3PO_4 does not appear in the urine.

In the plasma the concentration of the alkaline phosphate (Na_2HPO_4) exceeds that of the acid phosphate (NaH_2PO_4) by a ratio of 4:1.

Thus in the plasma:

$$pH = 6\cdot8 + \log_{10}\frac{[Na_2HPO_4]}{[NaH_2PO_4]}$$
$$= 6\cdot8 + \log_{10}4$$
$$= 6\cdot8 + 0\cdot6$$
$$= 7\cdot4$$

In highly acid urine the concentration of the acid phosphate may exceed that of the alkaline phosphate by as much as 200:1.

In this case the urine pH is given by:

$$\begin{aligned}
\text{Urine pH} &= 6\cdot8 + \log_{10}\tfrac{1}{200}\\
&= 6\cdot8 - \log_{10}\tfrac{200}{1}\\
&= 6\cdot8 - 2\cdot3\\
&= 4\cdot5
\end{aligned}$$

If large amounts of acid have to be excreted, the kidney tubule cells form ammonium salts. Ammonia is not extracted from the blood, but is manufactured in the kidney itself from the glutamine of the plasma, and from other circulating amino acids.

$$\underset{\text{Glutamine of plasma}}{\begin{matrix} CO\cdot NH_2 \\ | \\ CH_2 \\ | \\ CH_2 \\ | \\ CH\cdot NH_2 \\ | \\ COOH \end{matrix}} + H_2O \xrightarrow{\substack{\text{Kidney}\\ \text{glutaminase}}} \underset{\text{Glutamic acid}}{\begin{matrix} COOH \\ | \\ CH_2 \\ | \\ CH_2 \\ | \\ CH\cdot NH_2 \\ | \\ COOH \end{matrix}} + NH_3$$

The ammonia removes the hydrogen ions by forming ammonium salts.

Thus,

$$2\,NH_3 + H_2SO_4 \rightarrow \underset{\substack{\text{Ammonium}\\ \text{sulphate}}}{(NH_4)_2SO_4}$$

And,

$$NH_3 + H_3PO_4 \rightarrow NH_4H_2PO_4$$
$$2\,NH_3 + H_3PO_4 \rightarrow \underset{\substack{\text{Ammonium}\\ \text{phosphates}}}{(NH_4)_2HPO_4}$$

The urine also contains other inorganic salts such as calcium chloride and potassium chloride. In a mixture of these dissociated salts in solution, it is impossible to say whether we are dealing with, say, ammonium sulphate and sodium chloride, or ammonium chloride and sodium sulphate. For this reason it is usual to list the anions and cations in urine separately, and no attempt is made to pair them off [TABLE 15].

TABLE 15. PRINCIPAL IONS IN URINE

Cations	Anions
Na^+	Cl^-
K^+	SO_4^{--}
Ca^{++}	$H_2PO_4^-$
Mg^{++}	HPO_4^{--}
NH_4^+	HCO_3^- (alkaline urine only)

The only relatively insoluble combinations of these ions are those between the alkaline-earth elements (calcium and magnesium) and the phosphate radicals.

In acid urine the corresponding acid phosphates, $CaH_4(PO_4)_2$ and $MgH_4(PO_4)_2$, are fairly soluble in water, but in alkaline urine the alkaline phosphates are formed. These are $Ca_2H_2(PO_4)_2$ and $Mg_2H_2(PO_4)_2$ which are only slightly soluble in water. When urine is allowed to stand, the urea is converted to ammonia by bacterial action and the urine becomes alkaline. When this occurs the earthy phosphates may be precipitated making the urine cloudy. A further insoluble compound is formed. This is the triple phosphate NH_4MgPO_4 and this will also be precipitated.

Sodium Bicarbonate Reabsorption

It will be noted that although sodium bicarbonate is present in relatively large amounts in the blood, it is not present in the normal acid urine. It is completely reabsorbed in the tubule.

Carbonic anhydrase, the enzyme which acts as a catalyst for the conversion of carbon dioxide and water to carbonic acid and bicarbonate, plays an important part in the excretion of hydrogen ions in the distal tubule. At the same time the sodium ions are reabsorbed [FIG. 144]. Thus sodium ions are exchanged for hydrogen ions.

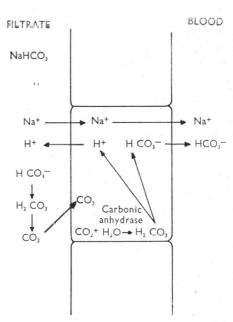

FIG. 144. The reabsorption of sodium bicarbonate by the kidney tubules. The sodium ions are absorbed in exchange for hydrogen ions. The carbonic acid formed in the tubule becomes CO_2 and water. The CO_2 enters the tubule cell. The hydrogen ion for the sodium exchange comes from carbonic acid made with the aid of carbonic anhydrase. Thus although the sodium ions are absorbed as such, new bicarbonate radicals are made from carbon dioxide to form the reabsorbed blood sodium bicarbonate.

This exchange converts the sodium bicarbonate in the distal tubule to carbonic acid, which splits into carbon dioxide and water. The sodium ions are taken into the cell and excreted in the blood capillary along with bicarbonate that has been made in the cell by the carbonic anhydrase. Thus although sodium bicarbonate disappears from the tubule and sodium bicarbonate appears in the capillary it is not the same bicarbonate radical in the two cases.

If a carbonic anhydrase inhibitor, such as acetazolamide (*Diamox*) is given, this interchange does not take place and the urine remains alkaline with the excretion of a large amount of sodium bicarbonate.

Potassium ions appear to be completely reabsorbed in the proximal tubule and to be re-excreted again in the distal tubule.

Alkalaemia (Alkalosis)

A vegetarian diet may lead to alkalaemia. As has been seen in CHAPTER 8, this is because the citrate and tartrate radicals of the sodium and potassium citrate and the sodium and potassium tartrate present are oxidized to carbon dioxide and water leaving the sodium and potassium ions and hydroxyl ions in the form of sodium and potassium hydroxide. These substances are buffered by the bicarbonate/CO_2 buffer system in the blood and converted to bicarbonate. But the hydroxyl ions have ultimately to be eliminated by the kidneys.

To achieve this the acid sodium phosphate is converted to the alkaline sodium phosphate. Thus:

$$NaH_2PO_4 + NaOH \rightarrow Na_2HPO_4 + H_2O$$

and the $\dfrac{Na_2HPO_4}{NaH_2PO_4}$ ratio is reversed.

Thus in highly alkaline urine, the ratio of alkaline phosphate to acid phosphate may rise to 40:1.
The pH would then be:

$$= 6\cdot8 + \log_{10} \tfrac{40}{1}$$
$$= 6\cdot8 + 1\cdot6$$
$$= 8\cdot4$$

In addition bicarbonate itself is excreted.

THE CONCENTRATION OF SOLUTES IN URINE

Specific Gravity

Urine is usually more concentrated than the plasma filtrate. The concentration of the solutes in the urine is given approximately by the specific gravity, which may be determined easily using a hydrometer which when used for this purpose is frequently called a urinometer [FIG. 145].

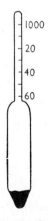

FIG. 145. Urinometer for the measurement of the specific gravity of urine. 1,000 on the scale corresponds to distilled water (= 1·000). 20 on the scale corresponds to 1,020 (or 1·020).

The calibration of the urinometer is checked by placing it in distilled water and noting the zero error, if any. The urinometer is then placed in the urine and allowed to float. The level of the fluid along the calibrated stem gives the specific gravity. It is usual to take the specific gravity of distilled water as being 1,000 instead of 1·000. The specific gravity of urine varies between 1,002 and 1,040 with an average value of 1,015.

When only a small volume of urine is available the concentration may be deduced from the optical refractive index of the urine. Portable instruments are available which will record directly both the refractive index and specific gravity using only one drop of urine [FIG. 146].

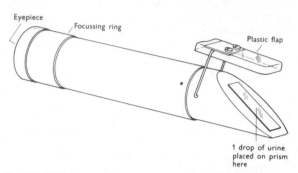

FIG. 146. Refractometer for measurement of specific gravity using one drop of urine. The flap is closed and the instrument is held up to the light. The position of the air-fluid interface on the scale gives the specific gravity.

The urine is placed on the prism at one end and the plastic flap closed. When held up to the light and viewed through the eyepiece, a line marking the limit of the zone of internal reflection is seen on the scale. The position of this line on the scale depends on the refractive index (and the specific gravity) of the urine. The instrument is calibrated in both specific gravity and refractive index at room temperature.

Osmolarity and Osmolality

Another convenient way of estimating the concentration of urine is to determine the depression of the freezing point of the water caused by the presence of the constituents. The freezing point of urine varies from between −0·25 °C. when the urine is very dilute to −2·6 °C. when the urine is concentrated. 1 g. molecule of a solute (the molecular weight in grams) dissolved in 1 litre of water depresses the freezing point by 1·86 °C. It also has an osmotic pressure of 22·4 atmospheres. Thus a freezing point depression of 1·86 °C. is equivalent to an osmotic pressure of 22·4 × 760 mm. Hg. If the solute molecule is ionized into two ions then 1 g. molecule of it will have twice this effect on the freezing point and the osmotic pressure. It is, therefore, more accurate to speak of 1 osmole of a solute per litre as depressing the freezing point by 1·86 °C. and having an osmotic pressure of 22·4 atmospheres, where an osmole is the molecular weight in grams divided by the number of ions into which the molecule dissociates.

Plasma and therefore the glomerular filtrate have a concentration of 0·3 osmoles per litre. Their freezing point is, therefore, −0·3 × 1·86 = −0·56 °C.

The final urine has a concentration, or osmolarity of up to 1·4 osmoles per litre of urine (or osmoles per kg. water = *osmolality*).

Since the *osmole* is a large unit, osmolarity is alternately expressed in *milli-osmoles*. Plasma has an osmolarity of 300 milliosmoles. The maximum osmolarity (and osmolality) of urine is 1400 milliosmoles.

TABLE 16 gives the relationship between specific gravity, osmolarity and depression of the freezing point.

TABLE 16

Sp. gr.	1,000	1,005	1,007	1,010	1,015	1,020	1,035
Osmoles/litre	0·0	0·2	0·3	0·4	0·6	0·8	1·4
Freezing point (°C.)	0	−0·37	−0·56	−0·74	−1·1	−1·5	−2·6

GLOMERULAR FILTRATE

A FURTHER CONSIDERATION OF THE MECHANISM OF WATER REABSORPTION

It is difficult to explain how the urine can become more concentrated than the plasma without postulating that water is actively reabsorbed by the tubule cells. *NO* But from experimental evidence it appears unlikely that active water transport occurs and it appears that the movement of water is passively determined by the concentration gradients of the solutes and particularly the sodium salts.

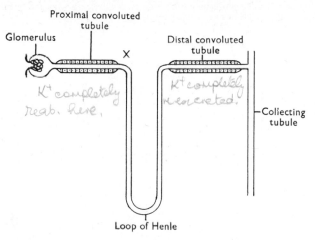

K⁺ completely reab. here,

K⁺ completely reexcreted.

FIG. 147. The constituent parts of the nephron (see text).

Micropipette techniques have enabled the composition and osmolarity of the tubular filtrate to be measured at different distances along the tubule. These results have shown that by the time the end of the proximal tubule has been reached [FIG. 147] seven-eighths of the sodium chloride and bicarbonate has been reabsorbed back into the blood by the active transport of the sodium ions from the lumen to the blood by the tubule cells. Seven-eighths of the water has followed passively. The rate of filtrate flow has thus decreased from 120 ml. per minute to 15 ml. per minute. If no further change in urine volume took place, the urine passed would be 22·5 litres per day, the volume voided in diabetes insipidus.

The composition of the filtrate at this point (point *X*, FIGURE 147) has altered slightly. All the glucose has disappeared due to reabsorption, whilst the concentration of the waste products such as urea, sulphate and creatinine has risen due to the limited reabsorption of these substances. But the osmolarity is unchanged at 0·3 osmoles per litre.

The filtrate now passes down the descending loop of Henle into the medulla of the kidney and returns via the ascending loop of Henle to reach the distal convoluted tubule and collecting tubule.

Micropipette estimations made at the end of the distal tubule show the filtrate to be still iso-osmotic. It appears, therefore, that the concentration of the urine takes place in the collecting tubules.

Kidney slice studies have shown that the osmotic pressure in the kidney increases with depth. The collecting tubule therefore runs through a region of increasing osmolarity and water is extracted from the collecting tubule by osmosis. FIGURE 148 shows how the osmolarity may be increased to 1·4 osmoles per litre which is the maximum concentration found in man.

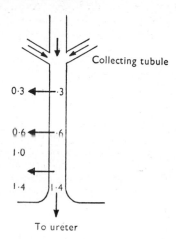

FIG. 148. Mechanism of concentration of urine. The collecting tubule passes through tissue spaces which have an increasing osmolarity. As a result water is absorbed and the urine becomes more concentrated.

Counter-current Multiplier Hypothesis

Some of the loops of Henle and their associated capillaries run into the region of high osmolarity. Wirtz has postulated that sodium ions are actively transported from the lumen out into the interstitial fluid by the cells of the ascending loop. These cells are also thought to be impermeable to water. Such an arrangement would enable the

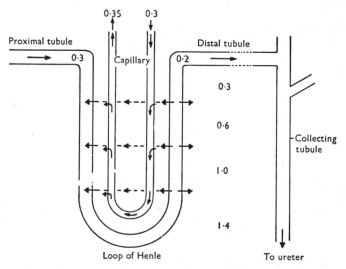

FIG. 149. The counter-current multiplier system for the formation of urine.

osmotic gradients to be set up and maintained by a counter-current multiplier system [FIG. 149].

In the ascending loop of Henle the arrows represent the movement of the sodium salts as they are actively transported from the lumen to the tissue fluid. They enter the descending loop by diffusion, only to be carried round to the ascending loop again, where they pass into the tissue fluid once more. The sodium salts are thus trapped in the region of the loop both in the lumen and in the tissue spaces.

The blood in the accompanying capillary also engages in this counter-current exchange. Sodium is taken up by simple diffusion at the arterial end of the capillary. Its concentration increases as the blood flows towards the loop and then decreases again as the capillary leaves the region of high osmolarity.

Although the osmolarity in the capillary may reach 1·4 osmoles in the loop, the blood leaves with an osmolarity only slightly greater than when it arrived (0·35 as against 0·3). The over-all effect is the removal of a small quantity of sodium salts without the corresponding water and the filtrate passing on to the distal tubule will be slightly hypotonic (0·2).

The osmolarity of the tissue fluid surrounding the distal tubule is the same as that of plasma, 0·3 that is, hypertonic compared with the filtrate in this part of the tubule. Thus water can now be absorbed from the distal tubule by osmosis as well as from the collecting tubule, when the circulating ADH makes the epithelial wall permeable to water.

A FURTHER CONSIDERATION OF THE REABSORPTION OF SODIUM IONS

It has been seen that seven-eighths of the sodium ions are normally reabsorbed in the proximal tubule and that some further sodium ions are actively reabsorbed from the loop of Henle in the countercurrent system. The reabsorption of sodium is 'time limited' in the proximal tubule, and with a low glomerular filtration rate proportionally more sodium is reabsorbed.

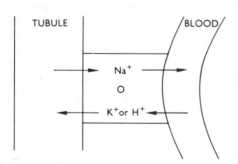

FIG. 150. Some of the Na^+ in the distal tubule is absorbed in exchange for K^+ or H^+.

Further reabsorption of sodium takes place in the distal tubule, partly as Na^+ and Cl^- ions, and partly as the cation exchange of Na^+ for K^+ or H^+ (NH_4^+) [FIG. 150]. It would appear that potassium and hydrogen ions share the same transport system for their excretion and that either one hydrogen ion or one potassium ion is excreted in exchange for every sodium which is reabsorbed by this mechanism.

This fact has several important implications.

1. Sodium appears to be so important to the body that in a sodium depleted subject, it is retained (by complete reabsorption) at the expense of the hydrogen ions which are excreted instead. This leads to alkalaemia. This form of alkalaemia is treated by raising the plasma sodium level.

2. With normal sodium levels and reabsorption, potassium ions and hydrogen ions are excreted in exchange for the sodium ions. In a potassium depleted subject (hypokalaemia), insufficient potassium ions will be available for this exchange. Hydrogen ions will be excreted instead leading to alkalaemia (hypokalaemic alkalosis). This form of alkalaemia is treated by giving potassium ions to raise the plasma potassium level.

3. With a high plasma potassium level (hyperkalaemia), potassium will be excreted in preference to hydrogen ions leading to hydrogen ion retention and acidaemia (hyperkalaemic acidosis).

4. Sodium ion excretion is reduced when hydrogen ions are being excreted as in metabolic acidosis [p. 102].

Although a severe metabolic acidosis is treated using an alkali such as sodium bicarbonate (or sodium citrate, tartrate or lactate which will be converted to bicarbonate by metabolism), a metabolic alkalosis may be treated by giving a neutral sodium chloride solution. Presumably the sodium ions are excreted in the urine at the expense of hydrogen ions which are retained along with the chloride ions. This will be equivalent to gaining hydrochloric acid.

Juxtaglomerular Apparatus

The reabsorption of sodium ions in the distal tubule is under the control of the hormone **aldosterone** which is secreted by the *zona glomerulosa* region of the adrenal cortex [p. 166].

(It should be easy to remember that it is the zona glomerulosa of the adrenal gland that is altering the composition of urine since, as its name implies, this part of the gland looks something like the kidney under the microscope.)

This hormone changes the permeability of the distal tubules to sodium ions.

The control system for the release of aldosterone [p. 167] is not fully understood, but the juxtaglomerular apparatus (JGA) of the kidney appears to play an important part.

The juxtaglomerular apparatus consists of a group of secretory cells situated around the afferent arteriole of the glomerulus which sense the sodium ion concentration in the distal tubule. This at first sight seems rather surprising but it should be remembered that the convoluted tubules surround the glomeruli, and that the *macula densa* area of the distal tubule is very close to the glomerulus and to these cells.

When the sodium ion concentration in the distal tubule region is low, the JGA cells increase their release of renin. As has been seen, renin is an enzyme which breaks down α2 globulin plasma protein to form an inactive decapeptide, angiotensin I, which is rapidly converted into an active octapeptide, angiotensin II [p. 45]. Angiotensin II stimulates the release of aldosterone from the adrenal cortex. More sodium ions are reabsorbed from the distal tubule.

A low level of sodium in the distal tubule usually implies a low plasma sodium level, and the increased sodium reabsorption acts in such a way as to restore the plasma sodium level to its correct value.

Conversely a high plasma sodium level will reduce the renin production and the lowered angiotensin and aldosterone levels will lead to a greater sodium ion excretion.

An increased renin production may also be the result of a low glomerular filtration rate which will allow more time for the sodium reabsorption in the proximal tubule.

This accounts for the increased renin production found when the kidney blood flow is reduced (ischaemic kidney). Since angiotensin in large quantities constricts arterioles, hypertension may result [see p. 46].

It also accounts for the excessive sodium retension seen in congestive heart failure. The low cardiac output state will give a low kidney blood flow, a reduced glomerular filtration rate, and high renin, angiotensin and aldosterone levels which will cause an excessive sodium reabsorption in the kidney tubules. Such a patient with congestive heart failure has to be placed on a sodium chloride restricted diet and be given diuretics to reduce the salt and water retention.

The sequence of events in congestive heart failure may be represented as follows:

CONGESTIVE HEART FAILURE → Low kidney blood flow → Reduced GFR → Sodium filtered reduced

→ Reduced sodium level in distal tubule → JGA → Renin increased → Angiotensin increased

→ Aldosterone increased → Kidney → INCREASED SODIUM REABSORPTION

In addition to its action in releasing aldosterone, it is possible that angiotensin may act directly on the kidney by constricting the afferent arteriole, lowering the glomerular filtration rate, reducing the sodium filtered and with more time available for its reabsorption, reducing the rate of sodium excretion.

CLEARANCE

'Clearance' is a mathematical concept useful when considering kidney function. Instead of considering the quantity of a substance Y removed from the body in the urine in absolute units such as grams per minute, we consider it instead in terms of the volume of blood that contains this amount of substance Y. This volume is known as the 'clearance' and represents the volume of blood that has theoretically been completely cleared of substance Y in one minute.

If the concentration of substance Y in the urine is U g. per ml. and the volume of urine passed is V ml. per minute, then the quantity of substance Y leaving the body in the urine per minute will be $U \times V$ g.

If the concentration of substance Y in the plasma is P g. per ml., then the volume of plasma containing $U \times V$ g. of the substance will be:

$$\frac{U \times V}{P} \text{ ml.}$$

This is the clearance of substance Y.

The clearance of a substance such as glucose will be zero, provided that there is no glycosuria, since all is reabsorbed. The corresponding volume of plasma that is completely cleared of glucose is nil.

The clearance of water is of theoretical interest only. It is 1.

Inulin

The clearance of substances that are filtered but not reabsorbed or secreted is of great interest since it will equal the glomerular filtration rate (GFR). Such a substance is the sugar inulin $(C_6H_{10}O_5)_n$. Its clearance is 120 ml. per minute [FIG. 151]. Thus the corresponding

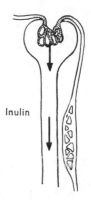

FIG. 151. Inulin clearance.

Example: P = 30 mg. inulin per cent.
V = 1 ml. urine per minute
U = 3,600 mg. inulin per 100 ml. urine

Thus clearance $C = \dfrac{3,600 \times 1}{30} = 120$ ml. per minute.

volume of plasma that is completely cleared is 120 ml. Since there is no change in the tubules this volume of plasma must have been filtered by the glomeruli.

PAH

Another group of substances of great interest are those that are both filtered and secreted by the tubules so that the plasma is completely cleared of these substances as it passes through the kidney. The clearance of such a substance is equal to the kidney plasma flow. As already mentioned, such a substance is para-aminohippuric acid (PAH). Its clearance is 650 ml. per minute. Thus the kidney plasma flow is 650 ml. per minute.

The kidney blood flow is calculated from the haematocrit. Thus:

$$\text{Kidney blood flow} = \text{Kidney plasma flow} \times \frac{100}{100 - \text{haematocrit}}$$

$$= 650 \times \frac{100}{100 - 45}$$

$$= 1,200 \text{ ml. per minute.}$$

The ratio of glomerular filtration rate to the kidney plasma flow gives the fraction of the plasma flow that is filtered and enters the tubules. It is termed the filtration fraction (FF).

Thus:

$$\text{Filtration fraction} = \tfrac{120}{650} = 0.19$$

$$= 19 \text{ per cent.}$$

The clearance of the various substances is represented diagrammatically in FIGURE 152.

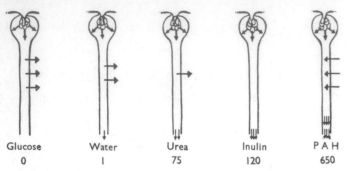

FIG. 152. The clearance of glucose, water, urea, inulin and para-aminohippuric acid (PAH).

A clearance of less than that of inulin (120 ml. per minute) indicates that reabsorption of this substance is taking place in the tubules. A clearance of over this figure means that the substance is being secreted into the filtrate by the tubule cells [FIG. 153].

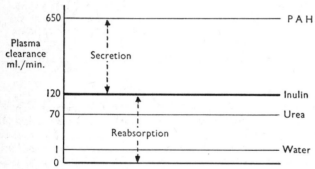

FIG. 153. If the clearance exceeds 120, secretion is taking place. If it is less than 120 then reabsorption is occurring.

Urea Clearance

The determination of urea clearance is used clinically to assess kidney function. It has the advantage that urea is a normal constituent of blood and urine and no injection of the test substance is therefore needed. Provided that the urine flow is high, that is, in excess of 2 ml. per minute, the urea clearance is a constant, independent of the urine flow. The normal value for this constant is 75 ml. per minute.

EXAMPLE

The urine flow is 3 ml. per minute, the plasma urea concentration is 30 mg. per 100 ml. and the urine urea concentration is 750 mg. per 100 ml.

Thus,

$$C = \frac{750 \times 3}{30} = 75 \text{ ml. per minute}$$

If the urine flow is increased to 5 ml. per minute, the plasma urea concentration is unchanged, but the urea concentration in the urine falls to 450 mg. per cent.

In this case,

$$C = \frac{450 \times 5}{30} = 75 \text{ ml. per minute}$$

Standard Urea Clearance

At normal urine flows (about 1 ml. per minute) the urea excreted and therefore the clearance depends upon the square root of the urine flow [FIG. 154]. In order to compare the clearance of one

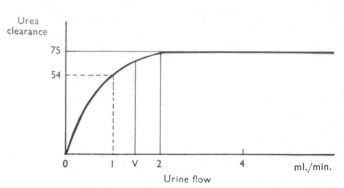

FIG. 154. If the urine flow exceeds 2 ml. per minute the urea clearance is a constant termed the maximum clearance (= 75). If the urine flow is less than 2 ml. per minute, the urea clearance is proportional to the square root of the urine flow.

The clearance at a flow of 1 ml. per minute is the standard clearance (= 54). This may be calculated from the clearance at urine flow V.

patient with another, an arbitrary flow of 1 ml. per minute is taken as the standard and the 'standard' clearance calculated. The standard clearance is an estimation of what the urea clearance would have been had the urine flow been 1 ml. per minute. It is determined as follows:

Let the urine flow be V ml. per minute and the urea concentration in the urine be U mg. per cent.

Then the quantity of urea excreted per minute is $U \times V$ when the flow is V ml. per minute.

If the flow were 1 ml. per minute, the amount excreted would have been

$$U \times V \times \frac{1}{\sqrt{V}} = U\sqrt{V}$$

If the plasma concentration is P

Then the standard clearance $= \dfrac{U\sqrt{V}}{P}$

The normal figure for standard clearance is 54 ml. per minute.

EXAMPLE

1. If the urine flow is 0·81 ml. per minute, the urea concentration is 1·8 g. per cent. in the urine and 30 mg. per cent. in the plasma. Then the true clearance C is given by:

$$C = \frac{1,800 \times 0·81}{30} = 49 \text{ ml. per minute}$$

But the standard clearance (with reference to a flow of 1 ml. urine per minute) C_s is given by:

$$C_s = \frac{1{,}800 \times \sqrt{0.81}}{30}$$

$$= \frac{1{,}800 \times 0.9}{30}$$

$$= 54 \text{ ml. per minute}$$

2. If the urine flow is reduced to 0·36 ml. per minute and the urea concentration is 2·7 g. per cent.
Then the true clearance is:

$$C = \frac{2{,}700 \times 0.36}{30} = 32 \text{ ml. per minute}$$

But $$C_s = \frac{2{,}700 \times 0.6}{30} = 54 \text{ ml. per minute}$$

Thus although the true clearances are different (49 and 32 ml. per minute respectively) the standard clearances, based on a urine flow of 1 ml. per minute, are the same (both 54 ml. per minute).

ESTIMATION OF UREA IN URINE

One method of estimating the quantity of urea in urine is the Dupré method. The urine is treated with an alkaline solution of sodium hypobromite and the urea content is calculated from the volume of nitrogen evolved.

$$CO(NH_2)_2 + 3\ NaOBr + 2\ NaOH$$
$$\rightarrow N_2 + 3\ NaBr + Na_2CO_3 + 3\ H_2O$$

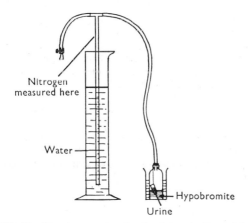

FIG. 155. The Dupré apparatus for the determination of the urea content of urine.

Hypobromite causes the evolution of nitrogen from the urea. The volume of nitrogen given off is measured and the urea calculated.

The apparatus is shown in FIGURE 155. 25 ml. of hypobromite solution, prepared by adding bromine to NaOH, are placed in the bottle. 4 ml. of urine are placed in the inner tube. The graduated burette is placed in a cylinder of water and after raising or lowering it until the level of water is the same outside and inside the burette, the reading is noted. The bottle is now tilted so that the urine runs into the hypobromite. The bottle is shaken and allowed to cool. After levelling up the increase in volume is noted.

1 g. of urea evolves 357 ml. (theoretically 373) N_2 at 0 °C. and 760 mm. Hg pressure. This is approximately equal to 400 ml. nitrogen at room temperature.

Let V = volume of gas evolved from 4 ml. of urine at room temperature.

Thus, 4 ml. urine must contain $\frac{V}{400}$ g. of urea. Therefore, 100 ml. urine contains $\frac{25\ V}{400}$ g. of urea.

ABNORMAL CONSTITUENTS OF URINE

TABLE 17 gives some of the commoner abnormal constituents of urine.

TABLE 17

Abnormal constituent	Possible cause
Protein	Kidney disease
Haemoglobin	Excessive breakdown of red cells
Glucose	High blood glucose level—diabetes mellitus
Ketone bodies	Ketosis—diabetes mellitus
Bile pigments	Jaundice
Bile salts	Obstructive jaundice
Galactose	Inborn error of metabolism
Phenylpyruvic acid	Inborn error of metabolism
Cystine	Inborn error of metabolism
Homogentisic acid	Inborn error of metabolism

Protein

The normal plasma proteins have too large a molecule to enter the tubule except in very small amounts which are reabsorbed. However, any abnormal protein in the plasma with a smaller molecular weight will enter the urine. Such proteins are haemoglobin from the haemolysis of red cells (molecular weight 67,000) and the Bence–Jones protein having a molecular weight of 35,000. Bence–Jones protein appears in the urine in the disease *myelomatosis*. This protein is probably the light chain of the γ-globulin antibody molecule [p. 22] which is produced in excessive amounts by abnormal antibody-producing cells in this condition. In kidney disease (nephritis) the normal plasma proteins, particularly the albumin, appear in the urine.

To detect the presence of albumin in urine, the simplest method is to acidify the urine with a few drops of acetic acid to dissolve the earthy phosphates [p. 139] and then to boil the urine. A precipitate indicates protein. In very severe cases the urine may solidify! Alternatively a protein precipitating agent such as sulphosalicylic acid may be added to the urine. A precipitate indicates protein.

Reagent strips are available which are coated with Tetrabromphenol Blue and a citrate buffer. These strips give a blue coloration when dipped into urine containing albumin.

Glucose

Glucose appears in the urine when the blood glucose level exceeds 180 mg. per cent. The return of the glucose molecules in the proximal tubule is an active process on the part of the tubule cells. These cells are limited in the quantity of glucose that they can transfer back to the blood in a given time.

When glucosuria is present, the urinary volume is increased. This is termed *polyuria*. There is, correspondingly, an excessive thirst termed *polydipsia* [osmotic diuresis, see below].

FIGURE 156 shows the excretion of glucose expressed graphically.

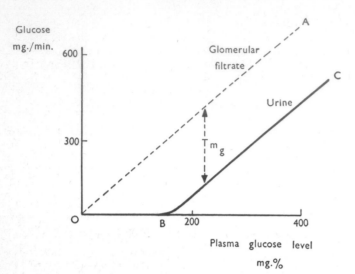

FIG. 156. The relationship between the blood glucose level and the glucose in the glomerular filtrate and in the urine.

As the plasma glucose level rises, the quantity of glucose in the glomerular filtrate rises linearly along line *OA*. The glucose reaching the bladder is zero until the plasma concentration reaches *B*. It then increases linearly along the line *BC* which is parallel to *OA*. All the time, between *B* and *C* glucose is being reabsorbed maximally. The vertical distance between the two parallel lines gives the maximum rate of tubular reabsorption for glucose. It is designated T_{m_g} and has a value of 300 mg. per minute.

Tests for Glucose

The standard test for glucose in the urine employs Benedict's qualitative sugar reagent which consists of copper sulphate, sodium citrate and sodium carbonate dissolved in water. Eight drops of urine are added to 5 ml. of the reagent and the solution is boiled vigorously for 3 minutes. A green, yellow or red precipitate indicates glucose.

In the reagent tablet method 5 drops of urine and 10 drops of water are placed in a tube and a tablet containing copper sulphate, citric acid, sodium carbonate and sodium hydroxide is added. The reaction between the citric acid and the sodium hydroxide produces sufficient heat to cause boiling. The resultant colours are compared with a test chart.

The above two tests give positive results with glucose, galactose, lactose, pentoses and homogentisic acid.

In the reagent strip method, a coated strip is dipped into the urine. A colour, produced by enzymic action, indicates the presence of glucose.

The enzyme glucose oxidase converts the glucose to gluconic acid and H_2O_2. The hydrogen peroxide oxidizes the o-tolidine on the reagent strip, in the presence of peroxidase to a blue compound. This colour change is specific to glucose and is not given by galactose. The colour should be noted after exactly 1 minute. Any colour appearing later should be ignored.

Ketone Bodies

The ketone bodies, acetoacetic acid, β-hydroxybutyric acid and acetone are always present in urine in traces, but the amount is greatly increased in ketosis.

Two tests are commonly employed to detect their presence. These are:

Rothera's Test

The urine is saturated with ammonium sulphate and then a few drops of a sodium nitroprusside solution are added in a test tube. Ammonium hydroxide is carefully run down the side of the tube. A purplish ring forming at the junction of the liquids indicates the presence of ketone bodies.

Diagnostic nitroprusside tablets are available that give a lavender to deep purple colour in the presence of ketone bodies when one drop of urine is placed on a tablet.

Gerhardt's Test

Ferric chloride is added to the urine and if acetoacetic acid is present a brownish-red colour is formed. This result is not given by the other two ketone bodies, and a false positive may be given by drugs such as aspirin.

Bile

Urine containing bile is more frothy than usual and the froth is stained by the bile pigments bilirubin and biliverdin. The bile salts, when present, lower the surface tension of the urine. One of the simplest ways of detecting their presence is to sprinkle powdered sulphur on the surface. With normal urine the sulphur will float, but if bile salts are present the sulphur will sink.

Fouchet's Test

Five ml. of 10 per cent. barium chloride solution is added to 10 ml. of acidified urine, and after shaking the moisture is filtered. One drop of Fouchet's reagent (1 per cent. ferric chloride in 25 per cent. trichloroacetic acid) is added to the precipitate on the filter paper. The presence of bile pigments is given by a greenish-blue coloration.

Galactose

Galactose appears in the urine in galactosaemia [p. 176]. It is a reducing sugar and like glucose reduces Benedict's reagent. It does not however give any reaction with enzymic reagent strips which detect glucose.

Phenylpyruvic Acid

Phenylpyruvic acid appears in the urine in phenylketonuria [p. 176]. It gives a blue-green colour with ferric chloride.

Homogentisic Acid

When homogentisic acid (alcapton) is present the urine darkens on exposure to air. Alcaptonuria results from an hereditary defect of metabolism in which there is incomplete oxidation of tyrosine [p. 114] and the excretion of homogentisic acid (dihydroxy-phenyl-acetic acid) in the urine.

OSMOTIC DIURESIS

If the urine contains glucose or an excessive amount of urea, the volume of urine is increased. This is because the osmotic pressure exerted by these substances in the filtrate opposes the reabsorption of water in the tubules, and as a result more water is excreted. There is thus an *osmotic obligation* which limits the minimum volume of urine passed. The sugar *mannitol*, which is not metabolized by the body and is excreted in the urine, is used to bring about a high rate of urine flow (*osmotic diuresis*) by this mechanism. An excess of potassium salts will have a similar effect.

RENAL FAILURE

If the kidneys fail, products of metabolism such as urea, uric acid, creatinine, and sulphuric acid (in buffered form) will accumulate in the blood. Urea is the substance which is usually monitored and in kidney failure the blood urea level rises from the *normal level of 5 mmol/l (30 milligrams per 100 ml.)* [p. 137] to 25 mmol/l (*150 milligrams per 100 ml.*) or more. The concentration of hydrogen ions in the blood, as indicated by the blood pH and plasma bicarbonate level [p. 102], is very important.

The glomerular filtration rate (GFR) falls and the kidneys are no longer able to control the excretion of water and sodium. Water and electrolyte balance can then only be maintained by controlling the intake. Provided that some kidney function remains, the accumulation of metabolites can be prevented by a careful choice of diet. Protein is severely restricted to minimize the production of sulphuric acid (from the surplus sulphur-containing amino acids, methionine and cysteine) which would otherwise lead to *acidaemia*.

Symptoms rarely appear before the glomerular filtration rate has fallen from 120 *millilitres per minute* to below 30 millilitres per minute, that is, whilst one-half of one kidney is still functioning. When the glomerular filtration rate has fallen to below 3 millilitres per minute, control by diet is no longer possible, and *dialysis* or a *kidney transplant* is needed to maintain life.

Haemodialysis

The fact that an 'artificial kidney' could be made, using a semipermeable membrane, such as cellophane, between the blood and a dialysing fluid, was first shown by Kolff in the early 1940s. Haemodialysis, as it is now termed, requires the cannulation of an artery and a vein to enable the blood to be passed through the dialysing apparatus before being returned to the patient. Usually the cannulae are left *in situ* and when they are not in use, they are joined together to form an arterio-venous shunt. This prevents clotting.

The composition of the dialysing fluid is adjusted so as to restore the blood to its correct composition. In order to remove water from the body the dialysing fluid is kept at a negative pressure with respect to the blood.

Such a technique enables a patient with kidney failure to be maintained in the normal state of health almost indefinitely provided that dialysis is carried out every few days.

The patient's own peritoneum may occasionally be used as a semi-permeable membrane in the technique known as *peritoneal dialysis*. In this case the dialysing fluid is run into the peritoneal cavity.

Kidney Transplants

The kidney of a donor is transplanted usually into the pelvis of the recipient. The internal iliac vessels on one side are used to provide a blood supply for this kidney. The transplanted ureter is inserted directly in the bladder.

The problem of a kidney transplantation is principally an immunological one; the transplanted kidney tends to be rejected by the body of the recipient (unless it is from an identical twin). To minimize the likelihood of such a rejection, red and white cell groupings are carried out between the recipient and possible donors. X-rays and anti-lymphocytic serum are used to suppress the antibody-forming cells [p. 157]. In addition the recipient has to be maintained permanently on immuno-suppressive drugs. Unfortunately these drugs will reduce the recipient's resistance to infection.

REFERENCES AND FURTHER READING

de Wardener, H. E. (1968) *The Kidney; An Outline of Normal and Abnormal Structure and Function*, 3rd ed., London.
Giebisch, G. (1962) Kidney, water and electrolyte metabolism, *Ann. Rev. Physiol.*, **24**, 357.
Gottschalk, C. W., and Mylle, M. (1959) Micropuncture study of the mammalian urinary concentration mechanism; evidence for the countercurrent hypothesis, *Amer. J. Physiol.*, **196**, 927.
Pappenheimer, J. R., and Kinter, W. B. (1956) Haematocrit ratio of blood within mammalian kidney and its significance in renal haemodynamics, *Amer. J. Physiol.*, **195**, 377.
Pitts, R. F. (1969) *Physiology of the Kidney and Body Fluids*, 2nd ed., Chicago.
Pitts, R. F., Gurd, R. F., Kessler, R. H., and Hierholzer, K. (1958) Localization of acidification of urine, potassium and ammonia secretion and phosphate reabsorption in the nephron of the dog, *Amer. J. Physiol.*, **194**, 125.
Richards, A. N. (1938) Processes of urine formation, Croonian Lecture, *Proc. roy. Soc. B*, **126**, 398.
Smith, Homer W. (1951) *The Kidney: Structure and Function in Health and Disease*, New York.
Smith, Homer W. (1956) *Principles of Renal Physiology*, New York.
Verney, E. B. (1947) The antidiuretic hormone and the factors which determine its release, Croonian Lecture, *Proc. roy. Soc. B*, **135**, 25.
Wirz, H. (1956) Der osmotische Druck in den corticalen Tubuli der Ratenniere, *Helv. physiol. pharmacol. Acta*, **14**, 353.

14. THE SKIN AND BODY TEMPERATURE

The skin acts as a covering for the body. It minimizes the loss of water from the underlying cells and tissue fluid. It contains many sensory endings which protect the body against injury. In addition, it is the main organ for the regulation of the body temperature.

Structure

The skin consists of a superficial layer, the *epidermis*, and a deeper vascular connective tissue layer termed the *dermis* or *corium*. The epidermis contains no blood or lymphatic vessel. It is composed of stratified squamous epithelium which varies in thickness in different parts of the body. On the soles of the feet and on the palms of the hands the epithelium is thick, hard and horny.

The superficial squames of the epidermal layer are continually being rubbed off and replaced from the underlying layers. The epidermal cells are continually being manufactured by the germinative layer (basal and prickle cell layers) which lie on the dermis. Between these layers is a clear layer (stratum lucidum) [FIG. 157(*B*)].

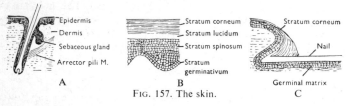

A   B   C

FIG. 157. The skin.

The rate of cell division is highest during sleep and lowest during muscular exercise and stress. Any increase in circulating adrenaline inhibits the cell division.

The dermis consists of dense connective tissue with blood vessels and lymphatics. It merges into the less dense subcutaneous tissue. It is the elastic fibres present in the dermis that give the skin its characteristic elasticity. These elastic fibres degenerate in old age and the skin becomes wrinkled.

The boundary between the dermis and the epidermis is undulating, with dermal papillae projecting into the epidermis. This prevents a separation of the two layers by shearing. In addition the epidermis dips deeply into the dermis at the hair follicles. If the skin is removed down to and including the germinative layer, the epidermis will still regrow by the proliferation of the epidermal germinative cells which still remain deep in the hair follicles and sweat glands. This fact is made use of in skin grafting. Skin, including the germinative layer, may be transferred to another part of the body without permanently denuding the donor area. However, if the skin is transferred to another person (except an identical twin [p. 173]), immunity reactions will cause the donated graft to be rejected and it will slough off after about 3 weeks [p. 157].

Hairs

Hairs are found everywhere except on the palms of the hands, the soles of the feet and parts of the external genitalia (penis and inner surfaces of the labia). These hairs vary considerably in external length. They are shortest on the skin of the eye-lids where they do not project above the surface and longest on the scalp. Hairs appear to be of two types; those that grow for a long time and rest for a short time, and those that grow for a short time and rest for a long time. Each hair consists of a root and a shaft. The root is embedded in an involution of the epidermis termed the hair follicle into which the duct of one or more sebaceous glands enter [FIG. 157(*A*)]. These glands secrete sebum which provides the skin with a covering of fat (lipid material). A similar secretion in the sheep, wool fat, is termed *lanolin* and is widely used in ointments, etc.

Sensory nerve endings around the base of the hair follicle are stimulated by movement of the hair which projects above the surface.

Minute bundles of smooth muscle fibres (*arrectores pilorum*) are connected to the hair follicle below the entrance of the sebaceous gland. These muscle fibres are innervated by the sympathetic nervous system. Contraction of these fibres causes the hair to 'stand on end' and the appearance of 'goose pimples'. At the same time sebum is squeezed out of the sebaceous gland.

Sweat Glands

Sweat glands consist of a single duct which leads from the body of the gland in the subcutaneous tissue to the surface. Where the epidermis is thick, such as on the palms and soles, the duct is spirally coiled as it passes through the epidermis.

There are two types of sweat gland, the *eccrine* and the *apocrine* glands. The eccrine glands are innervated by the sympathetic nervous system, but with acetylcholine as the chemical transmitter in place of noradrenaline. They secrete sweat that is mainly water but contains 0·1–0·4 per cent. of sodium chloride (NaCl). This percentage is under the control of the adrenal cortex hormones [p. 166].

The apocrine sweat glands are found in the axillae, around the nipples and vulva. These glands have no nervous control, but are stimulated by circulating adrenaline. They produce a milky odourless fluid which due to bacterial activity can subsequently develop a characteristic and sometimes offensive odour.

Aluminium compounds, such as aluminium chlorhydroxide and zinc compounds such as zinc phenosulphonate inhibit sweating from these glands and also act as bacteriostatic agents. Spread over the axilla in the form of a cream or paste, they act as an effective deodorant.

In the external meatus of the ear, modified sweat glands produce wax. These are the ceruminous glands.

Nails

Finger and toe nails consist of a modified and greatly thickened *clear layer* of the epidermis, with which it is in direct continuity [FIG. 157(*C*)]. The outer horny layer of the epidermis forms the thin cuticular fold (cuticle).

FUNCTIONS OF THE SKIN

The skins forms the boundary layer between the body and the external environment. It has five main functions: protection, sensation, storage, absorption and heat regulation.

Protection

The skin protects the internal structures of the body from external injury, and it protects the body from the invasion of harmful foreign organisms such as bacteria.

The outer layer of the epidermis is composed of a hard horny protein known as keratin. Keratin resists the action of all but the strongest acids and alkalis. It will, for example, withstand the action of acetic acid. A strong acid like nitric acid will, however, attack keratin in this case with the formation of yellow nitro-derivatives of the aromatic amino acids (phenylalanine, tyrosine and tryptophan). Hence the yellow stain when a drop of nitric acid falls on the skin.

The skin, nails and hair are continually being replaced. They are made by protein synthesis using the amino acids in the circulating blood.

Sensation

The sensory endings in the skin give warning, through the sensations of touch, pain and temperature [p. 191], of the presence of noxious stimuli, and in this way protect the body from injury. Anyone with sensory loss may readily develop severe injuries. Thus a patient with syringomyelia who had lost all sensation of heat and pain in his left arm, placed this arm on a hot surface and received a severe burn. He remarked that he smelt something burning and then noticed that it was himself.

Storage

The skin and subcutaneous tissues act as a store for water and fat. The adipose tissue under the skin is one of the main fat depots of the body. An accumulation of water under the skin leads to oedema [p. 59].

Absorption

The skin is not completely air-tight and a certain amount of gaseous exchange takes place. Some of the carbon dioxide is excreted in this way but the quantity is insignificant compared with that excreted via the lungs. An animal such as the frog with a moist skin can satisfy its requirements for oxygen through its skin.

The skin absorbs ultra-violet radiations which convert precursors in the skin such as 7-dehydrocholesterol to vitamin D.

7-Dehydrocholesterol

Vitamin D₃ (cholecalciferol)

TEMPERATURE REGULATION

Body Temperature

Man is a homoeothermal animal, that is he maintains a constant body temperature which is independent of the environmental temperature. In this respect he differs from the cold-blooded (poikilothermal) animals, such as the frog, whose body temperature depends upon the external temperature.

Man's body temperature is maintained between 97 99·5 °F. (36–37·5 °C.). The standard method of measuring this temperature is to use a clinical thermometer which is a form of maximum thermometer. Clinical thermometers are of two types, Fahrenheit and Centigrade. Fahrenheit thermometers have a scale which extends from 95° to 110° and are calibrated in 0·2° intervals. There is usually an arrow at 98·4 or 98·6 °F. to indicate the mean value of body temperature. Centigrade thermometers have a scale which extends from 36 °C. to 42 °C. in 0·1° intervals. The arrow, in this case, is placed at 37 °C.

Special thermometers are available with scales extending down to 80 °F. for use with patients who have excessively low body temperatures, such as babies and old people who are suffering from hypothermia.

Clinical thermometers are frequently marked '½ minute'. This indicates that the thermometer will reach its final reading in half a minute if placed in a water-bath. Used clinically the thermometer is placed under the tongue, in the axilla, in the groin or inserted into the rectum. The transfer of heat is much slower than in a water-bath, and if the thermometer is removed after the stated time, too low a reading will be obtained. It may take up to 5 minutes before the final temperature equilibrium has been reached. For an accurate determination of the temperature the thermometer should be replaced until constant readings are obtained. The reading obtained in the axilla is about ½ °F. lower than that in the mouth.

A Fahrenheit–Centigrade conversion scale is given in FIGURE 158. See Appendix [p. 223] for other ways of changing from one scale to the other.

The temperature recorded in the mouth may not be a true indication of the body temperature if a hot or cold drink has been taken within the previous half-hour. It will also be inaccurate if the subject is breathing through his mouth, for example with the common cold when the nasal passages are blocked.

The body temperature is not completely constant. It fluctuates throughout the day. It is commonly found that a maximum occurs in the evening and a minimum in the early hours of the morning

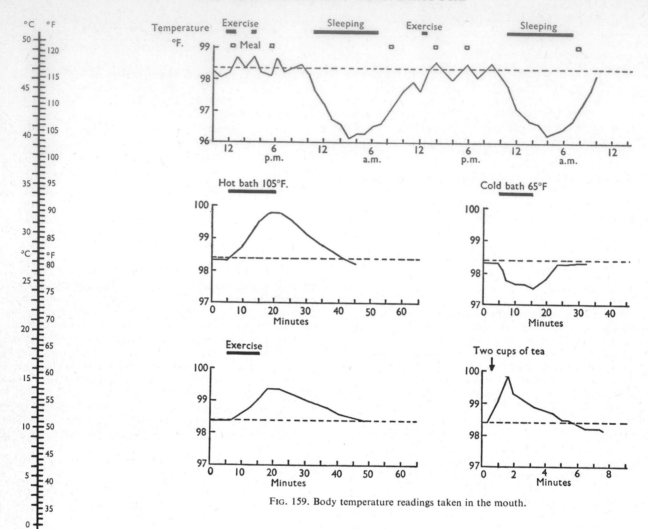

FIG. 159. Body temperature readings taken in the mouth.

FIG. 158.
Fahrenheit—Centigrade
conversion scale.

[FIG. 159]. In women there is also a monthly variation. The temperature in the second half of the menstrual cycle is higher than during the first half [p. 171].

The temperature rises during exercise and after a hot bath [FIG. 159]. The mouth temperature may not give an accurate indication of body temperature during exercise because of the mouth breathing.

Heat Balance

The body temperature is maintained at an approximately constant level because of the balance which exists between the heat gained and the heat lost.

If there were no heat loss, the metabolic processes of the body, even at the basal level of *300 kJ* per hour would produce sufficient heat to raise the body temperature by nearly 2 °F. every hour. This follows from the fact that a 70 kg. man produces 72 Calories per hour of heat due to the basal metabolism (40 Calories per m.² per hour multiplied by 1·8 m.² for the surface area) [p. 121].

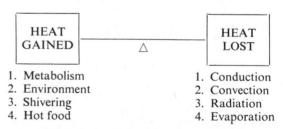

HEAT GAINED	△	HEAT LOST
1. Metabolism		1. Conduction
2. Environment		2. Convection
3. Shivering		3. Radiation
4. Hot food		4. Evaporation

The temperature rise is obtained from the equation:

$$\underset{\text{(in Calories)}}{\text{Heat gained}} = \underset{\text{(in Kg.)}}{\text{Mass}} \times \text{Specific heat} \times \underset{\text{(in °C.)}}{\text{Temperature change}}$$

The specific heat of the body is very close to that of water (=1) due to the high water content (over 70 per cent.). Thus the temperature rise in a 70 kg. man gaining 72 Cal. per hour with a specific heat of unity will be 1 °C. per hour. This will equal 1·8 °F. per hour.

At this rate of temperature rise it would only take a few hours for a fatal heat stroke to occur.

Heat Gained

Heat is gained by the body by internal metabolism and from the external environment if this is at a higher temperature than the body. There will also be a small gain of heat from any hot food or liquid that is ingested. If the heat production is insufficient to maintain the body temperature, further metabolism is brought about by the involuntary contraction of the skeletal muscles known as shivering.

Heat Lost

Heat is lost from the skin by conduction, convection, radiation and the evaporation of water. A small loss of heat occurs in the expired air which is at body temperature and also contains evaporated water; it is saturated with water vapour.

The heat lost by the skin depends on the temperature gradient between the skin and the environment. The skin temperature is lower than the mean body temperature and is regulated by the blood flow to the skin.

The skin temperature cannot be measured with a mercury in glass thermometer because the heat required to operate such a thermometer cools the skin. Instead a temperature sensing element with a very low thermal capacity is used. This often takes the form of a thermo-couple and galvanometer. Alternatively, an electronic temperature sensitive element such as the thermistor may be used. The electrical resistance of the thermistor changes with temperature, and after appropriate amplification of the current changes which this causes, the temperature is displayed on a meter or recorded on a chart.

With a low blood flow the skin temperature is low and heat loss to the environment is minimal. With a high blood flow, the skin temperature is high and approaches the body temperature. The heat loss to the environment is then maximal.

Heat is lost by conduction by physical contact with the environment. Thus when sitting, heat is conducted to the chair. Loss of heat by conduction is minimized by wearing clothes that are poor conductors of heat. Trapped air makes a good thermal insulator. A string vest worn as an under-garment is surprisingly warm because of the trapped pockets of air. The furs worn by Eskimos, with the fur facing inwards as well as outwards, is another example of the use of this principle.

Convection is similar to conduction in that the heat is transmitted from one molecule to another by physical contact, but this time the heat is transferred to the air which becomes less dense and rises taking the heat with it. Cold air comes in to take its place.

Radiation is the loss of heat by electro-magnetic waves. The wavelength of radiant heat is slightly longer than that of visible red light. The frequency, therefore, is slightly less, hence the term infrared radiation. Such radiation does not heat the medium through which it passes. It can be transmitted through a vacuum. The heat from the sun comes by this form of radiation.

The amount of heat lost by radiation depends not only on the temperature of an object but also on its colour and texture. Dark rough surfaces radiate maximally (black body radiators). Light shiny surfaces at the same temperature do not lose heat so rapidly. The same applies to the gaining of heat by radiation. A block of ice-cream melts far more rapidly if it is wrapped in black paper than if wrapped in shiny aluminium foil.

The human skin acts as a 'black body radiator' irrespective of the actual skin colour, but this does not apply to the clothes worn. White clothes are more suitable than black both in the tropics where it is very hot and in the Antarctic where it is very cold. In the tropics there is less heat gained, and in the Antarctic there is less heat lost.

Heat Lost by the Evaporation of Water

Heat is needed to convert water to water vapour. This heat is termed the latent heat of vaporization. The vaporization of 1 ml. of water needs 0·58 Calories (2·4 kJ). This is the amount of heat lost by the evaporation of the skin of 1 ml. of sweat. If the sweat falls off the skin without evaporating, there is no cooling effect.

The body loses about 900 ml. of water per day in temperature zones by evaporation, due to the fact that the skin is not a completely watertight covering, and via the sweat glands. This will result in the loss of 900 × 0·58 = 520 Calories (2200 kJ) of heat per day.

400 ml. of water are lost from the lungs each day as water vapour. The evaporation of this water will cause the loss of 400 × 0·58 = 230 Calories (1000 kJ) per day.

The sweat glands form the only mechanism available for the reduction of body temperature, when the environmental temperature is higher than that of the body. Under these conditions heat will be gained by conduction, convection and radiation. The heat lost by the evaporation of sweat will have to include the heat gained in these ways.

The evaporation of water does not take place if the humidity of the surrounding air is high. The body can withstand very high environmental temperatures if the air is dry but temperatures well below body temperature may be uncomfortable if the humidity is high.

The amount of water lost per day by sweating may be anything from zero up to 12 litres per day. The evaporation of 12 litres of water from the skin would represent a loss of 7,000 Calories (30,000 kJ) of heat.

Prolonged sweating leads to dehydration. Sweating then ceases and the body temperature may start to rise.

Sweat from the eccrine sweat glands is a dilute solution of NaCl. Since the sodium chloride in sweat is less than half that in blood from which the sweat is made, sweating will at first increase the concentration of NaCl in the blood. If only the water is replaced, then the NaCl concentration in the blood will fall, and a salt deficiency occurs. This may lead to painful muscular cramp, known as stoker's cramp because it was first noticed in stokers and other workers in high environmental temperatures. The remedy, as Haldane first demonstrated, is to take salt as well as water.

The dog is an animal without effective sweat glands. It lowers its body temperature by panting. Rapid shallow breathing moves the air rapidly over the moist tongue. Water evaporates and the blood flowing through the tongue is cooled.

Sweating occurs for many other reasons than a high environmental temperature. Emotional sweating takes place under conditions of mental stress. It is also a common accompaniment to seasickness, nausea, vomiting and fainting. Eating highly spiced foods leads to gustatory sweating.

Core Temperature and Temperature in the Extremities

The maintenance of a constant body temperature applies only to the 'core' of the body, that is, the interior of the skull, thorax and abdomen. In a cold environment the temperature in the extremities of the limbs may be well below the central 'core temperature'. A special mechanism is employed to ensure that cold blood from such a region does not arrive at the heart.

Skin Temperature Gradients Along Limbs

If the skin temperature is measured using an electronic thermometer at different points along a limb, it is found that there is a progressive fall in temperature as one proceeds away from the trunk. FIGURE 160 shows the skin temperature recorded with a thermistor skin thermometer at points along the lower limb when the environmental temperature was 19 °C. It will be seen that close to the trunk the skin temperature was 35 °C., but that by the time the tip of the big toe had been reached the temperature had fallen to room temperature.

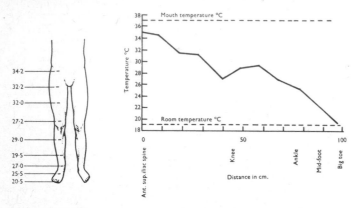

FIG. 160. Skin temperatures recorded along the anterior surface of the lower limb using thermistor thermometer. Limb exposed to environmental temperature of 19 °C. Note that at anterior superior spine of ilium skin temperature approached the body temperature as recorded in the mouth, whilst at tip of big toe temperature was the same as that of room temperature. Between these two points there was a progressive fall in temperature except over the patella which was relatively colder than the adjacent areas presumably due to its poor blood supply compared with muscles.

The blood returning from the extremities is cooler than the body temperature, but as it returns to the heat along the veins surrounding the arteries (venae comitantes) it is warmed by the arterial blood so that by the time it enters the trunk the temperature has risen to 37 °C. At the same time the temperature of the blood in the arteries will fall as the blood moves peripherally so that the arterial blood is already at a low temperature by the time the extremities are reached.

This constitutes a counter-current heat transfer system [FIG. 161] which allows the extremities to be at a very low temperature when necessary (such as walking through snow) without excessively cold blood returning to the heart.

HEAT REGULATING CENTRE

The temperature regulating centre is situated in the hypothalamus. It is in fact two centres. The anterior hypothalamus controls the heat loss by means of vasodilatation of the skin [p. 45] and sweating when the body temperature is rising. The posterior hypothalamus is the centre for conserving heat and for further heat production when the body temperature is falling.

The temperature regulating centre behaves like a thermostat. In many bacterial and viral infections this thermostat is set at a higher level. Profound vasoconstriction occurs and the patient feels cold. Shivering or rigors may occur. The body temperature rises to its new level, and the patient is said to have a fever. Once the new temperature has been reached the skin vessels will dilate and the patient will appear flushed.

When the infection is over and the temperature is falling to normal, marked sweating may occur to enable heat to be lost rapidly.

Drugs such as aspirin act as antipyretics and reduce the body temperature in a fever by their action on the heat regulating centre. They reset the thermostat to a lower level. Whilst the temperature is falling to the new level sweating may occur. They have no effect if the temperature is normal.

Sensation of Heat and Cold

The subjective sensation of heat and cold depends very largely on the skin temperature and thus on the skin blood flow. When the skin vessels are dilated one feels warm. When they are constricted one feels cold irrespective of the central body temperature. Alcohol taken on a cold night dilates the cutaneous blood vessels and gives a feeling of warmth, although the dilated skin vessels may cause the body temperature to fall profoundly.

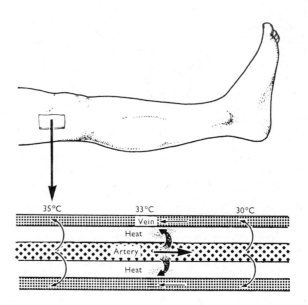

FIG. 161. Counter-current heat transfer. The main arteries to the limb are surrounded by veins (venae comitantes). Heat is transferred from the arteries to the veins which are bringing cold blood back from the extremities. The temperature of the blood in the arteries progressively falls as it passes to the extremities. This arrangement allows the extremities to be at a much lower temperature than the body temperature, without excessively cold blood coming back to the heart from these extremities along the veins.

REFERENCES AND FURTHER READING

Burton, A. C., and Edholm, O. G. (1955) *Man in a Cold Environment*, London.

Fry, F. E. J. (1958) Temperature compensation, *Ann. Rev. Physiol.*, **20**, 207.

Hardy, J. D. (1961) Physiology of temperature regulation, *Physiol. Rev.*, **41**, 521.

Kuno, Y. (1956) *Human Perspiration*, Oxford.

Medical Research Council (1960) Physiological Responses to Hot Environments, *Spec. Rep. Ser. med. Res. Coun. (Lond.)*, No. 298.

Montagna, W., and Ellis, R. A. (1958) *Biology of Hair Growth*, New York.

Pickering, G. (1958) Regulation of body temperature in health and disease, *Lancet*, i, 59.

Rose, J. E., and Mountcastle, V. B. (1959) Touch and kinesthesis, in *Handbook of Physiology*, Section 1, Vol. I, p. 387, Washington, D.C.

Sibinga, M. S. (1959) Observations on growth of finger nails in health and disease, *Pediatrics*, **24**, 225.

Smith, Audrey U. (1961) *Biological Effects of Freezing and Supercooling*, London.

Sweet, W. H. (1959) Pain, in *Handbook of Physiology*, Section 1, Vol. I, p. 459, Washington, D.C.

Uprus, V., Gaylor, J. B., and Carmichael, E. A. (1935) Shivering, *Brain*, **58**, 220.

Weddell, G., and Miller, S. (1962) Cutaneous sensibility, *Ann. Rev. Physiol.*, **24**, 199.

Weiner, J. S., and Hellman, K. (1960) The sweat glands, *Biol. Rev.*, **35**, 141.

Zotterman, Y. (1959) Thermal sensations, in *Handbook of Physiology*, Section 1, Vol. I, p. 431, Washington, D.C.

15. THE LIVER, THE SPLEEN AND THE THYMUS

THE LIVER

Blood Supply

The liver is the chemical factory of the body. It has a double blood supply from the *hepatic artery* and the *portal vein*. It receives 20 per cent. of its blood from the hepatic artery. This is a branch of the coeliac artery which arises from the front of the abdominal aorta just below the diaphragm. The hepatic artery supplies the liver with 'arterial' blood which is 95 per cent. saturated with oxygen. The remaining 80 per cent. of the liver blood flow is from the portal vein which drains the digestive tract, the spleen, pancreas and gall-bladder. This blood is only 85 per cent. saturated with oxygen.

The venous drainage of the liver is the *hepatic vein* which joins the inferior vena cava. There are no valves in the veins.

Structure

The liver is built up of lobules with liver cells lying between the blood sinusoids and the bile capillaries [FIG. 162]. There are no endothelial cells lining either the blood or bile spaces so that the blood comes into direct contact with the liver cells.

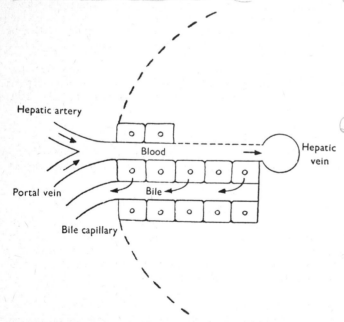

FIG. 162. Formation of bile by the liver. The bile canaliculi are arranged around the central vein (a tributary of the hepatic vein) like spokes of a wheel.

Kupffer cells, which are stellate cells belonging to the recticulo-endothelial system, are found at intervals along the vascular sinusoids. They are highly phagocytic and will ingest foreign substances.

FUNCTIONS OF THE LIVER

The functions of the liver are numerous and many have been discussed already. They may be conveniently grouped together under three main headings: (1) those concerned with blood; (2) those concerned with food; (3) those concerned with the removal of 'poisons'.

Blood

It has been seen that the liver plays an important part in the formation and destruction of the red blood cells [CHAPTER 2]. It is a site of formation of red cells in foetal life. It stores the haematinic principle for the normal maturation of the red cells. It removes from the blood the bilirubin formed when the red cells are broken down and excretes this bilirubin down the bile-duct into the duodenum, as bile pigments. If the liver is, for any reason, unable to keep the blood bilirubin at a level of less than 2 mg. per cent. then jaundice develops.

The liver manufactures the **plasma proteins** with the exception of some of the globulin fraction. The plasma albumin (A) exceeds the plasma globulin (G) and the normal A/G ratio is 1·7/1. In liver failure the plasma protein pattern is altered and the globulin fraction may exceed the albumin fraction. This change may be detected by a change in the electrophoretic pattern of the plasma proteins and by various flocculation tests which are based on the effect of proteins on colloidal solutions.

A fall in the over-all plasma protein content of the blood and in particular the loss of the relatively small but numerous albumin molecules will cause a drop in the plasma protein osmotic pressure and may result in the formation of oedema [p. 59].

The **blood clotting factors** such as prothrombin and fibrinogen are made in the liver. In liver disease blood clotting may be impaired leading to bleeding from the skin and mucous membranes. Vitamin K is used by the liver in the formation of prothrombin and a deficiency of this vitamin will lead to a delay in blood clotting.

Food

1. Carbohydrate

The liver stores glycogen from ingested carbohydrate and manufactures it from non-carbohydrate sources, i.e. proteins and fats. Circulating adrenaline and glucagon mobilize this liver glycogen and convert it to blood glucose. Only liver has the *phosphatase* (glucose-6-phosphatase) necessary for this breakdown [p. 113].

The liver maintains the normal blood sugar level at 60–100 mg. glucose per cent (3·3–5·5 mmol/l). Removal or failure of the liver leads to a fall in the blood glucose level leading to coma and ultimately death.

Galactose is only converted into glucose in the liver. This conversion is one of the tests employed in cases of suspected liver impairment. 40 g. of galactose are eaten and the blood galactose level is measured at half-hourly intervals. The blood galactose level remains low at an insignificant level except in liver disease.

2. Protein

The liver in common with many other tissues deaminates the surplus amino acids with the production of ammonia, but only the liver converts the ammonia to urea. In liver failure the amino acids in the blood are increased and they appear in the urine. The blood ammonia level is raised and may reach toxic levels. The blood urea level, on the other hand, falls.

3. Fats

Bile salts produced by the liver are essential for the digestion and absorption of fat [p. 109]. A decrease in bile salts leads to a reduction in the fat absorption and an increase of the fat in the faeces.

The fat-soluble vitamins are stored in the liver, hence liver oils such as cod-LIVER oil and halibut-LIVER oil are a potent source of vitamins A and D.

The depot fat is mobilized when required and is converted to the ketone bodies (acetoacetic acid, β-hydroxybutyric acid and acetone) by the liver. These ketone bodies are metabolized by other tissues with the production of heat and energy.

Ketosis, which is an excess of ketone bodies in the blood, does not occur in liver failure.

Poisons

Many normal physiological substances as well as drugs are modified by the liver in some way before being excreted by the kidneys. These *protective syntheses* are termed 'detoxication'. Some substances, such as the short-acting barbiturates, are completely destroyed by the liver. Other substances are conjugated with glucuronic acid (a derivative of glucose), glycine, sulphate or acetate.

Benzoic acid, which has been used as a food preservative, and is therefore eaten, is conjugated in the liver with the amino acid glycine to form benzoyl-glycine or hippuric acid which is excreted in the urine.

Care has to be taken when hypnotic drugs are given to anyone with liver damage as the duration of action of these drugs may be unduly prolonged.

The liver cells themselves may be damaged by the toxic action of such substances as carbon tetrachloride (used as a fat and grease solvent for dry cleaning), tetrachlorethane, chloroform, selenium and even alcohol. The liver appears to be best able to resist such toxic action when it has an adequate store of glycogen and amino acids, especially methionine.

Hepatic Insufficiency

If the liver fails (hepatic insufficiency) or is removed the following changes occur:
1. The blood glucose level falls leading to a fatal hypoglycaemia. This demonstrates the importance of the liver in maintaining the normal blood glucose level. The blood glucose level can be maintained by a continuous infusion of glucose (or fructose which is used in place of glucose outside the liver). Galactose, on the other hand, is not used if the liver is no longer able to convert it to glucose.
2. The blood urea falls, but the ammonia level rises, showing that the liver is the principal site for urea formation [p. 115].
3. Bilirubin accumulates in the blood and jaundice develops.
4. Blood clotting is impaired by a fall in the prothrombin and fibrinogen. These are made in the liver.
5. Detoxication ceases so that drug action is greatly prolonged.

The liver has, however, great powers of regeneration.

A Further Consideration of Jaundice

It has been seen in Chapter 2 [p. 11] that jaundice develops when the excretory pathway for bilirubin from broken-down red cells becomes blocked.

The bilirubin is transported to the liver combined with the plasma albumin. Although two molecules of bilirubin will combine with plasma albumin *in vitro*, the second molecule is only loosely bound. *In vivo* each molecule of plasma albumin will only carry one molecule of bilirubin, and when the ratio of bilirubin molecules to albumin molecules exceeds unity, some bilirubin remains unbound.

Unbound unconjugated bilirubin appears to interfere with cell metabolism and a high plasma bilirubin level leads to staining of the basal ganglia and brain damage. The condition is known as **kernicterus** (brain jaundice).

Thus with a rhesus mismatch (icterus gravis neonatorum) there is a risk of kernicterus in the baby if the plasma bilirubin exceeds 20 mg./100 ml. (350 μM./l.)

The bilirubin is conjugated with glucuronic acid for its passage down the bile-duct. For this conjugation the enzyme *bilirubin glucuronyl transferase* is needed. A deficiency of this enzyme could give rise to kernicterus.

Tests of Liver Function

Liver disease produces changes in the plasma proteins which were originally demonstrated by flocculation tests such as the 'thymol turbidity' test. These changes are now investigated by the determination of the total plasma protein level, the determination of the plasma albumin level (and hence the plasma globulin level by subtraction) and the electrophoretic pattern which usually shows five bands: albumin, α_1, α_2, β and γ-globulins.

If the patient is jaundiced, the plasma bilirubin level will be raised, whilst bile pigments and urobilinogen may be present in the urine.

Bilirubin in the plasma gives a reddish-purple colour when diazotized with sulphanilic acid, sodium nitrite and hydrochloric acid. This is known as the **Van den Bergh reaction**. It is carried out in two ways:

1. Direct Van den Bergh Reaction

Only bilirubin combined with glucuronic acid (post-hepatic bilirubin), which makes it water soluble, gives the above reaction directly.

2. Indirect Van den Bergh Reaction

As has been seen, the bilirubin en route from the red cell breakdown in the reticulo-endothelial system to the liver is combined with protein. Alcohol has to be added to remove the protein and to dissolve the bilirubin released before the colour reaction is obtained. This is termed an indirect reaction.

Liver Damage

Liver diseases may lead to a deficiency of the clotting factors prothrombin and fibrinogen. Such a deficiency will lead to a prolonged clotting time.

Liver cell damage will lead to a leak of enzymes from the liver cells into the plasma where they can be detected. These enzymes include the amino-transferases, alkaline phosphatase and lactate dehydrogenase.

Amino-transferases (Transaminases)

The liver cells contain enzymes for the formation of one amino acid using another. These enzymes are known as **amino-transferases** or **transaminases.** Two such enzymes are aspartate transaminase (SGOT) and alanine transaminase (SGPT).

Aspartate transaminase (AsT) catalyses the transamination:

Aspartic acid + 2–oxoglutaric acid
(amino group)   (no amino group)

$\rightleftharpoons$ Oxaloacetic acid + glutamic acid
(no amino group)   (amino group)

Since the same enzyme catalyses the reverse reaction, it is also known as glutamic-oxaloacetic transaminase or G-OT for short. The *serum glutamic-oxaloacetic transaminase* level is abbreviated to **SGOT.**

Alanine transaminase (AlT) catalyses the reversible transamination:

Alanine + 2–oxoglutaric acid $\rightleftharpoons$ pyruvic acid + glutamic acid
(amino group)   (no amino group)   (no amino group)   (amino group)

It is also known as glutamic-pyruvic transaminase (G-PT). **SGPT** stands for *serum glutamic-pyruvic transaminase.*

These two enzymes are also present in cardiac muscle and the plasma level rises shortly after a coronary thrombosis (myocardial infarction). There is thus a marked increase in the aspartate transaminase (SGOT) in blood taken from a patient 6–48 hours after the thrombosis.

Viral Jaundice

The virus of infectious (viral) hepatitis can be transmitted by a blood transfusion from a donor who is a carrier of this disease. It can also be transmitted faecal/orally and blood/orally. Hence precautions must be taken and the hands washed thoroughly after handling blood.

Anyone who has had jaundice is usually excluded from being a blood donor since the jaundice may have been of viral origin. Unfortunately many individuals have become carriers of this virus without having any knowledge of the fact and without any episode of jaundice in their life. Presumably at some time they have had a symptom-free subclinical attack of this virus disease. Such a person may become a blood donor and transmit viral hepatitis to a recipient.

In addition the virus can survive in the 'pooled' dried plasma which is made from blood from many donors after the time has expired for its use as a transfusion of red cells. In this case the virus may spread throughout many bottles of plasma. Dried plasma from a single donor is safer.

The virus can be transferred from person to person by the use, without proper sterilization, of the same syringe and needles, and for this reason is common amongst drug addicts.

The virus presents a problem in kidney dialysis units. Many of these patients have repeated blood transfusions so that the likelihood of at least one of them receiving this virus is greatly increased. Once it enters the dialysis apparatus it may be transmitted to other patients and even to the staff. The presence of the virus in the blood or plasma is very difficult to detect.

Australia Antigen

Patients who are receiving repeated blood transfusion develop antibodies to various antigens, such as abnormal plasma proteins, bacterial products and viruses, which are not normally present in their own blood. Although the production of an antibody has no apparent effect on the recipient, his plasma now provides a source of antibody for testing apparently normal people for the presence of the antigen. The antigen-antibody reaction often results in the precipitation of the antigen so its presence can be readily detected.

Using this technique an Australian aborigine was found to possess an antigen in his blood to which another patient had developed an antibody. The antigen is now named the Australia antigen or **Au antigen** for short. The antibody is designated **anti-Au.**

There is now strong evidence that the **Au antigen,** and the **anti-Au antibody,** are related to the hepatitis virus, particularly the strain of virus which has a long incubation period (serum hepatitis). The detection of one or other of these substances indicates that the person has been exposed to this hepatitis virus, and since he may still be a carrier of this virus, he, presumably, should not be used as a blood donor. The presence of the Au antigen is, however, very common in areas of the world where sanitation is poor.

Since the antibodies are carried in the γ-globulin fraction of the plasma proteins, this fraction could theoretically be used to prevent an infection after contact. However, normal human γ-globulin, which has been obtained from donor blood, will be ineffective as a prophylactic measure for preventing virus hepatitis, since persons with a history of viral jaundice will have been eliminated as blood donors, and, as a consequence, the viral hepatitis antibody content is likely to be low.

Measurement of Liver Blood Flow

In addition to containing the bile salts and the bile pigments the bile is a route for the excretion of cholesterol and drugs of the phenolphthalein type. Phenolphthalein itself is excreted by this route and acts as a laxative. Iodophenolphthalein is opaque to X-rays and is used to show up the bile passages and gall-bladder while it is being excreted. Another derivative, bromsulphthalein (BSP), is used to measure liver blood flow and its excretion is used to test liver function.

The liver blood flow is measured by a modification of the Fick principle. Bromsulphthalein is infused into the blood at a rate that just maintains the blood level constant. The rate of inflow is then equal to the rate of removal by the liver. The level in the blood arriving at the liver is equal to that in the arteries elsewhere in the body and is known. The rate of excretion is equal to the rate of infusion. All that is needed is the level in the blood leaving the liver. This is found by catheterizing the hepatic vein. A catheter is passed via a superficial vein in the arm round into the superior vena cava, into the right atrium and out again into the inferior vena cava and from there into the hepatic vein.

The liver blood flow is determined by dividing the amount of bromsulphthalein infused per minute by the concentration differences between the blood entering and leaving the liver. Such determinations give a liver blood flow of 1·5 litres per minute.

The **bromsulphthalein liver test** is usually carried out as follows. A dose of bromsulphthalein (5 mg. per kg.) is injected into the subject's circulation and blood samples are taken at 5 and 45 minutes afterwards. By 45 minutes all but 5 per cent. of the bromsulphthalein should have been removed by the liver and should therefore have disappeared from the blood. A level of more than 10 per cent. indicates liver damage.

THE SPLEEN

The spleen (lien) lies between the fundus of the stomach and the diaphragm on the left side of the abdominal cavity. Its blood supply is from the large splenic artery which is a branch of the coeliac artery which in turn arises from the abdominal aorta just below the diaphragm. The splenic vein joins the portal vein, so that the blood flows to the liver after having passed through the spleen.

The framework of the spleen consists of numerous fibrous bands known as *trabeculae*. In the spaces between the trabeculae is found the splenic pulp.

The splenic artery enters at the hilum of the spleen and breaks up into numerous small arteries. A sleeve of lymphoid tissue around these arteries (Malpighian bodies) is the site of lymphocyte formation. The blood flows through the splenic pulp to the small veins which unite to form the splenic vein. As the blood passes through the meshwork associated with the pulp, it is filtered and the numerous phagocytic cells associated with the pulp remove broken down cells and foreign particles.

The sympathetic nerve fibres to the spleen relay in the coeliac plexus. These fibres run to the blood vessels and to the smooth muscle associated with the capsule and trabeculae. In animals, such as the cat and dog, but not in man, rhythmical contractions of the spleen occur several times a minute, which force blood along the splenic vein to the liver.

FUNCTIONS OF THE SPLEEN

The spleen is not essential to life and it may be removed in both man and animals without ill effect.

Blood Cell Formation

In foetal life the spleen is a site of formation of red and white cells in much the same way as the red bone marrow. The lymphocyte production continues throughout life and the spleen may once again become a site of red cell formation in certain forms of anaemia.

As a Blood Store

In animals, such as the cat and dog, the spleen acts as a store of red blood cells. Sympathetic activity and the release of catecholamines by the adrenal medulla (adrenaline and noradrenaline) causes the spleen to contract and discharge its blood content into the circulation. This will increase the red cell count, the haematocrit and the viscosity of the blood. (Even without any alteration in the cardiac output or peripheral resistance this increase in viscosity will tend to increase the arterial blood pressure [see p. 46].)

In man, however, the spleen only weighs 150 g. and, even if it contained nothing but red blood cells, the effect of injecting the additional 150 g. of red cells into 5 litres of blood would be negligible. The spleen cannot, therefore, be considered to be an important blood storage organ in man.

Blood Destruction

The spleen is a site for the destruction of old red blood cells and the removal of their breakdown products. Thus there is a higher level of bilirubin in the splenic vein than in the splenic artery. Red cell breakdown occurs in other equally important sites such as in the bone marrow.

Red cells leaving the spleen are more spherical (spherocytosis) and haemolyse more easily.

The spleen also destroys platelets. In hypersplenism there may be excessive platelet destruction leading to purpura (spontaneous haemorrhages from capillaries such as under the skin). Such conditions are improved by removal of the spleen (splenectomy).

Other Functions

The action of the spleen as a blood filter has been referred to above. In addition, the spleen is an important part of the recticuloendothelial system engaged in the defence of the body against infection. It is the site of antibody formation, and enlarges in many infections. In malaria, for example, the spleen enlarges and may weigh many kilograms.

THE THYMUS

The thymus gland lies in the superior mediastinum of the thorax behind the sternum. It extends from the pericardium up into the neck to the lower border of the thyroid gland. The thymus consists of two fused asymmetrical lobes which have developed separately. The right lobe is larger than the left. Each lobe is composed of numerous lobules. Each lobule is made up of small follicles of about 1 mm. in diameter. The follicles have an inner medulla and an outer cortex and are surrounded by a connective tissue capsule. Blood vessels run from surrounding vascular plexuses into the interior. The cortex of the follicle is densely packed with small lymphocytes. The medulla contains fewer lymphoid cells and contains nest-like bodies termed Hassall's corpuscles.

At birth the thymus is relatively large weighing 13 g. and the cells show a high rate of cell division. Growth of the gland is active in the early years of life and it weighs 40 g. at puberty. Later in life it undergoes a gradual involution.

IMMUNOLOGICAL FUNCTION OF THE THYMUS

When an organ is transplanted from one animal to another of the same species, this transplant or homograft may be rejected by the recipient within a few hours (*immediate rejection*) or after a few weeks or even months (*delayed rejection*).

This rejection is due to the transplant acting as an antigen which stimulates the formation of an antibody in the host [p. 45]. The rejection is then brought about by the antigen-antibody or *immunological* reaction which then follows. The antibodies are made in the lymphoid tissues by the lymphocytes. Antibody formation may be depressed, for example, by exposure of the bone marrow and lymphoid tissue to high doses of X-rays, which will inhibit blood cell formation [p. 9]. Alternatively immuno-suppressive drugs, antilymphocytic serum and cortisol may be employed. Antilymphocytic serum is made by injecting human lymphocytes into an animal such as the horse.

In mammals the thymus is important in the immunological processes. It has been found that if mice have their thymus gland removed (thymectomy) at birth, they do not reject skin grafted on to them from other mice, or even rats [see p. 148]. Furthermore, there is defective development of lymphoid centres in lymph nodes, intestines (Peyer's patches) and the spleen, and the blood lymphocyte count remains low. They have a reduced resistance to infection, and unless kept under sterile conditions the animal wastes away in one to three months. These effects are not obtained if the thymus is removed later in life.

The effects of thymectomy at birth were originally thought to be due to the removal of lymphoid tissue. The period of growth of lymphoid tissue in the thymus precedes that of the other lymphoid tissue in the body (with the possible exception of that in the bone marrow). It has been suggested that thymic lymphocytes with antibody-forming properties (immunocytes) migrate via the thymic vein into the blood stream and form colonies in other lymphoid sites. It has also been considered possible that in addition thymectomy removes a thymus hormone which normally stimulates lymphocyte production in other centres and enables lymphocytes to react with antigens and form antibodies.

The thymus gland itself is not the site of antibody formation. In the thymus the antigen appears to be unable to pass through the *blood-thymus barrier* to reach the lymphoid tissue from the blood vessels.

Recent work has thrown further light on these findings and on the rôle of the thymus in the immunity response.

Immune System

The lymphocytes [p. 18] and their antibodies [p. 22] constitute the Immune System of the body. The majority of the lymphocytes at any given time are in the resting state. They are producing only a small amount of a particular antibody. This is displayed on the cell-surface of the lymphocyte. When stimulated by coming in contact with the matching antigen, a single small lymphocyte grows, divides repeatedly and produces a colony of offspring cells, termed a *clone*, all of which produce the same antibody. When this happens the lymphocyte is said to have responded positively. The lymphocyte is not thought to recognize the whole antigen, but only a specific part called an *epitope*. If the epitope concentration is too high or too low, the lymphocyte may respond negatively and become paralysed.

In embryonic life, lymphocytes which produce antibodies against one's own body tissues are destroyed or paralysed, so that after birth antibodies are only made against foreign cells and foreign substances. This self-tolerance enables lymphocytes to discriminate between *self* and *not-self*. Failure of this self-tolerance mechanism leads to auto-immune disease [p. 45].

T- and B-lymphocytes

Amongst the small lymphocytes in the body are populations of *T*-lymphocytes and populations of *B*-lymphocytes. Bone marrow lymphocytes are thought to be travelling continuously via the blood stream to the thymus gland. There they are *processed* to become thymus-dependent lymphocytes, or *T*-lymphocytes, for short.

B-lymphocytes also originate in the bone marrow, but are thymus independent. They are called *B*-lymphocytes (bursa-dependent) because in the chicken, where they were first recognized, they are *processed* in the Bursa of Fabricius. The site of processing of the *B*-lymphocytes in man is not known, but it is thought to be in the gut-associated lymphoid tissue, such as the Peyer's patches and the appendix.

Although both types of lymphocyte are thought to make antibody and to carry it on their cell-surface, only the *B*-lymphocytes and their derivatives, the plasma cells, release the antibodies into the blood [p. 18].

$$B\text{-lymphocyte} \xrightarrow{\substack{+\ \text{antigen} \\ \text{stimulation}}} \text{Plasmoblasts} \rightarrow \text{Plasma cells}$$

Circulating
antibody

These circulating antibodies=(immunoglobulins) are found mainly in the gamma globulin fraction of the plasma proteins (although some are found in the beta and alpha-2 globulin fractions). They are associated with the immediate rejection process.

On the other hand, the *T*-lymphocytes, when stimulated, divide and produce cells which destroy foreign cells by direct contact. These 'killer' cells damage the foreign cell's membrane, often with the aid of *complement*, so that cell's content leaks out and the cell dies. These *T*-lymphocytes are associated with cell immunity reactions and the delayed rejection process.

$$T\text{-lymphocyte} \xrightarrow{\substack{+\ \text{antigen} \\ \text{stimulation}}} \text{Lymphoblasts} \rightarrow \text{'Killer'}$$

lymphocytes

T-lymphocytes do not secrete antibodies into the blood. However, under suitable conditions, *T*-lymphocytes appear to be able to stimulate *B*-lymphocytes to produce circulating antibodies. This is important when the antigen is too weak to stimulate the *B*-lymphocyte directly. *T*-cells may, under other circumstances, paralyse *B*-cells.

Memory cells are lymphocytes of a clone which have reverted to the resting state, but which are ready to respond should the antigen reappear. Thus, the memory cells derived from the lymphocytes which responded positively to the measles virus in childhood may persist throughout one's lifetime.

REFERENCES AND FURTHER READING

Liver

Bollman, J. L. (1961) Liver, *Ann. Rev. Physiol.*, **23**, 183.

Maclagan, N. F. (Ed.) (1957) The liver, *Brit. med. Bull.*, **13**, No. 2.

Mann, F. C., and Magath, T. B. (1924) Die Wirkung der totalen Leberextirpation, *Ergebn. Physiol.*, **23**, 212.

Papper, H., and Schaffner, F. (1957) *Liver: Structure and Function*, New York.

Sherlock, S. (1968) *Diseases of the Liver and Biliary System*, 4th ed., Oxford.

Williams, R. T. (1959) *Detoxication Mechanisms*, 2nd ed., London.

Spleen

Knisely, M. H. (1936) Spleen studies I., *Anat. Rec.*, **65**, 23.

Mackenzie, D. W., Whipple, A. O., and Wintersteiner, M. P. (1941) Studies on the microscopical anatomy and physiology of living transilluminated mammalian spleens, *Amer. J. Anat.*, **68**, 397.

Thymus

Burnet, F. M. (1969) *Self and Non-self*, Cambridge.

Jerne, N. K. (1973) The immune system, *Scientific American*, **225**, No. 1, 52.

Miller, J. F. A. P., Marshall, A. H. E., and White, R. G. (1962) The immunological significance of the thymus, *Advanc. Immunol.*, **2**, 111.

Roitt, I. M. (1971) *Essential Immunology*, London.

16. THE ENDOCRINE GLANDS

The glands of the body may be divided into those with an external secretion (*exocrine glands*) and those with an internal secretion (*endocrine glands*). Examples of exocrine glands are the sweat, lacrimal and mammary glands which pass their secretion along ducts to the external surface of the body, and the glands of the mouth, stomach and intestines which pass their secretions along ducts into the alimentary tract. The endocrine or ductless glands on the other hand have no ducts or openings to the exterior. Their secretion is passed into the blood stream, and is transmitted via the circulation to modify the activity of some distant organ or tissue. This mechanism forms part of the communication system of the body, and provides the body with an alternative method of communication to the nervous system [CHAPTERS 19 and 20].

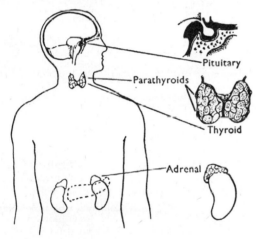

FIG. 163. The endocrine glands.

These 'chemical messengers' which are sent by the endocrine glands in this *humoral control system* are known as hormones. From an historical point of view the term hormone, which is derived from the Greek 'to excite', was first applied to secretin which, as has been seen, is released by the small intestine in response to the presence of food, and circulates in the blood to stimulate production of pancreatic juice. It is probably less confusing if secretin, gastrin and the other intestinal hormones are referred to as 'humoral agents', and the term hormone is restricted to the secretion of the endocrine glands.

The endocrine glands in the body are:

Pituitary gland (hypophysis)
Thyroid gland
4 Parathyroid glands
2 Adrenal glands (suprarenal glands)
2 Gonads (ovaries or testes)

The pancreas may be included in this list in respect of its production of *insulin* and *glucagon*, but it is also an exocrine gland in respect of its production of *pancreatic juice* which passes along the pancreatic duct to the duodenum.

The placenta, which is formed in pregnancy [p. 173], acts as an endocrine gland in addition to being the organ by which the baby receives its nourishment.

Methods of Study

Our knowledge of the action of the endocrine glands and their hormones has been derived, in a large measure, from a study of the changes which occur when the glands are over-active and under-active. To this knowledge has been added the effects of removal of all or part of the gland, and the results of the administration of either naturally occurring or synthetically prepared hormones.

Chemical Structure of Hormones

The hormones produced by the pituitary, parathyroid and pancreas are proteins (or peptides). They will be destroyed by the digestive juices if given by mouth, and for replacement therapy they must be given by injection.

The thyroid and adrenal medulla hormones are aromatic organic compounds (benzene derivatives). The remaining hormones belong to a group of substances known as *steroids*.

Steroids

In addition to these hormones many substances of physiological importance including cholesterol, bile salts and the precursors of vitamin D in the skin, are substances based on a chemical structure of carbon and hydrogen in the form of three six-carbon rings and a five-carbon ring fused together. They are known as *steroids*.

written briefly as:

The carbon atoms are numbered from 1 to 27 as follows:

When the parent structure is modified a series of suffixes is used to indicate the change. If the name of the compound ends in *-ane* then the molecule has single bonds between the carbon atoms in the rings (as above). If there are double bonds between the carbon atoms, the suffix *-ene* is used. If a hydrogen atom is replaced by an —OH group, the suffix *-ol* is used as in *cholesterol*, *cortisol* and *oestradiol*. If two hydrogen atoms are replaced by an oxygen atom, the suffix *-one* is employed, as in *aldosterone* and *progesterone*.

A TYPICAL STEROID (CHOLESTEROL)

THE PITUITARY GLAND

The pituitary gland is a small reddish-grey body 12 mm. × 8 mm. suspended from the base of the brain by a stalk or infundibulum. It lies in the pituitary fossa of the sphenoid bone above the back of the nasal cavity [FIG. 163].

From both a developmental and a physiological point of view it is convenient to consider the pituitary gland in two parts: the posterior pituitary (or neurohypophysis) which is derived from the brain, and the anterior pituitary (or adenohypophysis) which is derived from the mouth and nasal cavity. The pituitary gland produces many hormones.

Posterior Pituitary Gland (Neurohypophysis)

The neurohypophysis or posterior pituitary gland produces two hormones, the **antidiuretic hormone** (ADH, adiuretin, vasopressin) and the **oxytocic hormone** (oxytocin). The antidiuretic hormone has already been discussed in connexion with the formation of urine by the kidneys. Its function is to regulate the reabsorption of water from the distal and collecting tubules, and, by so doing, to regulate the volume of urine which passes to the bladder and is ultimately excreted. The term *antidiuretic hormone* is a more appropriate name for this hormone than the older name *vasopressin* or *vasopressor fraction*. Vasopressin was chosen when it was found that an injection of a relatively large amount of this hormone caused vasoconstriction and a rise of blood pressure. However, it is very doubtful whether the amount of circulating ADH is sufficient to have any effect on the blood pressure, whereas it has a profound effect on the kidney tubules.

A deficiency of this hormone leads to **diabetes insipidus.** Large volumes of urine are passed each day and the patient is excessively thirsty. The urine is very dilute. Before the days of chemical tests, the standard method of testing urine was to taste it. In diabetes insipidus the urine was tasteless or *insipid*, hence the name of this condition; in diabetes mellitus where there is also an increase in urine production, the urine was sweet to the taste since it contained glucose. Diabetes insipidus may be treated by administering ADH.

ADH has been synthesized and is a peptide made up of nine amino acids; it is an nonapeptide.

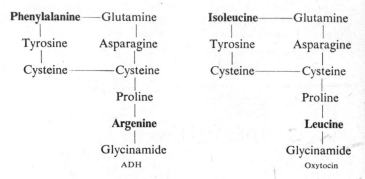

If two of the amino acids are changed, the second posterior pituitary hormone is obtained, the oxytocic hormone, or oxytocin. Oxytocin (Greek *oxys* = sharp, *tokos* = birth) is so named because of its ability to contract the pregnant uterus. Its release is one of the many factors causing *parturition* (birth of a baby). Injections of the hormone are used to induce uterine contractions and to cause the onset of labour and to enhance the rhythmic uterine contractions during labour. This action is probably due to the release of calcium ions (Ca²⁺) in the uterine muscle (myometrium) which bring about the smooth muscle contraction. It also causes milk ejection from the breasts after child-birth and the synthetic hormone is used in the treatment of disorders of lactation.

Although these hormones are released from the posterior pituitary, they are probably manufactured in the nerves which run from the hypothalamus to the gland via the pituitary stalk. These are **neuro-secretory**, that is, nerve impulses bring about the release of a hormone instead of a chemical transmitter substance.

Such neuro-secretory nerves form a link between the two control mechanisms of the body, the nervous system and the hormonal system.

Anterior Pituitary Gland (Adenohypophysis)

The anterior pituitary gland or adenohypophysis secretes a number of hormones. These are known as -trophic hormones (Greek *trophe* = nourishment) since they regulate the growth and activity either of the body as a whole or of other endocrine glands.

A series of portal vessels link capillary networks in the median eminence of the hypothalamus with sinusoids in the anterior pituitary. The release of the anterior pituitary hormones is controlled by the hypothalamus by a series of 'releasing factors', one for each of the anterior pituitary hormones, which pass along this portal system.

A feedback mechanism exists whereby the circulating blood hormones (from the endocrine glands controlled in this way) inhibit the hypothalamus and, to a greater or lesser extent, the anterior pituitary. Such a feedback system enables the level of hormones in the blood to be maintained within close limits. Any reduction in the level of a circulating hormone will reduce the inhibition on the hypothalamus. The increase in releasing factor and anterior pituitary stimulating hormone which results, will stimulate the endocrine gland concerned to release more of the hormone into the blood.

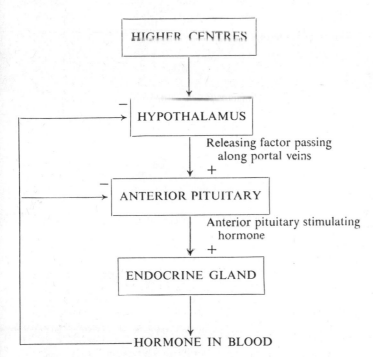

When stained histologically with both an acidic stain (eosin) and a basic stain (cf. staining white blood cells, page 17) three types of anterior pituitary cell are found: eosinophil cells (α cells), basophil cells (β cells) and chromophobe cells. The chromophobe cells are small cells which do not stain with the dyes. They may develop later into chromophil cells.

Human Growth Hormone (HGH)

This hormone is concerned with somatic growth. It is produced by the eosinophil or α cells of the anterior pituitary.

One's height depends on inherited factors. The children of tall parents tend to be taller than the average but not so tall as their parents. They *regress* or tend to be nearer the mean than their parents. (Hence the term 'regression line' for this mean in statistics.) A further factor, however, is the amount of circulating growth hormone.

Under-production of this hormone during the period of growth leads to retarded growth and dwarfism. Over-production leads to excessive growth of the bones of the body and also of the muscles and viscera leading to gigantism. Most giants 7–8 ft. tall are examples of pituitary overactivity.

Overactivity later in life does not increase the stature once the epiphyses of the long bones in the limbs have fused to the shafts (at an age of 18–24 years) and then no further increase in length is possible. The bones of the hands, feet and head, however, appear to retain their power of growth and the subject may first notice that his shoes no longer fit, and that his gloves and hats are becoming too small. This condition is known as **acromegaly** (from Greek *akron* = extremity and *megas* = great).

Human growth hormone is a protein consisting of a chain of 191 amino acids. It has a molecular weight of 21,500. The amino acids have sulphur-sulphur chemical bridges at two places in the chain, between amino acids 182 and 189 and between amino acids 53 and 165. This gives two loops, one containing 6 amino acids and the other loop 111 amino acids. Although the structural formula is now known and the hormone has been synthesized from its constituent amino acids, the principal source of the hormone is from human pituitary glands. It is species-specific and growth hormone from other animals, except possibly the monkey, has no effect in man.

Trials with the limited amount of human growth hormone available have shown that it has its expected effect on growth, and dwarf children have increased their rate of growth markedly when under treatment with this hormone.

The release of HGH is under the control of the releasing factor (HGH-RF) from the hypothalamus. This is a ten amino acid peptide (decapeptide), and being a smaller molecule is easier to synthesize than HGH itself. Its formula is given on page 209.

Thyrotrophic Hormone (Thyrotrophin, Thyroid Stimulating Hormone, TSH)

This hormone controls the activity of the thyroid gland. Increased production of the hormone leads to overactivity of the thyroid gland. A decreased production of TSH leads to underactivity of the thyroid.

Adrenocorticotrophic Hormone (Adrenocorticotrophin, Corticotrophin, ACTH)

This hormone is produced by the basophil or β cells. It is a straight chain peptide made up of 39 amino acids (molecular weight 4,500), but a peptide containing only the first 19 has the same activity. This hormone controls the activity of the adrenal cortex and the release of the corticoid steroids such as cortisol (hydrocortisone) and cortisone. It does not control the release of aldosterone.

Gonadotrophic Hormones (Gonadotrophins)

The three gonadotrophic hormones of the anterior pituitary control the activity of the gonads. They are the follicle stimulating hormone (FSH), the luteinizing hormone (LH) and the luteotrophic hormone (LTH, prolactin). These hormones are produced in the female in a cyclical pattern which repeats itself about every 28 days from puberty to the menopause except during pregnancy. In the male these hormones are produced at a steady rate without the cyclical variations. The action of these hormones will be considered in the next chapter.

THE THYROID GLAND

The thyroid gland produces three hormones in man, **thyroxine, triiodothyronine** and **calcitonin.** Only the hormones thyroxine and triiodothyronine will be considered in this section. The production of calcitonin by the thyroid gland will be considered in the next section [p. 165] following a discussion of the parathyroid glands.

The thyroid gland is situated in the neck. It consists of two lateral lobes on either side of the trachea and larynx, and a connecting isthmus which runs across the midline giving the gland a butterfly appearance [FIG. 163]. Occasionally the gland extends into the thorax behind the sternum. It has a rich blood supply from the thyroid arteries.

$$HO - \langle \begin{smallmatrix} 3 \\ 4 \\ 5 \end{smallmatrix} \rangle - CH_2 \cdot CH(NH_2)COOH$$

Tyrosine (4-hydroxyphenylalanine)

$$HO - \langle \rangle_{I}^{I} - CH_2 \cdot CH(NH_2)COOH$$

3:5 diiodotyrosine (DIT)

$$HO - \langle \rangle_{I}^{I} - CH_2 \cdot CH(NH_2)COOH$$

$$+ HO - \langle \rangle_{I}^{I} - CH_2 \cdot CH(NH_2)COOH$$

diiodotyrosine + diiodotyrosine

$$HO - \langle \rangle_{I}^{I} - O - \langle \rangle_{I}^{I} - CH_2 \cdot CH(NH_2)COOH$$

Thyroxine (tetraiodothyronine) (T₄)

$$HO - \langle \rangle_{I}^{I} - O - \langle \rangle_{}^{I} - CH_2 \cdot CH(NH_2)COOH$$

Triiodothyronine (T₃)

The thyroid gland weighs 25 g. in the healthy adult. It has developed from a median diverticulum known as the thyroglossal duct which arises from the floor of the embryonic mouth. The gland itself is composed of cuboidal cells which line a very large number of closed spherical vesicles of up to 0·1 mm. in diameter. These vesicles are filled with a semi-homogeneous material called colloid. This colloid contains the thyroid hormone in the storage form of a glycoprotein called *thyroglobulin.* The thyroid gland is unique amongst the endocrine glands in that it stores its active principle in the cavity of these vesicles; all the other endocrine glands store their hormones in the secreting cells. The thyroid

hormone thyroxine is split off from the thyroglobulin by a protein-splitting enzyme and circulates in the blood mainly combined with the α-globulin fraction of the plasma proteins (TBG, p. 22).

Formation of Thyroxine

Thyroxine is an iodine derivative of the amino acid thyronine. Thyroxine itself has four iodine atoms in the molecule and is referred to as T_4. The three-iodine compound triiodothyronine (T_3) is even more active and some of the thyroid hormone is circulating in this form.

Triiodothyronine circulates mainly in the free state and unlike thyroxine does not bind so readily with the plasma protein TBG.

Thyroxine and triiodothyronine are made in the body from the dietary amino acid tyrosine. For its synthesis inorganic iodide is concentrated in the cells of the thyroid. The iodide is then organically bound and oxidized to iodine by the cytochrome enzyme system forming mono and diiodo compounds of tyrosine. Oxidative coupling converts diiodotyrosine to triiodothyronine and thyroxine.

Control of Thyroxine Secretion

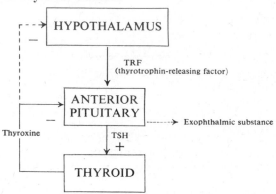

Thyroid hormone secretion starts in foetal life, reaches a maximum in childhood and declines in later life.

The secretion of thyroxine (and triiodothyronine) is regulated by the secretion of the thyrotrophic hormone (TSH) from the anterior pituitary gland. TSH is a glycoprotein containing 96 amino acids and 2 carbohydrate groups.

The release of TSH is, in turn, regulated by the thyrotrophin-releasing factor (TRF or TSH-RF) from the hypothalamus. TRF is a tripeptide. It has been synthesized and its formula is given on page 209. It raises the blood level of TSH and hence those of thyroxine and triiodothyronine. In addition to being effective by intravenous injection, it is also active when given by mouth. This is a surprising finding since it demonstrates that small peptides can be absorbed from the digestive tract without prior complete breakdown to amino acid by the peptidase enzymes of the digestive tract.

High levels of thyroxine and triiodothyronine inhibit the secretion of TSH by the anterior pituitary which in turn reduces the activity of the thyroid gland. They have little effect on the activity of the hypothalamus. The thyroid activity is thus self-regulated by a feedback system which is controlled by the level of thyroid hormones in the blood.

Action of Thyroxine

Thyroxine stimulates metabolism in the tissues generally. It raises the oxygen consumption and increases the heat produced. Metabolically it acts by accelerating the enzyme reactions responsible for the oxidative processes. It also promotes growth and development generally. The normal level of circulating thyroxine maintains a basal metabolic rate (B.M.R.) of 40 Calories per m.² body surface area per hour [p. 121] in a man and 37 Calories per m.² per hour in a woman.

Thyroxine and thyroid extracts are active when given by mouth. They differ in this respect from the hormones which are proteins or peptides and which will be completely broken down to amino acids and destroyed by the protein-splitting enzymes of the digestive tract. 1 mg. of thyroxine per day increases the metabolic requirements of a person from 2,500 to 3,500 Calories per day. If only 2,500 Calories are eaten the body reserves will be used up and there will be a loss of weight.

Goitre

A swelling of the thyroid gland in the neck is called a goitre. It may be associated with either underactivity or overactivity of the gland. The goitre which occurs when there is a deficiency of iodine in the diet is associated with underactivity of the gland. To prevent such a deficiency in areas of the world where iodine is absent from the soil, iodide is added to the table salt. Goitres due to oversecretion of TSH, on the other hand, are associated with overactivity of the thyroid gland.

Underactivity of the Thyroid Gland (Hypothyroidism) Myxoedema

Underactivity of the thyroid gland leads to a reduction in the metabolic processes in the body. There is a fall in the basal metabolic rate to as much as 50 per cent. of normal. The body temperature is subnormal. Both the heart rate and respiration are reduced. There is a rise in the blood cholesterol level.

A deficiency of thyroid activity at birth leads to a child becoming a cretin. The child is small, mentally retarded and usually has a large protruding tongue. Such a child responds to thyroxine by mouth, but as the child may be as much as 6 months old before the condition is diagnosed, it may be too late to restore completely the normal growth pattern.

A thyroid deficiency which occurs later in life leads to myxoedema. There is a reduction in the metabolic rate and a lowering of the body temperature and heart rate. Speech, thought and movement become very much slower and, in addition, there is an increase in weight due to the deposition of semi-fluid material which gives the name to this condition (G. *myxa*, mucus; *oidema*, swelling). The face and eyelids become puffy. The tongue swells and the skin becomes rough and course. The hair thins on the scalp and eyebrows. The menstrual cycle is disturbed in women. Thyroxine by mouth effectively treats myxoedema.

Overactivity (Hyperthyroidism, Thyrotoxicosis)

exophthalmic goitre Graves' disease ?

This may be due to a primary overactivity of the thyroid gland itself, a nodular goitre which actively produces T_3 and T_4, an excess of TSH from the anterior pituitary which causes overactivity of the thyroid gland or to LATS/LATSP (LATS-protector).

LATS is a **long-acting thyroid stimulator** found in the plasma of some patients with thyrotoxicosis. It is found in the IgG fraction of the γ-globulin and presumably has been made by antibody-forming tissues as the result of an auto-immune reaction [p. 45].

The metabolic processes of the body are speeded up. There is an increase in heart rate (tachycardia), an increase in pulmonary ventilation, oxygen consumption, carbon dioxide production and heat produced. The tachycardia persists during sleep but this is difficult to determine since the subjects become very light sleepers and tend to wake up if their pulse is taken.

The basal metabolic rate may be increased by as much as 50 per cent. As a result of the increased heat production the subject prefers the cold weather to the hot weather.

The subject becomes excessively nervous and irritable. The hands show a tremor particularly when held straight out.

Protrusion of the eyeballs (exophthalmos) gives an anxious staring expression. This is seen particularly when the thyroid activity is due to an excess of TSH, although the cause is probably not due to the thyroxine or the TSH but to some unidentified exophthalmos substance from the anterior pituitary.

When the hyperthyroidism is such that the subject becomes clinically ill the condition is termed thyrotoxicosis. For example, the increased excitability of the cardiac muscle may lead to *atrial fibrillation* [p. 35] and heart failure. (Atrial fibrillation may occur following the ingestion of an excess of thyroxine as a treatment for obesity.)

Thyrotoxicosis is treated by removal of most of the thyroid gland by surgery or by giving radio-active iodine. Thiocarbamide compounds such as *thiouracil*, *methimazole* and *carbimazole* prevent the incorporation of iodine in the tyrosine molecule. Thiocyanate and perchlorate compounds interfere with the uptake of iodine by the thyroid gland. Such compounds which reduce the formation of thyroxine are used to control thyrotoxicosis.

The ingestion of a large dose of iodine (or iodide) brings about a temporary remission of thyrotoxicosis which lasts for a few weeks. It is often given prior to a thyroidectomy operation to decrease the vascularity of the gland and hence to reduce bleeding at the operation. The mode of action of the iodine is not known. It does not bring about any long-term improvement.

Thyroid and Blood Cholesterol Level

$$CH(CH_3) \cdot CH_2 \cdot CH_2 \cdot CH_2 \cdot CH(CH_3)_2$$

Cholesterol

It has been known for over 50 years that an increase in thyroid activity reduces the level of cholesterol in the blood whereas a decrease in thyroid activity increases it.

The normal level of cholesterol is 200 mg. per 100 ml. plasma. In myxoedema the concentration may be as high as 600 mg. per 100 ml. (hypercholesterolaemia). In thyrotoxicosis it may fall to 100 mg. per 100 ml.

The blood cholesterol level represents the balance between ingestion and formation on the one hand, and excretion and utilization on the other.

Although cholesterol is present in the diet (egg yolk, liver, meat fats), the greater part is synthesized in the liver and other tissues from acetyl co-A in a series of steps (represented by the number of carbon atoms in the molecules):

Acetyl co-A (2C) → (6C) → (5C) → (15C) → (30C)
→ Cholesterol (27C)

Cholesterol is also destroyed by the liver and excreted in the bile (as cholesterol itself and as cholic acid). It is probably used as the precursor for the formation of steroid hormones. However, cholesterol deposited in the walls of arteries may lead to atherosclerosis and arterial thrombosis.

The action of the thyroid gland on cholesterol metabolism is complex. It stimulates the formation of cholesterol by the liver, but has a greater effect on increasing the excretion of cholesterol in the bile, and increasing the entry of cholesterol into cells. The net result is that the blood level falls.

Investigation of Thyroid Gland Activity

The traditional method of studying thyroid activity has been to investigate its effect on metabolism and the basal metabolic rate. The basal metabolic rate (BMR) is usually expressed as a percentage above or below normal for a person of the same sex and body surface area. The BMR is raised in cases of over-activity of the thyroid gland and lowered in cases of under-activity. A result more than 15 per cent. above or below this figure is considered abnormal.

This technique is now being replaced by methods based on a more direct determination of the level of the thyroid hormones in the blood and the ability of the gland to take up radioactive iodine from the blood.

The normal plasma thyroxine level (usually measured as *serum* thyroxine, that is, after the blood has been allowed to clot) is 7 micrograms per 100 ml. plasma. Such a person is said to be **euthyroid** (*eu* = normal). In overactivity (**hyperthyroid**) of the gland the plasma thyroxine level may rise to as high as 15 micro-grams per 100 ml. In underactivity (**hypothyroid**) the plasma thyroxine level may fall to 2 micrograms per 100 ml.

The plasma thyroxine and triiodothyronine levels are difficult to measure and, as an alternative, the protein bound iodine (PBI) level is often determined. This is a measure of the iodine compounds combined with protein in the plasma (mainly thyroxine, not triiodothyronine, but plus other iodine protein compounds). The normal level is 6 micrograms per 100 ml. plasma. A false high read-ing may be obtained if the patient has taken iodine-containing drugs (e.g. cough mixtures) or has had X-rays involving the use of iodine containing radio-opaque contrast media.

Radioactive iodine studies are based on the fact that the body cannot distinguish between normal 127-iodine and its two isotopes 131-iodine which has a half-life of 8 days and 132-iodine with a half-life of 2 hours. All are incorporated in thyroxine in propor-tion to their concentrations in blood.

An oral dose of radioactive iodine is given and the concentra-tion in the neck region measured 4 hours later by a radioactivity detector. The uptake is usually high in thyrotoxicosis and low in myxoedema.

Other tests include blood cholesterol level (see above) and the ankle jerk which is sluggish and prolonged in myxoedema and brisk in thyrotoxicosis.

THE PARATHYROID GLANDS

The four parathyroid glands are embedded in the thyroid gland [Fig. 163]. Each is oval in shape and approximately 6 mm. long. Microscopically the glands consist of columns of cells with vascular channels between the columns.

The hormone produced is the parathyroid hormone or para-thormone. It is a single chain peptide of 84 amino acids. The sequence of amino acids has been determined. A synthetic peptide consisting of 30 amino acids has the same biological action as the whole hormone. Since it is a peptide it is inactive by mouth.

Parathormone plays an important part in the maintenance of the plasma calcium level at 10 mg. per cent. (2·5 mmol per litre) [p. 128]. It is usual to consider the **plasma** calcium level rather than the **blood** calcium level because there is practically no calcium in

the red cells. The **blood** calcium level will be lower at 5·5 mg. per cent. since blood calcium level $= \dfrac{100 - H}{100} \times$ plasma calcium level, where H is the haematocrit [p. 5].

Unlike the thyroid gland, the parathyroid glands are not con-trolled by the pituitary gland, but probably by the plasma calcium level.

In order to study the actions of the parathyroid gland, it is necessary to consider the composition of bone.

Bone

Bone consists of a matrix of protein (mainly collagen) embedded in a mucopolysaccharide ground substance. The matrix is made hard and rigid by the deposition of mineral salts, principally the double salt of calcium phosphate and calcium hydroxide termed **calcium hydroxyapatite** $3Ca_3(PO_4)_2 \cdot Ca(OH)_2$.

The majority of bone is preceded by cartilage but some, such as the roof and sides of the skull, is preceded by membrane. The bone is formed by **osteoblast** cells which lay down calcifiable collagen fibrils. As ossification proceeds the original cartilage cells degenerate and disappear.

The ossification of bone is a complex process involving a local increase in the concentration of calcium and phosphate ions so that their solubility product is exceeded. The collagen in bone seems to differ from collagen in other tissues in that its molecular arrangement favours the deposition of calcium hydrogen phos-phate, $CaHPO_4 \cdot 2H_2O$, which is then converted into crystals of calcium hydroxyapatite.

Bone is in a state of dynamic equilibrium and is being continu-ally remoulded by the combined action of osteoblast cells which are forming bone, and large multi-nucleated **osteoclast** cells which are destroying it.

Actions of Parathyroid Hormone

The actions of parathormone (PTH) may be summarized as follows:
1. It has a rapid action on the kidney tubules causing an increased excretion of phosphate which leads to a mobiliza-tion of calcium and phosphate from bone.
2. It has a delayed effect on bone causing:
   (a) proliferation of the osteoclasts,
   (b) a decrease in osteoblastic activity leading to a decrease in bone matrix formation.
3. It has a delayed action on the digestive tract, reducing the calcium in the digestive secretions, and hence reducing the calcium lost in the faeces and increasing the calcium retained by the body.

Deficiency of Parathyroid Hormone (Hypoparathyroidism)

A deficiency of the circulating parathyroid hormone due to underactivity or removal of the parathyroid glands leads to a decrease in the phosphate excretion by the kidneys, and an increase in the phosphate level in the blood. At the same time the plasma calcium level falls (hypocalcaemia).

There is an inverse relationship between the plasma calcium and inorganic phosphate levels. Thus, when the plasma phosphate level rises from 6 mg. per cent. to 10 mg. per cent. the plasma calcium level tends to fall from 10 mg. to 6 mg. per cent.

Tetany

If the plasma calcium level falls to 6 mg. per cent. **tetany** occurs. There is an increased excitability of nerves and neuromuscular junctions which leads to muscular spasms particularly of the hands and feet which is known as *carpopedal spasm*. An increased excitability of the nervous system may lead to convulsions.

Before the onset of tetany, the increased excitability of nerves (*latent tetany*) may be demonstrated in a number of ways. Tapping the facial nerve (VII) as it crosses the angle of the jaw has no effect in a normal person, but in *latent tetany* the muscles on that side of the face will twitch and may go into spasm (Chvostek's sign of latent tetany). A tourniquet or blood pressure cuff around the upper arm in latent tetany will stimulate the underlying nerves and cause the onset of carpo-spasm (Trousseau's sign).

The plasma calcium level depends upon adequate absorption of calcium from the intestines as well as on the level of parathormone. Thus tetany may occur in rickets and osteomalacia due to a vitamin D deficiency since this vitamin is essential for adequate absorption of calcium from the intestines. Tetany may also occur, unrelated to the plasma calcium and parathyroid levels, in alkalosis resulting from the ingestion of alkalis and following overventilation. In this case the increased excitability of nerves is brought about by the fall in the hydrogen ion concentration.

Excess of Parathyroid Hormone (Hyperparathyroidism)

An excess of parathormone due to overactivity of the glands or to an injection of the hormone raises the plasma calcium level to 20 mg. per cent. or more (hypercalcaemia). The phosphate reabsorption by the kidney is inhibited and the phosphate is lost in the urine. Calcium and phosphate are mobilized from the bone leading to rarefaction, cyst formation and spontaneous fractures. There is an increased calcium excretion in the urine so that the 'skeleton is passed out through the urethra'. However, the calcium is excreted less rapidly than the phosphate and the blood calcium level rises.

The rarefaction of bone in hyperparathyroidism is confirmed by comparing the X-ray appearance of the bone of a limb with that of a normal limb taken simultaneously as a control.

Calcitonin = thyrocalcitonin

A hypocalcaemic factor, originally thought to be secreted by the parathyroid gland and named *calcitonin*, is now known to be produced mainly if not entirely by the thyroid gland in man and the alternative name of *thyrocalcitonin* has been suggested.

Further work has shown that this hypocalcaemic factor is produced by cells situated in the last gill arch of the fish (ultimobranchial arch). In the evolution process these cells have migrated in man to the thyroid gland where they lie alongside the thyroid vesicles. They are known as *parafollicular cells* or more simply C-cells (C for calcitonin). This hormone lowers the blood calcium when the level is too high. The mode of action is probably to suppress the mobilization of calcium from bone. Its importance in the regulation of the plasma calcium level has not yet been evaluated but a deficiency would presumably lead to a rarefaction of bones. Animal experiments suggest that calcitonin may be important in preventing the decalcification of a mother's bones by the foetus's need for calcium during pregnancy.

It has already been seen that in addition to the parathyroid and thyroid glands, the plasma calcium level is dependent upon the uptake of calcium from the digestive tract and that this is under the control of 1:25 dihydroxycholecalciferol (a vitamin D derivative) which is made by the kidney [p. 126]. The growth and development of bone is influenced by the pituitary gland [growth hormone, p. 161], thyroxine [p. 162] and the adrenal cortex [p. 166].

THE ADRENAL GLANDS

The two adrenal or *suprarenal* glands are situated immediately above and in front of the upper pole of each kidney [FIG. 163]. The adrenal gland is a double gland with a central part or medulla and an outer part or cortex. Each part behaves as a separate and independent endocrine gland.

Adrenal Medulla

The adrenal medulla releases hormones which are *amines* of *catechol* (dihydroxybenzene) and which are known collectively as *catecholamines*. The important hormones are noradrenaline (norepinephrine) and adrenaline (epinephrine). Noradrenaline is the chemical transmitter of the sympathetic nervous system.

The adrenal medulla is innervated by preganglionic sympathetic fibres and its activity in many respects augments that of the sympathetic nervous system throughout the body.

The action of these hormones is to dilate the pupils of the eyes, contract the *arrectores pilorum* muscles which cause the hair to stand on end, relax the bronchioles thereby increasing the size of the airways, inhibit digestion, contract the sphincters of the digestive tract, and inhibit the bladder musculature. With regard to the cardiovascular system, these hormones increase the blood pressure. They constrict the arterioles and veins by a direct action (with the exception of the coronary vessels and muscle vessels which are dilated by adrenaline). The hormones increase the activity of the heart by their direct action. However, since an injection of these

Noradrenaline (norepinephrine)

Adrenaline (epinephrine)

hormones increases the blood pressure and increases the baroreceptor activity, any tendency to vasoconstriction and cardiac acceleration will be opposed by the baroreceptor reflexes. As a result, under certain conditions, there may be a reflex cardiac slowing and/or a reflex vasodilatation, with increased blood pressure due to an increased cardiac output resulting from a more efficient emptying of the ventricles. (See also page 205 for α and β receptors.)

The vasoconstrictor action of noradrenaline and adrenaline is utilized when they are added to a local anaesthetic to localize its action by constricting the surrounding blood vessels.

Adrenaline mobilizes the liver glycogen and converts it to blood glucose [p. 113].

The adrenal medulla is active under a wide variety of stress conditions such as anger, fear, cold, low blood sugar, low blood pressure, cerebral anoxia and asphyxia. It has recently been shown that angiotensin [p. 45] stimulates the release of the adrenal medulla hormones.

Catecholamines are excreted in small amounts in the urine unchanged and as the metabolic product hydroxymethoxymandelic acid (HMMA) formerly known as vanillylmandelic acid.

$$HO \underset{3}{\overset{4 \quad 1}{\bigcirc}} CH(OH)COOH$$
$$CH_3O$$

HMMA (4-hydroxy-3-methoxymandelic acid)

This is formed by **monoamine oxidase (MAO)**, an enzyme which oxidizes the terminal $-CH_2 \cdot NH_2$ group of noradrenaline and the terminal $-CH_2 \cdot NH \cdot CH_3$ group of adrenaline to the carboxyl group $-COOH$ forming mandelic acid. In addition methylation occurs at position 3 of the benzene ring.

The presence of large amounts of these substances in the urine forms the basis of the test for a phaeochromocytoma, which is a tumour of the adrenal medulla which produces high blood levels of adrenaline and noradrenaline.

Adrenal Cortex

Unlike the medulla the adrenal cortex has no nerve supply. It consists of columns of cells which are divided into three zones from the capsules of the gland inwards; the zona glomerulosa, zona fasciculata and the zona reticularis. The adrenal cortex is rich in vitamin C and cholesterol. Cholesterol is probably the precursor of the hormones produced which are termed corticoids. They may be divided into three groups:

Group I. Mineralocorticoids—Aldosterone

Aldosterone is the important mineralocorticoid produced by the adrenal cortex, that is, a corticoid that regulates the mineral salts in the body.

Aldosterone acts mainly on the sodium ion (Na^+) and by so doing maintains the level of sodium chloride in the body. It does this by stimulating the reabsorption of sodium ions in the kidney tubules and by decreasing the sodium ion content of sweat. At the same time the excretion of potassium by the kidneys is increased.

Any fall in the level of circulating aldosterone will result in a

CH₂OH

CHO CO

HO CH₃ 11 13

O

Aldosterone

fall of sodium chloride in the extracellular fluid of the body and a corresponding loss of water so that the osmotic pressure is unchanged.

Aldosterone is produced by the zona glomerulosa. It is not controlled by the adrenocorticotrophic hormone (ACTH) from the pituitary gland, but its release is stimulated by circulating angiotensin [p. 45].

The level of angiotensin in the blood rises when there is an increased production of renin by the kidney as a result of a reduction in plasma sodium level or when there is a reduction in the blood flow to the kidney (see 'renin–angiotensin–aldosterone system', page 142).

Aldosterone and Blood Volume Regulation

This interrelationship gives a possible mechanism for the regulation of the blood and fluid volumes in the body. Should the blood volume be low, the reduced cardiac output will reduce the blood pressure and pulse pressure in the afferent arterioles of the glomeruli of the kidneys. Surrounding these arterioles are cells which release **renin** into the circulation under these conditions. They form the juxta-glomerular apparatus (JGA) [p. 142]. Renin, as has been seen, is an enzyme which acts on the α_2-globulin fraction of the plasma proteins (angiotensinogen), causing the formation of the peptide **angiotensin**. The angiotensin will act on the adrenal cortex causing an increase in the release of **aldosterone**. Aldosterone will increase the reabsorption of sodium by the kidney tubules and thus raise the sodium level in the blood and extracellular fluid.

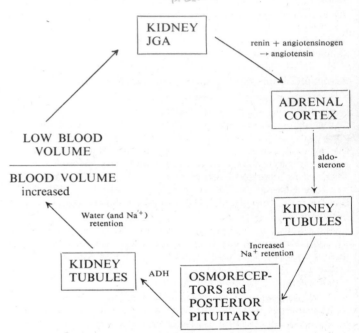

Possible mechanism for the restoration of blood volume by the renin–angiotensin–aldosterone–ADH mechanism.

The sodium retention will increase the osmotic pressure of the blood, and this will stimulate the osmoreceptors in the region of the hypothalamus. The osmoreceptor stimulation will increase the release of ADH from the posterior pituitary gland which will increase the reabsorption of water by the kidney tubules. Water is thus retained.

This retention of sodium and water will restore the blood volume to normal.

Group II. Glucocorticoids—Cortisol

Cortisol (hydrocortisone) is the chief member of the group of hormones known as glucocorticoids which are produced by the zona fasciculata and zona reticularis.

ACTH stimulates the production of cortisol by the adrenal cortex by activating the enzyme adenyl cyclase [p. 113]. This enzyme converts ATP to cyclic AMP which then activates phosphorylase which is necessary for the synthesis of cortisol.

NB

The control of the adrenal cortex may be represented diagrammatically:

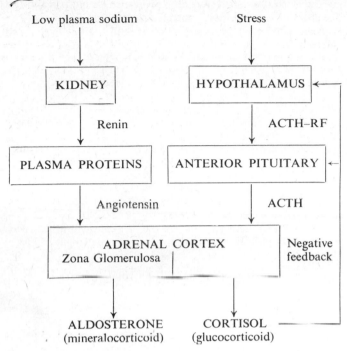

Cortisol = *Hydrocortisone*

The glucocorticoid hormones regulate the general metabolism of carbohydrates, proteins and fats on a long-term basis. They play a particularly important role in modifying the metabolism in times of mental and physical stress.

The blood level of cortisol shows a diurnal variation being highest in the morning and lowest about midnight.

Cortisol increases the breakdown of tissue protein to amino acids which are then converted to glycogen in the liver. The conversion of glucose to glucose-6-phosphate is inhibited. This is the first step in the formation of glycogen from glucose [p. 111]. There is thus a

Cortisol (Hydrocortisone)

decrease in carbohydrate metabolism and an increase in protein breakdown. Fat is mobilized from the fat depots, and transported to the liver for conversion into ketone bodies. An excess of cortisol will thus lead to a high blood glucose level and ketosis, i.e. *diabetes mellitus*.

Cortisol impedes the development of cartilage and decreases the absorption of calcium from the intestines by antagonizing vitamin

D. Excess cortisol will lead to rarefaction of bones, especially vertebrae.

Cortisol has other actions such as reducing the number of circulating eosinophils. It reduces the allergic response of the body. It reduces inflammation. It has some mineralocorticoid effect.

The cortisol production is regulated by ACTH from the anterior pituitary gland (which in turn is probably regulated by ACTH-RF from the hypothalamus). Like thyroxine, the corticoids feed back to the anterior pituitary and inhibit its activity. Thus a high level of blood cortisol reduces the ACTH production by the anterior pituitary.

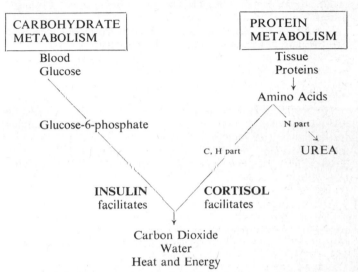

Cortisol increases the use of protein and decreases the use of carbohydrate for the production of heat and energy.

The **chief effects of cortisol** when given to a patient may be summarized as follows:

1. *It increases the use of protein, and decreases the use of carbohydrate for the production of heat and energy. In excess it causes diabetes mellitus.*

2. *It is anti-inflammatory, that is, it reduces the inflammatory response of the body.*

3. *It is anti-allergic, that is, it reduces the allergic responses of the body.*

4. *It causes some sodium (and hence water) retention.*

Synthetic Corticoids

Synthetic analogues of cortisol have been made in which one or more of these properties of cortisol predominates. Thus fluorine derivatives such as *dexamethasone* have been found to have a greatly enhanced anti-inflammatory effect, whilst *prednisolone* (cortisol with an extra double bond in the ring joined to the oxygen) has very much less sodium retaining properties than cortisol itself.

Group III. Sex Hormones

The adrenal cortex is the site of formation of sex hormones—androgens, oestrogens and progesterone. These appear to be relatively unimportant compared with the sex hormones produced by the gonads. However, tumours of the adrenal cortex may lead to precocious puberty in children, virilism in adult females, and feminization in adult males.

Underactivity of the Adrenal Gland

Addison's disease is a generalized underactivity of the whole adrenal gland. If untreated it may be fatal. There is muscle weakness, a low blood pressure (B.P. 80/50 is typical), and pigmentation of the skin. There is an excessive loss of sodium chloride in the urine which leads to a reduction in the tissue fluid and plasma volumes. It is treated by giving sodium chloride and cortisol.

Overactivity of the Adrenal Gland

Overactivity of the adrenal cortex leads to **Cushing's syndrome.** If it is due to overactivity on the part of the anterior pituitary in producing an excess of ACTH, the condition is termed **Cushing's disease.**

In both cases the level of circulating cortisol is greatly increased, and the diurnal variation in level no longer occurs.

The retention of sodium leads to oedema and a high blood pressure (hypertension). The face becomes fat like a full moon (moon-shaped) and the ears are no longer visible when viewed directly from the front.

The breakdown of muscle protein to amino acids leads to wasting and weakness of the skeletal muscles. The skin becomes thin due to the removal of protein and shows purple striations. Diabetes mellitus develops. There is a decrease in the number of circulating eosinophils. Other changes include abnormal growth of hair on the face, chest and abdomen (hirsutism) and the absence of menstrual periods (amenorrhoea) in the female.

All these changes are produced by excessive doses of cortisol.

Over-production of aldosterone usually due to a tumour of the zona glomerulosa tissue is termed **primary aldosteronism** or **Conn's disease.** It is a rare disease associated with sodium retention, muscular weakness and an excessive potassium and water loss.

PANCREAS AS AN ORGAN OF INTERNAL SECRETION

The islet cells of the pancreas were described by Langerhans in 1869. The cells of the **islets of Langerhans** are of two types. The alpha cells are now known to produce the hormone **glucagon** whilst the beta cells produce **insulin.** In 1889 Von Mering and Minkowski demonstrated that the removal of the pancreas in the dog led to sugar appearing in the urine, that is diabetes mellitus.

Early attempts to isolate **insulin** were unsuccessful because insulin is a protein and is destroyed by the trypsinogen present in the pancreatic juice-secreting cells. However, if the pancreatic duct is tied, these cells degenerate and then insulin can be extracted from the gland.

Paulescu in 1916 reported the isolation of insulin, but it was not until the work of Banting and Best in 1921 that insulin became commercially available for the treatment of diabetic patients. It is extracted from the pancreas of animals.

Insulin has been synthesized. It is a protein built up of 51 amino acid units in two coupled chains. Chain A contains 21 amino acid units and has a molecular weight of 2,750. Chain B contains 30 amino acid units and has a molecular weight of 3,700.

Insulin is produced by the β-cells of the islet tissue. It is thought to be stored as a zinc compound. In the plasma insulin is bound to β-globulin [p. 22]. Insulin *facilitates the entry of glucose into cells.* A deficiency leads to a high blood sugar, fatigue and loss of weight. An excess leads to a low blood sugar, irritability, double vision, sweating and a sensation of hunger, and ultimately to coma due to the low blood sugar.

Insulin is the antidiabetic hormone and is used to treat diabetes mellitus. Being a protein it is inactive by mouth and has to be given by injection. It restores the ability to use glucose and fats. The excessive breakdown of protein ceases. The blood glucose level is lowered as the glucose is converted to liver and muscle glycogen and utilized as a source of heat and energy. The ketosis disappears as the ketone bodies are metabolized in the presence of the carbohydrate metabolism [p. 114]. Glucose and ketone bodies disappear from the urine.

Glucagon

Glucagon is a peptide made up of 29 amino acids. It has a molecular weight of 3450.

Glucagon is produced by the α-cells of the islet tissue. **It raises blood glucose by mobilizing liver glycogen.** In this connexion it is more potent than adrenaline. Glucagon also brings about an increase in the formation of glycogen in the liver from proteins and fats (neoglucogenesis). In the liver, glucagon activates the enzyme *adenyl cyclase* which converts ATP to cyclic AMP [p. 113]. The cyclic AMP then activates the enzyme *phosphorylase* which plays an important part in the breakdown of glycogen to glucose phosphate and then to glucose.

Glucagon has no action on muscle glycogen.

Glucagon acts directly on the β-cells of the islets of Langerhans and stimulates them to release insulin.

Ingested glucose brings about a greater increase in plasma insulin (and hence a smaller blood glucose increase) than glucose given intravenously. Glucose in the duodenum appears to release a substance (enteroglucagon) which stimulates the release of insulin probably by the release first of glucagon which will then stimulate the release of insulin. Both secretin and pancreozymin [p. 109] bring about the release of insulin but enteroglucagon appears to be a different peptide.

REFERENCES AND FURTHER READING

Albright, F., and Reifenstein, E. C. (1948) *Parathyroid Glands and Metabolic Bone Disease*, Baltimore.

Banting, F. G., Best, C. H., and Macleod, J. J. R. (1922) The internal secretion of the pancreas, *Amer. J. Physiol.*, **59**, 479.

Cannon, W. B. (1915) *Bodily Changes in Pain, Hunger, Fear and Rage*, London.

Chester Jones, I. (1957) *The Adrenal Cortex*, Cambridge.

Cushing, H. (1912) *The Pituitary Body and its Disorders*, Philadelphia.

Du Vigneaud, V. (1954) Hormones of the posterior pituitary gland: oxytocin and vasopressin, *Harvey Lect.*, **50**, 1.

Fortier, C. (1962) Adenohypophysis and adrenal cortex, *Ann. Rev. Physiol.*, **24**, 223.

Grundy, H. M., Simpson, S. A., and Tait, J. F. (1952). Isolation of a highly active mineralo-corticoid from beef adrenal extract, *Nature (Lond.)*, **169**, 795.

Harington, C. R., and Barger, G. (1927) Chemistry of thyroxine, *Biochem. J.*, **21**, 169.

Harris, G. W. (1955) *The Neural Control of the Pituitary Gland*, London.

Heller, H. (Ed.) (1956) *The Neurohypophysis*, London.

Hubble, D. V. (1957) Endocrines and growth, *Brit. med. J.*, **1**, 601.

Matsuzaki, F., and Raben, M. S. (1965) Growth hormone, *Ann. Rev. Pharmacol.*, **5**, 137.

Myant, N. B. (Ed.) (1960) The thyroid gland, *Brit. med. Bull.*, **16,** No. 2.

Oliver, G., and Schäfer, E. A. (1895) The physiological effects of extracts of the suprarenal capsules, *J. Physiol. (Lond.)*, **18,** 230.

Pitt-Rivers, R., and Tata, J. R. (1959) *The Thyroid Hormones*, London.

Prunty, F. T. G (Ed.) (1962) The adrenal cortex, *Brit. med. Bull.*, **18** No. 2.

Raven, H. M. (1924) The life history of a case of myxoedema, *Brit. med. J.*, **2,** 622.

Robbins, J., and Rall, J. E. (1960) Proteins associated with thyroid hormones, *Physiol. Rev.*, **40,** 415.

Selye, H. (1955) *Stress*, Montreal.

Smith, R. W., Gaebler, O. H., and Long, C. N. H. (1955) *The Hypophyseal Growth Hormone: Nature and Actions*, New York.

Sonenberg, M. (1958) Chemistry and physiology of thyroid stimulating hormone, *Vitam. and Horm.*, **16,** 205.

Spence, A. W. (1953) *Clinical Endocrinology*, London.

Young, F. G. (Ed.) (1960) Insulin, *Brit. med. Bull.*, **16,** No. 3.

17. THE PHYSIOLOGY OF REPRODUCTION

FSH
LH
LTH

Hormones play a dominant part in the physiology of reproduction. Under the influence of releasing factors from the hypothalamus [p. 209] the anterior pituitary gland produces gonadotrophic hormones which control the activity of the gonads, the ovaries in the female and the testes in the male. These hormones are produced in a cyclical manner in the female but at a constant rate in the male. The gonadotrophic hormones have been named according to their function in the female, and although the same hormones are produced in the male, it will be convenient to consider the female system first.

FEMALE REPRODUCTIVE SYSTEM

The reproductive organs in a woman consist of the ovaries, uterine (Fallopian) tubes, the uterus and vagina which are internal organs situated in the pelvis [FIG. 164] and the vulva which includes

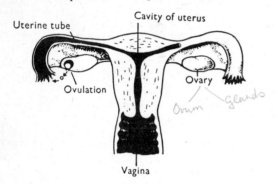

FIG. 164. The ovary and uterus.

At ovulation the ovum is discharged into the abdominal cavity. It enters the open end of the uterine (Fallopian) tube and passes along this tube to the cavity of the uterus.

the mons pubis, the labia, the clitoris, the bulbus vestibuli and the greater vestibular glands which are the external organs (external genitalia) situated below and in front of the pubic arch.

The uterus is a hollow thick-walled muscular organ which is situated in the mid-line behind the bladder and in front of the rectum. It is 7·5 cm. in length, 5 cm. in breadth (about the size of a pear). It consists of a body and cervix. The inner lining of the uterus is the endometrium.

The cavity of the uterus is connected to the outside by the vagina. The vagina extends from the cleft between the labia minora to the uterus and lies between the urethra and bladder in front and the rectum behind. It is about 8 cm. long. It runs upwards and backwards and joins the uterus at right angles.

The cleft between the labia minora is termed the vestibule of the vagina. In it is situated the vaginal and external urethral orifices. The vaginal orifice is partially closed by a thin fold of mucous

membrane known as the hymen. When it has been ruptured small elevations known as *carunculae hymenales* remain. The clitoris, which is homologous with the male penis is an erectile structure. It is partially covered by the anterior ends of the labia minora.

Menstrual Cycle

Starting at puberty, a sequence of events occurs in the internal organs, controlled by the cyclical release of gonadotrophic hormones by the anterior pituitary gland. This sequence of events is repeated monthly and is associated with the passage of 50–250 ml. of blood via the vagina during a period of 3–5 days of the cycle. This is known as the menstrual flow, and the sequence of events associated with it, the menstrual cycle. These cycles continue at a more or less regular interval of 28 days until the menopause at the age of 45–55.

By convention the day of onset of the menstrual flow is taken as day 1 of the menstrual cycle.

Ovulation

The two ovaries are situated in the pelvis on either side of the uterus [FIG. 164]. They are 3 cm. long, 1·5 cm. wide and 1 cm. thick (about the size of an almond). These ovaries have a double function. They produce the female egg-cell or ovum and they are endocrine glands. The ovaries produce the female sex hormone **oestrogen.** Oestrogen first appears in the blood at puberty and causes the development of the female secondary sexual characteristics, the development of the breasts, the female distribution of hair in the pubic region with the upper margin concave upwards and under the arms (axilla) but absent from the face. It also causes the female distribution of fat and the emotional and psychological changes associated with womanhood. The naturally occurring oestrogen is oestradiol:

Natural oestrogen (Oestradiol)

Every month one or other of the ovaries produces an ovum. This ovum has been present from birth, but has developed in the previous few weeks in an ovisac known as a Graafian follicle (vesicular ovarian follicle) [FIG. 165(A)]. The maturation commences under the stimulus of the first anterior pituitary gonadotrophic hormone, the follicle stimulating hormone (FSH). Under the combined influence of FSH and increasing levels in the blood of the second gonadotrophic hormone known as the luteinizing hormone (LH), maturation is completed by the 14th day of the menstrual cycle.

The mature follicle 10 mm. in diameter approaches the surface of the ovary and on the day of ovulation (day 14) the ovum is expelled into the abdominal cavity. Ovulation may be associated with a recognizable abdominal pain.

The ovum enters one or other of the Fallopian (ovarian) tubes and passes along the tube to the cavity of the uterus [FIG. 164].

Under the influence of the luteinizing hormone (LH), the vesicular ovarian follicle from which the ovum has been expelled is converted into **a corpus luteum** (yellow body) [FIG. 165(B)].

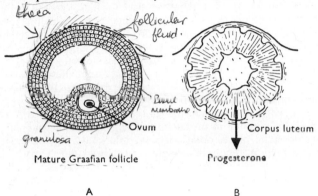

FIG. 165

A. Mature ovarian (Graafian) follicle.
B. Corpus luteum. After ovulation the Graafian follicle is converted into a corpus luteum (yellow body) which secretes the hormone progesterone.

The corpus luteum is itself an endocrine gland, and releases the hormone **progesterone.** This circulates in the blood and prevents any further ovulation from occurring. The release of progesterone

Progesterone

is possibly also under the control of the third gonadotrophic hormone, the luteotrophic hormone (LTH) but the release of this third hormone does not appear to be essential in women (unlike some animals such as the rat).

If the ovum is not fertilized with a male sperm, it only remains in the uterine cavity for 14 days. Thus on the 28th day [FIG. 166] it is shed via the vagina together with the endometrium and a certain amount of blood. This constitutes the menstrual flow and denotes the commencement of another cycle. If we revert back to the first 5 days of this cycle we find that the menstruation which occurs, results in the removal of the unfertilized ovum from the previous month's ovulation.

The regrowth of the endometrium from day 5 onwards is due to the action of oestrogen until day 14 when progesterone appears and aids the growth. The endometrium is being prepared for gestation (pregnancy) as the name pro-gesterone implies.

It will be noted [FIG. 166] that progesterone is only present in the circulation between days 14 and 28 and is absent between days 1 and 14. It is the withdrawal of progesterone which follows on the degeneration of the corpus luteum a few days earlier that leads to the shedding of the endometrium on day 28 and the onset of menstruation.

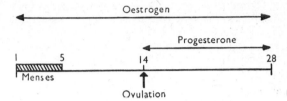

FIG. 166. The menstrual cycle.

The onset of menstruation (menses) is taken as the first day of the cycle. The duration of the cycles may vary from 23–35 days, with a mean of 28 days. The menstrual cycle ceases during pregnancy.

A slight rise in body temperature occurs shortly after ovulation and persists until the next menstrual flow. Daily records of body temperature may be used to determine whether ovulation has occurred.

Progesterone causes retention of fluid which leads to a slight generalized oedema towards the end of each menstrual cycle and may be associated with the premenstrual tension experienced by many women at this time.

MALE REPRODUCTIVE SYSTEM

In the male there is no monthly sex-cycle as in the female. The male gonads or testes lie in the scrotum where they are exposed to a lower environmental temperature than the body temperature.

The testis has two functions. It produces the male sex hormone testosterone and it produces the male reproductive cells, the spermatozoa or sperms.

Male Sex Hormone Production

Testosterone is formed by the interstitial cells of the testis under the influence of the luteinizing hormone of the anterior pituitary gland. Since the term 'luteinizing hormone' is now inappropriate, the same pituitary hormone is often referred to as the interstitial cell stimulating hormone (ICSH).

The production of testosterone commences at puberty and leads to the development of the male secondary sexual characteristics.

Testosterone

Hair starts to grow on the face and the male distribution of hair develops on the trunk, pubic region and in the axillae. Testosterone, like the pituitary growth hormone, stimulates growth and increases the conversion of amino acids to tissue protein particularly in the skeletal muscles. The increased muscle development leads to the male physique. Growth of the larynx deepens the voice—the voice breaks. The organs of reproduction, the testes, epididymis, seminal vesicles, prostate and penis undergo a rapid development.

Testosterone production continues throughout life.

Spermatozoa Production

The coiled tubes of the testis known as seminiferous tubules form the male reproductive cells, the spermatozoa or sperms. These are much smaller than the ova and are highly motile due to the presence of a propulsive tail. The reproductive material is carried in the head. Sperms are produced in the hundreds of millions, unlike the ova which are produced singly each month.

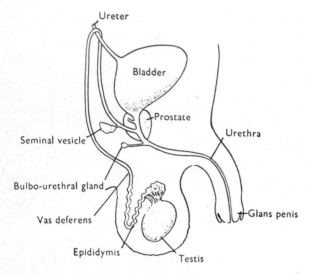

FIG. 167. The male genital system.

The spermatocytes, from which the sperms are derived, undergo a reduction division by which the number of their chromosomes is reduced from 46 to 23. (A similar reduction division has taken place in the formation of the ovum.) Since these chromosomes contain the genetic material, only one half of the parent's characteristics are present in the sperm and will pass on to the offspring. The same applies to the ovum. Further, since two of the 46 chromosomes in the spermatocyte were the sex chromosomes X and Y, and only one of these is passed on to the sperms, it follows that there will be two kinds of sperms, those carrying the Y chromosome and those carrying the X chromosome. The Y chromosome sperms produce boys and the X chromosome sperms produce girls. The mother is neutral in determining the sex of the baby since all the ova carry the X chromosome in the total of 23. If the ovum is fertilized with a Y-chromosome-carrying sperm, the resultant offspring will have XY and will be a male. If the fertilization is by an X-chromosome-carrying sperm, the offspring will be XX and will be female. All the offspring have a total of 46 chromosomes and will have received 23 from each parent.

The formation of sperms (spermatogenesis) starts at puberty and continues throughout life. The sperms are non-motile in the

seminiferous tubules and are pushed onwards into the vasa efferentia of the testis by the formation of fresh cells. These channels are lined with cilia which propel the sperms into the epididymis at the base of the testis. Here they become motile.

The sperm production is probably under the control of the follicle stimulating hormone of the anterior pituitary gland. The seminiferous tubules which produce the sperms atrophy if the pituitary is removed, and sperm formation ceases.

The sperms are stored in the epididymis before passing up along the seminal duct (vas deferens) and along the inguinal canal into the abdomen where they pass through the prostate gland to the urethra [FIG. 167]. The seminal vesicles which join the seminal duct shortly before the prostate is reached add an alkaline yellow viscid fluid. The prostate gland adds a thin opalescent secretion which gives the seminal fluid (semen) its characteristic odour. The bulbo-urethral glands add a mucoid secretion. Thus the seminal fluid is made by the activity of the seminiferous tubules, seminal vesicles, the prostate and the bulbo-urethral glands. Approximately 3 ml. of semen containing 200 million sperms is released with each ejaculation.

With the continuous production of semen, if sexual intercourse does not take place, occasional erections and the discharge of seminal fluid, particularly at night (wet dreams), may take place. This is entirely physiological and should in no way be associated with a feeling of guilt.

Physiology of Sexual Intercourse (Coitus)

When sexual intercourse takes place the erect male penis is inserted into the female vagina, and as a result of friction between the glans penis and the vaginal wall, seminal fluid is ejected into the vagina. The motile sperms enter the cavity of the uterus via the cervix and pass up to the uterine tubes. It appears that fertilization of the ovum is most likely to occur whilst the ovum is passing along the tube.

The usual position is for the woman to lie on her back with her legs apart and knees slightly raised, and for the man to lie face downwards on her taking some of his weight on his arms. But there are alternative positions. Movement of one or both of the partners causes the friction between the glans penis and the vaginal wall.

Erection of the penis is brought about by the distension with blood of the venous sinuses in the erectile tissue, in response to sexual excitement. It is essential for satisfactory intercourse. Erection is caused by an increased activity on the part of the sacral parasympathetic nerves (the nervi erigentes) which cause vasodilatation of the arterioles of the penis and constriction of the main dorsal vein. The corpora cavernosa and spongiosa of the penis fill with blood under high pressure and the penis changes from a small flabby organ covered with wrinkled skin to a rigid elongated organ which will allow semen to be deposited high up in the vagina near the entrance to the uterus.

Friction between the glans penis and the vagina reflexly stimulates sympathetic activity to the smooth muscle of the epididymis, vas deferens, seminal vesicle and prostate, causing the discharge of seminal fluid into the urethra. The internal sphincter of the bladder closes so that the seminal fluid cannot enter the bladder nor can urine be voided. The internal sphincter is closed by the sympathetic nervous system activity [p. 206]. The ejaculation of the seminal fluid is brought about by the somatic (voluntary) nervous system which causes rhythmical contractions of the muscles associated with the penis (bulbo- and ischiocavernosus muscles).

If the sacral parasympathetic outflow is damaged erection of the penis no longer occurs. If the lumbar sympathetic outflow is interrupted such as will occur after a bilateral lumbar sympathectomy operation, ejaculation no longer occurs although erection of the penis is still present.

It will be noted that during sexual intercourse both the sympathetic and parasympathetic divisions of the autonomic nervous system are active at the same time. The increased sympathetic activity, which is associated with the release of noradrenaline and adrenaline from the adrenal medulla causes an increase in heart rate, stroke volume and cardiac volume. The increase in cardiac activity may lead to awareness of the heart's activity (palpitations). The blood pressure is raised. Breathing becomes rapid, the face flushes and sweating may occur.

Although the emission of seminal fluid can occur as a simple physiological reflex, the act of sexual intercourse in humans is associated with emotional and psychological influences at the highest cortical levels, which may reinforce or inhibit the basic reflexes. The emission of semen in the male is associated with a pleasurable sensation called an orgasm. In general, sexual intercourse is most satisfactory when both partners are physically and mentally concentrating on the sexual act and are free from extraneous worries.

In the female sexual excitement is not essential for conception, but is highly desirable from a psychological and reflex point of view. Similar cardiovascular and respiratory changes to those in the male occur in the female in sexual excitement. There is an increase in the heart's activity and respiration is stimulated. The flushing may extend to the chest and produce a patchy rash resembling that of measles. Changes in the breasts include enlargement, erection of the nipples and distension of the surrounding areolar tissue. The vulva becomes engorged and the clitoris becomes erect. The adductor muscles of the thighs are relaxed and the legs are separated. The secretion of mucus in the vagina and vulva facilitates the entry of the penis into the vagina. Stimulation of the clitoris and surrounding structures heightens the physical excitement caused by the vaginal stimulation, and provided that sufficient 'love play' has been employed, an orgasm occurs in the female. Ideally the orgasm should occur simultaneously in the two sexes, but this requires much practice and adjustment. It is possible that during orgasm the uterus may contract and suck the seminal fluid into its cavity. The female is capable of repeated orgasms but the male needs a period of relaxation before a further orgasm can be achieved.

PREGNANCY

The fertilized ovum starts to divide as it passes along the uterine tube to the uterus where it embeds itself in the uterine wall and develops into a baby [FIG. 168]. The corpus luteum persists as the *corpus luteum of pregnancy*. As a result progesterone is maintained in the circulation and no menstrual flow occurs on the 28th day of the cycle. The menstrual cycles now cease until after the baby has been born, and no further ovulations occur during this time. The absence of the menstrual period in a woman who has previously been completely regular, is strongly indicative of pregnancy. At this time (day 28) the embryo is only 14 days old. There are other causes of missed or delayed periods. Young women may find that a change in environment may cause such a delay, presumably due to the effect of higher centres on the hypothalamus [p. 209] which affects the cyclical release of the anterior pituitary hormones.

Twins and Multiple Births

Should two ova be released at ovulation and both be fertilized, twins will result. These twins may be of the same or opposite sex and will be simply two children of the family having a common birthday. (Such twins may be blamed on the wife for having ovulated twice instead of once!) Another form of twins may occur from a single ovulation. At the early stages of development the daughter-cells derived from the fusion of the ovum and sperm can differentiate into an entire baby. Should the cell mass split into two at this early stage, two babies will develop. They will be of the same sex and have the same genetic structure. They will be alike and are termed identical twins. Identical twins have one great advantage over the rest of the population, which is that tissue grafts may be carried out from one twin to the other without any fear of antibody formation, immunity reaction and rejection of the graft.

A double or triple ovulation, associated with one or more of the fertilized ova giving rise to identical twins, will lead to triplets, quadruplets or even quintuplets. Twins occur in approximately 1 in 80 births about one-third being identical twins, triplets 1 in 80^2, quadruplets 1 in 80^3 and quintuplets 1 in 80^4.

Placenta

The fertilized ovum embedded in the uterine wall has developed through the morula stage to the trophoblast stage [FIG. 168].

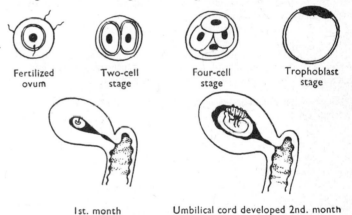

FIG. 168. Development of the fertilized ovum (not to scale).

The developing embryo forms two enveloping membranes. The outer is termed the chorion and the inner the amnion. Chorionic villi invade the uterine wall and serve as a means of transferring nutrients from the maternal blood to the embryo. Later a part of the chorion is formed into the placenta which feeds the embryo for the rest of the pregnancy. The embryo develops in the amnion completely immersed in the amniotic fluid. It is connected to the placenta by the umbilical cord along which it receives its food and sends its waste products.

The placenta is detached from the uterine wall during the birth and is expelled a short time after the baby (the afterbirth).

The chorion, and later the placenta, acts as an endocrine gland. It forms oestrogen, progesterone and a gonadotrophic hormone similar to LH formed by the anterior pituitary. This is termed **human chorionic gonadotrophin (HCG)**. Once the placenta has formed, the progesterone it produces will maintain the pregnancy and the corpus luteum of pregnancy is no longer so important. However, it persists throughout pregnancy. Prior to the formation of the placenta, failure of progesterone production by the corpus luteum or removal of the ovaries will cause the pregnancy to be terminated with a miscarriage.

The high level of gonadotrophic hormone in the blood during pregnancy once the placenta has formed, leads to its appearance in the urine, where it may be detected. This forms the basis of the pregnancy tests. Originally urine, after suitable treatment, was injected into an animal and the effect of any gonadotrophic hormone present in the urine on the sex organs noted. In the **Aschheim– Zondek test** immature mice were employed and corpus luteum formation was looked for as an indication that ovulation had occurred. The **Friedman test** employed virgin rabbits and when gonadotrophin hormone was present ovulation occurred. The most rapid answers were obtained from tests using the rat and the Xenopus toad. Urine containing chorionic gonadotrophin when injected into an immature female rat produced a marked hyperaemia of the ovaries (distension of the blood vessels) in 16 hours. Normal urine did not. This was known as the *rat ovarian hyperaemia* test. When urine containing chorionic gonadotrophin was injected into the dorsal lymph sac of the male toad, there was a discharge of spermatozoa within 3 hours of the injection.

The above tests have now been superseded by *agglutination inhibition tests* in which gonadotrophin-coated red cells or latex particles are added to the urine and an agglutinating antiserum. If gonadotrophin is present in the urine it will react with the antiserum and prevent the agglutination. Thus an absence of agglutination indicates pregnancy.

Positive results are not usually obtained in the above tests until 6 weeks after the last menstrual period. The foetus is then about 28 days old (since ovulation is 14 days after the menstrual period). However, using radio-immunoassay techniques, HCG can be detected in the urine one week after conception.

Chromosome Abnormalities

The 46 human chromosomes are only visible under the light microscope when a cell is dividing. The female sex chromosomes (XX) are, however, often visible as a drum-stick chromatin body attached to the multilobed nucleus of neutrophils, as a Barr body in mucosal cells of the mouth, and they can be demonstrated by a special fluorescent staining technique in a section of hair cut 1 cm. from the root. Such observations are used as the basis of a sex test in international athletic competitions.

For a more detailed study of the chromosome pattern of an individual, body cells, such as lymphocytes, are grown in tissue culture, and their development stopped at the *metaphase* stage by the addition of the drug *colchicine*. At this stage, prior to cell division, the chromosomes are clearly visible. Hypotonic saline is added to make the cells swell and disperse. The chromosomes are cut out from a photomicrograph and arranged in pairs according to the length of the short arm. This process is termed **karyotyping.**

Abnormal patterns occasionally found:

XO   45 chromosomes   female   Turner's syndrome
XXY   47 chromosomes   male   Klinefelter's syndrome
XYY   47 chromosomes   male
YO is not compatible with life
Mongolism—47 chromosomes (additional chromosome on 21st Pair)
              46 chromosomes (translocation—chromosome 21 attached to one of the 13–15 group pairs)

These abnormalities may be due to a defect in either the sperm or the ovum.

CHILD-BIRTH (PARTURITION)

The baby is born by the intermittent contraction of the smooth muscle of the uterine wall aided by a bearing down action on the part of the mother. When labour commences the waves of uterine contraction speed up until the contractions of the uterine muscle occur every 2 minutes. Between each contraction the uterine muscle relaxes. Each contraction raises the pressure in the uterus to 40 mm. Hg. The baby is expelled through the dilated cervix of the uterus and out through the vagina [FIG. 169].

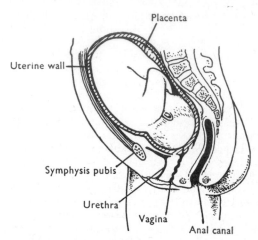

FIG. 169. The pregnant uterus at term.

The onset of labour is complex. Changes occur in the placenta and corpus luteum which lead to a fall in blood progesterone level. It seems that it is the imbalance between the oestrogen and progesterone levels that is important.

Women paralysed by spinal injury in the mid-thoracic region have had normal births so it would appear that the voluntary nervous system plays a relatively small part.

Oxytocic hormone from the posterior pituitary gland may be important; an infusion of this hormone is one of the ways used to promote the onset of labour. It increases the frequency and power of rhythmical contraction of the uterus. The baby suckling at the breast causes the release of the oxytocic hormone and contraction of the uterus. This is important in aiding the return of the uterus to its original size after child-birth.

Mammary Glands

The breasts or mammary glands secrete milk to feed the new-born baby. These glands exist in both sexes, but in the male they remain in a rudimentary state. In the female the breasts start to enlarge with the onset of puberty under the influence of the circulating hormones oestrogen and progesterone. They increase in size during pregnancy and especially during lactation. In old age they become atrophied.

In each breast there are about fifteen separate milk producing systems arranged radially around the nipple. Each terminates in its own orifice at the nipple [FIG. 170].

The milk is made by the epithelial cells lining the milk alveoli and is secreted into the lumen of the gland. Clusters of alveoli supply the small lactiferous ducts which unite with other ducts to form one large excretory duct for each system. Each excretory duct has a dilatation termed the lactiferous sinus close to its nipple orifice. This sinus acts as a reservoir for the milk.

The nipple has a pink or brownish hue. The base of the nipple is surrounded by a coloured area of skin termed the areola. In the virgin this has a delicate rosy colour. In early pregnancy the areola becomes larger and darker. Towards the end of pregnancy it is dark brown in colour. The colour decreases once lactation has ceased, but it never regains its original colour entirely. The colour changes may help in the diagnosis of a first pregnancy.

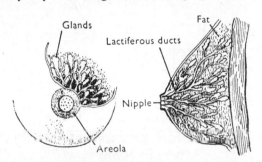

FIG. 170. The mammary gland.

The gland consists of fifteen to twenty subcutaneous duct systems each with its own orifice at the nipple.

Lactation

The production of milk on a large scale does not occur until a few days after the baby has been born. The initiation of milk production (lactogenesis) is under hormonal control of the anterior pituitary gland via the hormone *prolactin*. This is a 200 amino acid peptide with 3 S-S bridges. Prolactin is also important for the maintenance of lactation. Surges of prolactin secretion following each suckling stimulate the breast for the next feed. The placenta also produces a lactogenic substance *human placental lactogen*.

The formation of prolactin is inhibited by progesterone during pregnancy. With the disappearance of the corpus luteum of pregnancy and the expulsion of the placenta, the progesterone level in the blood falls. The oestrogen which still remains stimulates the prolactin formation and the onset of lactation. An excessively high level of oestrogen will inhibit milk secretion and artificial oestrogens are used to prevent the onset of lactation [see also *prolactin inhibiting factor* page 209].

The continuation of the milk secretion (galactopoiesis) depends on adequate levels of the growth hormone and the thyroid hormone. It also depends on the breasts being emptied at frequent intervals. In the absence of suckling, milk production ceases.

The baby by its sucking and champing action can express milk from the ducts, but it cannot extract the milk deep down in the mammary alveoli. Stimulation of the nipples by the sucking sends nerve impulses to the supra-optic nucleus of the hypothalamus and thence to the posterior pituitary gland which cause the release of the *oxytocic hormone*. This hormone reaches the mammary glands via the blood and causes contraction of the myoepithelial cells surrounding the alveoli so that milk is moved into the more superficial ducts.

At one time it was thought that conception was not possible during lactation, but this has been disproved. There is a delay in the return of the menstrual periods after the birth of the baby (3–4 months), but often ovulation recommences first. If this ovum is fertilized, another pregnancy will commence without any intervening menstrual period. If the ovum is not fertilized menstrual periods recommence a fortnight after the recommencement of ovulation.

Colostrum

The secretion from the breast for the first few days after the birth is yellow in colour and is termed *colostrum*. It contains cells from the centre of the alveoli which are the colostrum corpuscles. It is rich in the protein globulin but low in fat. It has been suggested that this fluid is a means of transferring protective antibodies from the mother to the baby

Milk

During the first few weeks of life a baby gains in weight by 25 to 30 grams per day with milk as its only source of food for growth and repair of tissues and for heat and energy. Milk, therefore, contains all the dietary requirements.

Milk contains fat, milk sugar (lactose), and the proteins lactalbumin and casein. In human milk the main protein is lactalbumin which is easily digested by the baby. In cows' milk the main protein is caseinogen which forms a relatively insoluble mass of calcium caseinate under the action of pepsin and calcium salts in the stomach. Cows' milk also contains less carbohydrate than human milk. To make cows' milk suitable for a new-born baby it has to be diluted and sugar added.

In the mother the milk is made from the circulating amino acids and the plasma proteins, the lactose is made from the blood glucose, and the milk fat is synthesized from the blood fat and the acetate residues.

The only cells in the body to synthesize the disaccharide lactose are those in the mammary gland. The enzyme involved is termed *lactose synthetase* and its formation is stimulated by circulating prolactin.

Certain drugs taken by the mother appear in the milk. These include morphine, sulphonamides, barbiturates and the laxative phenolphthalein.

THE NEW-BORN BABY

At birth the baby's environment changes from water at 37 °C. (amniotic fluid in the uterus) to room air at a very variable environmental temperature. The baby's heat regulating centre is poorly developed at birth and it therefore has to be kept warm.

Brown Adipose Tissue

When subjected to cold, the new-born infant has the ability to increase its heat production by an additional mechanism to shivering. It does so by increasing metabolism in a special type of adipose tissue which is widely distributed about the body of the infant and is known as 'brown body fat'.

This tissue is highly vascular and consists of multilocular fat cells with a profusion of mitochondria. Similar adipose tissue is found between the shoulders of hibernating animals.

When subjected to cold, the infant increases its oxygen uptake and at the same time uses up its brown body fat. The stimulus which brings this about is probably an increase in the circulating noradrenaline. The temperature in the fat rises, and this tissue supplies heat to the rest of the body. The process is termed 'non-shivering thermogenesis'.

Circulatory Changes at Birth

In utero the baby (foetus) received its oxygen supply from the maternal blood via the placenta and the umbilical vein. Large

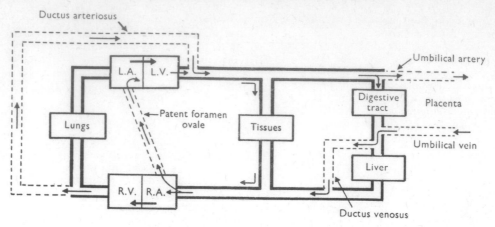

FIG. 171. The circulation before and after birth.

The oxygen and food supply for the foetus comes from the maternal circulation via the placenta, but the foetal and maternal circulations are independent.

The ductus venosus, the ductus arteriosus, foramen ovale and umbilical vessels (all shown in dotted lines) close after birth leaving the normal postnatal circulation (shown in heavy lines). The ductus arteriosus may not close completely until several days after birth.

L.A. = left atrium; R.A. = right atrium; L.V. = left ventricle; R.V. = right ventricle.

maternal sinuses, fed from the uterine arteries, supply oxygen to the foetal blood in the chorionic villi which dip into the sinuses. This blood has been pumped to the placenta along the umbilical arteries by the foetal heart. It returns to the foetus along the umbilical vein 80 per cent. saturated with oxygen. Most of this blood by-passes the liver by the ductus venosus and enters the right atrium of the heart [FIG. 171]. It is diluted by poorly oxygenated blood returning from the rest of the body. The lungs are inoperative for gaseous exchange and are by-passed in two ways. Firstly, the foramen ovale is open between the right and left atria so that particularly the blood from the inferior vena cava by-passes the right heart and lungs by entering the left atrium. Secondly, the ductus arteriosus transfers much of the blood which reaches the pulmonary artery directly to the aorta.

It will be noted that for this to occur the pulmonary artery pressure in the foetus must exceed that in the aorta. This is in contrast to the state of affairs after birth where the aortic pressure is 120/80 mm. Hg and the pulmonary artery pressure only 25/8 mm. Hg.

At birth the stimulation of the skin by the cold environment causes the baby to cry and by so doing it inflates its lungs with air. A dramatic fall in the pulmonary resistance to blood flow occurs and the pulmonary artery pressure falls. Simultaneously the aortic pressure rises. The ductus venosus, and the foramen ovale close. The ductus arteriosus constricts and closes within the next few days. Flow in the umbilical cord ceases and the cord is usually cut and tied a few inches from the abdominal wall. The remnant of the cord becomes detached after a few days leaving a scar (the umbilicus or navel).

METABOLIC DISORDERS OF THE NEW-BORN

Galactosaemia and phenylketonuria are two examples of metabolic disorders which are present at birth. Although these conditions are comparatively rare, their early detection is of great importance.

Galactosaemia

A baby suffering from galactosaemia has an *inborn error of metabolism* which prevents the utilization of galactose. It is an inherited recessive disorder which appears shortly after birth.

The sugar in milk (lactose) is converted by the enzyme lactase in the small intestine into the monosaccharides glucose and galactose [p. 105]. Normally the galactose is changed into glucose by the liver [p. 154], and is metabolized with the production of heat and energy. In this condition the galactose conversion is impaired and galactose accumulates in the blood. This causes toxic effects which include diarrhoea and vomiting, and later, mental retardation, cataract and liver damage. Galactose appears in the urine. It reduces Benedict's reagent, but does not give any reaction with reagent strips specific to glucose [p. 146]. A baby with this disorder must be fed on a galactose-free diet, and it will be 'poisoned' by milk.

Phenylketonuria

Phenylketonuria is another inborn error of metabolism which appears shortly after birth. It is caused by the baby's inability to convert surplus phenylalanine to tyrosine and thence to thyroxine [p. 162], adrenaline and noradrenaline [p. 165] and melanin the dark pigment in hair and skin. The lack of melanin is shown by the fair hair and a fair complexion of the infant.

$CH_2COCOOH$ $CH_2CH(NH_2)COOH$

Phenylpyruvic acid
(in urine)
a 'phenyl ketone'

Phenylalanine

$CH_2CH(NH_2)COOH$

OH
Tyrosine

Phenylalanine and its derivative phenylpyruvic acid accumulate in the blood. They are toxic to the brain and lead to mental deficiency. The 'phenylketone' phenylpyruvic acid appears in the urine—hence the name given to this condition. It gives a blue-green colour with ferric chloride [p. 146].

Since phenylalanine is an amino acid essential for growth and repair of tissues [p. 115], a phenylalanine-free diet cannot be employed. Instead a diet with a low but controlled level of phenylalanine is given.

PROSTAGLANDINS

The name 'prostaglandin' was given by von Euler in 1935 to a group of hormone-like substances found in the seminal fluid of man and animals. These substances were originally thought to be produced by the prostate gland (hence their name), but further work has shown that they are produced by the seminal vesicles. More recently it has been found that they are produced by many other tissues of the body.

Chemically the prostaglandins consist of a chain of 19-carbon atoms with a carboxyl group (—COOH) attached to one end of the chain making a total of 20-carbon atoms. They thus fall into the category of fatty acids. The carbon atoms are usually numbered starting with the carboxyl-carbon as number 1.

A bond between C—atom 8 and C—atom 12 makes the molecule ⊂ shaped by forming a 5-carbon ring (cyclopentene ring) near the centre of the chain.

Most of the carbon atoms are saturated with hydrogen but all members of the group have one or more double bonds between certain carbon atoms. Prostaglandins are thus 'unsaturated' fatty acids.

$$-CH_2-CH_2-CH_2-CH_2-$$ saturated
$$7654$$

$$-CH_2-CH=CH-CH_2-$$ unsaturated
$$7654$$

In addition the molecule contains 2 or 3 additional oxygen atoms in the form of side chain —OH or =O groups.

Formation of Prostaglandins

The prostaglandins are synthesized in the body from ingested **unsaturated fatty acids.** Such unsaturated fatty acids are 'essential' in the diet, unlike saturated fatty acids which can be made in the body from acetyl co-A [p. 111].

It has been known for some time that rats fed on a fat-free diet have a stunted growth, develop skin lesions and that reproduction ceases. Only the unsaturated fatty acids, such as are needed for prostaglandin synthesis, will remedy this deficiency.

Classification of Prostaglandins

The classification of prostaglandins has become progressively more confusing as new members of the group are isolated. Initially the prostaglandins were subdivided into groups **A, B, C, D,** etc., but to save confusion with blood groups and vitamins, these are written as **PGA, PGB, PGC, PGD,** etc. (PG stands for prostaglandin). Each group is subdivided and given a suffix 1, 2, 3 to indicate the number of double bonds in the molecule. Each subgroup is further subdivided (where necessary) into α, β, γ, etc. Thus the two prostaglandins of current interest are referred to as 'PGE$_2$' and 'PGF$_2\alpha$'.

Both PGE and PGF groups have —OH groups at positions 11 and 15. PGE has an =O group at position 9 whereas PGF has an —OH group at this position. The double bonds in both PGE$_2$

and PGF$_2\alpha$ are between carbon atoms 5–6 and 13–14. Set out in full the structural formula of PGE$_2$ becomes:

Prostaglandin PGE$_2$

Actions of Prostaglandins

Prostaglandins are extremely potent substances. They bring about their effects when present in very low concentrations. They are also destroyed very rapidly after production and this has made their detection and isolation in man very difficult. Their actions are complex and difficult to predict. The first reported actions in animal experiments was a fall in blood pressure, and contraction and sometimes relaxation of uterine muscle. It is now known that the results vary from species to species and depend upon the structural formula of the prostaglandin used.

Prostaglandins have been detected in the central nervous system.

Prostaglandins also release adenyl cyclase which converts ATP to cyclic AMP [p. 113].

The principal action of prostaglandin appears to be on smooth muscle. It may be either as a direct effect on the muscle itself or an indirect one by the inhibition of the release of noradrenaline at the sympathetic postganglionic terminations [p. 204].

Thus prostaglandin PGE_1 (formula as PGE_2 but with only one double bond at 13–14) relaxes the smooth muscle of the bronchi. It has been used as an alternative to *isoprenaline* [p. 205] in an aerosol inhaler for the treatment of asthma.

Prostaglandin PGE_1 constricts the blood vessels of the nose and makes breathing easier in cases of nasal congestion.

In the dog, an intravenous infusion of prostaglandin PGE_1 and PGA_1 (formula as PGE_1 but with no —OH group at position 11 and double bond between 10–11) acts as a diuretic and increases both the volume of urine and quantity of sodium excreted. Prostaglandin PGE_1 and PGF_2 inhibits gastric secretion in the dog.

Prostaglandins and the Female Reproductive System

An intravenous infusion of PGE_2 or $PGF_2\alpha$ (as low as a few micrograms per minute) increases the force and frequency of contraction of the uterus in pregnant women. Since prostaglandins have been detected in the venous blood of the woman during labour, it is possible that naturally produced prostaglandins may be playing a role in parturition.

Intravenous infusion of PGE_2 is used in place of oxytocin to induce labour.

Work on female monkeys indicates that prostaglandin $PGF_2\alpha$ may have another action on the female reproductive system. It appears to reduce the amount of progesterone released by the corpus luteum in the early stages of pregnancy. Since a high level of progesterone is essential for the implantation of the fertilized ovum in the endometrium of the uterus, this fact may provide a possible means for limiting the world's population.

REFERENCES AND FURTHER READING

Barclay, A. E., Franklin, K. J., and Prichard, M. M. L. (1944) *The Foetal Circulation and Cardiovascular System, and the Changes that they Undergo at Birth*, Oxford.
Barcroft, J. (1946) *Researches on Prenatal Life*, Oxford.
Born, G. V. R., Dawes, G. S., Mott, J. C., and Widdicombe, J. G. (1954) Changes in the heart and lungs at birth, *Cold Spr. Harb. Symp. quant. Biol.*, **19**, 102.
Caldeyro-Barcia, R., and Heller, H. (Eds.) (1961) *Symposium on Oxytocin*, London.
Corner, G. W. (1942) *Hormones in Human Reproduction*, Princeton.
Corner, G. W. (1952) The primate ovarian cycle, *Brit. med. J.*, **2**, 403.
Cross, K. W. (Ed.) (1961) Foetal and neonatal physiology, *Brit. med. Bull.*, **17**, No. 2.
Davies, M. E., and Plotz, E. J. (1957) Progesterone metabolism, *Recent Progr. Hormone Res.*, **13**, 347.
Dorfman, R. I., and Shipley, R. A. (1956) *Androgens: Biochemistry, Physiology and Clinical Significance*, London.
Folley, S. J. (1956) *The Physiology and Biochemistry of Lactation*, Edinburgh.
Harris, H. (1963) *Garrod's Inborn Errors of Metabolism*, London.
Linzell, J. L. (1959) Physiology of the mammary glands, *Physiol. Rev.*, **39**, 534.
Lloyd, C. W. (Ed.) (1959) *Recent Progress in the Endocrinology of Reproduction*, New York.
Lyons, W. R., Li, C. H., and Johnson, R. E. (1958) Hormonal control of lactation, *Recent Progr. Hormone Res.*, **14**, 219.
Pike, J. E. (1971) Prostaglandins, *Scientific American*, **225**, 5.
Ramwell, P. W., and Shaw, J. E. (Eds.) (1970) Prostaglandins, *Ann. N.Y. Acad. Sci.*, **180**.
Reynolds, S. R. M. (1939) *Physiology of the Uterus*, London.
Velardo, J. T. (1958) *Essentials of Human Reproduction*, New York.
Velardo, J. T. (1961) Reproduction, *Ann. Rev. Physiol.*, **23**, 263.
Villee, C. A. (Ed.) (1961) *The Control of Ovulation*, London.

18. MUSCLES AND NERVES

MUSCLE

Skeletal muscle is composed of long thin cells, known as muscle fibres. These are 1–50 mm. long, but only 10–100 μ in diameter. These fibres are enclosed in a structureless outer coat known as the sarcolemma.

Muscle fibres are striated, that is, under the microscope they show alternate light and dark bands along their length. Each fibre is seen to be made up of a large number of myofibrils (1–2 μ diameter). These show the same striation.

The light and dark bands change as the microscope is racked up and down. The band which appears dark when the microscope is slightly out of focus (due to the microscope objective being too close to the muscle) has a high refractive index to light. This refractive index is different for polarized light oscillating parallel to the axis from that oscillating at right angles. This is termed *birefringence* or anisotropy. Hence these bands are termed A bands. The alternate band is the I band. At the centre of the dark band A is a light area, H, and at the centre of the light I band is a dark Z band [FIG. 172].

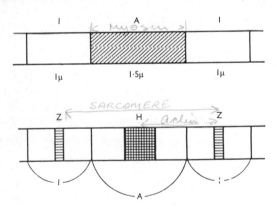

FIG. 172. Diagram showing the microscopical appearance of the striations of voluntary muscle.

The area between the Z bands is the contractile unit or *sarcomere*; the Z bands become closer together during a contraction.

When the muscle fibres contract the I band shortens, but the A band remains constant. The H band, however, narrows.

Muscle is composed of the proteins actin (with tropomyosin and troponin) and myosin. Myosin is confined to A band. Actin filaments run from the Z band to terminate in the H zone [FIG. 173].

Electron microscope studies have shown that striped muscle is built up of two overlapping series of filaments.

The myosin filaments are thick filaments. The actin filaments are thin. The thin filaments are attached together at the Z bands.

Cross-bridges extending from the myosin to the actin filaments provide the force which causes the filaments to move relative to one another. Detachment and re-attachment further on of each cross-bridge presumably occurs as the shortening proceeds.

CHANGES IN MUSCLE DURING CONTRACTION

Mechanical

When stimulated a muscle fibre contracts. The natural stimulus arrives via the motor neurone, but the muscle will respond to an external stimulus such as an electric current.

As will be seen later the muscle contraction may be static or **isometric**, that is, the length remains constant, but the tension increases, or it may be dynamic or **isotonic** with the muscle becoming shorter and thicker.

There is a time interval between the application of a stimulus and the onset of the contraction. This interval is known as the *latent period*. There then follows a *contraction phase* followed by a *relaxation phase*.

Chemical

When a muscle is active, the glycogen stores are depleted, oxygen is used up and carbon dioxide is formed. It has been seen

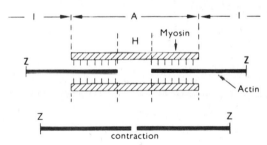

FIG. 173. The contractile elements in striated muscle.

[p. 112] that the muscle glycogen breaks down to pyruvic acid with the release of energy which is stored by the formation of high energy ATP molecules, and that the pyruvic acid is further oxidized to CO_2 and H_2O in the citric acid cycle with the formation of many more ATP molecules. If the oxygen supply is inadequate, an oxygen debt develops, and lactic acid is formed. After the exercise this oxygen debt is repaid by maintaining an increased pulmonary ventilation and raised oxygen intake. About one-fifth of the lactic acid is oxidized to CO_2 and H_2O and the energy from this reaction is used to reconvert the rest of the lactic acid back to glycogen.

The formation of pyruvic and lactic acids is not essential for muscular contraction. A muscle still contracts after monoiodo-acetic acid has been added, but no pyruvic or lactic acid is now formed.

179

The contractile substances in the muscle are myosin and actin.

$$\text{Actin} + \text{myosin} \rightarrow \text{actomyosin}$$

Actomyosin threads shorten when ATP is added. Actin has a molecular weight of 70,000. Myosin has a higher molecular weight of 450,000. It is adenosine triphosphatase, that is, it splits ATP.

These changes involve the release of calcium ions (Ca^{2+}) which activate the myosin to split ATP to ADP. The released energy causes the actomysin to contract. The ATP is rebuilt by the breakdown of creatine phosphate (phosphorylcreatine). Thus:

$$Ca^{2+} + \text{myosin} \longrightarrow \text{activated myosin}$$

$$ATP \longrightarrow ADP + \text{phosphate} + \text{energy}$$

$$\text{Creatine phosphate} + ADP \longrightarrow \text{creatine} + ATP$$

Heat

The energy developed by an active muscle is converted in part to mechanical work and in part to heat. Muscles have a maximum efficiency of 25 per cent. and this implies that at least three-quarters of the energy is lost as heat.

Sound

Sound is produced when muscle contracts. The sound of the masseter muscle contracting may occasionally be heard when clenching the jaw.

NERVES

A nerve cell together with its processes, *axon* and *dendrites*, forms a **neurone.** The long process of a nerve cell is termed a nerve fibre or simply a nerve. The nerve cell is the cell of origin of the nerve fibre, that is the nerve fibre grows out from this cell in the embryo. It is essential for its nutrition. The nerve fibre will cease to function and degenerate if the connexion with its cell of origin is severed.

One of the striking features of the nerve cell stained with Methylene Blue is the presence in the cytoplasm of numerous Nissl granules which stain with this basic dye [FIG. 174]. They have been shown

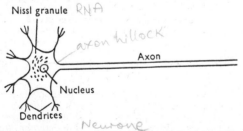

FIG. 174. The nerve cell with its axon and dendrites. The Nissl granules contain RNA and are probably associated with protein synthesis.

to contain RNA [p. 116] and are probably associated with protein synthesis in the cell. They are absent from the part of the cell giving rise to the axon which is known as the axon hillock.

The Nissl granules disintegrate and disappear if the axon is sectioned. This is known as *chromatolysis.*

Two types of nerve fibre are found, termed medullated and non-medullated.

Medullated Nerve Fibres	Non-Medullated Nerve Fibres
1. Axon	1. Axon
2. Medullary sheath (myelin)	2. Neurilemma
3. Neurilemma	

In the medullated nerve fibre, the central axon (axis cylinder) is surrounded by both a medullary (myelin) sheath and an outer neurilemmal sheath [FIG. 175]. In the non-medullated nerve, the

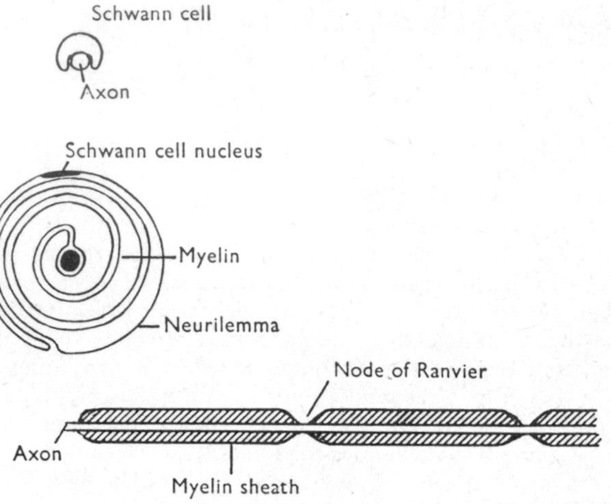

FIG. 175. The Schwann cell spins the myelin (medullary) sheath around the axon. The myelin sheath is interrupted at intervals known as nodes of Ranvier.

medullary sheath is absent and the neurilemmal sheath is the only covering to the axon.

In both cases, a cell nucleus may be seen lying under the neurilemma and this has been termed the cell of Schwann. Modern electron microscope studies have shown that the difference between medullated and non-medullated fibres is not as great as at first seemed.

In both cases the axon is enveloped along its length by Schwann cells; the outer membrane of this cell is the neurilemma. In the case of medullated nerves, the Schwann cell has rotated many times round the axon, wrapping a spiral of lipid-protein round it rather like a roll of carpet. This is the myelin sheath. In the non-medullated nerve the Schwann cell encloses the axon, but has not formed a myelin sheath.

Most of the nerves are medullated, but the postganglionic autonomic fibres and other small fibres of less than 1 μ in diameter are non-medullated.

The myelin sheath is interrupted at intervals at the nodes of Ranvier which represents intervals between the Schwann cells.

Each bundle of nerve fibres is enclosed by a coat of connective tissue known as the perineurium. A number of bundles are bound together by a sheath known as the epineurium.

NERVE IMPULSES

A nerve fibre conducts *nerve impulses*. These impulses form the basis by which communication information is transmitted through the nervous system. An impulse is difficult to define. The usual definition is a 'physico-chemical change transmitted by nerves'. It is associated with electrical changes in the nerve (action potential), with an increased oxygen uptake and carbon dioxide production and an increased heat production.

A **stimulus** is the external force which sets up the impulse. It may be mechanical, electrical or chemical. For experimental purposes, electrical stimuli are frequently employed, because they are easily reproducible and do not cause permanent damage to the nerve.

The Membrane Potential

The cytoplasm inside the nerve cell and axon contains potassium proteinate in the form of positively charged potassium ions and negatively charged protein ions. Although the potassium ions are able to diffuse through the surrounding membranes, the protein ions cannot.

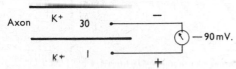

FIG. 176. Although the membrane surrounding the axon is permeable to potassium ions, the potassium ion level in the axon is thirty times higher than that in the surrounding tissue fluid. The positively-charged potassium ions are held by the electrostatic attraction of the surplus of the negatively-charged ions in the axon. This excess of negative ions over positive ions gives the interior of the axon an electrical potential which is −90 mV. with respect to the outside.

Outside, in the extracellular fluid, the sodium chloride is in the form of positively charged sodium ions and negatively charged chloride ions. Both sodium and chloride ions are able to enter through the membrane into the cell but the sodium is immediately extruded again by the sodium pump mechanism. Only the chloride is freely diffusible.

The only diffusible ions are therefore potassium from inside the membrane and chloride from outside. In such a case the ions arrange themselves according to the Donnan Equilibrium

$$[K^+]_{IN} . [Cl^-]_{IN} = [K^+]_{OUT} . [Cl^-]_{OUT} \qquad (1)$$

high   low   low   high

where the square brackets represent concentrations. Inside, since the K^+ concentration is high (from the potassium proteinate), the Cl^- concentration will be low. Outside, since the Cl^- concentration is high (from the NaCl), the K^+ concentration will be low. It will be noted that neither the sodium concentration outside nor the ionized protein concentration inside appear in this equation (1).

Equation (1) may be rewritten:

$$\frac{[K^+]_{OUT}}{[K^+]_{IN}} = \frac{[Cl^-]_{IN}}{[Cl^-]_{OUT}} \qquad (2)$$

The high concentration of potassium ions will bombard the inside of the membrane much more intensely than the low concentration of potassium ions outside and potassium will tend to diffuse outwards. The high concentration of positively charged potassium ions inside can only be maintained if they are being held there by an excess of negatively charged ions, i.e. an electrical potential exists whereby the inside is negative to the outside.

The magnitude of this potential is given by the formula:

$$E = \text{constant} \times T \times \log_{10} \frac{[K^+]_{OUT}}{[K^+]_{IN}}$$

where T is the absolute temperature.

At 37 °C. this becomes

$$E = 61 \log_{10} \frac{[K^+]_{OUT}}{[K^+]_{IN}}$$

With a potassium ratio of 1:30 [FIG. 176] this becomes:

$$E = 61 \times \log_{10} \tfrac{1}{30}$$
$$= -61 \log_{10} 30 = -61 \times 1.48 = -90 \text{ mV.}$$

Direct measurements have shown potentials of 50–70 mV. negative between the inside and outside of nerve fibres and nerve cells.

Action Potential

When an impulse is passing along a nerve the 'active' part of the nerve changes its electrical state. The inside now becomes positive with respect to the outside [FIG. 177].

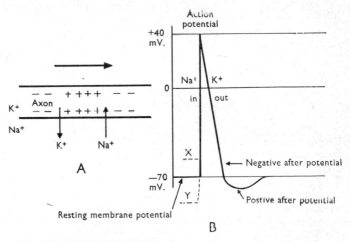

FIG. 177. The nerve impulse is associated with the entry of sodium ions into the axon which changes the potential from −70 mV. to +40 mV. This is followed shortly afterwards by the exit of potassium ions so that the electrical state of −70 mV. is restored. The voltage change associated with this transient reversal of potential is known as the *spike* or action potential. Similar 'depolarization' changes take place in the nerve cell.

If the resting potential in the nerve cell is reduced towards X, the nerve cell becomes more excitable. When the potential reaches the critical level X, a propagated spike potential is produced which travels down the axon (nerve impulse). If the potential is increased towards Y, the nerve cell becomes less excitable.

Excitatory nerves acting on the nerve cell tend to move the cell potential towards X. Inhibitory nerves tend to move the cell potential towards Y.

When the impulse reaches the region, the membrane becomes suddenly permeable to sodium ions which rush into the inside, making the inside positive (+40 mV.) with respect to the outside. The potential inside has thus changed from −70 mV. to +40 mV., a change of 110 mV. This potential change is termed the action or spike potential.

On theoretical grounds, if the membrane becomes temporarily impermeable to K$^+$, and the Na$^+$ concentration outside is ten times that inside, then $E = 61 \log_{10} [\text{Na}^+]_{\text{OUT}}/[\text{Na}^+]_{\text{IN}} = +61 \log_{10} \frac{10}{1} = +61$ mV.

The membrane then ceases to be permeable to sodium and becomes permeable to potassium so that the membrane potential rapidly returns to its resting level.

Thus the nerve has gained a little sodium and lost a little potassium, but has reverted to its previous electrical state and is ready for another impulse. The sodium and potassium ions are re-exchanged later during a slow recovery period.

Nerve Cell Potentials

Similar 'depolarization' changes take place in the nerve cell.

If the resting potential of a nerve cell is reduced from -70 mV. towards zero, the nerve cell becomes more excitable. If the reduction in potential reaches a critical level X [FIG. 177] a propagated spike potential is produced which travels down the axon as a nerve impulse.

If the resting potential of the cell is increased towards Y [FIG. 177] the nerve cell becomes less excitable.

Both excitatory and inhibitory nerves synapse with the nerve cell [FIG. 178]. The excitatory nerves tend to move the nerve cell's potential towards X. Such a change from the resting level is termed an **excitatory postsynaptic potential** (EPSP).

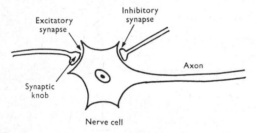

FIG. 178. A large number of neurones synapse with each nerve cell in the central nervous system. One excitatory and one inhibitory synapse is shown. The gap between the synaptic knob and the nerve cell is bridged by a chemical transmitter. The excitatory neurones produce excitatory postsynaptic potentials in the nerve cell. The inhibitory neurones produce inhibitory postsynaptic potentials which tend to cancel out the excitatory postsynaptic potentials. When the excitatory postsynaptic potentials have changed the cell potential sufficiently to reach X [FIG. 177] a propagated spike potential is produced which travels down the axon as a nerve impulse.

The inhibitory nerves tend to move the cell's potential towards Y. Such a change is termed an **inhibitory postsynaptic potential** (IPSP). It will tend to cancel out any EPSP and to make the cell less excitable.

There is a small gap, known as the *synaptic cleft*, between the synaptic knob of the excitatory (or inhibitory) neurone and the nerve cell [FIG. 179]. When a nerve impulse arrives at the synaptic knob it causes the release of a small quantity of chemical transmitter from the synaptic vesicles. This chemical transmitter bridges the gap and brings about the potential changes in the nerve. The chemical transmitter is then rapidly destroyed or removed. Two different transmitter substances have been postulated for the excitatory and inhibitory synapses. Sherrington suggested that they be designated **E** and **I**.

Presynaptic Inhibition

Inhibition also occurs in a slightly different manner. Some inhibitory nerve fibres do not terminate on the body of a nerve cell. Instead they terminate on the synaptic knob of an excitatory nerve fibre which is itself in contact with the body of the nerve cell.

Such inhibitory neurones prevent the release of chemical transmitter at the excitatory synaptic knob and thus prevent excitation of the nerve cell. This form of inhibition is termed *presynaptic inhibition* to distinguish it from *synaptic inhibition* [FIG. 178].

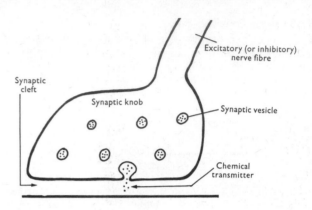

FIG. 179. The synaptic cleft is the small gap between the synaptic knob and the nerve cell. When a nerve impulse reaches the synaptic knob it causes the release of transmitter substance from the synaptic vesicles which bridges the gap and brings about potential changes in the nerve cell. The chemical transmitter is destroyed shortly after its release.

PROPAGATION OF THE NERVE IMPULSE

When the nerve cell is depolarized, a nerve impulse passes along the axon. The propagated action potential is associated with a wave of depolarization which passes along the length of the nerve fibre. The mechanism is as follows:

Electric currents flow from the active area to the inactive area ahead [FIG. 180]. These currents depolarize the region ahead and

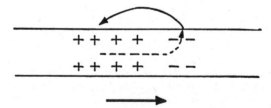

FIG. 180. Propagation of the nerve impulse. The region of the nerve where the nerve impulse has reached is 'depolarized'. The flow of electric current from this depolarized region to the region of axon in advance of the nerve impulse results in depolarization of this new region and propagation of the nerve impulse.

cause propagation of the impulse. In the myelinated nerve the active process leaps from one node of Ranvier to the next. This is known as *saltatory conduction* (L. *saltare* to leap) and gives an increase in the conduction velocity [FIG. 181].

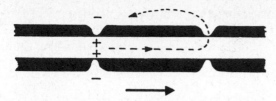

FIG. 181. Saltatory conduction. In medullated nerves the depolarization leaps from node to node. This speeds up the velocity of conduction.

Conduction Velocity

Nerve fibres vary considerably in size and conduction velocity. Nerve fibres are divided into A, B and C groups [FIG. 182]. The

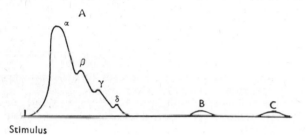

FIG. 182. The propagation time varies with the type of nerve fibre. Thus when a single stimulus is applied to a mixed nerve and the compound action potential recorded some distance away, a series of peaks are seen corresponding to groups of fibres with differing velocities. The Aα fibres have the fastest conduction velocity. The C fibres have the slowest.

group A fibres are the medullated fibres found in sensory and motor nerves. This group is further subdivided into α, β, γ and δ on the basis of the nerve fibre diameter. The larger Aα fibres have a diameter of 15–20 μ. The diameter of the Aδ may be as small as 1 μ.

The **group B** fibres are the preganglionic autonomic fibres.

The **group C** are the non-medullated fibres.

The Aα fibres are the most easily stimulated.

Nerve fibre	Velocity
A 20 μ α	120 m. per sec. (most easily stimulated)
β	6 × (fibre size in μ)
γ	≃ velocity in m. per sec.
1 μ δ	3 m. per sec.
B < 3 μ	4 m. per sec.
C 0·4 μ–1·2 μ	1–3 m. per sec.

The velocity of conduction of medullated fibres is proportional to the fibre diameter; the conduction rate in m. per second is approximately six times the fibre diameter in microns in fibres larger than 3 μ. The largest medullated fibres with a diameter of 20 μ conduct at 120 m. per second.

The non-medullated fibres have a conduction velocity proportional to the square root of the fibre diameter. With a diameter of 1 μ the conduction velocities are approximately the same. Below 1 μ non-medullated fibres have a faster conduction rate than medullated fibres.

Refractory Period

During the spike potential a nerve fibre is *absolutely refractory*. During this time no stimulus, no matter how strong, can initiate a fresh impulse. The nerve is inexcitable. This is followed by a *partial refractory* state when only a very strong stimulus can excite and this produces a subnormal spike. Large fibres recover in 1 millisecond and therefore could theoretically propagate 1,000 impulses per second.

The refractory state is followed by the negative after-potential state, during which the nerve is hyperexcitable. This is followed by the positive after-potential state during which time the nerve is in a subnormal state of excitability.

NEUROMUSCULAR TRANSMISSION

When a striated nerve reaches its muscle fibre it loses its medullary sheath and the bare axon ramifies with the muscle sarcoplasm forming a motor end-plate [FIG. 183]. When the nerve impulse

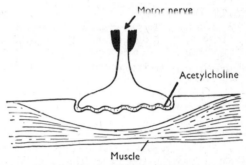

FIG. 183. The motor end-plate and neuromuscular transmission. Nerve impulses cause the release of acetylcholine which causes a motor end-plate potential, a propagated muscle action potential and contraction of the muscle fibre. The action is blocked by curare.

reaches the motor nerve termination *acetylcholine* is released to bridge the gap between the nerve and the muscle. This acetylcholine becomes attached to the end-plate receptor of the muscle, allows Na^+ to enter and generates a local end-plate potential. When this end-plate potential reaches a critical magnitude the rest of the muscle fibre becomes depolarized giving rise to the propagated muscle action potential which precedes the development of the mechanical contraction.

Acetylcholine is the acetyl derivative of choline which may be considered to be a substituted ammonium hydroxide compound. Thus:

$$H-\underset{\underset{\displaystyle H}{|}}{\overset{\overset{\displaystyle H}{|}}{N^+}}-H \quad OH^- \qquad CH_3-\underset{\underset{\displaystyle CH_3}{|}}{\overset{\overset{\displaystyle CH_3}{|}}{N^+}}-CH_2\cdot CH_2\cdot OH \quad OH^-$$

Ammonium hydroxide   Choline

$$CH_3-\underset{\underset{\displaystyle CH_3}{|}}{\overset{\overset{\displaystyle CH_3}{|}}{N^+}}-CH_2\cdot CH_2\cdot O\cdot CO\cdot CH_3 \quad OH^-$$

Acetylcholine

The release of acetylcholine is favoured by the presence of calcium ions (Ca^{2+}) and antagonized by magnesium ions (Mg^{2+}).

The action potential is due to sodium ions entering and potassium ions leaving the muscle (as with nerves FIG. 177). It releases calcium ions in the muscle which activate the myosin [p. 180].

The events at the neuromuscular junction may be summarized as follows:

Nerve impulse $\xrightarrow[\underset{Mg^{2+} \text{ antagonizes}}{Ca^{2+} \text{ aids}}]{1 \text{ millisecond delay}}$ Release of transmitter

Acetylcholine $\rightarrow$ end-plate depolarization $\xrightarrow{\text{muscle cell action potential}}$ transverse tubular system transmission

$\xrightarrow[\text{reticulum}]{\text{coupling to sarcoplasmic}}$ Ca^{2+} released from sarcoplasmic reticulum $\xrightarrow[\text{protein}]{\text{activation of contractile}}$ muscle contraction

The acetylcholine released by a nerve impulse has been removed before the next nerve impulse arrives. This is partly due to diffusion, and partly due to the action of the enzyme **cholinesterase** (true cholinesterase or acetylcholinesterase). It breaks acetylcholine down to choline and acetic acid.

Anticholinesterases are drugs which inactivate cholinesterase, and thus enhance the action of the acetylcholine released. Such anticholinesterase drugs include *eserine* (physostigmine) which is obtained from Calabar beans which grow in West Africa, *neostigmine* (prostigmin) a synthetic compound, and certain organophosphate compounds.

Neuromuscular Blocking Agents

The transmission across the neuromuscular junction is blocked by *curare* and its derivatives which therefore act as paralytic agents on the neuromuscular junction in voluntary muscle. Tubocurarine is the active principle from one of the forms of curare. Curare was used by the South American Indians as an arrow-poison. It is used as a muscle relaxant in operative surgery, and electroconvulsion therapy.

Since it paralyses the respiratory muscles, which are voluntary (striated) muscles, artificial respiration may have to be employed whilst the patient is curarized.

After curare the end-plate currents are reduced to too small a value to trigger action potentials in the adjacent muscle fibres.

Gallamine (flaxedil) is a synthetic compound which acts in a similar manner to tubocurarine.

These *competitive blocking agents* are antagonized by anticholinesterase.

Neuromuscular transmission may be blocked in a different manner. *Succinylcholine* (suxamethonium) is a drug which acts like acetylcholine in depolarizing the motor end-plate, but since it is destroyed more slowly, the depolarization persists and this leads to paralysis. Such drugs are known as **depolarizing blocking agents.** Succinylcholine is destroyed mainly by a plasma cholinesterase termed butyryl cholinesterase or **pseudocholinesterase.**

The toxin from the botulinum bacillus (botulinum toxin) produces paralysis in another way. It prevents the release of acetylcholine from the motor nerve terminations.

It will be noted that although acetylcholine is also the chemical transmitter released at autonomic ganglia (between the preganglionic and postganglionic fibre) in both the sympathetic and parasympathetic nervous systems [p. 204] and at the termination of cholinergic postganglionic fibres [p. 206], different types of receptors are involved. The acetylcholine released at autonomic ganglia is blocked by hexamethonium [p. 204] and that at the cholinergic ending by atropine [p. 206].

ELECTRICAL STIMULATION OF NERVE AND MUSCLE

An electric current provides a convenient and reproducible method of applying a stimulus to a nerve or muscle. Galvani in 1791 discovered such a method of stimulation when he noted that frogs' legs, which were being prepared for culinary purposes, twitched when suspended from a metal fence and touched the metal balcony. He thought that this was due to animal electricity. Galvani's name has been given to direct current electricity produced by a chemical action in a battery. A **galvanic current** in physiology is used to mean a direct current, as would be produced from such a source.

The electrical change set up in a nerve in the neighbourhood of the stimulating electrodes is termed *electrotonus*. There is an increased excitability near the cathode (catelectrotonus) and a decreased excitability near the anode (anelectrotonus) on make (i.e. when the current is switched on) and during the flow of current. When the electrotonus at the cathode (catelectrotonus) reaches a certain magnitude an impulse is generated, but on the other hand, when the electrotonus at the anode reaches a certain magnitude the nerve propagation may be blocked. Thus, a muscle on the anode side will not be stimulated by this (ascending current), whereas a muscle on the cathode side (descending current) will still be stimulated [FIG. 184].

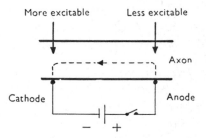

FIG. 184. The stimulation of a nerve by a direct current (galvanic stimulation). On 'make' and when the current is flowing the anode region of the nerve becomes less excitable and the cathode region becomes more excitable. Under certain conditions an impulse generated at the cathode will be blocked at the anode and thus the nerve impulse will only be propagated in one direction, that is, away from the anode. A muscle on the cathode side will be stimulated whereas a muscle on the anode side will not.
On 'break' the above conditions are reversed. The anode becomes more excitable and the cathode less excitable.

On break (i.e. when the current is switched off), the situation is reversed; the excitation occurs at the anode and the blocking at the cathode.

The alternative method of stimulation is to use the induction coil [FIG. 187, p. 186]. A current change in the primary coil is used to induce an e.m.f. in the secondary coil which is applied to the preparation electrodes. Due to the self-inductance of the primary coil the rate of rise of current in the primary on switching on is less than its rate of fall when switching off. As a result an induction coil gives a larger voltage across the secondary when *breaking* the primary circuit than when *making* it.

The primary circuit may be made or broken giving single shocks, or it may be continuously switched on and off using an interrupter. Such a stimulus which consists of a series of shocks is termed a **faradic current.**

The stimulus strength is varied by changing the separation between the coils. When the primary and secondary coils are at their maximum distance from one another, the strength of the stimulus may be further reduced by rotating the secondary coil. There will be no induced e.m.f. when the secondary coil is at right angles to the primary coil.

Both these methods of stimulation are tending to be replaced by electronic stimulators which are usually arranged to give a series of square pulses of known duration and with a known interval between the pulses. By varying this interval, the repetition rate is varied. The amplitude of these pulses is adjustable so that the stimulus strength may be varied.

RELATIONSHIP BETWEEN STIMULUS AND RESPONSE IN A NERVE

Single Neurone

Provided that the stimulus is adequate the magnitude of the spike potential is independent of the strength of the stimulus. There is an 'all or none' relationship between the stimulus and response. The magnitude of the spike potential depends only on the state of the nerve fibre and not on the magnitude of the exciting stimulus. Thus, under normal conditions in the body, all action potentials are identical in size in a neurone, and the only variable for the conveyance of information will be the number of impulses per second.

Nerve Trunk

As the stimulus increases there appears to be a graded response until the maximum is reached. Each neurone is still acting on an 'all or none' basis, but the nerve trunk contains fibres having a range of excitability. The fibres with the lower excitability are only brought in when the stimulus strength has increased considerably.

Time Factor

The effectiveness of an electrical stimulus depends on the strength of the current and the time for which it is applied. A weak current has to be applied for a longer time than a stronger current to be an effective stimulus. If the current is below a certain minimum in strength, it will be ineffective even if applied for infinite time. If the time of application is too short, no current, no matter how strong will be effective. FIGURE 185 shows a typical strength-duration curve.

The curve is a hyperbola asymptotic to the lines $x =$ minimum time and $y =$ minimum current. An additional point is required to fix the curve. This is obtained in the following manner.

The minimum current that will stimulate the preparation is the **rheobase** ($= R$). Twice this current is taken and the minimum duration necessary for this current to stimulate is termed the **chronaxie** ($= C$). It is an index of the relative excitability of the tissue. The point P is now fixed by its co-ordinates C and **2R**.

The current must have a minimum gradient to excite. If the current rises slowly it may have no effect no matter what its ultimate strength is. This is due to adaptation on the part of the nerve.

MOTOR NEURONES

FIGURE 186 shows a cross-section of the spinal cord. The central H-shaped grey matter contains the nerve cells. Its anterior and posterior projections on each side are termed 'horns'. The motor neurone cells supplying the skeletal muscle fibres lie in the anterior horns and are known as **anterior horn cells.**

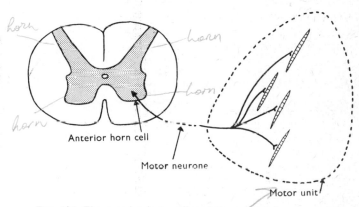

Anterior horn cell

Motor neurone

Motor unit

FIG. 186. The anterior horn cell and its associated motor unit. Each motor neurone running to voluntary (skeletal or striated) muscle branches and supplies from 5 (eye muscles) to 150 muscle fibres.

Each anterior horn cell gives rise to a single motor neurone which leaves via the ventral (anterior) nerve root. On reaching the muscle supplied, this neurone branches and supplies a number of muscle fibres. The motor neurone and its muscle fibres constitute a **motor unit** which is the smallest group of muscle fibres employed either voluntary or in a reflex action.

Isometric and Isotonic Contractions

If the contracting muscle is allowed to shorten under a steady load, the contraction is termed **isotonic.** Such a contraction may be recorded by fixing one end of the muscle and by attaching the other to a lever which records on a kymograph drum. On the other hand, a muscle will still contract even though it is not allowed to shorten. In this case it develops an increased tension. This tension change may be recorded by fixing one end of the muscle and attaching the other to a strong spring fitted with a strain gauge

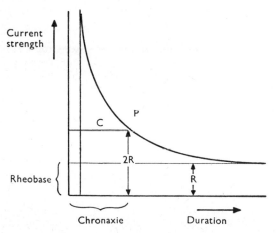

Current strength

P

C

2R

Rheobase

R

Chronaxie   Duration

FIG. 185. The strength-duration curve for the electrical excitation of nerve and muscle. If the current is too weak (less than the rheobase) it is ineffective even if applied for infinite time. If the time of application is too short, no current, no matter how strong, will stimulate. The minimum time for which a current, of strength equal to twice the rheobase, must be applied to stimulate is termed the *chronaxie.*

or other form of tension recorder. This type of contraction is termed **isometric.**

Voluntary movements are a mixture of isotonic and isometric contractions. Lifting a heavy weight is mainly an isotonic contraction, but holding it in mid-air involves an isometric contraction of the muscles. Two equally balanced tug-of-war teams are exerting isometric contraction which only becomes isotonic when one side starts to win.

NERVE-MUSCLE PREPARATION

The response of a muscle to the stimulation of its motor nerve may be studied in the nerve-muscle preparation. In its simplest

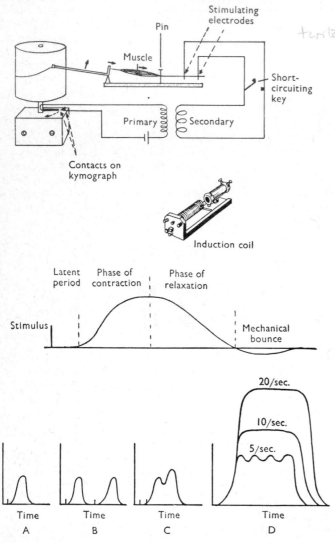

FIG. 187. The frog nerve-muscle preparation.

form this consists of a muscle (such as the frog gastrocnemius muscle) with the motor nerve still attached. The muscle is fixed at one end, and the other is connected via a thread to a lever writing on a kymograph [FIG. 187]. Although the mechanical inertia of the recording system leads to some inaccuracies in the results,

nevertheless, such a preparation demonstrates some of the important factors affecting skeletal muscle contraction.

The nerve is stimulated electrically. The stimulus is applied to the nerve via a pair of electrodes connected to the secondary circuit of an induction coil. The strength is restricted so that only break shocks are effective. Alternatively, an electronic stimulator may be employed. A contact on the kymograph drum enables the stimulus to be initiated at a known point on the record. A second contact on the drum enables a second stimulus to be applied at a known time after the first.

Strength of Single Stimulus

A single stimulus is applied and its strength is slowly increased. When the stimulus is too weak it will fail to stimulate any of the nerve fibres. Such a stimulus is said to be *subliminal.*

As the stimulus is increased then after a short latent period a weak contraction of the muscle occurs. This takes the form of a brief contraction followed by a relaxation known as a twitch. With such a weak stimulus only a few nerve fibres (the most excitable) are stimulated and it is their motor units that are contracting. This is a *liminal* stimulus. As the stimulus progressively increases, more and more nerve fibres are stimulated giving increased contraction. The stimulus is *submaximal.*

When all nerve fibres are stimulated the maximum twitch response is obtained. Any further increase in the stimulus strength (*supramaximal*) will have no additional effect.

A supramaximal stimulus is employed to study the effects of a succession of stimuli.

Two Stimuli

If a second stimulus is applied after the first the effect will depend on the time interval between the stimuli. If there is a long interval between the stimuli, two identical twitches will result [FIG. 187(B)]. On the other hand, if the interval is very short (within a few milliseconds), the second stimulus may have no effect; it has occurred during the refractory period. Between these two extremes, the second stimulus may give rise to varying degrees of summation whereby the resultant contraction may be considerably greater than that of a single twitch [FIG. 187(C)].

Series of Stimuli

If a series of stimuli are applied the force of contraction increases as the stimulus rate increases up to a maximum known as **tetanus.** This is a sustained contraction. At lower frequencies the force of contraction, although greater than that of a single twitch may fluctuate at the stimulus frequency giving rise to a tremulous response or **clonus** [FIG. 187(D)].

Fatigue

With repeated stimulation the nerve-muscle preparation shows fatigue. The latent period becomes longer, the contraction smaller and more prolonged with slow relaxation and ultimately the muscle does not return to its original length.

The fatigue associated with prolonged muscular exercise is probably a central phenomenon in the brain and spinal cord and not in the muscles.

In this preparation, all the motor neurones are firing synchronously with each electrical stimulus. At low stimulus frequencies the preparation shows a tremulous response. When the spinal cord is active at such low rates, the anterior horn cells fire asynchronously, that is, out of step with one another, so that although the individual motor units may show clonus, the combined effect of a large number of such units will be a steady contraction.

ELECTRICAL CHANGES IN SKELETAL MUSCLE

When a muscle is being stimulated via its motor nerve a localized potential appears at the motor end-plate with each nerve impulse. As has been seen, when this end-plate potential reaches a critical magnitude, a propagated muscle action potential is produced which spreads over the muscle fibre.

A stimulus giving rise to a single nerve impulse gives a single muscle action potential which occurs at the onset of the contraction phase of the muscle [FIG. 188(*A*)].

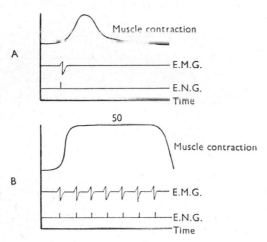

FIG. 188. The muscular contraction, electromyogram (E.M.G.) and electroneurogram (E.N.G.) in the nerve-muscle preparation.

A. A single nerve impulse gives a single muscle action potential and a muscle twitch.
B. A series of nerve impulses at a frequency sufficient to give a sustained contraction (tetanus) still gives discrete muscle action potentials.

A recording of this muscle action potential is an electromyogram (E.M.G.). The recording of the nerve action potential is an electroneurogram (E.N.G.).

With repetitive stimuli a muscle action potential arises in response to each nerve impulse even though the muscle is in a state of sustained contraction (tetanus) [FIG. 188(*B*)].

HUMAN MOTOR NERVES

Very similar results to those described above are obtained in human experiments. FIGURE 189, for example, shows the stimulation of the motor nerve supplying a voluntary muscle in man at different stimulation rates. A large neutral electrode is applied to the arm, and the nerve supplying the superficial flexor muscles of the fingers is stimulated by a probe electrode applied over the motor point. It will be seen that with low stimulation frequencies of the motor neurones a twitch results whereas at high frequency these twitches fuse into a sustained tetanus [FIG. 190].

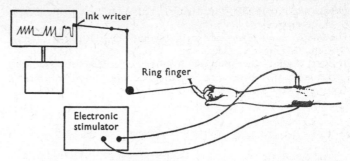

FIG. 189. Recording the contraction of a human muscle in response to electrical stimulation. Contraction of the flexor digitorum superficialis muscle causes flexion of the ring finger which is recorded on the kymograph. The muscle is stimulated at different frequencies using an electronic stimulator. A probe electrode is used to localize the motor point. The second electrode consists of a saline pad placed on the other side of the arm. The area of this pad is large so that the local current density is too low to cause stimulation at this electrode. For records see FIG. 190.

The electromyogram associated with the activity of voluntary muscles may be recorded using concentric-needle electrodes which are inserted through the skin into the muscle. After suitable amplification the electromyogram may be displayed on a cathode ray oscilloscope or on a suitable recorder. Under suitable conditions, it is possible to record the electromyogram, showing activity in the underlying muscles, using surface electrodes applied to the skin.

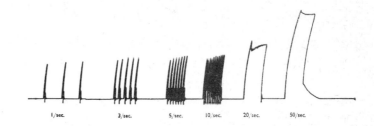

Impulses per second

FIG. 190. Contraction of the flexor digitorum superficialis following stimulation of the motor point at frequencies ranging from 1 per second to 50 per second. At low frequencies twitches are recorded. At the higher frequencies these fuse into a sustained contraction or tetanus.

With this method of stimulation all the motor neurones are transmitting synchronous nerve impulses to their motor units. When the same muscle is stimulated naturally for voluntary movements, the anterior horn cells discharge asynchronously and a steady muscular contraction results at low frequencies of discharge as well as at the high frequencies. Although individual motor units are twitching, their asynchronous activity summates into a steady contraction.

REFERENCES AND FURTHER READING

Bourne, G. H. (1960) *The Structure and Function of Muscle*, New York.

Ehrenpreis, S. (Ed.) (1967) Cholinergic mechanisms, *Ann. N.Y. Acad. Sci.*, **144**, 2.

Hodgkin, A. L. (1951) The ionic basis of electrical activity in nerve and muscle, *Biol. Rev.*, **26**, 339.

Huxley, A. F. (1957) Muscle structure and theories of contraction, *Progr. Biophys.*, **7**, 255.

Huxley, A. F., and Stämpfli, R. (1949) Evidence for saltatory conduction in peripheral myelinated nerve fibres, *J. Physiol. (Lond.)*, **108**, 315.

Huxley, H. E. (1957) The double array of filaments in cross-striated muscle, *J. biophys. biochem. Cytol.*, **3**, 631.

Murray, J. M., and Weber, A. (1974) The cooperative action of muscle proteins, *Scientific American*, **230**, No. 2, 59.

Paton, W. D. M. (Ed.) (1956) Physiology of voluntary muscle, *Brit. med. Bull.*, **12**, No. 3.

Tasaki, I. (1953) *Nervous Transmission*, Springfield.

19. THE NERVOUS SYSTEM

The nervous system is the important control and communication system of the body. It consists of a central part, the *brain* and *spinal cord* which together constitute the central nervous system (C.N.S.), and a peripheral part, the *nerves* of the body.

The nerves which are carrying nerve impulses towards the central nervous system are termed *afferent* (L. *afferre*, to carry to) or sensory nerves. Those nerves which are carrying nerve impulses away from the central nervous system are termed efferent (L. *efferre*, to carry away) or motor nerves.

The nervous system develops in the embryo from the neural tube. Three expansions of the neural tube appear at the head-end due to unequal rates of growth. These expansions develop into the forebrain, midbrain and hindbrain. The rest of the neural tube becomes the spinal cord.

TABLE 18
Development of the neural tube into the brain and spinal cord

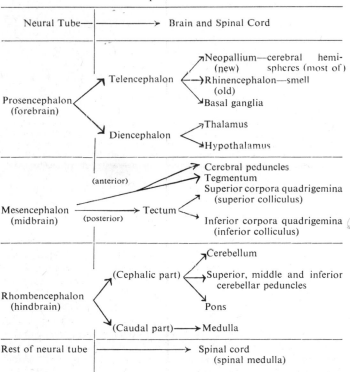

The forebrain (*prosencephalon*) develops into the cerebral hemispheres and basal ganglia (*telencephalon*) and the thalamus and hypothalamus (*diencephalon*). The midbrain (mesencephalon)

forms the tectum, tegmentum and cerebral peduncles, whilst the hindbrain (rhombencephalon) develops into the cerebellum, pons and medulla [TABLE 18].

This part of the central nervous system lies in, and is protected by the skull [FIG. 191]. It will be seen that the cerebral cortex

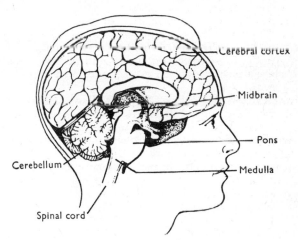

FIG. 191. A sagittal section of the brain.

occupies the major part of the cranium. The two cerebral hemispheres are connected by their peduncles to the midbrain which is in continuity with the pons, medulla and spinal cord. The cerebellum lies posteriorly to the pons and medulla, and is connected to the brain stem by three cerebellar peduncles.

Twelve pairs of cranial nerves originate from this region of the central nervous system [p. 3].

The rest of the neural tube develops into the spinal cord. The spinal cord is protected by the vertebral column in which it lies. Thirty-one pairs of spinal nerves develop on a segmental basis in association with the spinal cord [p. 2]. Each nerve has a ventral (anterior) and a dorsal (posterior) root attaching it to the spinal cord.

MENINGES AND CEREBROSPINAL FLUID (C.S.F.)

The brain and spinal cord are covered by three membranes known as the *meninges*. Meningitis is the name given to inflammation of these membranes. From within outwards the membranes are termed the **pia,** the **arachnoid** and the **dura.** The pia is closely adherent to the nervous tissue. There is a space of variable depth between the pia and the arachnoid which is known as the subarachnoid space. It is filled with cerebrospinal fluid (C.S.F.). The same fluid is found in the hollow cavities (ventricles) of the cerebral hemispheres and brain stem. There are four ventricles. The two ventricles, one in each cerebral hemisphere, are termed the lateral ventricles, whilst the two in the midline extending down to the central canal of the spinal cord are the third and fourth ventricles.

C.S.F. is secreted by the cells of the *choroid plexuses* in the lateral ventricles and third and fourth ventricles. These ventricles are in continuity with one another and are in communication with the subarachnoid space via the median and two lateral apertures in the fourth ventricle (foramina of Magendie and Luschka, respectively).

C.S.F. is absorbed by the *arachnoid villi* associated with the superior sagittal sinus and the spinal cord. Excessive production of C.S.F., or failure of reabsorption due to obstruction of the fourth ventricle foramina, leads to an accumulation of fluid in the brain ventricles leading to *hydrocephalus.*

The C.S.F. is a clear slightly alkaline fluid, sp. gr. 1·005. Its composition is similar to protein-free plasma, but there is less glucose, potassium and calcium than in plasma. Certain ions such as sulphate, iodide and thiocyanate, which pass readily through capillary walls into the tissue fluid, will not pass into the C.S.F. Thus C.S.F. formation is a selective secretion on the part of the choroid plexus cells, and not a simple filtration [see page 103].

The total volume of C.S.F. is 125 ml. It acts as a support for the brain and spinal cord, and maintains a uniform pressure on the nervous structures. Since the skull may be considered to be a watertight box containing the brain, blood and C.S.F., any increase in blood volume in the skull must be associated with a decrease in the volume of C.S.F. or else the nerve tissue will be compressed.

The fact that the amount of blood in the brain is more or less constant was first pointed out by Monro in 1783 and by Kellie in 1824. It is known as the Monro–Kellie hypothesis.

If C.S.F. is allowed to escape, it is formed at the rate of 500 ml. per day.

The spinal cord in the adult terminates opposite the spine of the first lumbar vertebra, but the subarachnoid space extends down to the sacral region. A sample of C.S.F. for diagnostic purposes may be obtained by inserting a hollow needle into the space below the termination of the spinal cord. For example, the needle could be inserted between the fourth and fifth lumbar vertebrae. Such a procedure is termed a *lumbar puncture.*

NERVE FIBRE TRACTS

It has already been seen that a section through the spinal cord shows white and grey areas, and that the central grey H-shaped area contains nerve cells. The white area is occupied by nerve fibres and it is the myelin sheath of these fibres that gives this part of the spinal cord its white appearance.

Many of the nerve fibres are running up and down the spinal cord in tracts which are situated in the white 'columns'. These columns are named anterior, lateral, and posterior, according to their position in the white matter of the spinal cord [FIG. 192].

The majority of these fibre tracts extend from the spinal cord up into the brain stem where their position is different from that in the spinal cord, and depends on the region of the brain stem reached.

The fibre tracts running between the brain and spinal cord are named according to their origin and destination. Thus, the spino-cerebellar tracts run from the spinal cord to the cerebellum and the corticospinal tracts run from the cerebral cortex to the spinal

cord. In this connexion the medulla is sometimes referred to by its older name, the bulb; hence corticomedullary tracts may be referred to as the corticobulbar tracts.

These nerve pathways have been plotted by noting the results of nerve degeneration. When a nerve fibre is cut off from its cell of origin, the fibre degenerates, and its staining properties alter. The pathway of a degenerated tract may be traced through the various levels of the brain stem by successive sections. The staining methods commonly employed are the Weigert-Pal method which leaves the degenerated fibres unstained but stains the other fibres black, and the Marchi method which stains degenerated fibres black and normal fibres yellow.

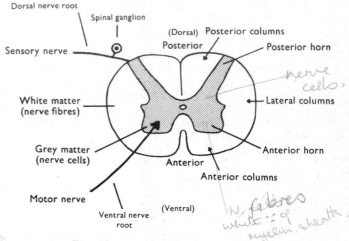

FIG. 192. A cross-section of the spinal cord.

Left. Sensory nerves have their cells of origin outside the spinal cord in the spinal ganglia. These nerves enter the spinal cord via the dorsal nerve roots. Motor nerves have their cells of origin in the anterior horn of grey matter of the spinal cord. These nerves leave via the ventral nerve roots.

Right. The columns of nerve fibres running up and down the spinal cord are named according to their position in the white matter of the spinal cord. Nerve tracts in these columns are named according to their origin and destination. Thus the spino-cerebellar fibres are running from the spinal cord to the cerebellum.

SENSORY NERVES

The sensory or afferent nerves bring information into the central nervous system in the form of coded nerve impulses. This information is, in part, protective and gives warning of possible injury, and is, in part, related to the complex control mechanism associated with voluntary movement.

The sensory nerves enter the spinal cord via the dorsal (posterior) nerve roots [FIG. 192]. Their cells of origin lie outside the spinal cord in the spinal ganglia. Although the sensory and motor nerves have separate spinal cord roots, the nerves join, and run for the greater part of their route as a mixed nerve with motor and sensory fibres in adjacent bundles.

Sensory nerve impulses arise from sensory receptors in the skin, deeper structures and viscera, in addition to the special sense organs. Each sensory neurone works on the 'all or none' principle; all impulses in a given neurone are identical to one another, and the only variables will be the number of impulses transmitted along the neurone per second, and the pattern of this discharge.

Most sensory receptors when stimulated cause, at first, a rapid succession of impulses in the associated sensory neurone, but due to adaptation this rate of firing decreases and may cease even though the stimulus is still being applied [FIG. 193].

Action
potential

Stimulus

FIG. 193. The relationship between a sustained stimulus and the action potential in a sensory nerve. Due to adaptation the impulse frequency decreases with time until either the activity ceases altogether, or the firing rate is reduced to a steady rate.

Since neurones do not vary the amplitude or size of the nerve impulse, an increase in a sensory stimulus is transmitted to the C.N.S. as more impulses per second (a higher impulse frequency) in the sensory neurone. In addition, a stronger stimulus will cause a spread of activity to adjacent sensory receptors and neurones, so that more sensory neurones are active.

touch; pain; temp. itch?   sensory sym.

SKIN RECEPTORS

The skin behaves as if it has several distinct types of sensory receptors each with its own discrete sensory nerve. These receptors are classified as touch receptors, pain receptors, and temperature receptors (both heat and cold).[13] Stimulation of any one type of receptor gives rise to its own specific sensation. It is possible that the sensation of itching is due to a further type of receptor, the 'itch' receptor.

Tactile Sensation

A certain minimum stimulus is necessary to give a sensation of touch. The tactile acuity or sensitivity varies in different parts of the body. It is high on the tongue, lips and finger tips. In certain parts of the body the acuity is greatly increased by the presence of hairs on the skin. These hairs have sensory receptors around the base of the hair follicle which give a sensation of touch when the hair is moved. Tactile acuity is tested by arranging for the subject to close his eyes and to say when he feels a sensation. He should state the type of sensation without being asked. Cotton wool lightly touching the skin is employed as the stimulus. For a quantitative examination Frey's hairs are employed. These are hairs of different thicknesses mounted in a holder and graduated according to the pressure necessary to bend the hair.

Localization of Sensation

Each sensory receptor has its own nervous pathway leading to the central nervous system; from early childhood, we learn to associate a sensation arriving at consciousness along a given path-

[13] Occasionally two types of sensation are found in the same neurone. This may be due either to the receptor responding to more than one type of stimulus or to the neurone being shared between receptors, rather like the two subscribers on a party-line telephone system.
Other types of sensation such as sharpness, roughness, texture in general, are deduced by the brain from the information received via the nerves from the above receptors.

way with a particular part of the body where the sensory receptor is situated. We have a mental picture of our body and sensations are related to this mental picture. Thus if someone treads on our toe, we are aware, without looking, which foot it is and which toe on that foot. This process of associating sensation with a part of the body is known as 'localization'.

Referred Pain

We always assume that any nerve impulse arriving along a sensory nerve will have been generated by the usual sensory receptor. Thus, even if the nerve impulse has been generated by pressure or mechanical irritation of the sensory nerve at some point along its pathway, we associate the sensation with the receptor at the distal end of the nerve. The sensation is said to be 'referred' to the skin area or muscle concerned. A patient with a prolapsed intervertebral disc (a slipped disc) may complain of pain down the back of the legs rather than in the lumbar region where the prolapsed protrusion has occurred. Even after amputation when a leg has been removed, mechanical stimulation of the nerves in the stump may give rise to a sensation of touch, pain and temperature which appears to come from the absent limb—a phantom limb.

In the abnormal circumstances of nerves being stimulated at some intermediate point along their length, impulses will be transmitted in both directions. An example of this is seen in *herpes zoster* or shingles. This is an infection in which the dorsal root ganglia are attacked by a virus very similar to the chickenpox virus. The irritation caused may generate impulses. The impulses passing centrally give a sensation usually of pain. Those passing peripherally arrive antidromically at the skin, and produce a typical triple response reaction [see CHAPTER 4], with redness and blister formation of the skin along the area of distribution of the sensory nerve. Thus an infection of the dorsal root ganglia associated with the fifth thoracic (T.5) nerve on the right, will give an inflamed area of skin extending half-way round the chest on the right side at the level of the cutaneous branch of T.5, i.e. the T.5 skin dermatome which lies just below the nipple.

Punctate Nature of Skin Receptors

If an area of skin (1 cm.^{2}) is subdivided into small squares (1 mm.^{2}), and each square tested for touch, pain, heat and cold, it is found that the receptors to the different modalities are distinct and separate. Each is localized to a small area. When the identical skin site is tested the next day it is found that the relative position of the receptors has changed, and an area that had previously responded to heat, now responds, say, to pain. It appears that there are more receptors than are in operation at any given time, and that these receptors alternate with one another.

Discrimination

This is the ability to distinguish two sensations as distinct when they are close together. This is of importance in Braille writing for the blind. If the dots are too close they cannot be deciphered. Try an experiment. Take two pins with the points 1 mm. apart and touch the skin simultaneously with the two pins. They will feel like a single pin everywhere except on the tongue, which has the highest discrimination. If the points are 2½ mm. apart they can be discriminated as two by the finger tips. On the back the points may have to be as far as 6 cm. apart before they are felt as two.

Adaptation *not accomodation.*

With a continuously applied stimulus, the sensation becomes reduced with time. This is known as adaptation. This is in part due to adaptation in the receptor and in part due to consciousness appearing to ignore sensory information. One may be completely unaware of a clock ticking as a sound, but notice immediately if the clock stops. This is termed the 'off-effect'. Due to adaptation we are unaware of the stimulation of the skin by our clothes. A wrist watch or bracelet may be noticed for the first few days but then one becomes unaware of its presence.

An experiment showing adaptation of the temperature receptors may be carried out as follows. Place the left hand in a bowl of hot water and the right hand in a bowl of cold water. The left hand has a sensation of heat. The right hand has a sensation of cold. After a few minutes a certain amount of adaptation will have taken place. Next plunge both hands into a bowl of tepid water. To the left hand this water feels cold, to the right hand this water feels hot.

MUSCLE AND JOINT RECEPTORS

Receptors in muscles, ligaments, joints and tendons are known collectively as *proprioceptors*.

The corresponding words which are not used so frequently are exteroceptors on the surface of the body and interoceptors in the viscera.

The information from the proprioceptors is of two types. Firstly, the degree of contraction of the muscles and secondly, the position of joints in space.

Information concerning the degree of contraction of muscles comes from muscle spindle receptors in muscles themselves and Golgi receptors in the muscle tendons. These form part of a servo-mechanism controlling muscle contraction that will be considered later. This information does not reach consciousness.

The conscious information concerning the degree of joint movement whether it be flexion or extension, abduction or adduction or circumrotation, comes from the receptors in the joint capsule. These receptors form part of a remarkable mechanism whereby the position of our limbs in space is computed by the brain from this information. It is termed 'position sense'. The efficiency and accuracy of this mechanism can be tested in the following manner. Place the index finger of the right hand anywhere in the air. With the eyes shut take the index finger of the left hand and touch the index finger of the right hand. Most people can do this (you may be an inch or two out). Try it again with the right index finger behind the back.

Consider the number of joints involved; interphalangeal, metacarpophalangeal, intermetacarpal, carpometacarpal, carpal, wrist, elbows and shoulder, all with a wide range of possible movements. Not only can the resultant position be calculated, but another limb, connected to a different part of the trunk, can be taken to an identical position in space. It would be difficult, if not impossible, to construct a machine that could do just this.

The ability to touch one's nose with the tip of a finger with the eyes closed is the clinical test frequently employed to test proprioception in the upper limbs.

Loss of proprioception can be compensated for to a large degree by vision. But any defect becomes more evident when the eyes are shut or at night time.

Loss of proprioception in the legs is one of the late manifestations of syphilis which is called *tabes dorsalis*. A patient with such a loss of proprioception is able to stand and walk about in daytime, but he loses his balance and falls over if he shuts his eyes. Such patients do not go walking at night time. A normal person when walking lifts the feet only a very short distance from the ground. A tabetic patient is unaware of where his feet are with reference to the ground and tends to lift them high off the ground and to hit the ground with a bang when he puts the foot down again. This gives the stamping gait that is so characteristic of the condition.

Nerve endings responding to pain are found in the muscles and deeper structures.

SENSORY PATHWAYS

The various nerve fibres found in a sensory nerve may be classified as follows:

Class	Origin	Type	Size (μ)
Group I$_A$	Muscle spindles	Aα	12–20
Group I$_B$	Golgi tendon end organs	Aα	12–20
Group II	Touch and pressure receptors in skin	Aβ and Aγ	5–12
Group III	Pain receptors	Aδ	2–5
Group IV	Pain receptors	C	0·5–1

All the sensory nerves enter the spinal cord via the dorsal (posterior) nerve roots, but the pathways then depend on the type of

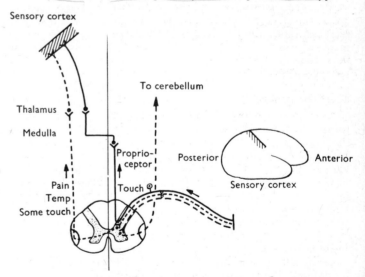

FIG. 194. The three main sensory pathways from sensory receptors in the skin and proprioceptors to the thalamus and sensory cortex and to the cerebellum.

sensation. The proprioceptors concerned with the position of joints in space, and the touch fibres which give accurate localization of the area touched, pass up in the posterior columns of white matter at the rear of the spinal cord. These fibres terminate in the cuneatus or gracile nuclei in the medulla. Second neurones originate in these nuclei and cross to the midline to run in the medial lemniscus up to the thalamus of the opposite side [FIG. 194]. Third neurones relay the impulses to areas 3, 1 and 2 of the sensory cortex [FIG. 195].

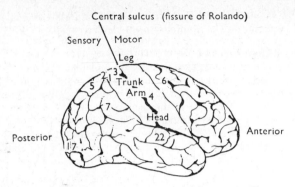

FIG. 195. The motor and sensory areas of the cortex. The motor area is situated in front of the central sulcus (areas 4 and 6). The sensory area is situated behind the central sulcus (areas 3, 1, 2).

Area 17 is the visual area. Area 22 is the auditory area.

Pain, temperature and the remaining touch fibres relay shortly after entering the dorsal (posterior) nerve root. The second neurone crosses to the opposite side in the spinal cord and runs up in the anterolateral column of white matter to the thalamus of the opposite side.

Thus, by either pathway the sensory information reaches the thalamus of the opposite side and some of the information reaches the sensory cortex of the opposite side.

The proprioceptor impulses concerned with the co-ordination of muscular contraction enter via the dorsal nerve roots along with the other fibres. They then relay and enter the spinocerebellar tracts of the same side to run to the cerebellum.

These three main sensory pathways may be summarized as follows:

1. Touch receptor $\xrightarrow{\text{Neurone 1}}$ Cuneatus and gracile nuclei in Proprioceptor medulla
$\xrightarrow{\text{Neurone 2}}$ Thalamus $\xrightarrow{\text{Neurone 3}}$ Sensory cortex
×
———————————————————

2. Pain receptor
Temperature receptor $\xrightarrow{\text{Neurone 1}}$ Spinal cord
Touch receptor
$\xrightarrow{\text{Neurone 2}}$ Thalamus $\xrightarrow{\text{Neurone 3}}$ Sensory cortex
×
———————————————————

3. Proprioceptor $\xrightarrow{\text{Neurone 1}}$ Spinal cord
$\xrightarrow{\text{Neurone 2}}$ Cerebellum for co-ordination of movement
———————————————————

× = cross to other side

Cerebral Cortex

Unlike the spinal cord which has its grey matter (nerve cells) inside the white matter (nerve fibres) [FIG. 192, p. 190], the cerebral cortex has its grey matter on the surface.

The complete 'wiring diagram' of the cortex (cortical connexion) is far from being worked out, but certain connexions are known to exist. Many cortical areas have been found to have a more or less specific function and to facilitate their localization, a numbering system of cortical areas has been adopted [FIG. 195].

The surface of the cerebral hemisphere is convoluted. Each convolution is termed a *gyrus* and the depression between the gyri is termed a *sulcus*. The sulcus separating the frontal from the parietal lobe is the **central sulcus** or fissure of Rolando. The sensory cortex lies in the *postcentral gyrus* (ascending parietal gyrus) which lies immediately posterior to the central sulcus (areas 1, 2 and 3, FIG. 195). The whole body is represented along this gyrus in an upside-down fashion. The sensory impulses from the head arrive at the head area in the lower part of the gyrus. The impulses from the legs arrive at the upper part of the gyrus [FIG. 195].

ELECTROENCEPHALOGRAM (E.E.G.)

In 1875 Caton demonstrated the electrical activity of the brain using electrodes on the exposed surface of the cerebral cortex of animals and a sensitive galvanometer. Such a recording from the exposed surface is termed an *electrocorticogram*. The analysis of the recording obtained when electrodes are applied to the scalp of man (*an electroencephalogram*) was first made by the German neurologist Berger in the early part of this century.

The most prominent rhythm seen in an electroencephalogram is the alpha rhythm which has a frequency of 8–13 cycles per second. It is present when the eyes are closed but disappears when the eyes are opened or when the subject uses his brain for mental arithmetic. It is recorded with electrodes usually in the form of small silver discs applied to the scalp and with electrode jelly or fluid to ensure a good electrical connexion. The waves recorded have an amplitude of only 50 microvolts so that a large amount of electrical amplification is required before the signal is of sufficient strength to operate a pen recorder. Transistor amplifiers are now employed. Precautions must be taken to screen the subject from other electrical apparatus in view of the low signal level.

When the electroencephalogram is analysed other frequencies are sometimes found to be present. The waves have now been classified according to their frequencies:

Delta	½– 3 cycles per second
Theta	4– 7 cycles per second
Alpha	8–13 cycles per second
Beta	14–30 cycles per second

Under anaesthesia the alpha rhythm is replaced by the beta rhythm of 14–30 cycles per second. Cerebral anoxia leads to large amplitude slow waves and then electrical silence.

The origin of the electrical activity is obscure. The waves are only recorded when a large number of cortical neurones are firing synchronously. If they are firing asynchronously the electrical potentials produced by adjacent cells will tend to cancel out and there will be no over-all voltage detected at the scalp electrodes. It would appear that when the eyes are closed synchronous firing occurs involving the thalamus and cortical neurones.

During sleep the electroencephalogram shows large waves with a delta rhythm. When these waves rhythmically increase and decrease in amplitude, they are termed *spindle waves*.

The word 'spindle' denotes a 'tapering towards each end'. Care should be taken not to confuse 'spindle' when used in connexion with an electroencephalogram record with 'muscle spindle' which, as has already been seen [p. 192] is a sensory receptor in muscle.

Dreaming during sleep is frequently associated with rapid movements of the eyes (**rapid eye movements, R.E.M.**). The movements are brought about by contractions of the external ocular muscles [p. 212]. The electromyogram activity of these muscles is frequently picked up by the electrodes and the activity is super-imposed on the electroencephalogram recording.

It has been suggested that activity in these muscles when the eyes are closed is the origin of the alpha rhythm in the waking state.

ASCENDING RETICULAR FORMATION

The ascending sensory neurones not only run to the thalamus, sensory cortex and cerebellum, but collaterals are sent to the reticular formation which extends throughout the brain stem from the medulla to the thalamus. Nerve fibres from the reticular formation and its associated thalamic nuclei run diffusely through the cerebral cortex

Spinal sensory nerve → Ascending reticular formation

→ Cortex

Impulses in the nerves 'arouse' the cortex to activity.[14] This is termed the reticular activating system (R.A.S.). Destruction of the reticular formation leads to deep sleep, insensibility to sensory stimuli and immobility even when the main nerve tracts from the spinal cord to the cortex are still intact.

On this basis sleep would appear to be the natural state, and we awake because of the activation of the reticular formation by sensory impulses. Anaesthetics probably act by depressing the activity of the reticular formation. This could be due to the blocking of transmission at the large number of synapses found in the reticular system.

The reticular formation is the important integrating mechanism of the central nervous system. Here the sensory information is sorted and filtered; some is sent to 'consciousness' and other information is rejected.

MOTOR NERVES

Corticospinal Tract

Immediately in front of the central sulcus is the *precentral gyrus* (ascending frontal gyrus) which is the motor area (cortex areas 4 and 6) [FIG. 195]. Electrical stimulation of this area brings about movement of the opposite side of the body. As in the case of the sensory cortex the body is represented upside-down along this gyrus, with the head area in the lowest part, and the feet in the upper part.

The corticospinal tract (pyramidal tract) is the pathway for rapid voluntary movement. The fibres cross to the other side in the medulla (pyramidal decussation) and run down in the lateral columns of white matter. There may be a short internuncial neurone before the anterior horn cell is reached [FIG. 196].

Thus, the right side of the body is controlled by the left cerebral hemisphere, and an injury to the left motor cortex will cause muscular weakness on the right side of the body. Correspondingly, the left side of the body is controlled by the right cerebral hemisphere. In actions involving muscles on both sides of the body and, therefore, both hemispheres, it is the left cerebral hemisphere that

[14] Bremer (1935) found that the brain isolated by a transection at the midbrain level (cerveau isolé) showed the behavioural and electroencephalogram pattern of sleep. A brain transected at the spinal level is termed encéphale isolé. This does not show the sleep pattern.

usually dominates. Such a person would be right-handed. Left-handed people have a dominant right cerebral hemisphere.

Anterior Horn Cell and Motor Neurone

The voluntary muscle fibres are completely relaxed (flaccid) unless impulses are arriving at the muscle along the motor nerve. The motor nerves have their cells of origin inside the spinal cord in the anterior horns of grey matter.

If we count up the number of muscle fibres in an anatomical muscle and the number of neurones in the motor nerve, we find that the muscle fibres exceed the neurones by a ratio that may be as large as 150 to 1. It follows therefore that muscle fibres must share motor neurones. Each anterior horn cell gives rise to one motor neurone which branches on entering the muscle and supplies a number of muscle fibres. All these fibres will contract simultaneously when an impulse generated by the anterior horn cell arrives via the motor neurone. The fibres form what is termed a 'motor unit'. As has been seen in the previous chapter it is the smallest unit employed in a voluntary movement or a reflex action. We never employ a half, or any other fraction, of a motor unit.

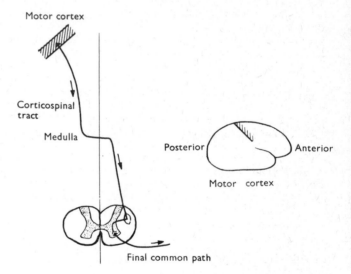

FIG. 196. The corticospinal pathway (pyramidal tract) for voluntary movement. An interruption to the pathway results in paralysis of the appropriate motor units to voluntary movement although involuntary and reflex movements may still occur. The fibres decussate in the medulla. Thus the left motor cortex controls the muscles on the right side of the body, and the right motor cortex controls the muscles on the left side of the body. There may be a short internuncial neurone between the corticospinal fibres and the anterior horn cell which supplies the motor unit along the motor neurone or 'final common path'. An interruption of the final common path leads to a flaccid paralysis with the loss of both voluntary and reflex contraction.

An anatomical muscle is made up of several thousand motor units each with its own anterior horn cell in the spinal cord. The force of contraction of a muscle used in a voluntary movement depends upon:

1. The number of motor units employed.
2. The impulse frequency in the neurone supplying each individual motor unit.

Not all the motor units in a given muscle are used when a voluntary movement takes place. As a greater force of contraction is needed, more and more motor units are brought into play (recruited).

The force of contraction of each motor unit depends upon the rate of firing of the anterior horn cells. If no impulses are sent along the motor nerve, the motor unit will be completely relaxed. At an impulse frequency of 5 per second weak contractions will occur. At an impulse frequency of 100–200 per second the maximum contraction of the unit is obtained, which may be many times more powerful than that obtained at low frequencies. Typical tensions developed by motor units are 5 g. at low frequencies increasing to 30 g. at high frequencies of discharge.

Although at low frequencies of discharge of the anterior horn cell the muscle fibres of each motor unit may be producing twitches rather than a sustained contraction, this is not noticeable in the whole anatomical muscle, because of the asynchronous discharge of individual anterior horn cells. Thus, when one group of muscle fibres is relaxing, another is contracting giving an over-all sustained contraction.

Should all the anterior horn cells to an anatomical muscle fire synchronously due to some abnormal state, then at low firing rates, twitching of the whole muscle may occur giving rise to a tremor.

It is very rare in a voluntary manoeuvre to employ all the motor units in a muscle at their maximum capacity. If one did the force of contraction would be enormous. There seems to be some central psychological block preventing all the anterior horn cells from being brought into play. It is well known that psychological factors play a big part in an athletic performance. The cheering crowds who urge the competitors on, play an important part in facilitating this recruitment of anterior horn cells. You may have played the party trick whereby four people try to lift a person using only their index fingers under the arms and legs. This is found to be impossible. Then following some mumbo-jumbo, such as putting the hand on the person's head and imagining that he has become lighter, you find to your surprise that the person can be lifted easily. Only on the last attempt, when psychological preparation has caused sufficient anterior horn cells to be employed, does this lifting become possible.

The superhuman strength developed when in a tight spot or emergency, is just the use of strength that is always available, but which is rarely employed.

Final Common Path (Algebraic summation at anterior horn cell)

The anterior horn cell and its motor neurone are the only route by which nerve impulses can cause contraction of the muscle fibres constituting the motor unit. It is an excitatory nerve. There

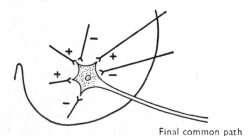

Final common path

FIG. 197. The final common path. All motor neurones outside the spinal cord to striated muscle are excitatory. There are no inhibitory neurones. Inside the spinal cord both excitatory (+) and inhibitory (−) neurones impinge on the anterior horn cell. Algebraic summation occurs.

are no inhibitory nerves supplying the voluntary muscle fibres (cf. involuntary muscle fibre, page 3).

The expression 'final common path' is applied to the motor neurone. An alternative expression is 'lower motor neurone'.

Inside the spinal cord there are many paths by which the anterior horn cell may be stimulated to produce muscular contraction. Excitatory neurones synapse with the anterior horn cells, and by their activity tend to increase the discharge frequency of the cell (a *plus* effect) [FIG. 197]. However, other neurones synapsing with the anterior horn cell inhibit its activity (a *minus* effect). Algebraic summation occurs at this anterior horn cell (and other parts of the central nervous system). The pluses and minuses will tend to cancel one another out and the resultant activity of the anterior horn cell depends on the relative magnitude of these two opposing effects.

A moment's thought will reveal examples in everyday life. If something too hot is picked up, the normal response from the temperature receptors is to cause the object to be dropped. However, if the hot object happens to be a hot cup of tea in a rather valuable tea service, over an expensive carpet, one opposes the normal response and places the cup as quickly as possible on the table.   = conditioned reflexes.'

The desire to cough and blink can be opposed, at any rate for a limited period.

In addition to the corticospinal (pyramidal) tract, neurones from other parts of the brain and spinal cord impinge upon the anterior horn cell. These will now be considered.

Renshaw Cells

The motor neurone gives off a collateral branch which synapses with a cell known as a Renshaw cell. When active the Renshaw cell inhibits the anterior horn cell [FIG. 198]. This system provides

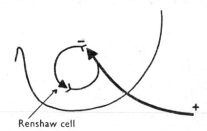

Renshaw cell

FIG. 198. The anterior horn cell is inhibited by the activity of the Renshaw cell.

'negative feedback' to the anterior horn cell and stabilizes its activity by making it less excitable.

Reflexes

Sensory nerves from the same and nearby segments of the spinal cord give off branches which synapse with the anterior horn cells. Often there are one or more relay neurones (internuncial neurones) along the route [FIG. 199]. Such synapses give rise to spinal reflexes which will be capable of causing muscular contraction even if the nerve tracts from the brain are interrupted by injury as a result of section of the spinal cord. Such movement would be involuntary.

We may define a spinal reflex as the involuntary response to a stimulus involving nerve pathways restricted to the spinal cord.

Reflexes are not restricted to the spinal cord. The baroreceptor [p. 40] and chemoreceptor [p. 42] reflexes are examples which involve the medulla. Some reflexes are inborn, others develop as a result of habit (conditioned reflex). The production of saliva is both an inborn reflex due to the presence of food in the mouth, and a conditioned reflex due to the thought, sight or smell of food. Many procedures carried out regularly become automatic and reflex in character. Even complex manoeuvres such as driving a car become to a very large extent a reflex act.

The muscle reflexes are the easiest to assess quantitatively.

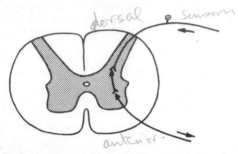

FIG. 199. The reflex arc. One or more internuncial neurones may be present between the sensory and motor neurones.

The Stretch Reflex

The simplest spinal reflex is the stretch reflex which involves only two neurones and one synapse. When a muscle is stretched, the stretch receptors in the muscle spindles are stimulated and nerve impulses enter the dorsal (posterior) nerve root of the spinal cord via the IA sensory nerves. These nerves synapse with the anterior horn cells supplying the same muscle. The anterior horn cell activity causes a muscular contraction which removes the tension from the muscle spindle.

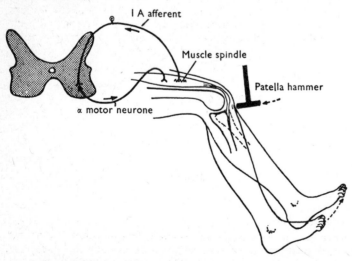

FIG. 200. The knee jerk. Tapping the patellar tendon stimulates the muscle spindles (IA afferents). The IA afferent reflex arc is monosynaptic, that is, it does not involve any internuncial neurones.

The existence of such reflexes may be demonstrated in a dramatic, although somewhat artificial, manner by applying a sudden jerk to the muscle instead of the steady stretch which is the normal stimulus. After a brief interval, the muscle responds with a short contraction. A convenient way of stretching the muscle is to apply a short sharp tap to its tendon with a rubber-covered hammer[15]. For example, in the case of the quadriceps muscle of the thigh, the muscle may be stretched by striking the patellar tendon quickly with a percussion hammer. The resultant involuntary jerk of the leg is due to the sudden contraction of the quadriceps muscle. It is termed the *knee-jerk reflex* [FIG. 200].

A similar reflex, this time resulting in plantar flexion of the foot produced by contraction of the calf muscles, occurs when the Achilles tendon is struck (*ankle-jerk reflex*).

The stretch reflex is an example of reflex excitation, that is, the sensory nerve is a plus nerve and increases the activity of the anterior horn cell. Reflex inhibition also occurs at a spinal level. In general, when the extensor muscles to a joint contract the flexor muscles relax, and when the flexor muscles contract the extensor muscles are relaxed. An example of a spinal reflex involving the reciprocal action of flexors or extensors with both excitation and inhibition of muscles is the crossed-extensor reflex. This involves several spinal segments and muscle groups. It may be demonstrated in animals after the influence of higher centres has been removed by section of the brain stem.

Stimulation of a sensory nerve on one side causes contraction of the flexors and relaxation of the extensors of the knee on the same side (ipsilateral side) whilst on the other side (contralateral side) the flexors relax and the extensors contract giving extension on the other side [FIG. 201].

This reflex is shown in man after a spinal injury which isolates the spinal cord from the brain. Even though the legs are paralysed and no voluntary movement is possible, when a nocuous stimulus is applied to the left leg or foot the left leg is withdrawn and the right leg pushed out although no conscious pain has been felt.

Some General Features of Reflex Action

From an analysis of such reflexes it is possible to list certain general features. Some of the more important are:

Synaptic Delay

The total reflex time is greater than the propagation time in the nerves involved due to the time taken for the impulse to be transmitted across the synapses in the reflex arc. The delay is of the order of 0·5–1·0 milli-seconds at each synapse.

Summation

Two or more stimuli, each of which is subliminal, may summate and produce an enhanced response on a motor neurone. If the impulses resulting from the stimuli arrive via different neurones this effect is termed spatial summation. If they arrive as a repetitive stimulation of the same neurone, it is termed temporal summation.

Occlusion and Subliminal Fringe

When two or more sensory nerves are stimulated simultaneously and are acting on the same motor neurones, the combined responses may be greater or less than the sum of the individual responses. If the response is greater, it is because of the additional activity of a subliminal fringe of motor neurones which are only stimulated when all the sensory nerves are active. If the response is less than expected it is due to occlusion, which occurs when some of the motor neurones are common to more than one sensory nerve.

[15] Alternatively the application of a vibrator to the tendon stimulates the muscle spindles and often brings about a sustained contraction of the muscle.

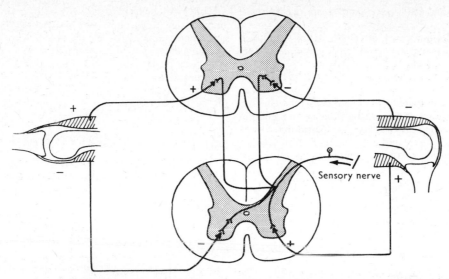

FIG. 201. The crossed-extensor reflex. Stimulation of the sensory neurone causes flexion
on the same side (ipsilateral side) and extension on the other side (contralateral side).
Many segments of the spinal cord are involved.

Recruitment and After-discharge

With the continued excitation of the sensory nerve it is often found that the motor response continues to increase (until a plateau is reached) due to the increasing number of motor nerves which are brought into play. This phenomenon is termed recruitment. After the stimulation of the sensory nerves has ceased, the motor response may persist for a time. This is termed after-discharge. Recruitment and after-discharge are seen in the crossed-extensor reflex.

EXTRAPYRAMIDAL SYSTEM

In addition to the corticospinal tract, and local spinal neurones, the anterior horn cell is under the influence of neurones from higher centres known collectively as the extrapyramidal fibres.

The extrapyramidal tracts include the reticulospinal tracts from the reticular formation, the vestibulospinal tracts from the vestibular nuclei, the tectospinal tract from the tectum (superior and inferior corpora quadrigemina for optic and auditory reflexes), olivospinal and rubrospinal tracts from the olive and red nucleus respectively.

The cerebellum plays an important part in regulating the activity of the extrapyramidal system, but there are no direct cerebellospinal tracts.

THE CEREBELLUM

The cerebellum lies below the cerebrum and above the pons and medulla. It consists of a small central part, the *vermis* and two large, laterally placed cerebellar hemispheres. It is connected to the brain stem via three peduncles known as the superior, middle and inferior peduncles. The inferior peduncle is also known as the *restiform body*.

From a functional point of view it is convenient to divide the cerebellum into the phylogenetically older part, the palaeocerebellum, and the newer neocerebellum.

Flocculonodular Lobe

The flocculonodular lobe of the palaeocerebellum forms part of the relay system associated with the vestibular apparatus for controlling body posture. Some of the nerve fibres from the organs of balance and acceleration (otolith organs and the semicircular canals, page 217) run to the flocculonodular lobe of the cerebellum before returning to the vestibular nuclei. This part of the cerebellum is concerned with the maintenance of one's balance. Removal of this lobe of the cerebellum leads to disturbances of equilibrium and difficulty in maintaining the erect posture, whilst a child with a medulloblastoma brain tumour which affects the flocculonodular lobe is unsteady when trying to stand or walk.

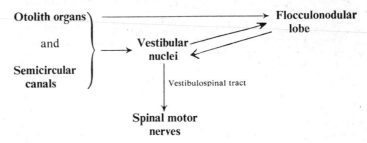

Anterior Lobe

The rest of the palaeocerebellum is made up of the physiological anterior lobe (central lobe, culmen and lobulus simplex), and the posterior part of the posterior lobe (pyramis, uvula and paraflocculus).

This region of the cerebellum receives the spinocerebellar tracts, and has been termed by Sherrington 'the head ganglion of the proprioceptive system'. The anterior spinocerebellar tracts enter the cerebellum via the superior peduncle. The posterior spinocerebellar tracts enter via the inferior peduncle. The majority of these fibres are uncrossed and have come from proprioceptors on the same side of the body.

Some of the posterior column fibres, after relaying in the nuclei gracilis and cuneatus in the medulla, send nerve impulses to the cerebellum via the inferior peduncles.

In addition, the cerebellum receives proprioceptor fibres from the head via the cranial nerves.

The nerve fibres from the cerebellar cortical cells relay in the deep nuclei of the cerebellum and run in the inferior and middle peduncles to the reticular nuclei of the pons and medulla which send reticulospinal fibres to the spinal cord. Other fibres run to the vestibular nuclei which send vestibulospinal fibres to the spinal cord.

Neocerebellum

The remainder of the cerebellum including most of the cerebellar hemispheres, is a more recently developed structure in the evolutionary process and is termed the neocerebellum. Each cerebellar

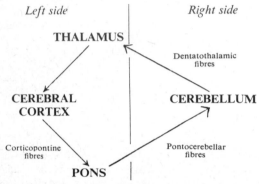

hemisphere is closely associated with the cerebral hemisphere of the opposite side. Since the corticospinal fibres cross and the cortex controls the opposite side of the body, the cerebellum will be co-ordinating movements on the same side of the body.

The cerebellum receives information from the cerebral hemisphere, particularly from the sensory and motor cortex and from the temporal lobe via relays in the pons nuclei (nuclei pontis) of the other side. The fibres cross the midline and reach the cerebellum via the middle cerebellar peduncle.

The fibres from the dentate nucleus of the cerebellum leave via the superior peduncle and cross to the thalamus of the other side. Here they relay and pass to the cerebral cortex.

Unilateral lesions of the neocerebellum lead to disturbances in posture and voluntary movement on the same side of the body. The disturbances of posture include a reduction in muscle tone on the affected side so that, for example, the hand becomes flail-like. The head is frequently turned to the opposite side. If the hands are held straight out there is a tremor with a slow downward and quick upward movement. If the eyes are closed, the arm on the affected side sways outwards. The eyes when open tend to deviate to the opposite side and then quickly return giving nystagmus. There is a reluctance to move the limbs on the affected side, so that the arm does not swing when walking. The subject when walking sways irregularly and tends to fall to the affected side. He may be taken for a drunken man. Rapid repetitive movements, such as a rapid alternate pronation and supination of the forearm, cannot be carried out (dysdiadochokinesia). He finds it particularly difficult to walk along a straight line marked on the ground by placing one foot directly in front of the other.

The movements are clumsy with poor co-ordination of the muscles (asynergia) and with the incorrect force (dysmetria). The movements appear to be broken down into the constituent parts known as decomposition of movement.

Superimposed upon the voluntary movement is a tremor which becomes worse as the subject concentrates on the action—an intention tremor. It may be impossible to drink from a cup, since all the liquid is spilled by the shaking before the lips are reached.

DESCENDING RETICULAR FORMATION

The reticulospinal fibres forming part of the extrapyramidal system are of two types, inhibitory and facilitatory [FIG. 202].

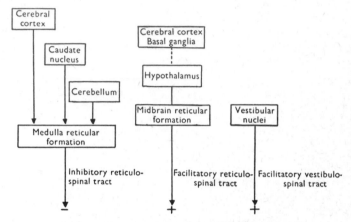

FIG. 202. The main extrapyramidal inhibitory and facilitatory pathways in the spinal cord.

Stimulation of the medullary reticular formation inhibits the anterior horn cells and abolishes movements originating in the cortex or from the spinal reflexes. The descending reticulospinal fibres from this region are diffusely scattered throughout the anterior and lateral columns of white matter of the spinal cord. This inhibitory activity originates from the cerebellum, basal ganglia and cerebral cortex. An important inhibitory cortical area lies in the suppressor band just in front of the motor cortex. Fibres from these areas relay in the reticular formation in the medulla.

Stimulation of the reticular formation higher in the brain stem causes facilitation of both cortically and reflexly induced movements. These facilitatory reticulospinal fibres lie more posteriorly than the inhibitory fibres in the spinal cord. The facilitatory activity originates in the hypothalamus and surrounding structures, and in the midbrain.

The vestibulospinal tract, although not strictly a part of the reticulospinal system, is an important facilitatory pathway that is normally inhibited by the action of the cerebellum on the vestibular nuclei.

Influence of Higher Centres on Spinal Reflexes

We have now seen that the anterior horn cells involved in, say, the stretch reflex, are also under the influence of the pyramidal and extrapyramidal tracts. The extrapyramidal system has both a facilitatory and an inhibitory component. Under normal circumstances the inhibitory component is dominant and 'damps down' any spinal reflexes. In some subjects this inhibition may be so great that it is difficult to elicit a knee jerk. This central inhibition can, in this case, be reduced if the subject concentrates on some other voluntary movement whilst the test is being performed. A typical manoeuvre, known as *reinforcement*, is to ask the subject to clasp his hands and then pull outwards. Whilst he is engaged in doing this, the patellar tendon is tapped, and often a larger knee-jerk response is obtained.

γ-EFFERENT ACTIVITY TO MUSCLE SPINDLES

The muscle spindles contain intrafusal (L. *intra*, inside; *fusus*, spindle) muscle fibres which are in series with the annulospiral sensory endings in the nuclear bag. These intrafusal fibres have their own nerve supply from small anterior horn cells. The motor neurones leading to the intrafusal fibres are smaller in diameter than those running to the main muscle mass and are in the Aγ group (2–8 μ in diameter). The main motor neurones are Aα fibres (12–20 μ in diameter).

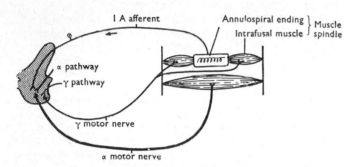

Fig. 203. The muscle spindle and γ-motor activity.

Activity on the part of the gamma (γ) motor neurones (γ-efferents or fusiform neurones) will cause contraction of the intrafusal fibres. This will apply tension to the muscle spindle sensory ending and increase activity in the IA sensory nerve running from the spindle to the spinal cord. Due to the stretch reflex arc this activity will increase the firing of anterior horn cells and cause contraction of the extrafusal muscle fibres of the main muscle mass. Since the muscle spindles are in parallel with the extrafusal muscle fibres, this contraction will decrease the muscle spindle activity. The contraction of the main muscle continues until sufficient shortening has occurred to remove the tension from the muscle spindle endings [FIG. 203].

α-Motor neurone activity decreases the rate of firing of the muscle spindle sensory elements. γ-Motor neurone activity to the intrafusal fibres, on the other hand, increases it. Thus the muscle spindle signals the discrepancy between the two systems [FIG. 204].

The following generalization may now be made:

1. α-motor neurone stimulation *decreases* muscle spindle activity.
2. γ-motor neurone stimulation *increases* muscle spindle activity.

It follows that if both the alpha and gamma systems are active simultaneously to the same extent the annulospiral ending stimulation of the muscle spindle will be unchanged.

The γ-efferent system provides the body with a servo-mechanism for altering the degree of contraction of the voluntary muscles. The muscle spindles are re-set by a change in the γ activity and the main muscle mass alters its degree of contraction to the new equilibrium position.

This is similar to changing the temperature of a room by re-setting a thermostat which will then turn the heating system on and off until the appropriate temperature is reached.

The γ-efferent activity is controlled by the extrapyramidal system. The cerebellum is apparently able to transfer activity from the α system to the γ system so that changes in the position of the limbs are brought about by the γ system.

Recent work indicates that respiration is brought about by the simultaneous discharge of α and γ activity to the respiratory muscles during inspiration. Any resistance to breathing which tends to prevent shortening of the respiratory extrafusal muscle fibres will produce an inbalance between the two systems and an increase in the power of the respiratory muscles (via the IA afferent stretch reflex) to overcome the resistance.

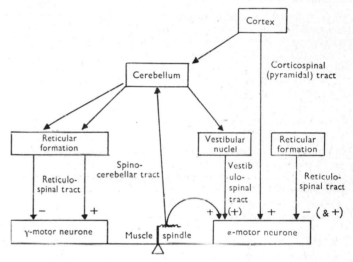

Fig. 204. The muscle spindle signals the imbalance between the α- and γ-motor systems.

MUSCLE TONE

In a conscious human being, many of the skeletal muscles are, at any one time, in a state of reflex contraction known as muscle tone. This applies particularly to the muscles which are opposing the effects of gravity. It is the degree and distribution of this tone that determines the body posture. This muscle tone is maintained involuntarily without fatigue by the activity of the anterior horn cells in the spinal cord. The discharge frequency of each of these cells is low and cells supplying adjacent motor units fire asynchronously, so that the muscle concerned exerts a steady pull. Should synchronous firing occur, the twitching nature of the muscular contraction will give rise to a tremor.

Muscle tone is basically of two types. It is usually due to the reflex stimulation of the α-motor neurones by the muscle spindles in the muscle concerned, resulting from an increased γ-efferent activity. Alternatively it may be due to a direct stimulation of α-motor neurones without the stretch reflex being involved.

The clinician judges muscle tone by the ease with which the joints may be passively flexed and extended. This tests not only the degree of γ- and α-motor neurone activity, but also, to a certain extent, the physical state or pliability of the muscles themselves.

Decerebrate Rigidity

If the brain stem of an animal is sectioned at the level of the midbrain the animal is decerebrate and has no power of voluntary movement. The γ-efferent activity in the spinal cord, however, increases to such a degree, particularly in the anti-gravity muscle, that these muscles are able to support the weight of an animal. It will stand rather like a table will stand, but will fall over if given a push. This is termed decerebrate rigidity. It is a release phenomenon due to the interruption of inhibitory pathways from above.

The γ-motor neurone activity appears to originate from the anterior lobe of the cerebellum, since it disappears when the activity of this lobe is suppressed by surface cooling, and reappears when the lobe is warmed.

Removal of the anterior lobe of the cerebellum in a decerebrate animal abolishes the γ-efferent discharge, but the decerebrate rigidity persists. It has changed from being due to an excessive γ discharge to an excessive α discharge, due to vestibulospinal activity no longer being inhibited by the cerebellum [FIG. 204].

A similar type of decerebrate rigidity involving direct activation of the α-neurones is produced by tying off the blood supply to the brain (ischaemic decerebration; Pollock and Davis, 1930). It is not abolished by cutting the dorsal nerve roots.

Other Factors Influencing Muscle Tone

The distribution of muscle tone is influenced by the neck muscles and position of the head in space. The neck reflexes originate from the proprioceptors in the neck muscles. In an animal such as the cat the tonic neck reflexes may be summarized as follows:

1. Ventroflexion of the head causes the fore limbs to flex and the hind limbs to extend.
2. Dorsiflexion of the head causes the fore limbs to extend and the hind limbs to flex.
3. Pressure on the lower cervical vertebrae causes flexion of all four limbs.
4. Rotation of the head causes the 'jaw' limbs (i.e. on the side to which the head is facing) to extend and the vertex limbs to flex.

Such reflexes can only be clearly demonstrated when vestibular mechanism and higher centres have been destroyed.

In a man whose cortex has been destroyed, the legs are extended and the arms adducted at the shoulder and semiflexed at the elbow with pronation and flexion of the fingers and wrist. Rotation of the head to one side causes the arm and leg on the side of the jaw to extend, and the opposite arm and leg to flex.

The vestibular reflexes from the otolith organs which signal the position of the head in space [p. 217] may be demonstrated in an animal after a plaster of Paris collar has been fitted to eliminate the neck reflexes. With the animal in the supine position (on its back) the tone in the limb muscles is maximal. In the prone position, the tone is minimal.

The vestibulospinal tracts are essential for decerebrate rigidity to occur.

The Erect Posture in Man

In the normal standing position both extensors and flexors are contracted to make the limbs into pillars—the positive supporting reaction.

Man in the standing position is in a dynamic rather than static equilibrium. He can maintain the upright posture with surprisingly little muscular activity in the trunk and legs as has been shown by electromyograph studies. However, any tendency to fall forward is counteracted by the contraction of the muscles at the back of the legs (increased tone). As he sways backwards, the muscles at the front of the legs contract causing him to sway forwards. Similarly, lateral muscles prevent him falling sideways. Thus, when standing up we sway forwards and backwards and from side to side. Each movement is counteracted by the appropriate muscular contraction.

The balancing, in daytime, is carried out mainly by the eyes. Blind people find it more difficult to maintain their balance when standing in a bus or train than a sighted person. Without sight, the balance is maintained by the proprioceptors in the legs and by the vestibular mechanism. The vestibular mechanism is poorly developed in man, but it is the only mechanism available for the orientation of the body when diving and when swimming under water.

Cerebral Vascular Accidents

The small blood vessels in the brain, such as the branches of the lenticulostriate artery, which supply the fibre tracts running through the internal capsule from the motor cortex to the midbrain, may burst if the blood pressure becomes too high. The haemorrhage which follows is commonly called a 'stroke'. Alternatively a cerebral thrombosis may occur. In either case the interruption of the corticospinal tracts will lead to weakness or paralysis on the opposite side of the body (hemiplegia). The interruption of the inhibitory reticulospinal fibres will remove the inhibition from the spinal reflexes and increase the γ-motor activity. The stretch reflexes may be enhanced to such an extent that passive movements of the joints become difficult. This condition is termed *spasticity*. Usually an 'upper motor neurone' lesion, such as this, leads to a **spastic paralysis.** However, should the facilitatory fibres be affected to a greater extent than the inhibitory fibres, a **flaccid paralysis** will result. A lesion of the lower motor neurone (final common path) always leads to a flaccid paralysis since the muscles have lost their nerve supply and are completely relaxed once the motor nerve has been cut.

When spasticity is present and an attempt is made to stretch a muscle passively, it extends in a series of short jerks and between each increment of lengthening, stretching is resisted.

If the force applied to the muscle is progressively increased, a point is reached where the spastic muscle suddenly relaxes completely. The corresponding joint will flex (or extend) rapidly with a clasp-knife like action. This is due to the inhibition of the γ-motor activity by the Ib sensory fibre from the Golgi tendon organs in response to the increased tension in the tendon. It is termed the lengthening reaction.

BASAL GANGLIA

The telencephalon develops into the cerebral hemispheres and the basal ganglia. The basal ganglia consist of the globus pallidus (or pallidum), the corpus striatum (or striatum) which is made up of the caudate nucleus and the putamen, substantia nigra, and the subthalamic nucleus (corpus Luysii). They form a group with the globus pallidus at the centre.

The nerve pathway from the corpus striatum to the globus pallidus (striopallidal tract) carries inhibitory reticular fibres from the suppressor areas of the cortex and the caudate nucleus

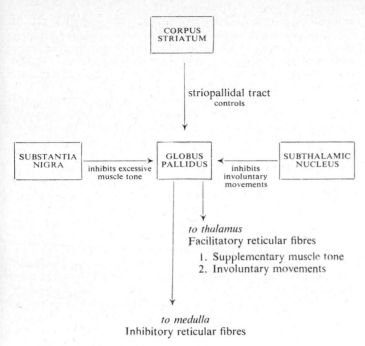

CORPUS
STRIATUM

striopallidal tract
controls

SUBSTANTIA
NIGRA → inhibits excessive muscle tone → GLOBUS PALLIDUS ← inhibits involuntary movements ← SUBTHALAMIC NUCLEUS

to thalamus
Facilitatory reticular fibres

1. Supplementary muscle tone
2. Involuntary movements

to medulla
Inhibitory reticular fibres

NERVOUS SYSTEM LESIONS

Effects of Mixed Nerve Section

The effects of section of a peripheral nerve such as the ulnar nerve in the arm may be summarized as follows:

1. Flaccid paralysis of the muscles supplied.
2. Loss of sensation over the area exclusively supplied.
3. Loss of reflexes.
4. Paralysis of sweat glands and pilomotor nerves.
5. Vasodilatation (later vasoconstriction).
6. Wasting of the muscles.
7. Trophic changes in the skin.
8. Wallerian degeneration and later regeneration.

The interruption of the motor nerves paralyses the muscles supplied to both voluntary and reflex stimulation. The muscles lose their tone and become flaccid.

Due to the overlap of the sensory nerves the area of complete loss of sensation is often less than expected. For example, section of a single thoracic nerve (say T.6) gives no sensory loss because of the overlap of T.5 and T.7 nerves.

The absence of sweating, the paralysis of the pilomotor nerves which when active cause the hairs 'to stand on end' and the vasodilatation are all the result of the interruption of the sympathetic fibres which are present in the mixed nerve. The arterioles later regain some degree of tone.

The muscles waste and unless the nerve regenerates will ultimately change to fibrous tissue. The contracture of this fibrous tissue may produce gross deformity at the joint supplied.

The skin over the area without sensation becomes thin, red and shiny. This skin may break down leading to ulceration. Any skin sores heal very slowly. Pressure leads to local tissue anoxia and capillary damage. In a normal person the continual movements even when asleep in bed, prevent severe tissue anoxia from occurring. In a paralysed limb, pressure may be present for an extended period of time leading to bed-sores. At one time it was thought that these 'trophic changes' were due to the interruption of nerves which played some part in the nutrition of the skin. This now appears unlikely and the changes are probably due to the loss of protective sensory information, immobility of the area, and the loss of vasomotor reflexes.

of the corpus striatum. These fibres run on down to the medullary reticular formation and the inhibitory reticulospinal tract. Facilitatory reticular fibres from the cortex run via the putamen to the globus pallidus, then to the thalamus, and down to the midbrain reticular formation and facilitatory reticulospinal tract.

Disease of the basal ganglia may lead to an imbalance in the activity of the two systems, with the result that there is a marked increase in the γ-efferent activity, and an increased muscle tone in both the flexor and extensor muscles. This condition is termed **muscular rigidity** and it is one of the features of Parkinson's disease or *paralysis agitans*. Notwithstanding this latter name, the subject is not *paralysed* but his voluntary movements are impeded by the muscular rigidity. He has a shuffling gait and a fixed mask-like expression. He is said to be imprisoned by the rigidity of his muscles. The rigidity disappears during sleep. The augmentation of the γ-activity is thought to be due to a failure on the part of the substantia nigra to provide sufficient inhibition of the activity of the globus pallidus.

Disorders of the basal ganglia also lead to involuntary movements presumably due to abnormal activity on the part of the extrapyramidal system and γ-efferent activity. The subject with Parkinson's disease has a fine tremor of the hands and gives the impression that he is rolling pills with his fingers—pill rolling tremor. The tremor and rigidity is relieved by destruction of the globus pallidus or its afferent fibres.

The symptoms of some patients with Parkinson's disease are relieved when the level of *dopamine* (noradrenaline without an hydroxyl group) in the brain is raised. *l*-dopa (dihydroxyphenylalanine) is a substance which crosses the *blood–brain barrier* and is decarboxylated in the brain to form dopamine.

Hemiballismus is another disorder of the basal ganglia. In this case, the involuntary movements predominate with sudden irregular movements of the limbs on one side of the body. These movements disappear during sleep, and are thought to originate in the basal ganglia due to a failure of inhibition by the subthalamic nucleus.

Degeneration and Regeneration of Nerves

The changes which occur following nerve section were described in 1850 by the English physiologist Waller, and for this reason the process is often referred to as Wallerian degeneration.

When the nerve is cut, its cell of origin shows changes. The Nissl granules disappear and the nucleus may become eccentric. The nerve itself degenerates distal to the site of injury and back for a short distance, usually to the previous node of Ranvier. The axon breaks up within a few days. The myelin sheath degenerates leaving only the Schwann cells and the neurilemmal tubes. The debris is removed by phagocytic cells from the endoneurium. The Schwann cells proliferate and if the cut ends of the nerves are brought close together, these cells will bridge the gap.

The central axon grows out as a series of neurofibrils in varying directions. These are guided by the Schwann cells into the neurilemmal tubes. Once a fibril has entered a tube, all the other branches degenerate and the axon regrows down that tube. The daily rate of growth is of the order of 1 mm. per day. A new medullary sheath is formed round the regrowing axon by the Schwann cells.

Except when a nerve is injured by crushing it is unlikely that the axon will grow down its old neurilemmal tube. Many axons never re-establish connexions at all and many grow down the wrong tubes, i.e. sensory nerves in motor tubes. Other sensory nerves grow down tubes leading to a different area. New central connexions will have to be made to enable the information from these receptors to be interpreted correctly. It is not surprising, therefore, that usually only partial recovery occurs.

Transection of the Spinal Cord

Spinal Shock

If the spinal cord is cut through, there is at first a state of spinal shock. All the muscles below the level of the transection are paralysed and all muscle tone is lost (flaccid paralysis). There is a complete loss of all sensation in the parts of the body innervated by spinal segment below the transection. All reflexes are lost and the bladder and rectum are paralysed. The internal sphincter of the bladder is often not affected and this leads to retention of urine.

The vasomotor centre activity maintaining arteriolar and venous tone runs down in the lateral columns of white matter of the spinal cord and leaves via the sympathetic outflows arising from the first thoracic to the second lumbar segments of the spinal cord. A high thoracic spinal section will interrupt this vasoconstrictor tone causing a marked fall in blood pressure. A section below L.2, on the other hand, will have very little effect on blood pressure.

The venous return from the lower part of the body is poor, due to the arteriolar and venous dilatation and the absence of muscular activity. The lower limbs become cold and blue and bed-sores may develop.

The duration of spinal shock is greater in animals high up the evolutionary scale. It lasts for a matter of minutes in the cat, days in the monkey and about 3 weeks in man.

Reflex Activity following Spinal Transection

As the spinal shock passes off the smooth muscle recovers first. The bladder muscle regains its power of contraction and reflex micturition occurs without conscious effort when the bladder is full. Spinal patients learn to detect fullness of the bladder by such devices as noting a flushing of the face, due to a rise in blood pressure, which results from an autonomic reflex in the paralysed region. Then by pressing on the bladder through the abdominal wall the bladder-emptying reflex may be initiated.

The muscle tone returns, but since all γ-efferent activity has been lost the tone is less than before the injury. The flexor tone often exceeds the extensor tone giving paraplegia in flexion.

Although no voluntary movements are possible the paralysed muscles will still contract reflexly. Stimulation below the lesion may give rise to a mass reflex with flexor spasm of the legs, involuntary evacuation of the bladder and sweating below the level of the lesion. Stimulation of the skin of the sole of the foot causes an upward movement of the big toe (dorsiflexion) and a spreading of the other toes (abduction). These movements constitute the **Babinski response.** When the pyramidal tracts are intact this spinal reflex is suppressed, and the stimulus now causes a downward movement of the toes (plantar flexion)—the **normal plantar response.**

With a complete spinal transection there is no regeneration of the interrupted nerve pathways, but the paraplegic patient may

be educated to use the muscles that are still innervated, and in very many cases can be taught to dress himself and to perform a useful occupation. A muscle such as latissimus dorsi which is innervated from the cervical region (C.6, 7, 8) will probably still be innervated and may be developed sufficiently to move the pelvis so that walking becomes possible.

Spinal transection high in the thoracic region will paralyse the intercostal muscles used in breathing, but the diaphragm is innervated from the cervical region (C.3, 4, 5) and diaphragmatic respiration will continue.

Hemisection of the Spinal Cord (Brown-Séquard Syndrome)

In 1851 Brown–Séquard described a lesion of the lateral half of the spinal cord with paralysis on the same side and some sensory loss on the other side. This has since been called the Brown–Séquard syndrome. It is usually due to a tumour in the spinal cord which involves several segments. It is of gradual onset and is not associated with spinal shock.

Below the level of the lesion there is paralysis of the muscles on the same side. Muscle sense and touch are lost but pain and temperature remain. On the opposite side there is loss of pain and temperature. At the level of the lesion, there is flaccid paralysis of the muscles due to destruction of the anterior horn cells, anaesthesia due to a sensory loss, a loss of reflexes, and loss of sympathetic outflow, leading to vasodilatation of the segments concerned.

Memory Loss

Following head injuries, severe anoxia and electro-shock therapy, recent memories may be lost whilst old memories are retained. This is termed **retrograde amnesia.** Thus a driver having a bad motor-car accident may never remember what happened. This amnesia for recent events but not past events suggests that some method of consolidating memory exists in the brain.

Recent memories are probably stored electrically by the establishment of synaptic connexions. Circuits are set up round which neurone-firing may reverberate. But animals surviving 'deep freeze', where all electrical activity in the brain has ceased, retain their conditioned reflexes. Experiments have been carried out with the *Planaria* worm in which simple conditioned reflexes with light can be established. If cut in two, both halves grow into intact worms and both retain the conditioned reflexes.

Such experiments suggest that memory may be consolidated chemically, possibly by encoding DNA and RNA molecules.

CEREBRAL BLOOD FLOW

The brain needs a high continuous blood flow for its activity. It was not until 1945 that Kety and Schmidt described a method that has enabled cerebral blood flow to be measured in man. The original method was based on the inhalation of nitrous oxide and its uptake by brain tissues. More recently radioactive krypton 85 has been used to measure total cerebral blood flow, and the blood flow to different parts of the brain (regional cerebral blood flow) has been determined using xenon 133.

The radioactive inert gas, dissolved in saline, is injected rapidly into the internal carotid artery. The blood flow is determined from the changes in radioactivity in regions of the brain with time.

The weight of the brain of an adult man is about 1·5 kg., and its blood flow is about 800 ml. blood per minute or 16 per cent. of the cardiac output at rest. The blood flow is thus 55 ml. blood per 100 g. brain tissues per minute. This figure may be compared with that of skeletal muscle [p. 56] where the flow is 2 ml./100 g. muscle/minute at rest and 40 ml./100 g. muscle/minute in exercise.

Mental activity has little effect on the cerebral blood flow or its oxygen requirements.

In addition to this high blood flow, the brain extracts a higher than average amount of oxygen from each 100 ml. blood. Each 100 ml. blood arrives carrying 19 ml. O_2 and leaves with only 13 ml. O_2. The A-V oxygen difference is thus 6 ml. O_2/100 ml. blood compared with an average of 5 ml. O_2/100 ml. for the rest of the body [p. 82].

The carbon dioxide production of the brain equals the oxygen uptake and the R.Q is therefore 1. This implies that the principal food for the nutrition of the brain is glucose although there is evidence that it can metabolize small quantities of fat, especially during starvation.

Metabolic Regulation of Cerebral Blood Flow

Since the brain is situated above the heart in the sitting and standing position, the first requirement for an adequate cerebral blood flow is an adequate blood pressure. Under normal conditions however, the cerebral blood flow is independent of the perfusion blood pressure since the vessels change their size to match the changes in pressure. Such a mechanism, which is also found in the kidney, is termed 'autoregulation'. However, a profound fall of blood pressure resulting say, from a haemorrhage, will lead to a reduced blood flow and loss of consciousness.

Although the brain blood vessels have a sympathetic innervation, the autonomic nervous system does not seem to be important in the regulation of cerebral blood flow. The cerebral blood vessels are, however, very sensitive to metabolites and in particular to CO_2 and pH changes.

A fall in pH (acidosis) dilates cerebral blood vessels. A rise in pH (alkalosis) constricts these vessels. For this reason, small pieces of cotton wool soaked in bicarbonate are used by brain surgeons to stem bleeding.

Changes in brain pH are most frequently the result of changes in blood and cerebrospinal fluid carbon dioxide. Breathing a gas mixture containing 6 per cent. CO_2, for example, will increase the cerebral blood flow by as much as 75 per cent.

Overventilation [p. 98], which leads to a fall in alveolar and blood CO_2, constricts the cerebral blood vessels and reduces the cerebral blood flow. The flow may fall to 35 ml. blood/100 g. brain tissue/minute. Vision is affected and the subject becomes dizzy and light-headed. The tetany (carpopedal spasm) which ensues may be partly cerebral in origin [p. 98].

This sensitivity to carbon dioxide, which is also a metabolite of the brain, may account for the autoregulation. Any tendency for the cerebral blood flow to be reduced will be opposed by the vasodilatation which results from the accumulation of carbon dioxide. Any increase in cerebral blood flow will wash out carbon dioxide from the brain and the lowered brain carbon dioxide will bring about cerebral vasoconstriction.

Cerebral blood vessels are slightly dilated by oxygen lack and slightly constricted by oxygen excess. Breathing 10 per cent. oxygen will increase the cerebral blood flow to 75 ml. blood/100 g. brain tissue/minute. Breathing 100 per cent. oxygen reduces the cerebral blood flow to 45 ml. blood/100 ml. brain tissue/minute.

The cerebral blood flow is reduced by a raised venous pressure which impedes the venous outflow, and by a rise in intracranial pressure to over 30 mm. Hg which compresses the cerebral capillaries and veins.

In senile cerebral atherosclerosis there is an increase in cerebral vascular resistance and the reduction in cerebral blood flow may lead to senile dementia. In addition the 'autoregulation' may be lost so that a fall in blood pressure will reduce the blood flow.

The mental state, however, appears to be related more to the brain's oxygen consumption than its blood flow. Thus the coma of diabetic ketoacidosis [p. 114] is associated with a reduced brain oxygen uptake, but there is no reduction in cerebral blood flow.

It has already been seen [p. 190] that the cranium is a rigid box containing the brain, blood vessels and cerebral spinal fluid and that any increase in size of the blood vessels will be associated with a reduction in the volume of cerebrospinal fluid in the skull.

REFERENCES AND FURTHER READING

Adrian, E. D. (1928) *Basis of Sensation*, London.

Bremer, F. (1953) *Some Problems in Neurophysiology*, London.

Brookhart, J. M. (1960) The reticular formation, in *Handbook of Physiology*, Section 1, Vol. 2, p. 1245, Washington, D.C.

Bucy, P. (Ed.) (1949) *The Precentral Motor Cortex*, Illinois.

Caton, R. (1875) The electric currents of the brain, *Brit. med. J.*, **2**, 278.

Delafresnaye, J. F. (Ed.) (1954) *Brain Mechanisms and Consciousness*, Oxford.

Dow, R. S., and Moruzzi, G. (1958) *The Physiology and Pathology of the Cerebellum*, Minneapolis.

Eccles, J. C. (1957) *The Physiology of Nerve Cells*, Baltimore.

French, J. D. (1960) The reticular formation, in *Handbook of Physiology*, Section 1, Vol. 2, p. 1281, Washington, D.C.

Granit, R. (1955) *Receptors and Sensory Perception*, New Haven.

Guttman, L., and Whitteridge, D. (1947) Effects of bladder distension on autonomic mechanisms after spinal cord injuries, *Brain*, **70**, 361.

Jung, R., and Hassler, R. (1960) The extrapyramidal motor system, in *Handbook of Physiology*, Section 1, Vol. 2, p. 863, Washington, D.C.

Magoun, H. W. (1950) Caudal and cephalic influences of the brain stem reticular formation, *Physiol. Rev.*, **30**, 459.

Pavlov, I. P. (1927) *Conditioned Reflexes*, London.

Penfield, W., and Rasmussen, T. (1950) *The Cerebral Cortex of Man*, New York.

Sherrington, C. S. (1906) *The Integrative Action of the Nervous System*, New Haven.

Terzuolo, C. A., and Adey, W. R. (1960) Sensorimotor cortical activities, in *Handbook of Physiology*, Section 1, Vol. 2, p. 797, Washington, D.C.

Zanchetti, A. (1962) Somatic functions of the nervous system, *Ann. Rev. Physiol.*, **24**, 287.

FILM

Cerebral cortex of the monkey, I.C.I. Film Library, London (C. S. Sherrington).

20. THE AUTONOMIC NERVOUS SYSTEM

A preliminary discussion of the role of the autonomic nervous system has already been given in CHAPTER 1. It will now be considered in more detail so that its over-all function as part of the controlling systems of the body may be more clearly seen.

The autonomic nervous system was divided into the sympathetic nervous system and the parasympathetic nervous system by Langley in 1911. Both of these are efferent, that is, motor systems, but it is convenient to consider also under this heading the visceral afferents which form the sensory system regulating and modifying the sympathetic and parasympathetic activity.

The autonomic nervous system supplies:

(a) Involuntary muscles.
(b) Secreting glands.
(c) The heart.

THE SYMPATHETIC NERVOUS SYSTEM

This is a two neurone system. The neurone which leaves the spinal cord is not the neurone which arrives at the organ supplied. Somewhere along the pathway the first neurone terminates in a *synapse* with a second neurone which conveys the nerve impulses to their destination. The place where the synapse is situated is termed a *ganglion*. The fibre before the synapse is designated the **preganglionic fibre** and that after the synapse the **postganglionic fibre** [FIG. 205].

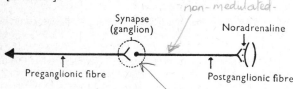

non-medulated-

FIG. 205. The pre- and postganglionic sympathetic nerve fibres.

neurohumoral transmitter

The chemical transmitter at the ganglion is acetylcholine. Transmission of the impulse across the ganglion may be prevented by drugs such as hexamethonium and its derivatives which are known as 'ganglionic blocking agents'.

The chemical transmitter at the termination of the postganglionic fibre is usually **noradrenaline.** Such fibres are said to be *adrenergic*.

$$HO-\bigcirc\!\!-CH \cdot CH_2 \cdot NH_2$$
$$HO \qquad\qquad OH$$

Noradrenaline

Some postganglionic sympathetic fibres, such as those to the sweat glands, are *cholinergic* and release *acetylcholine*. Other sympathetic fibres are thought to release acetylcholine which in turn causes the

skin sweat glands
skin blood vessels — *symp. cholinergic*
adr. medulla

release of noradrenaline. The entry of calcium ions is thought to precede the noradrenaline release.

$$CH_3 \qquad\qquad OH^-$$
$$\qquad\quad +$$
$$CH_3-N-CH_2 \cdot CH_2 \cdot O \cdot CO \cdot CH_3$$
$$CH_3$$

Acetylcholine

The released noradrenaline is inactivated shortly after its release mainly by its re-uptake back into the postganglionic sympathetic nerve termination. The released acetylcholine, on the other hand, is inactivated mainly by the presence of *acetylcholinesterase* [p. 184] which splits it into choline and acetic acid. The sympathetic outflow is limited to the thoracic and upper lumbar segments of the spinal cord. Between the first thoracic and second lumbar segments inclusive the grey matter of the spinal cord has lateral horns in addition to the anterior and posterior horns. The sympathetic neurones have their cells of origin in these horns and the preganglionic fibres pass out via the ventral nerve roots [FIG. 206]. These preganglionic

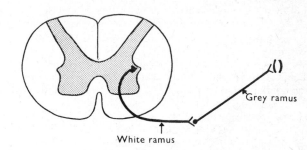

FIG. 206. The sympathetic fibres arise in the lateral horn of the grey matter of the spinal cord in the segments T.1 to L.2. The preganglionic fibres are medullated and white in colour. The postganglionic fibres are non-medullated and grey in colour.

fibres are medullated (with a myelin sheath) which gives them a whitish appearance. For this reason they are known as the white rami communicantes. These fibres enter the sympathetic trunk (sympathetic chain) which runs on either side of the spinal cord. The sympathetic trunk extends for the whole length of the spinal cord. It has a swelling or ganglion associated with each spinal segment with the exception of the cervical region where there are only three ganglia. These cervical ganglia are known as the superior, middle and inferior cervical ganglia. The preganglionic fibres may run up or down the sympathetic trunk before synapsing with a postganglionic fibre in one of the ganglia [FIG. 207].

Postganglionic fibres leave from the whole length of the sympathetic chain, and one preganglionic fibre may give rise to many postganglionic fibres. Each postganglionic fibre leaves the chain as a non-medullated grey ramus communicans and runs to its destination in a mixed nerve. Thus the sympathetic fibres to the forearm arrive in the three main nerves of the arm, the radial, median and ulnar nerves. These fibres will be vasoconstrictor to the blood vessels, secretomotor to the sweat glands and motor to the arrectores pilorum muscles.

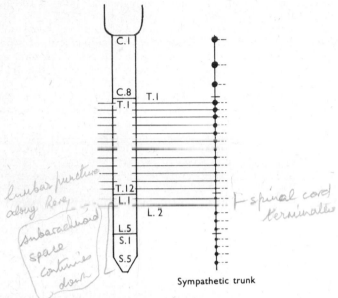

lumbar puncture
along here —
subarachnoid
space
continues
down

= spinal cord
terminalis

Sympathetic trunk

FIG. 207. The sympathetic outflow.

The sympathetic fibres to the head are an exception. There are no mixed nerves running in a suitable direction to convey sympathetic fibres from, say, the superior cervical ganglion to the eye. The nerves adjacent to this ganglion are the IX and X cranial nerves and these pass into the skull to terminate in the medulla. The cranial sympathetic fibres run in the outer coat (adventitia) of the blood vessels which are the internal and external carotid arteries and their branches.

The sympathetic fibres to the digestive tract do not relay in the sympathetic trunk. The preganglionic fibres leave the chain again as the greater, lesser and lowest splanchnic nerves in the thorax. These nerves pass through the crura of the diaphragm to relay in the sympathetic plexuses which lie in front of the abdominal aorta. These midline plexuses are named according to the neighbouring branch of the abdominal aorta. Thus the plexus around the coeliac artery is termed the coeliac plexus. That around the superior mesenteric artery is the superior mesenteric plexus. Other plexuses include the inferior mesenteric plexus and the hypogastric plexus.

Adrenal Medulla

The activity of the sympathetic nervous system is augmented by the release of the catecholamines noradrenaline and adrenaline by the medulla of the adrenal (suprarenal) gland [see p. 165]. The noradrenaline and adrenaline circulate in the blood as hormones and reach their target structures by diffusing from the blood through the tissue fluid.

α and β Receptors and Blocking Agents

The catecholamine, whether released from the sympathetic post-ganglionic terminations or arriving via the blood, acts on two types of receptor in the tissues. These have been designated alpha (α) and beta (β) receptors.

The α receptors are associated with the stimulation (contraction) of smooth muscle. They are found, for example, at the termination of the postganglionic fibres to the arterioles. These receptors are blocked by the ergot alkaloids, *phenoxybenzamine* (dibenzyline) and *phentolamine*. Noradrenaline is particularly active on these receptors.

The β receptors are associated with the inhibition (relaxation) of smooth muscle. They are also the receptors which bring about an increase in the force of contraction and rate of the heart. These receptors are blocked by the blocking agents such as *propranolol* (isopropylamine derivative of propanol). The isopropyl derivatives of noradrenaline (*isoprenaline*) are particularly active on these receptors.

To avoid confusion it has been suggested that the receptors in heart muscle should be termed β1 receptors, and those in smooth muscle β2 receptors. Since blocking agents specific to cardiac muscles are being developed, it would be better to restrict the term β receptor to smooth muscle receptors, and to introduce an entirely new term (such as γ or C receptor) for the cardiac receptors.

Conduction of the nerve impulse along the postganglionic adrenergic fibres is impaired by drugs such as *guanethidine*. These drugs are used to reduce the vasoconstrictor tone in cases of high blood pressure (hypertension).

The tissue receptors associated with the cholinergic post-ganglionic sympathetic fibres are blocked by *atropine* and *hyoscine*.

Immuno-sympathectomy

Levi-Montalcini (1964) found a nerve growth factor (NGF) that controls the growth of sympathetic nerve cells in mice and rats. It is a protein which may be extracted from the salivary glands and other tissues.

Injection of NGF into new-born mice and rats stimulates the growth of sympathetic ganglia, whereas the injection of an anti-serum which destroys the normal circulating NGF leads to the destruction of sympathetic ganglia and nerve fibres bringing about what has been termed an 'immuno-sympathectomy'.

It is possible that NGF affects the growth of nervous tissue in general, and that these effects on the sympathetic nervous system are due to the stage of development that it has reached.

Prostaglandins and Noradrenaline Release

It has recently been shown that the release of noradrenaline at the sympathetic postganglionic nerve terminations as a result of sympathetic stimulation in an isolated rabbit's heart preparation is inhibited by prostaglandin PGE_2 [p. 177]. Prostaglandin inhibitors, on the other hand, increase the noradrenaline released.

The importance of these findings has not yet been evaluated but it has been suggested that prostaglandins may be acting as a negative feedback mechanism in the regulation of the sympathetic nervous system activity.

THE PARASYMPATHETIC NERVOUS SYSTEM

This is also a two neurone system. The preganglionic fibre is medullated whilst the postganglionic fibre is non-medullated [FIG. 208].

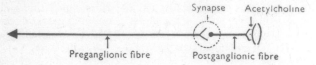

FIG. 208. The pre- and postganglionic parasympathetic nerve fibres.

The ciliary, otic and submandibular ganglia are relay stations for parasympathetic fibres. Elsewhere the synapse is situated in the organ supplied. Thus in this case the postganglionic fibres will be very short and will lie entirely in the organ supplied.

The chemical transmitter at both the synapse and at the postganglionic termination is acetylcholine. These are cholinergic nerves. Transmission across the synapse is blocked by the same ganglionic blocking agents (hexamethonium and its derivatives) as for the sympathetic nervous system. The receptors in the tissues are blocked by atropine and hyoscine.

Parasympathetic Cranial Outflow

Parasympathetic fibres are found in the cranial nerves III, VII, IX, and X [FIG. 209].

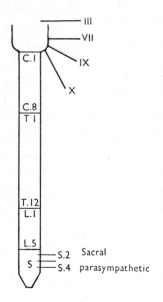

Parasympathetic outflow

FIG. 209. The parasympathetic outflow.

The parasympathetic fibres in the oculomotor nerve (III) arise in the Edinger–Westphal nucleus in the midbrain. The preganglionic fibres reach the ciliary ganglion via the nerve to the inferior oblique muscle of the eye. Here they relay and the postganglionic fibres run in the short ciliary nerves to supply the ciliary muscle and iris of the eye [p. 211].

The parasympathetic fibres in the facial nerve (VII) reach the submandibular and sublingual salivary glands by a complex pathway. These fibres arise in the superior salivary nucleus which lies in the reticular formation in the lower pons. The fibres travel in the seventh nerve until they leave in the chorda tympani nerve just above the stylomastoid foramen. The chorda tympani runs across the ear-drum to reach the lingual nerve which is a branch of the fifth nerve. As its name implies the lingual nerve runs to the tongue but a branch runs to the submandibular ganglion and the submandibular salivary gland. Postganglionic fibres arise from the ganglion and from synapses in the gland itself.

The parasympathetic fibres in the glossopharyngeal nerve (IX) supply the parotid gland. These fibres arise in the inferior salivary nucleus which lies in the reticular formation below the superior nucleus. The fibres travel in the tympanic branch of the ninth nerve to the lesser superficial petrosal nerve to reach the otic ganglion. Here they relay and the postganglionic fibres run in the auriculotemporal nerve (a branch of the fifth nerve) to the parotid gland.

The parasympathetic fibres to the salivary glands are vasodilator as well as secretomotor, that is they cause an increase in blood flow in addition to a flow of saliva.

The vagus (X) is very largely a parasympathetic nerve. It supplies the thorax and abdomen with parasympathetic fibres. The synapses are situated in small ganglia in the organs supplied.

The action of the vagus on the heart has already been considered. Tonic vagal activity maintains the normal heart rate. A reduction in vagal tone causes the heart to speed up.

The vagal fibres constrict the circular muscle of the bronchi and air-passages and are, therefore, bronchoconstrictor.

The vagus supplies the digestive tract as far as the transverse colon. The branches to the stomach are secretomotor and increase the motility as well as initiating the production of gastric juice [p. 107]. The parasympathetic fibres relax the pyloric sphincter.

The branches to the pancreas bring about pancreatic secretion in association with secretin and pancreozymin [p. 108].

The intestinal branches are secretomotor to the glands and increase the motility of the duodenum, small intestine, caecum, ascending and transverse colons. They relax the ileocaecal sphincter between the small and large intestine. The intestinal synapses are found in Meissner's plexus in the submucous layer of the intestine and in Auerbach's plexus which is between the circular and longitudinal muscle layers.

Sacral Outflow

The sacral parasympathetic outflow arises from the second, third and fourth segments of the spinal cord and forms the pelvic splanchnic nerves or the *nervi erigentes*. The pudendal nerve of the somatic nervous system has the same nerve roots.

The sacral parasympathetic outflow supplies the rest of the digestive tract, that is the descending colon, the pelvic colon, the rectum and the anal canal.

The fibres to the rectum and anal canal are important in defaecation as they contract the muscular coats and inhibit the internal anal sphincter. The external sphincter, which also has to be released before defaecation occurs, is under voluntary control and is supplied by the pudendal nerve.

The fibres to the bladder are important in micturition as they are visceromotor to the bladder musculature and inhibitory to the internal sphincter. As in the case of the rectum, there is also a second sphincter, the external sphincter, which is under voluntary control [p. 110].

The parasympathetic fibres to the external genitalia are vaso-dilator to the erectile tissue [p. 172].

VISCERAL AFFERENT FIBRES

The afferent or sensory fibres in the vagus have their cells of origin in the inferior (nodose) ganglion which lies in the neck in close proximity to the superior cervical ganglion of the sympathetic system. These afferent fibres include baroreceptor fibres from the aortic arch, pulmonary artery and common carotid arteries, chemoreceptor fibres from the aortic and pulmonary bodies, stretch receptors from the lungs and air-passages (the Hering–Breuer fibres), atrial and ventricular receptors from the heart, and fibres from the digestive tract.

The afferent fibres associated with the spinal outflow part of the autonomic nervous system have their cells of origin in the dorsal root ganglia, and the fibres enter the dorsal roots of the spinal cord in just the same way as the sensory nerves of the somatic nervous system.

Some of the sensory fibres associated with the sympathetic system run for part of their route in the sympathetic trunk, but they do so without relaying.

HIGHER CENTRE CONTROL OF THE AUTONOMIC NERVOUS SYSTEM

Many parts of the autonomic nervous system are able to function on a spinal basis and produce complex responses to afferent stimuli [see spinal transection, page 202]. However, the activity of the autonomic nervous system is normally under the control of centres in the medulla, hypothalamus and cerebral cortex. Many of the medullary autonomic centres such as the cardiac centre and vaso-motor centre have already been discussed, so also has the role of the hypothalamus in the regulation of body temperature. The hypothalamus and the limbic lobe of the cerebral cortex form the *limbic system* or visceral 'brain' for the higher control of autonomic activity particularly that associated with the emotions fear, grief, etc.

Micturition

The emptying of the bladder (micturition) may be taken as an example of an act which normally involves the co-ordination of both the autonomic and voluntary nervous systems. In spinal man, it reverts to a spinal autonomic reflex.

The contraction of the bladder musculature (detrusor muscle) is brought about by parasympathetic activity [FIG. 210]. The internal sphincter leading to the urethra is normally closed by the sympathetic activity. During micturition this sympathetic activity is decreased and the concurrent increase in parasympathetic activity causes relaxation of the sphincter. The external sphincter is under voluntary control. It is supplied by the pudendal nerve (S.2, 3, 4).

The sympathetic supply to the bladder arises from the first and second lumbar segments of the spinal cord. The preganglionic fibres pass through the sympathetic chain and run directly or via the coeliac and superior mesenteric ganglia, without relaying, to the presacral nerve. The presacral nerve divides into the two hypogastric nerves which run to the hypogastric ganglia on either side of the rectum. The postganglionic fibres which run to the bladder and the internal sphincter arise in these ganglia.

The parasympathetic fibres arise from the second, third (and fourth) sacral segments. The preganglionic fibres run in the nervi erigentes to the same hypogastric ganglia. The postganglionic parasympathetic fibres to the bladder and internal sphincter also arise in this ganglia.

The afferent or sensory fibres from the bladder run along the sympathetic pathways to the dorsal nerve roots of L.1 and L.2 and also along the parasympathetic pathways to the dorsal nerve roots of S.2, 3, and 4. These sensory fibres convey information concerning the degree of distension of the bladder. They also contain pain fibres which allow bladder pain to be accurately localized.

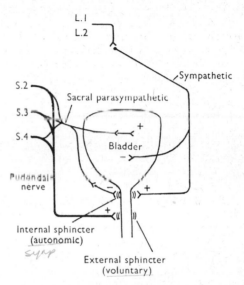

FIG. 210. The nervous control of micturition. In man both the sympathetic and parasympathetic fibres relay in the hypogastric plexuses situated on either side of the rectum.

As the bladder fills the pressure slowly rises, but the pressure rise is small (less than 10 cm. H_2O until the bladder contains 500 ml. urine). As more and more urine enters the bladder the muscle relaxes to accommodate the increased volume of urine. Thus, if a further 50 ml. of fluid are introduced into the bladder rapidly via a catheter, there is an initial rise in pressure, but after a short time this initial increase disappears, and the pressure falls to a value only slightly in excess of the previous value [see also page 223].

When the bladder is full, contraction waves appear which stimulate the pressure receptors in the bladder wall. The afferent nerve impulses entering via the spinal cord run to higher centres and give the desire to micturate. Nerve impulses from the cerebral cortex increase the parasympathetic activity and decrease the sympathetic activity causing relaxation of the internal sphincter and contraction of the bladder muscle. At the same time the external sphincter is relaxed by inhibition of the activity in the pudendal nerve. The intra-abdominal pressure is raised by holding the breath and making a forced expiration against a closed glottis. This causes the diaphragm to descend. At the same time the abdominal musculature is contracted and the increased abdominal pressure presses on the bladder, and aids the expulsion of the urine.

As has been seen [p. 172] the internal bladder sphincter closes during sexual intercourse thus preventing seminal fluid entering the bladder or urine being voided.

SUMMARY OF THE EFFECT OF SYMPATHETIC AND PARASYMPATHETIC ACTIVITY

The organs of the body supplied by the autonomic nervous system will be divided into: (1) those supplied by both the sympathetic and parasympathetic systems; (2) those supplied predominantly by the sympathetic system; and (3) those supplied predominantly by the parasympathetic system.

Organs with a Double Innervation (1)

ORGAN SUPPLIED	EFFECT OF INCREASED SYMPATHETIC ACTIVITY	EFFECT OF INCREASED PARASYMPATHETIC ACTIVITY
Pupil of the eye	Dilates	Constricts
Bronchi	Dilates	Constricts
Heart	Increases heart rate	Decreases heart rate
Digestive tract	Inhibits motility	Increases motility
Sphincters of digestive tract	Constricts	Relaxes
Bladder	Allows filling	Empties

It should be noted that the effects of parasympathetic underactivity are the same as those of sympathetic overactivity and vice versa. For example, if the parasympathetic nerves to the pupil of the eye are paralysed by homatropine drops, the result will be the same as that of sympathetic overactivity, namely, dilatation of the pupils.

Organs Supplied mainly by the Sympathetic System (2)

ORGAN	NO SYMPATHETIC ACTIVITY	INCREASED SYMPATHETIC ACTIVITY
Blood vessels	Vasodilatation	Vasoconstriction
Sweat glands	Nil	Production of sweat

Organs Supplied mainly by the Parasympathetic System (3)

ORGAN	INCREASED PARASYMPATHETIC ACTIVITY
Salivary glands	Secretion of saliva and an increased blood flow
Gastric glands	Secretion of gastric juice rich in enzymes
Pancreatic glands	Secretion of pancreatic juice rich in enzymes
External genitalia	Vasodilatation of erectile tissue

Two rules may act as a guide for remembering the effects of activity of the two systems:

1. The sympathetic system *is active in times of stress*, i.e. the 'fight or flight' reaction. The typical response in such a case is a fast heart, dilatation of the pupils, dilatation of the bronchi, contraction of the arrectores pilorum muscles causing the 'hair to stand on end', sweating and a cessation of digestion. Parasympathetic activity, in general, has the opposite effect. An increase in respiration will probably also occur under these conditions of sympathetic activity but this has not been included because the respiratory muscles are controlled by the somatic nervous system.

2. The parasympathetic system is a more discrete system than the sympathetic and widely differing activities may be occurring simultaneously in different parts of the body. However, we may generalize by saying that the parasympathetic system is *the emptying mechanism of the body*. Its activity speeds up the passage of food along the digestive tract. It brings about defaecation. It brings about micturition. Sympathetic activity has the opposite effect.

HYPOTHALAMUS

The functions of the hypothalamus have been studied in animals by the techniques of electrical stimulation and removal (ablation) of specific areas. Such methods have shown that, in addition to controlling the release of the pituitary hormones, the hypothalamus plays an important role in the regulation of activities which involve the sympathetic and parasympathetic nervous systems.

These experimental techniques have shown that the areas for a specific function are not so highly localized as in some other parts of the brain and that there is a great deal of overlap between the areas. Stimulation of the posterior hypothalamus leads in some animals to a *rage reaction* which is associated with signs of increased sympathetic activity, such as dilated pupils, and contraction of the *arrectores pilorum* muscles which cause the hairs to stand on end. Electrical stimulation of other areas has been shown to bring about blood pressure and respiratory changes (both increases and decreases), drowsiness, and changes associated with parasympathetic activity such as salivation, vomiting, defaecation and micturition.

The principal functions of the hypothalamus may now be summarized:

1. Limbic System

The hypothalamus is part of the limbic system for the higher control of autonomic activity.

2. Controls Posterior Pituitary Activity—Hypothalamus—Posterior Pituitary Neurosecretory System

(a) Antidiuretic Hormone Secretion

The osmoreceptors, which are sensitive to the electrolyte content of the plasma, are situated in the region of the supra-optic nucleus of the hypothalamus. Nerve fibres run from this hypothalamic nucleus to the posterior lobe of the pituitary gland (supra-optico-hypophyseal tract). An increase in the osmotic pressure of the blood increases the ADH secretion from this system. The increase in ADH level causes water retention by increasing the kidney tubular reabsorption of water [p. 136]. Failure of this system leads to diabetes insipidus [p. 136]. Removal of the pituitary gland, however, does not usually lead to diabetes insipidus (although the other signs of pituitary insufficiency occur) presumably because of the release of sufficient ADH from the vesicles associated with the supra-optico-hypophyseal tract which still remain.

This is a neuro-secretory system [p. 160].

The release of ADH usually shows a *diurnal variation*, that is, it has a 24 hour cycle. As a result less urine is formed during the night.

There is an increased release of ADH in emotional states, exercise, fainting and cigarette smoking (due to the action of nicotine on the hypothalamus). In all these cases there will be a temporary reduction in the volume of urine produced. So much ADH may be released in fear that the skin blood vessels are constricted by the 'vasopressin' action of ADH that pallor results.

It should be noted that ADH controls the amount of urine *entering the bladder* per minute. In the above emotional states, there may, quite independently, be a desire to pass urine. This is due to an increase in parasympathetic activity, which, by raising the bladder pressure, stimulates the bladder sensory receptors.

(b) Oxytocic Hormone Secretion

Like ADH the oxytocic hormone is thought to be made by the nerve cells of the hypothalamo–hypophyseal tract and released from the posterior pituitary when nerve impulses pass along these nerves. This hormone brings about contraction of the pregnant uterus and milk ejection in women [p. 175]. Its function, if any, in the male is unknown. The nerve (and hormonal) pathways which act on the hypothalamus to bring about the release of this hormone have not been fully worked out, but since stimulation of the nipples by sucking stimulates the release of this hormone during lactation, presumably the afferent nerve pathway is from sensory endings at the nipple to the supra-optic nucleus of the hypothalamus and thence to the posterior pituitary.

3. Controls Anterior Pituitary Activity—Hypothalamo–Hypophyseal Portal System

Unlike the posterior pituitary, the anterior pituitary does not have a nervous connexion with the hypothalamus. The hypothalamus exerts its control on the secretions of the anterior pituitary by means of *releasing factors* and *inhibiting factors* (release inhibiting factors) which pass from the hypothalmus to the anterior pituitary via the portal venous system [p. 161].

It seems likely that one releasing and one inhibiting factor will be found for each of the anterior pituitary hormones. This will treble the number of hormones to be considered. There will then be HGH-RF, TSH-RF, ACTH-RF, FSH-RF, LH-RF and P-RF, plus HGH-IF, TSH-IF, ACTH-IF, FSH-IF, LH-IF and P-IF, in addition to HGH, TSH, ACTH, FSH, LH and Prolactin (LTH).

Work on the isolation and elucidation of the structure of the releasing and inhibiting factors is proceeding rapidly. The releasing factors so far isolated appear to be peptides of a comparatively low molecular weight, and should be relatively easy to synthesize once the formula is known.

Once these releasing factors become readily available and more is known about their physiological action, it seems likely that they will be included in the category of hormones and their names will be changes to *releasing hormones* and *inhibiting hormones*.

(a) TSH-RF (Thyrotrophin Releasing Hormone)

The thyroid stimulating hormone releasing factor has been isolated and synthesized. It is a small peptide made up of the amino acids: glutamic acid, histidine and proline.

The terminal glutamic acid has undergone a cyclic structure change with the removal of water from the terminal —NH$_2$ group and the side chain —COOH group. The resultant residue is termed pyroglutamic acid. This change makes the resultant molecule more stable.

Glutamic Acid → Pyroglutamic Acid

The arrangement of peptides in TSH–RF is:

$$\text{pyroglu—his—proNH}_2$$

It seems probable that this, and possibly other releasing factors are made by brain cells remote from the hypothalamus, secreted into the lateral ventricles, transported via the CSF to the third ventricle and the hypothalamus. They then pass down the hypothalamo–hypophyseal portal system to the anterior pituitary.

TSH-RF promotes the synthesis as well as the release of TSH. An injection of 50 micrograms of TSH-RF intravenously raises the level of circulating TSH to a peak in twenty minutes. Rather surprisingly, it is also active when given, in larger doses (20 milligrams), by mouth. This implies that sufficient escapes destruction by the digestive enzymes, and that small peptides can be absorbed without prior break down to amino acids!

The structural formula of TSH-RF is:

Pyroglutamic Acid — Histidine — Proline-Amide

If synthetic peptides can be found that act as antagonists to this natural releasing factor, they may be of great value in the treatment of thyroid overactivity.

(b) LH-RF (Gonadotrophin Releasing Hormone)

The luteinizing hormone releasing factor is a decapeptide having the following amino acid sequence:

$$\text{pyroglu—his—try—ser—tyr—gly—leu—arg—pro—glyNH}_2$$

It has already been synthesized. In the doses used (10^{-9} grams) it appears to release FSH as well as LH.

LH-RF may prove to be of value in bringing about ovulation in women who do not normally ovulate (due to a failure of the pituitary to release FSH and LH) so that these women can become pregnant and have children.

Synthetic analogues of LH-RF, by blocking the action of the natural releasing factor, may prove to be of use as an oral contraceptive.

(c) HGH-RF

The human growth hormone releasing factor is also a decapeptide. Its amino acid sequence is:

val—his—leu—ser—ala—glu—glu—lys—glu—ala

When available, this releasing factor may prove to be of value in the treatment of dwarf children; to enable growth to be stimulated so that they can reach a normal stature. As has been seen [page 161], the treatment of this condition is at present restricted by the limited supply of HGH, which being an 191 amino acid peptide, is unlikely to be synthesized on a commercial scale in the foreseeable future.

Inhibiting factors

Work on the isolation and elucidation of inhibiting factors has not proceeded as rapidly as that on releasing factors.

Section of the pituitary stalk in animals leads to galactorrhoea (excessive milk production) which is not present before the pituitary stalk is cut. It appears therefore that there is a tonically active *inhibiting factor* P-IF (prolactin-inhibiting factor) normally present, which inhibits the release of prolactin and hence inhibits lactation.

Should this inhibiting factor become available it might prove to be of use in inhibiting post-partum lactation in a woman when the baby is being transferred to bottle-feeding.

A fourteen amino-acid peptide has been isolated which inhibits the release of human growth hormone. This HGH-IF may prove to be of value in the treatment of gigantism and acromegaly.

The melanocyte stimulating hormone is secreted by the same cells of the anterior pituitary as secrete ACTH, and brings about pigmentation of the skin in animals. A three amino acid peptide, melanocyte stimulating hormone inhibiting factor MSH-IF has been isolated but it has no known function.

4. Thirst

The water intake of the body is safeguarded by the sensation of thirst. Electrical stimulation of the dorsal hypothalamus in the goat produces excessive drinking (polydipsia) [p. 129]. The **thirst centre** (or drinking centre) probably lies in this region. Angiotensin stimulates this centre.

5. Hunger and Satiety

The food intake of the body is probably controlled by two centres in the hypothalamus known as the **appetite centre** (or feeding centre) and the **satiety centre.** Their presence has been demonstrated in animals by electrical stimulation [p. 123].

Stimulation of the appetite centre causes an animal to eat excessively (hyperphagia or bulimia), whilst stimulation of the satiety causes an animal to cease eating.

6. Temperature Regulation

The temperature regulating centre in the hypothalamus consists of two parts. The centre in the anterior hypothalamus regulates heat loss by sweating and skin vasodilatation whilst the centre in the posterior hypothalamus increases heat production and conserves heat [p. 152].

7. Vasodilator Nerves

Some vasodilator nerves to blood vessels such as those which bring about vasodilatation in muscle at the onset of a haemorrhagic faint [p. 43] are thought to be controlled directly from the hypothalamus instead of the medullary vasomotor centre.

8. Facilitatory Reticular Formation

The facilitatory reticulo-spinal tract activity is thought to originate in the hypothalamus and surrounding structures as well as in the cortex and basal ganglia [p. 201]. These fibres facilitate both cortical and reflexly induced movements.

REFERENCES AND FURTHER READING

Burn, J. H., and Rand, M. J. (1965) Acetylcholine in adrenergic transmission, *Ann. Rev. Pharmacol.*, **5,** 163.

Cannon, W. B., and De La Paz, D. (1911) Emotional stimulation of adrenal secretion, *Amer. J. Physiol.*, **28,** 64.

Cannon, W. B., and Rosenblueth, A. (1937) *Autonomic Neuro-effector Systems*, New York.

Ciba Foundation Symposium (1960) *Adrenergic Mechanisms*, London.

Denny-Brown, D., and Robertson, E. G. (1933) On the physiology of micturition, *Brain*, **56,** 149.

Elliott, T. R. (1904) On the action of adrenalin, *J. Physiol.* (*Lond.*), **31,** 20P.

Hess, W. R. (1957) *The Functional Organization of the Diencephalon*, New York.

Ingram, W. R. (1960) Central autonomic mechanisms, in *Handbook of Physiology*, Section 1, Vol. 2, p. 951, Washington, D.C.

Langley, J. N. (1901) Observations on the physiological action of extracts of the suprarenal bodies, *J. Physiol.* (*Lond.*), **27,** 237.

Langley, J. N. (1911) The effect of various poisons upon the response to nervous stimuli chiefly in relation to the bladder. [First use of terms sympathetic and parasympathetic], *J. Physiol.* (*Lond.*), **43,** 125.

Levi-Montalcini, R. (1964) Growth control of nerve cells by a protein factor and its antiserum, *Science*, **143,** 105.

Loewi, O. (1921) Uber humorale Ubertragbarkeit der Herzner-venwirkung, *Pflügers Arch. ges. Physiol.*, **189,** 239.

Symposium on Catecholamines (1959) *Pharmacol. Rev.*, **11,** 381.

von Euler, U. S. (1959) Autonomic neuroeffector transmission, in *Handbook of Physiology*, Section 1, Vol. 1, p. 215, Washington, D.C.

21. SPECIAL SENSES

THE EYE

The eye is like a camera. Light enters via the transparent cornea and by means of a lens system an inverted image is formed, not on a sensitive photographic film as in a camera, but on a sensitive structure known as the **retina** which converts the light into nerve impulses. These impulses are relayed to the optical cortex in the occipital lobe of the brain and are interpreted as sight.

The human eye is a globe about 1 inch in diameter [FIG. 211].

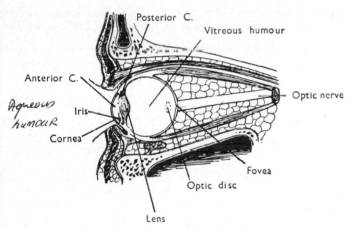

FIG. 211. The eye.

Anterior C = anterior chamber of the eye;
Posterior C = posterior chamber of the eye.

Note that the posterior chamber is situated in front of the lens. The anterior and posterior chambers are filled with aqueous humour. The chamber behind the lens is filled with vitreous humour.
The optic disc is the blind-spot. The optic nerve enters at this point and there are no light sensitive rods or cones.

The space in front of the **lens** is filled with *aqueous humour*. The part in front of the **iris** is termed the anterior chamber and that behind the iris the posterior chamber. The space behind the lens is filled with *vitreous humour*.

Formation of Aqueous Humour

Aqueous humour is secreted by the ciliary process of the ciliary body in the posterior chamber [FIG. 212]. It passes through the opening in the iris and enters the anterior chamber where it is absorbed in the *filtration angle* into the canal of Schlemm which leads to the venous blood.

The pressure of the aqueous humour is 25 mm. Hg. An upset in the balance of formation over absorption may lead to an increased intra-ocular tension, a condition known as *glaucoma*.

Iris—Size of the Pupil

The iris is a diaphragm with a circular aperture known as the *pupil* through which light passes.

The size of this aperture regulates the amount of light entering the eye. Just as with a camera, a small aperture gives a greater depth of focus (near objects and distant objects are in focus at the same time) and lens distortion (spherical and chromatic aberration) is reduced.

The size of the pupil is regulated by circular and radial smooth muscle fibres in the iris itself. This muscle is supplied by the autonomic nervous system and is one of the few places in the body where this system can be seen in action. Stand in front of a mirror. Cover your eyes with your hands. The darkness causes the pupils to dilate. Remove your hands and you can see the pupils constrict. Shining a light from a torch is the method employed clinically to test pupillary reaction.

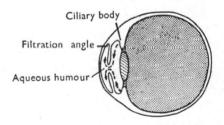

FIG. 212. The formation and reabsorption of aqueous humour. Aqueous humour is secreted from the ciliary body in the posterior chamber. It passes through the aperture in the iris (the pupil) and enters the anterior chamber. It is reabsorbed into the canal of Schlemm in the filtration angle.

The parasympathetic fibres constrict the pupil. They arise from the parasympathetic part of the cranial third nerve nucleus (Edinger-Westphal nucleus, page 206). The fibres run with the third nerve, via the nerve to the inferior oblique muscle and relay to the ciliary ganglion.

The sympathetic fibres dilate the pupil. They arise from the first thoracic segment and relay in the superior cervical ganglion.

The pathway of the light reflex—constriction of the pupil to bright light—is from the retina via the optic tract to the superior corpus quadrigeminum (superior colliculus) via the colliculonuclear fibres to both third nerve nuclei and back via the ciliary ganglia to the iris [FIG. 213]. Since the fibres run to the third nerve nuclei on both sides, shining a light on one eye causes both pupils to become smaller.

Constriction of the pupils also occurs when looking at a near object.

The size of the pupil depends on the balance between sympathetic and parasympathetic activity. Excessive constriction may be due to excessive parasympathetic activity or insufficient sympathetic activity. Conversely, dilatation may be due to excessive sympathetic activity or insufficient parasympathetic activity. Eye drops for dilating the pupil contain an atropine[16] derivative to block the parasympathetic activity; the unopposed sympathetic activity dilates the pupil.

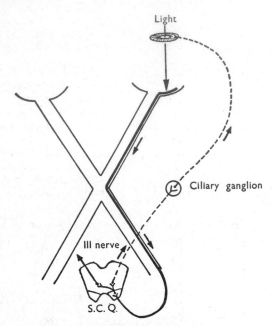

FIG. 213. Pathway for the light reflex. Fibres from the superior corpus quadrigeminum (S.C.Q.) run to both third nerve nuclei (para-sympathetic nuclei) so that light shining in one eye causes the con-striction of both pupils.

CONVERGENCE-ACCOMMODATION REFLEX

The eye at rest is focused at infinity. When we look at a close object, three things happen, the eyes converge, the pupils constrict and the eyes are focused.

1. Convergence

The eyes converge on their visual axes, so that both are looking at the object. This is brought about by the six extrinsic eye muscles, the *superior*, *inferior*, *lateral* and *medial* recti and the *superior and inferior* oblique muscles, particularly the *medial rectus*.

All are supplied by the third cranial nerve, except the *lateral rectus*, which is supplied by the sixth cranial nerve and the *superior oblique*, which is supplied by the fourth cranial nerve. As a memory aid the following mnemonic based on a *pseudo*-chemical formula may be employed:

$$LR_6(SO_4)_3$$

where LR_6 stands for lateral rectus supplied by the sixth,
    SO_4 stands for the superior oblique supplied by the fourth
    and all the rest are supplied by the third.

[16] Hence the name for the plant belladonna (= beautiful lady) dating from the time when it was fashionable for ladies to put the juice of this plant in their eyes to dilate the pupils.

2. Constriction of the Pupils

The pupils constrict due to an increase in parasympathetic activity and a reduction in the sympathetic activity.

3. Focusing—Ability of the Eye to Alter its Lens Power

In humans the object is brought into sharp focus on the retina by an alteration in the curvature of the front surface of the lens. For distant vision the lens is flattened by the outward pull of ligaments attached to the ciliary body. For near vision the ciliary muscles contract and relax the suspensory ligaments. The lens has an elastic capsule and it becomes fatter.

The ciliary muscle is supplied with parasympathetic fibres from the third cranial nerve. The pathway for the accommodation reflex is from the retina to the occipital cortex to the frontal lobes and to the third nerve nucleus. Accommodation is paralysed by atropine and homatropine. Such a paralysis of the ciliary muscle is termed *cycloplegia*.

Amplitude of Accommodation

In optics it is convenient to consider the power of lenses, not in terms of their focal lengths, but as the reciprocal of this. The usual unit is the dioptre (D) and is defined as $\dfrac{1}{\text{focal length in metres}}$. The advantage of this system is that when two or more lenses are used together their combined power may be determined by algebraic summation. Thus, if a 2D lens and a 3D lens are placed together the combined power is 5D. Further, if a reducing lens power −1D is now added, the combined power of all three would be 4D.

The eye focuses an object at infinity on to the retina lying only 1·7 cm. (= 0·017 m.) behind the lens. This means that it has a lens power of $\dfrac{1}{0·017} = 59$ dioptres. Most of the refraction takes place at the cornea (43D) and only 16D occurs in the lens proper.

To change the focus from infinity to 1 metre needs an additional 1D. A pair of spectacles of this strength would do just that. To focus at 10 cm. (= $\frac{1}{10}$ m.) would require a lens of 10D.

Young people can increase the power of their eyes from 59D to 70D, an increase of 11D. This is the *amplitude of accommodation*. Hence if the 59D enabled them to focus at infinity, the 70D would enable them to focus at $\frac{1}{11}$ metre = 9 cm.

If the lens is removed at a cataract operation, the power of the eye will have fallen from 59D to 43D. A lens is placed in front of the eye to make up the difference.

The nearest distance at which an object can be seen clearly is known as the *near point*. The farthest distance is the *far point*. With increase in age, the near point recedes so that ultimately only distant objects are in focus. Two pairs of glasses may be required, one for close work and reading, and one for middle distance vision such as watching television. Such a loss of accommodation is termed *presbyopia*.

The alteration in curvature at the front surface of the lens can be seen by means of the Purkinje–Sanson images. In a darkened room the images of a candle can be seen reflected from the surfaces of the eye [FIG. 214]. Neither the image from the anterior surface of the cornea, nor that from the posterior surface of the lens change with accommodation. Only that from the anterior surface of the lens changes.

In a short-sighted person (**myope**) the cornea and lens form too powerful an optical system for the size of the eye. As a result, when looking at a distant object the image falls in front of the retina. This is corrected by diminishing the power of the eye; a negative lens (concave) is introduced in front of the eye in the form of spectacles. Near objects can, however, be brought to a sharp focus on the retina by a myopic subject without optical correction.

In a long-sighted person (**hypermetrope**) the lens power is insufficient and an optical correction in the form of a positive lens (convex) is needed.

If the cornea and lens form a cylindrical rather than a circular lens system, horizontal rays will be brought to a different focus from the vertical rays, and it will be impossible to see both clearly at the same time. This condition is known as **astigmatism.** It is corrected by introducing an equal and opposite cylindrical lens to cancel out the defect.

A normal sighted person is termed an *emmetrope*.

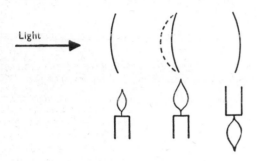

FIG. 214. The Purkinje–Sanson images. In a darkened room three reflected images of a light, such as a candle, can be seen from the surfaces of the eye. The reflections are from the surface of the cornea, from the anterior surface of the lens and from the posterior surface of the lens. When the eye is focused on a close object only the reflected image from the front surface of the lens changes size. This experiment confirms that focusing involves a change in the curvature of the anterior surface of the lens.

Retina

The light sensitive elements forming the retina are the rods and cones. The rods are used for night vision and the cones for day vision. When using only the rods we are colour-blind since these give only the sensation of shades of grey. With the cones colours are perceived.

Structural Arrangement

The light-sensitive rods and cones lie adjacent to the outer pigment layer at the back of the retina [FIG. 215]. The light has to pass through the nerve fibres and bipolar cells to reach these elements, except at the central yellow spot or fovea. This region is tightly packed with cones only. The bipolar cells and nerve fibres have been displaced to one side [FIG. 215].

Visual Acuity

Under a magnifying glass a picture in a newspaper appears as a series of dots. With the naked eye the dots are not visible. This is because the visual acuity of the eye is not great enough to discriminate between the individual dots. The standard visual acuity corresponds to 1 minute of an arc, that is, two points can be seen to be separate if they subtend an angle of at least 1 minute at the

eye. The *acuity* is the reciprocal of the angle expressed in minutes [TABLE 19].

TABLE 19

MINIMUM ANGLE	ACUITY
2 minutes	0·5
1 minute	1
30 second	2
15 seconds	4

Too great a visual acuity could mean, for example, a desire to sit farther away from the television set in order not to see the scanning lines.

Test types are employed to determine the visual acuity. Two sets are used, one for near vision (30 cm.) and one for distant vision (6 m.). They consist of black letters on a white background of such a size and shape that the detail of each letter would subtend an angle of 1 minute if the type were viewed from the distance specified for each line of type. These distances range from 60 metres to 5 metres for the distant vision type. If the small 5 metre line of type can be read from the standard distance of 6 metres, the acuity is said to be 6/5. If only the larger 18 metre type can be read from this distance the acuity is only 6/18. Normal vision is 6/6.

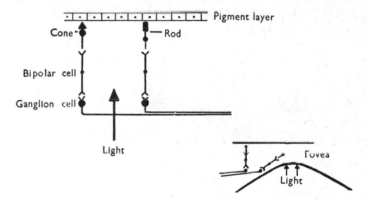

FIG. 215. The layers of the retina. Note that the light passes through the ganglion cell layer and the bipolar cell layer to reach the cones and rods except at the fovea where they are displaced to one side. Several rods share the same bipolar cell. Several bipolar cells share the same ganglion cell.

The Visual Pathway

Although the nerve impulses from the fovea pass to both occipital cortices, the outer fields are unilaterally represented. The fibres from the lateral parts of the retina (i.e. nasal fields of view) pass to the occipital cortex of the same side. The fibres from the medial part of the retina (i.e. temporal fields of view) decussate and cross to the occipital cortex of the other side [FIG. 216]. The fibres relay in the lateral geniculate body.

Cones (Rods) $\xrightarrow{\text{1st neurone}}$ bipolar cells $\xrightarrow{\text{2nd neurone}}$ ganglion cells ⟶ optic

chiasma ⟶ lateral geniculate body ⟶ cells in lateral geniculate

body ⟶ optic radiation $\xrightarrow{\text{3rd neurone}}$ calcarine cortex

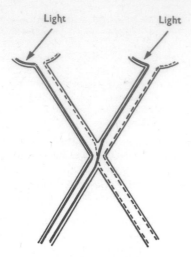

FIG. 216. The optic chiasma.

There is no decussation in a vertical plane. The superior fibres from the retina run to the superior aspect of the calcarine cortex [FIG. 217] and the inferior fibres from the retina to the inferior aspect.

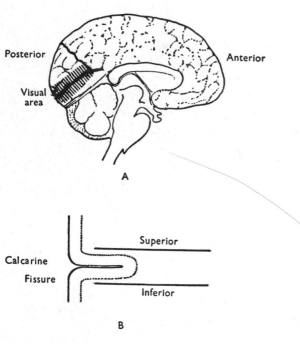

FIG. 217

A. Sagittal section of the brain showing the visual cortex. The visual cortex lies on either side of the calcarine fissure on the inner (medial) side of each cerebral hemisphere.

B. The medial retinal fibres decussate in the horizontal plane [FIG. 216] but there is no decussation in a vertical plane. Thus the fibres from the superior part of the retina arrive at the superior aspect of the calcarine fissures. The inferior fibres arrive at the inferior aspect.

Rod and Cones—Duplicity Theory

When the intensity of illumination is low, scotopic vision using rods is employed. At higher levels of illumination photopic (colour) vision using cones is used. The luminosity curve for the rods and cones is different [FIG. 218]. At dusk blues appear unusually bright compared with reds. This is called the Purkinje shift. Both cones and rods are now being used and the rods register blue as white, and red as black. The bleaching curve for visual purple (rhodopsin) is identical with the scotopic luminosity curve. Using cones the most sensitive part of the spectrum is 5,500 Å yellow-green. Using rods the most sensitive part is 5,050 Å, which is green-blue.

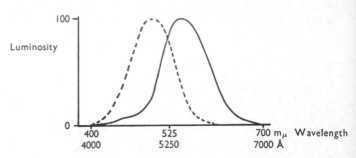

FIG. 218. The luminosity curves for cone and rod vision adjusted so that the maximum luminosity is 100 in each case.

--, The scotopic luminosity curve (rods) for different colours. All these colours are seen as shades of grey.

—, The photopic luminosity curve (cones). When using the cones, colours are seen (colour vision). 4,000 Å = violet, 4,500 Å = blue, 5,500 Å = green-yellow, 6,000 Å = orange, 7,000 Å = red. 1 Å = 10^{-1} mμ = 10^{-10} m. = 10^{-8} cm. = 10^{-1} nanometres.

The change from light-adapted (photopic) vision to night—or scotopic vision is termed *dark adaptation*. It takes 30 minutes before the maximum sensitivity of the dark-adapted eye is obtained (see vitamin A, page 124).

Binocular and Monocular Vision

Binocular vision has many advantages over monocular vision. It gives a larger visual field, and a lesion in one eye is not important particularly if that part of the field is covered by the other eye. This is illustrated clearly in connexion with the blind spot. We are blind in each eye over the area of retina where the optic nerve enters, but we are not aware of this defect in our vision since the blind spot of each eye is covered by the other eye.

The great advantage of binocular vision is that the retinal images of the two eyes are slightly different, that is it gives stereoscopic vision. This enables us to appreciate size, shape and distance of objects. In addition, the effort of convergence allows distance to be judged.

Even with one eye, distance may be judged, but not so accurately. We use apparent size, overlapping of distant objects by near objects, intersection of objects with horizontal plane, parallax, modification of colour (far away objects tend to appear more blue), shadows and reflected light to give the shape and distance of an object.

The eye is unable to detect the polarization of light. That is the plane in which vibrations are occurring.

Colour Vision

When white light is separated out into the individual wavelengths a spectrum or rainbow effect is obtained. Although the eye is sensitive to all these colours, that is, light ranging in wavelength from 4,000 to 7,500 Å (400–750 mμ), it is possible to produce the sensation of all colours using only three wavelengths, one in the red, one in the green, one in the blue. This principle is widely used in colour television and colour photography where only three emulsions, sensitive to these spectral bands, are used to reproduce all colours. The Young–Helmholtz theory of colour vision is based on trichromatic or three-colour theory. Three different types of cone are postulated which are sensitive to these three primary colours. Three cone pigments erythrolabe, chlorolabe and cyanolabe have also been postulated. Although there are many objections to such a theory it explains the various types of colour-blindness.

Dichromats can only see two of the three colours. If the missing colour is red the condition is called *protanopia*, if it is green *deuteranopia* and if blue *tritanopia* (very rare).

Granit has suggested an alternative hypothesis, namely, that some receptors are dominators responding to all wavelengths, whereas others are modulators responding to narrow bands of colours. Six or seven modulators have been postulated in this polychromatic theory of colour vision.

Adequate Illumination

It is now realized that an adequate level of illumination is important in order to create the correct working conditions, in offices, factories, libraries, etc.

For comparison purposes, the original standard was the **standard candle**, a wax candle of fixed dimensions which produced a luminous intensity of one candle-power. The present standard is approximately the same, but is termed the **candela**. It is defined as $\frac{1}{60}$th of the luminous intensity per cm.^{2} of platinum at 1773 °C. The luminous intensity of source compared with one candela is expressed in *candle-power*.

Illumination from a point source is determined by the inverse square law. Thus, if the distance from a source of illumination is doubled, the illumination is reduced to one quarter. If the distance is trebled, the illumination is reduced to one-ninth. Thus:

$$\text{Illumination} \propto \frac{\text{Luminous intensity of source}}{\text{Distance}^2}$$

If the distance is measured in feet, the illumination is measured in **foot-candles**. If the distance is measured in metres the illumination is measured in **metre-candles** or **lux**.

The amount of light energy given off by one candela is **4π lumens**. This light is given off in all directions and will illuminate equally the surface of an imaginary sphere radius r from the point source. Since the total surface of the sphere is $4\pi r^2$, the illumination when r = 1 foot will be 1 foot-candle, and the illumination when r = 1 metre will be 1 lux.

Thus   1 foot-candle = 1 lumen/ft^2

           1 lux   = 1 lumen/m^2

Since 10 square feet approximately equal 1 square metre 10 lumens/m^2. $\simeq$ 1 lumen/ft^2, i.e.

$$10 \text{ lux} \simeq 1 \text{ foot-candle}$$

The inverse square law only applies to a point source. If a parallel beam of light is produced by a lens or mirror system (search-light or car headlamp) the light is not radiated equally in all directions and the illumination will not fall off so rapidly with distance.

With parallel beam of light produced by a laser, the illumination will be theoretically constant over large distances and will only fall off due to dust particles in atmosphere.

Typical levels of illumination are:

Laboratories
20 lumens/ft^2 (*20 foot-candles*) = 200 lumens/m.^{2} (*200 lux*)

Reading rooms and libraries
30 lumens/ft^2 (*30 foot-candles*) = 300 lumens/m.^{2} (*300 lux*)

Outdoor daylight
10,000 lumens/ft^2 (*10,000 foot-candles*) = 100,000 lumens/m^2 (*100,000 lux*)

THE EAR

The External Ear

The external ear collects sound waves and transmits them to the tympanic membrane (ear-drum) [FIG. 219].

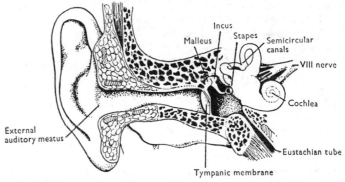

FIG. 219. The ear.

The external ear in man consists of a fixed auricle composed of elastic cartilage covered with skin. The external auditory (acoustic) meatus is an S-shaped tube leading down to the tympanic membrane. Numerous wax-secreting glands (ceruminous glands) are present in this meatus; accumulation of this wax is a common cause of deafness. The lateral part of this meatus is cartilaginous, the medial part is bony.

The Middle Ear

The middle ear lies on the inner side of the tympanic membrane or ear-drum. Three small ossicles transmit the vibrations of the tympanic membrane due to sound to the oval window of the inner ear. The inner ear is fluid-filled and the two membranes and ossicles act as an impedance-matching device changing the energy of sound in air into energy of sound in water.

The ear drum has twenty times the area of the oval window and the ossicles cause a 3:2 reduction in movement [FIG. 220].

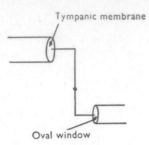

FIG. 220. The ossicles act as a lever between the tympanic membrane and the oval window, and reduce the movement in the ratio of 3:2 in man.

The impedance-matching can be further improved by connecting an exponential horn to the ear, an idea used for centuries in the form of an ear trumpet.

The middle ear is air-filled. It communicates with the nasopharynx via the auditory (pharyngotympanic or Eustachian) tube. Swallowing, sneezing, or yawning opens the pharyngeal end of this tube and allows the pressure inside and outside to be equilibrated. Swallowing is important when ascending to or descending from high altitudes, hence the sweet issued by airline companies.

This tube can be blocked by a cold in the head, in which case the air in the middle ear is absorbed into the blood. As a result the ear-drum is drawn in and this leads to deafness which disappears when the tube becomes unblocked again.

The Inner Ear

The inner ear converts the vibrations in the fluid into nerve impulses which are interpreted as sound.

The inner ear consists of a complicated series of cavities in the temporal bone filled with *perilymph*. Inside these bony cavities is a series of tubes filled with *endolymph*. The perilymph has a composition similar to cerebrospinal fluid. The endolymph, on the other hand, is like the fluid in cells with high potassium ion concentration and low sodium ion concentration.

The **cochlea** consists of three tubes wound two and three-quarter times round the central pillar called the modiolus.

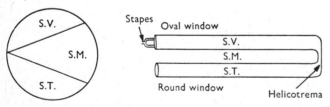

FIG. 221. Diagram of the cochlea.
S.V. = scala vestibuli; S.M. = scala media; S.T. = scala tympani.

The oval window leads to the scala vestibuli filled with perilymph [FIG. 221]. Reissner's membrane separates it from the scala media (ductus cochlearis). The basilar membrane separates it from the scala tympani. The scala tympani has a round window closing it off from the middle ear.

The basilar membrane is short (0·13 mm.) at its base near the oval window and increases in length to 0·4 mm. near the helicotrema. The hair cells of the organ of Corti form a spiral tunnel on top of the basilar membrane [FIG. 222].

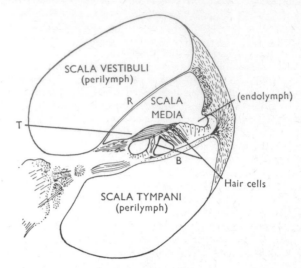

FIG. 222. Cross-section of the cochlea.
T = tectorial membrane; B = basilar membrane;
R = Reissner's membrane.

Audible sound consists of frequencies ranging from 20 cycles per second to 20,000 cycles per second. The upper limit of hearing is reduced with age, so by the time one is able to afford a super high-fidelity record reproducer one can no longer hear the high notes reproduced!

The high frequencies cause the base of the basilar membrane to vibrate, lower frequencies cause the whole membrane to vibrate.

Cochlear Microphonics

The resting endolymph electrical potential in the scala media (ductus cochlearis) is +80 mV. The cellular structures bordering the scala media are at a potential of −40 mV.

Movement of the basilar membrane towards the scala vestibuli reduces the 80 mV. positive potential in the scala media. Movement of the basilar membrane away from the scala vestibuli increases this positive potential. Thus, sound vibrations of the basilar membrane cause an alternating voltage of the same frequency to be superimposed on the steady endolymph potential. This is termed *cochlear microphonics*. Nerve impulses are set up in the cochlear nerve when the scala media becomes less positive with respect to the scala tympani.

Frequency Discrimination

With low frequencies the sense of pitch is conveyed by the frequency of nerve impulses. With higher frequencies, only parts of the basilar membrane close to the base are stimulated and the frequency is conveyed by the pattern of this vibration. The hair cells lie between the basilar membrane and the tectorial membrane [FIG. 222]. The vibration of the basilar membrane causes a shearing strain on these hair cells of the organ of Corti and the stimulation of the nerve endings at their base.

Auditory Pathways

The pathway from cochlear nerve to the cortex (superior temporal gyrus) is as follows:

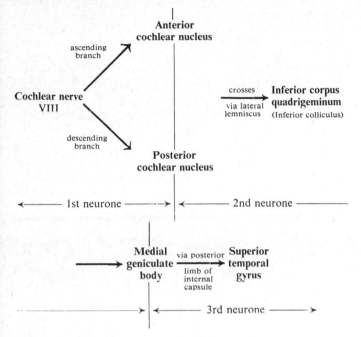

During the induction of anaesthesia, hearing is the last sensation to be lost. It is the first to return during recovery.

VESTIBULAR APPARATUS

The vestibular apparatus consists of the three **semicircular canals**, the **saccule** and the **utricle.** The saccule and utricle respond to the position of the head in space, whilst the semicircular canals respond to any change in the velocity of the head (acceleration). Neither respond to a steady velocity.

The three semicircular canals are the lateral, the superior (anterior) and the posterior.

The left superior canal is in the same plane as the right posterior canal and vice versa. The membranous canals are filled with endolymph and lie in a bony canal from which they are separated by perilymph. The three canals open into the utricle by five apertures, the superior and posterior canals share an opening. A dilatation (ampulla) in each canal contains the *crista.*

The utricle and saccule contain a projecting ridge called the *macula.*

The *ductus endolymphaticus* joins the utricle with the saccule, and the *canalis reuniens* unites the saccule with the scala media of the cochlea.

Crista and Macula

The vestibular fibres of the eighth cranial nerve terminate in nerve endings in the crista and macula. In each of these structures hairs project into firm gelatinous material, *cupula terminalis.* In the semicircular canals the cupula extends to the roof of the ampulla and acts as a movable partition [Fig. 223(C)]. The hair cells at its base are stimulated by movement of the endolymph fluid with reference to the canal. This will be brought about by acceleration and deceleration in the plane of the canal, but not by a steady velocity.

In the saccule and utricle the cupula contains chalky particles called *otoliths* or *statoconia.* Maximum stimulation occurs when the hairs are pointing downwards [Fig. 223(B)]. Minimum stimulation occurs when the hairs are pointing upwards [Fig. 223(A)].

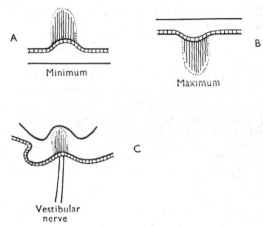

FIG. 223. The vestibular apparatus.

With the head erect the hairs of the saccules are in the horizontal plane and pointing laterally, thus the saccules are affected by the lateral tilt of the head. The head on the left shoulder gives the minimum stimulation to the right saccule and the maximum stimulation to the left saccule.

With the head erect the hairs of the utricles are pointing upwards. The utricles respond to fore and aft tilt and are maximally stimulated by standing on one's head.

Vestibular Connexions

Fibres from the maculae and cristae run to the vestibular nuclei in the pons and medulla and to the flocculonodular lobe of the cerebellum [p. 197].

The vestibular nuclei send fibres to:

1. Cerebellum.
2. Reticular nuclei.
3. Oculomotor nuclei of both sides.
4. Opposite thalamus.
5. Down vestibulospinal tracts to anterior horn cells.

The importance of many of these pathways has already been discussed in CHAPTER 19.

THE SENSATIONS OF TASTE AND SMELL

Taste

The taste receptors (taste buds) are found in the walls of the papillae on the upper surface of the tongue. There are four sets of taste buds which give the sensations of salt, sweet, sour and bitter tastes. A salt taste is given by a substance such as sodium chloride, a sweet taste by substances like sugar or saccharine, a sour taste by an acid such as hydrochloric acid and a bitter taste by a substance such as quinine.

The nerve supply to these receptors is the lingual nerve for the taste buds lying on the anterior two-thirds of the tongue and the glossopharyngeal nerve (IX) for the posterior third of the tongue. Although the lingual nerve is a branch of the fifth cranial nerve, electrophysiological studies have shown that many of the taste fibres reach the seventh cranial nerve via the chorda tympani (cf. nerve supply to the submandibular salivary gland, page 206).

There seems to be little connexion between the chemical structure of a substance and the taste sensation it produces. Often a small change in the chemical structure will completely alter the taste. The methyl derivative of the two amino acid peptide, aspartic acid-phenylalanine is 180 times sweeter than sucrose and may replace saccharin as a sweetener.

p-Ethoxyphenylthiourea is an example of a substance whose taste varies from person to person. About one-third of the population are unable to taste this substance. The remaining two-thirds find that it has a bitter taste. The ability to taste it is inherited according to the Mendelian laws.

Taste is important as a stimulus for the initiation of gastric and pancreatic secretions (in addition to saliva) It is also a protection against the ingestion of 'distasteful' substances which include many of the poisons.

Smell

Many so-called 'tastes' are, in fact, smells. This is demonstrated by the alteration in the taste of food which occurs when one is suffering from the common cold. The increase in nasal secretions in this condition prevents the stimulation of the olfactory mucosa.

The olfactory epithelium, containing the nerve cells which give the sensation of smell, are localized, in man, to a very small area of mucosa situated high in the nasal cavity. It is above the main air-stream and the sensation of smell is aroused by eddy-currents which carry the odoriferous substances in the air to the receptors. A sniff, on the other hand, directs the air-stream directly to the receptors.

The nerve supply is the olfactory nerve (I). There is a synapse between the olfactory cell and the olfactory nerve. The fibres pass from the receptors through the *cribriform plate* of the ethmoid bone to the olfactory bulb.

The sense of smell shows a rapid adaptation to one odour without impairing the sensitivity to another. However, some odours mask the effect of others, a fact made use of in deodorant and 'air freshener' preparations. Formalin is an example of a substance which paralyses the nasal mucosa and thus tends to eliminate other smells.

Olfactory sensitivity is modified by feedback loops which act on the olfactory nerve cells and alter this sensitivity. Thus the sense of smell is influenced by hunger. Olfactory sensitivity may become abnormally acute when there is underactivity of the adrenal cortex.

REFERENCES AND FURTHER READING

The Eye

Brindley, G. S. (1957) Human colour vision, *Progr. Biophys.*, **8**, 49.

Brindley, G. S. (1960) *Physiology of the Retina and the Visual Pathway*, London.

Davson, H. (1962) *Physiology of the Eye*, 2nd ed., London.

Granit, R. (1943) A physiological theory of colour perception, *Nature (Lond.)*, **151**, 11.

Granit, R. (1947) *Sensory Mechanisms of the Retina*, London.

Ishihara, Sh (1960) *Tests for Colour Blindness*, 15th ed., London.

Langham, M. E. (1958) Aqueous humour and control of intra-ocular pressure, *Physiol. Rev.*, **38**, 215.

The Ear

Békésy, G. von (1956) Current status of theories of hearing, *Science*, **123**, 779.

Galambos, R., and Davis, H. (1943) The response of single auditory nerve fibres to acoustic stimulation, *J. Neurophysiol.*, **6**, 39.

Tasaki, I. (1957) Hearing, *Ann. Rev. Physiol.*, **19**, 417.

Wever, E. G., and Bray, C. W. (1930) Action currents in the auditory nerve in response to acoustical stimulation, *Proc. nat. Acad. Sci. (Wash.)*, **16**, 344.

Whitfield, I. C. (1957) The physiology of hearing, *Progr. Biophys.*, **8**, 1.

Taste and Smell

Adey, E. R. (1959) The sense of smell, in *Handbook of Physiology*, Section 1, Vol. 1, p. 535, Washington, D.C.

Adrian, E. D. (1956) The mammalian olfactory organ, *J. Laryng.*, **70**, 1.

Heimer, L. (1971) Pathways in the brain, *Scientific American*, **225**, 1.

Pfaffmann, C. (1959) The sense of taste, in *Handbook of Physiology*, Section 1, Vol. 1, p. 507, Washington, D.C.

Zotterman, Y. (1958) Studies in the nervous mechanism of taste, *Exp. Cell Res.*, Suppl. **5**, 520.

APPENDIX I

S.I. UNITS

The system of units based on the metre, kilogram and second has now been adopted internationally and is known as the 'Système International d'Unités' (S.I. unit system). This system will, in the near future, replace all other systems for scientific and medical purposes.

The basic S.I. units are:

PHYSICAL QUANTITY	NAME OF UNIT	SYMBOL
mass	kilogram	kg
length	metre	m
time	second	s
electric current	ampere	A
temperature	degree Kelvin	°K
luminous intensity	candela	cd
amount of substance	mole	mol

They are defined as follows:

The **International Kilogram** (kg) (adopted 1875) is a cylinder of platinum-iridium kept at Sèvres near Paris.

The **International Metre** (m) (adopted 1962) is 1650763·73 times the wavelength of the orange line in the krypton spectrum.

The **Second** (s) was until recently $\frac{1}{31,556,925·9747}$ part of the year of 1900. It has been replaced (1972) by a standard based on a frequency of radiation of the caesium atom (9, 192, 631, 770 Hz).

1 cycle per second = 1 Hertz (Hz)

The **Ampere** is now defined as the electric current which when flowing through two straight and parallel conductors one metre apart produces an attractive force of 2×10^{-7} newtons per metre length.

Temperature is measured with reference to absolute zero using the centigrade (Celsius) degrees. It is expressed as degrees Kelvin (°K). The degree Celsius (°C.) may also be used. Absolute zero = $0 °K = -273·15 °C$.

The **Candela** has been defined on p. 215.

Multiples

10^{-12}	pico	p		10^{3}	kilo	k
10^{-9}	nano	n		10^{6}	mega	M
10^{-6}	micro	μ		10^{9}	giga	G
10^{-3}	milli	m		10^{12}	tera	T

The **unit of force** in the S.I. system is the **newton** (N). One newton is defined as the force which will accelerate a mass of 1 kilogram by 1 metre per second^2. It is independent of g (acceleration under gravity).

The principal advantage of using the kilogram metre, rather than the gram and centimetre, is that the unit of work, the **joule,** fits directly into the units already in use in electricity.

Thus 1 joule (J) is the potential energy lost when a weight of 1 kg. falls under gravity through 1 metre.

1 joule is the energy produced by 1 watt of electricity in 1 second, *i.e.* 1 watt = 1 joule per second.

1 joule is the work done when a force of 1 newton moves its point of application through 1 metre in the direction of the force.

The value of g (acceleration under gravity) in the S.I. system is $9·80665$ m.sec.$^{-2}$. For practical purposes this is taken as $9·81$ m.sec.$^{-2}$

The unit of pressure (= force/area) is the newton per metre^2, Nm^{-2} [also known as the **pascal** (Pa)].

$$1 \text{ Pa} = 1 \text{ Nm}^{-2}$$

1 mm. Hg = $13·595 \times 9·81 = 133·3$ pascals = $133·3$ Nm^{-2}

1 mm. Hg = $0·1333$ kilo-pascals = $0·1333$ kNm^{-2}

120 mm. Hg = $16·00$ kilo-pascals = $16·00$ kNm^{-2}

760 mm. Hg = $101·3$ kilo-pascals = $101·3$ kNm^{-2}

TABLE 20

Millimetres of mercury to S.I. Units (kilo-pascals or Nm$^{-2} \times 10^3$)
1 mm. Hg = $13·595 \times 9·81$ Nm^{-2}

KILO-PASCALS or NEWTON METRES$^{-2} \times 10^3$

mm. Hg	0	1	2	3	4	5	6	7	8	9
0	0·00	0·13	0·27	0·40	0·53	0·67	0·80	0·93	1·07	1·20
10	1·33	1·47	1·60	1·73	1·87	2·00	2·13	2·27	2·40	2·53
20	2·67	2·80	2·93	3·07	3·20	3·33	3·47	3·60	3·73	3·87
30	4·00	4·13	4·27	4·40	4·53	4·67	4·80	4·93	5·07	5·20
40	5·33	5·47	5·60	5·73	5·87	6·00	6·13	6·27	6·40	6·53
50	6·67	6·80	6·93	7·07	7·20	7·33	7·47	7·60	7·73	7·87
60	8·00	8·13	8·27	8·40	8·53	8·67	8·80	8·93	9·07	9·20
70	9·33	9·47	9·60	9·73	9·87	10·00	10·13	10·27	10·40	10·53
80	10·67	10·80	10·93	11·07	11·20	11·33	11·47	11·60	11·73	11·87
90	12·00	12·13	12·27	12·40	12·53	12·67	12·80	12·93	13·07	13·20
100	13·33	13·47	13·60	13·73	13·87	14·00	14·13	14·27	14·40	14·53
110	14·67	14·80	14·93	15·07	15·20	15·33	15·47	15·60	15·73	15·87
120	16·00	16·13	16·27	16·40	16·53	16·67	16·80	16·90	17·07	17·20
130	17·33	17·47	17·60	17·73	17·87	18·00	18·13	18·27	18·40	18·53
140	18·67	18·80	18·93	19·07	19·20	19·33	19·47	19·60	19·73	19·87
150	20·00	20·13	20·27	20·40	20·53	20·66	20·80	20·93	21·06	21·20
160	21·33	21·46	21·60	21·73	21·86	22·00	22·13	22·26	22·40	22·53
170	22·66	22·80	22·93	23·06	23·20	23·33	23·46	23·60	23·73	23·86
180	24·00	24·13	24·26	24·40	24·53	24·66	24·80	24·93	25·06	25·20
190	25·33	25·46	25·60	25·73	25·86	26·00	26·13	26·26	26·40	26·53
200	26·66	26·80	26·93	26·06	27·20	27·33	27·46	27·60	27·73	27·86

mm. Hg

100	200	300	400	500	600	700	800	900
13·33	26·66	40·00	53·33	66·66	79·99	93·33	106·7	120·0

Example: (1) 152 mm. Hg = 20·27 kilo-pascals

(2) 742 mm. Hg = 700 + 42 mm. Hg

= 93·33 + 5·60 = 98·93 kPa

The **Calorie** is not included in the S.I. system. The joule is used for both energy and heat. To convert from one system to the other remember:

$$4185 \cdot 5 \text{ joules} = 1 \text{ Calorie } (15 °C.)$$
$$[4184 \text{ joules} = 1 \text{ Calorie (thermochemical)}]$$

Heat given out or taken in:

$$= \text{mass (kg.)} \times \text{sp. heat} \times \text{temperature change.}$$

If the specific heat of water is expressed in Cal/kg. degree C. (= 1) the result will be in Calories.

If the specific heat of water is expressed in joules/kg. degree C. (= 4186) the result will be in joules.

Latent heat of vaporization of water at 100 °C. = 0·54 Cals/gram = 2260 joules/gram.

The daily intake of 2800 Calories becomes $2800 \times 4186 = 11 \cdot 715 \times 10^6$ joules (approximately 12 million joules per day).

A BMR of 40 Calories/m.^{2}/hr becomes 40×4186

$$= 167,000 \text{ joules/m.}^2\text{/hr [p. 123]}$$
$$= 167 \text{ kilo-joules/m.}^2\text{/hr.}$$

Carbohydrate gives 17,000 joules/gram (17 kilo-joules/g.)
Fats give 38,000 joules/gram (38 kilo-joules/g.)
Proteins give 17,000 joules/gram when used for heat

and energy (17 kilo-joules/g.).

TABLE 21

Calories to S.I. Units (joules)

1 Calorie = 1 k calorie = 1000 calories = 4185·5 joules
= 4·186 kilo-joules

KILO-JOULES (kJ or J × 10^3)

Calories

	0	10	20	30	40	50	60	70	80	90
0	0000	41·86	83·71	125·6	167·4	209·3	251·1	293·0	334·8	376·7
100	418·6	460·4	502·2	544·1	586·0	627·8	669·7	711·5	753·3	795·2
200	837·1	879·0	920·8	962·7	1004	1046	1088	1130	1172	1214
300	1256	1297	1339	1381	1423	1465	1507	1549	1590	1623
400	1674	1716	1758	1800	1842	1883	1925	1967	2009	2051
500	2093	2135	2176	2218	2260	2302	2344	2386	2427	2469
600	2511	2553	2595	2637	2679	2720	2762	2804	2846	2888
700	2930	2972	3013	3055	3097	3139	3180	3223	3265	3306
800	3348	3390	3432	3474	3516	3557	3599	3641	3683	3725
900	3767	3809	3851	3892	3934	3976	4018	4060	4102	4144
1000	4186	4227	4269	4311	4353	4395	4437	4478	4520	4562
2000	8371	8413	8455	8496	8538	8580	8622	8664	8706	8748
3000	12556	12598	12640	12682	12724	12766	12808	12849	12891	12933

Example: 2480 Cals = 2000 + 480 Calories
≡ 8371 + 2009 kilo-joules
= 10,380,000 joules

THE MOLE—THE S.I. UNIT OF QUANTITY

The table of atomic weights on page 221 is based on the atomic weight of carbon which is taken to be exactly 12·00. There are $6 \cdot 022 \times 10^{23}$ atoms in 12 g. of carbon and this number forms the basis for the unit of quantity in the S.I. system. The term *mole* is used, but its scope has been extended beyond that envisaged by Ostwald [p. 131].

The *mole* is now defined as the amount of a substance which contains the same number of elementary units (atoms, molecules, ions, electrons, etc.) as there are atoms in 0·012 kg. (12 g.) of carbon.

Care must be taken when using this term in its new form to be clear which elementary unit is being considered. A *mole of molecules* is not necessarily the same as a *mole of ions*. For example, 2 g. represents one mole of hydrogen molecules ($6 \cdot 022 \times 10^{23}$ molecules) whilst 1 g. represents one mole of hydrogen ions ($6 \cdot 022 \times 10^{23}$ ions).

When molecules are being considered, the sum of the constituent atomic weights in grams represents one mole. Thus 180 g. glucose ($C_6H_{12}O_6 : 6 \times 12 + 12 \times 1 + 6 \times 16 = 180$) and 58·5 g. sodium chloride ($NaCl : 1 \times 23 + 1 \times 35 \cdot 5 = 58 \cdot 5$) both contain the same number of molecules ($6 \cdot 022 \times 10^{23}$) and each represents one mole of the substance. One mole of potassium chloride ($39 \cdot 1 + 35 \cdot 5 = 74 \cdot 6$ g.) will contain the same number of molecules.

Special care must be taken when considering ions. One mole of sodium chloride molecules will give rise to two moles of ions, one mole of sodium ions plus one mole of chloride ions. One mole of sulphuric acid will give two moles of hydrogen ions plus one mole of sulphate ions.

With body fluids, whenever the molecular weight of a constituent is known, it is proposed that in future only the molar concentration (in moles per litre, millimoles per litre, micromoles per litre or nanomoles per litre) will be used. This will give an indication of the number of molecules (or ions) per unit volume of fluid. Thus 1 mmol/l of NaCl and 1 mmol/l of KCl both have the same number of molecules and ions (since both are monovalent) per unit volume of fluid whereas solutions of NaCl and KCl with the same mass concentration, say, 1 mg./100 ml., do not.

It is proposed that the *mole* should replace the older terms 'gram-molecular weight', 'gram-ionic weight' and 'gram-equivalent weight', and that the use of the term 'equivalent' should be discontinued. However, as has been discussed on page 130, there are occasions when it is advantageous to continue to use *milliequivalents per litre* rather than the newer millimoles per litre since the former enables the positive and negative ions to be equated as in FIG. 134, p. 131.

The **plasma calcium level** requires special mention since it underlines the problem that has arisen with the use of milliequivalents per litre in the past. Calcium has a molecular weight of 40 and, when ionized with a valency of 2, an equivalent weight of 20. The plasma calcium level is 10 mg./100 ml. plasma. If all the calcium were ionized, then the level in milli-equivalents per litre would be 5 mEq. per litre. However, only about half the plasma calcium is ionized so the level of ionized calcium in plasma is only 2·5 mEq. per litre. The remainder of the calcium is unionized and this fraction cannot, strictly speaking, be expressed as milli-equivalents per litre since equivalent weight (equals molecular weight divided by valency) has no meaning unless the substance is ionized.

With the newer molar concentration there is no problem. The total calcium level is 2·5 mmol per litre, of which 1·25 mmol per litre is ionized.

When the exact molecular weight of a substance is not known, the molar system cannot be used, and the concentration of such a substance will continue to be expressed in grams per litre, milligrams per litre, etc. In such cases it has been proposed that the litre (or the cubic decimetre, dm^3) should be the standard volume rather than 100 ml. Thus the traditional plasma protein level of 7 g./100 ml. becomes 70 g. per litre [p. 20].

APPENDIX II

ATOMIC WEIGHTS

Hydrogen	H	1·0
Helium	He	4·0
Carbon	C	12·0
Nitrogen	N	14·0
Oxygen	O	16·0
Fluorine	F	19·0
Sodium	Na	23·0
Magnesium	Mg	24·3
Aluminium	Al	27·0
Silicon	Si	28·0
Phosphorus	P	31·0
Sulphur	S	32·1
Chlorine	Cl	35·5
Argon	A	39·9
Potassium	K	39·1
Calcium	Ca	40·1
Chromium	Cr	52·0
Manganese	Mn	54·9
Iron	Fe	55·9
Cobalt	Co	58·9
Nickel	Ni	58·7
Copper	Cu	63·5
Zinc	Zn	65·4
Bromine	Br	79·9
Strontium	Sr	87·6
Silver	Ag	107·9
Cadmium	Cd	112·4
Tin	Sn	118·7
Iodine	I	126·9
Barium	Ba	137·4
Platinum	Pt	195·2
Gold	Au	197·2
Mercury	Hg	200·6
Lead	Pb	207·2

MOLECULAR WEIGHTS

MOLECULAR WEIGHT

Hydrogen	2·0
Nitrogen	28·0
Oxygen	32·0
Carbon dioxide	44·0
Water	18·0
Sodium bicarbonate	84·0
Sodium chloride	58·5
Urea	60·0

MILLI-MOLES PER LITRE

(milli-equivalents per litre in brackets if different)

1 milli-mole per litre

$= 23\cdot0$ mg. Na^+ per litre

$= 58\cdot5$ mg. sodium chloride per litre

$= 84\cdot0$ mg. sodium bicarbonate per litre

$= 39\cdot1$ mg. K^+ per litre

$= 74\cdot6$ mg. potassium chloride per litre

$= 100\cdot11$ mg. potassium bicarbonate per litre

$= 35\cdot5$ mg. Cl^- per litre

$= 40\cdot0$ mg. Ca^{2+} per litre (20·0 mg.)

$= 24\cdot4$ mg. Mg^{2+} per litre (12·2 mg.)

$= 61\cdot0$ mg. HCO_3^- per litre

$= 44\cdot0$ mg. CO_2 per litre

$= 22\cdot3$ ml. CO_2 per litre at 0 °C. and 760 mm. Hg

$(= 2\cdot2$ ml. CO_2 per cent.)

For an ideal gas:

One mole occupies 22·4 litres at 0 °C. and 760 mm. Hg (101·3 kPa).

For carbon dioxide the figure is 22·26 litres at 0 °C. and 760 mm. Hg (101·3 kPa).

HEART RATE

DURATION OF CARDIAC CYCLE—HEART RATE PER MINUTE

secs.	H.R.	secs.	H.R.	secs.	H.R.
0·30	200·0	0·64	93·8	1·00	60·0
		0·65	92·3		
0·31	193·5	0·66	90·9	1·02	58·8
0·32	187·5	0·67	89·6	1·04	57·7
0·33	180·0	0·68	88·2	1·06	56·6
0·34	176·5	0·69	87·0	1·08	55·6
0·35	171·4				
0·36	166·7	0·70	85·7	1·10	54·5
0·37	162·2				
0·38	157·9	0·71	84·5	1·12	53·6
0·39	153·8	0·72	83·3	1·14	52·6
		0·73	82·2	1·16	51·7
0·40	150·0	0·74	81·1	1·18	50·8
		0·75	80·0		
0·41	146·3	0·76	78·9	1·20	50·0
0·42	142·8	0·77	77·9		
0·43	139·5	0·78	76·9	1·22	49·2
0·44	136·4	0·79	75·9	1·24	48·4
0·45	133·3			1·26	47·6
0·46	130·4	0·80	75·0	1·28	46·9
0·47	127·7				
0·48	125·0	0·81	74·1	1·30	46·2
0·49	122·4	0·82	73·2		
		0·83	72·3	1·32	45·5
0·50	120·0	0·84	71·4	1·34	44·8
		0·85	70·6	1·36	44·1
0·51	117·6	0·86	69·8	1·38	43·5
0·52	115·4	0·87	69·0		
0·53	113·2	0·88	68·2	1·40	42·9
0·54	111·1	0·89	67·4		
0·55	109·1			1·42	42·3
0·56	107·1	0·90	66·7	1·44	41·7
0·57	105·3			1·46	41·1
0·58	103·4	0·91	65·9	1·48	40·5
0·59	101·7	0·92	65·2		
		0·93	64·5	1·50	40·0
0·60	100·0	0·94	63·8		
		0·95	63·2	1·52	39·5
0·61	98·4	0·96	62·5	1·54	39·0
0·62	96·8	0·97	61·9	1·56	38·5
0·63	95·2	0·98	61·2		
		0·99	60·6		

221

CONSTANTS

	$\log_{10}$
$\pi = 3\cdot1415927$	$0\cdot4972$
$\simeq \frac{22}{7}$	
$e = 2\cdot7183$	$0\cdot4343$
$g = 9\cdot80665$ m. sec^{-2}	
$\simeq 9\cdot81$ m. sec^{-2}	$0\cdot9917$
$= 981$ cm. sec^{-2}	$2\cdot9917$
$= 32\cdot2$ ft. sec.$^{-2}$	$\bar{1}\cdot5079$

1 joule	$= 7\cdot371 \times 10^{-1}$ ft. lb.	$\bar{1}\cdot8676$
1 newton	$= 2\cdot247 \times 10^{-1}$ lb. wt.	$\bar{1}\cdot3516$
1 kg. m.	$= 7\cdot233$ ft. lb.	$0\cdot8593$
1 ft. lb.	$= 0\cdot1383$ kg. m.	$\bar{1}\cdot1407$
	$= 1\cdot356$ joules	$0\cdot1325$
1 lb. wt.	$= 4\cdot45$ newtons	$0\cdot6484$

1 atmosphere of pressure $= 760$ mm. Hg

$= 1\cdot0133 \times 10^6$ dynes cm.$^{-2}$ $6\cdot0056$

$= 14\cdot7$ lb. wt. in.$^{-2}$ $1\cdot1673$

$= 101\cdot3$ kPa $2\cdot0056$

1 bar $= 750\cdot062$ mm. Hg $= 100$ kPa $2\cdot8752$

Pressure of 1 mm. column of water

$= 7\cdot3554 \times 10^{-2}$ mm. Hg $\bar{2}\cdot8669$

1 mm. column of Hg $= 13\cdot5954 \times 9\cdot80665$ Pa

$= 0\cdot133325$ kPa $\bar{1}\cdot1249$

1 metre $= 39\cdot370113$ inches	$1\cdot59517$
$= 3\cdot2808$	$0\cdot5160$
$= 1\cdot0936$ yards	$0\cdot0389$

1 square metre $= 10\cdot7639$ feet^2	$1\cdot0320$
$= 1\cdot196$ yards^2	$0\cdot0777$

1 cubic metre $= 1\cdot308$ yards^3 $0\cdot1166$

1 litre $= 61\cdot0239$ inches^3	$1\cdot7855$
$= 1\cdot76$ pints	$0\cdot2455$
$= 0\cdot22$ gallons	$\bar{1}\cdot3424$

1 gram $= 15\cdot43236$	$1\cdot1884$
$= 0\cdot03527$ oz. (avoir)	$\bar{2}\cdot5474$

1 kilogram $= 2\cdot20462$ pounds	$0\cdot3433$
1 grain $= 0\cdot0648$ grams	$\bar{2}\cdot8116$
1 ounce (avoir) $= 28\cdot350$ grams	$1\cdot4526$
1 pound $= 0\cdot4536$ kilograms	$\bar{1}\cdot6567$
1 pint (U.K.) $= 0\cdot568$ litres	$\bar{1}\cdot7545$
1 pint (U.S.A.) $= 0\cdot4544$ litres	$\bar{1}\cdot6575$
1 gallon (U.K.) $= 4\cdot546$ litres	$0\cdot6576$
1 gallon (U.S.A.) $= 3\cdot651$ litres	$0\cdot5607$
1 inch $= 2\cdot5400$ centimetres (exactly)	$0\cdot40483$
1 foot $= 30\cdot4800$ centimetres (exactly)	$1\cdot48401$
1 yard $= 0\cdot9144$ metres (exactly)	$\bar{1}\cdot96114$
1 mile $= 1\cdot6093$ kilometres	$0\cdot20665$
1 square inch $= 6\cdot4516$ cm.^{2}	$0\cdot8097$
1 cubic inch $= 16\cdot387$ cm.^{3}	$1\cdot2145$

SATURATED WATER VAPOUR PRESSURE

Temperature (°C.)	mm. Hg	kPa
0	4·6	0·61
10	9·2	1·23
11	9·8	1·31
12	10·5	1·40
13	11·2	1·49
14	12·0	1·60
15	12·8	1·71
16	13·6	1·81
17	14·5	1·93
18	15·5	2·07
19	16·5	2·20
20	17·5	2·33
21	18·7	2·49
22	19·8	2·64
23	21·1	2·81
24	22·4	2·99
25	23·8	3·17
26	25·2	3·36
27	26·7	3·56
28	28·3	3·77
29	30·0	4·00
30	31·7	4·23
31	33·7	4·49
32	35·7	4·76
33	37·7	5·03
34	39·9	5·32
35	42·2	5·63
36	44·6	5·95
37	47·1	6·28
38	49·7	6·63
39	52·4	6·99
40	55·1	7·35
50	92·3	12·31
60	149	19·87
70	233	31·07
80	355	47·33
90	525	70·00
100	760	101·33

Before 1964 1 litre was the volume of 1 kilogram of water at its maximum density (3·98 °C.) and at 760 mm. Hg atmospheric pressure:

= 1·000027 cubic decimetres (dm³)

Since 1964 1 litre (1964)· = 1 cubic decimetre exactly.

1 yard (1963) = 3 feet is now defined as 0·9144 metre.

1 pound (1963) is now defined as 0·453 592 37 kilogram.

LAPLACE'S LAW

The inflation of a rubber balloon is difficult at first, but once the balloon has started to expand, a much lower pressure is required to continue the inflation. This is because a given pressure inside the balloon produces a greater tension in the wall of the balloon as the radius increases. The same effect may be demonstrated using a soap-bubble. The relationship between the wall tension, the internal pressure and the radius is given by Laplace's Law:

Wall Tension = Internal Pressure × Radius

This law has several important physiological implications:

1. A small blood vessel, such as a capillary, although thin-walled, can withstand a very high internal pressure without rupturing because its radius is small.

2. A dilatation in a large artery (an aneurysm) tends to get bigger because the same blood pressure has a greater stretching effect on the wall as the radius increases (as with the balloon above).

3. A dilated ventricle of the heart will have to produce a greater wall tension in order to produce the same blood pressure. Although in the early stages of a dilatation, this increased tension can be obtained by the increased force of contraction which results from the *initial stretching of the cardiac muscle fibres* ('Starling's Law of the Heart [p. 46]), a point is reached in progressive heart failure when the mechanical disadvantage of the dilatation outweighs the Starling effect, and the blood pressure falls.

4. The bladder is able to accommodate more and more urine with only a small increase in internal bladder pressure (although the wall tension is increasing) as long as the radius continues to increase.

5. If two inflated balloons (or soap bubbles) are arranged to communicate with one another, the one with the smaller radius collapses and empties its contents into the larger one. Were it not for surfactant [p. 70], the surface tension of the fluid lining the lungs would cause smaller alveoli to collapse and empty their contents into the larger alveoli. This explains why babies with *hyaline membrane disease*, who have a deficiency of surfactant in their lungs, have areas of alveolar collapse.

TEMPERATURE CONVERSION

0 °C. = 32 °F.   100 °C. = 212 °F.

To inter-convert °C. and °F. mentally remember:

−40 °C. = −40 °F.
10 °C. = 50 °F.
40 °C. = 104 °F.

Choose one of these equivalents and multiply (or divide) degrees above or below by 1·8.

To multiply by 1·8, double and take away a tenth.

To divide by 1·8, halve and add a ninth (or a succession of tenths).

Thus:

37 °C. = **10** °C. + 27 °C.
= **50** °F. + (54 − 5·4) °F. = 98·6 °F.

And:

68 °F. = **50** °F. + 18 °F.
= **10** °C. + (9 + 1) °C. or 9·999 . . . °C. = 20 °C.

Alternatively, add 40°, multiply or divide by 1·8 (as above), take away 40°.

PRESSURE CONVERSION

75 mm. Hg = 10 kPa

To convert mm. Hg to kilopascals mentally

Double the pressure
Double it again
Divide by 3
Divide by 10

This is equivalent to dividing by 7·5.

Thus:

63 mm. Hg × 2 = 126
126   × 2 = 252
252   ÷ 3 = 84
84   ÷ 10 = 8·4 kPa

To convert kilopascals to mm. Hg mentally

Multiply the pressure by 10
Halve the answer
Halve it again
Multiply by 3

This is equivalent to multiplying by 7·5.

Thus:

8·4 kPa × 10 = 84
84   ÷ 2 = 42
42   ÷ 2 = 12
21   × 3 = 63 mm. Hg

Thus 8·4 kPa = 63 mm. Hg

APPENDIX III

COMMON LOGARITHMS
(Logarithms to Base 10)

	0	1	2	3	4	5	6	7	8	9	Add: 1	2	3	4	5	6	7	8	9
10	0·0000	0043	0086	0128	0170	0212	0253	0294	0334	0374	4	8	12	17	21	25	29	33	37
11	0·0414	0453	0492	0531	0569	0607	0645	0682	0719	0755	4	8	11	15	19	23	26	30	34
12	0·0792	0828	0864	0899	0934	0969	1004	1038	1072	1106	3	7	10	14	17	21	24	28	31
13	0·1139	1173	1206	1239	1271	1303	1335	1367	1399	1430	3	6	10	13	16	19	23	26	29
14	0·1461	1492	1523	1553	1584	1614	1644	1673	1703	1732	3	6	9	12	15	18	21	24	27
15	0·1761	1790	1818	1847	1875	1903	1931	1959	1987	2014	3	6	8	11	14	17	20	22	25
16	0·2041	2068	2095	2122	2148	2175	2201	2227	2253	2279	3	5	8	11	13	16	18	21	24
17	0·2304	2330	2355	2380	2405	2430	2455	2480	2504	2529	2	5	7	10	12	15	17	20	22
18	0·2553	2577	2601	2625	2648	2672	2695	2718	2742	2765	2	5	7	9	12	14	16	19	21
19	0·2788	2810	2833	2856	2878	2900	2923	2945	2967	2989	2	4	7	9	11	13	16	18	20
20	0·3010	3032	3054	3075	3096	3118	3139	3160	3181	3201	2	4	6	8	11	13	15	17	19
21	0·3222	3243	3263	3284	3304	3324	3345	3365	3385	3404	2	4	6	8	10	12	14	16	18
22	0·3424	3444	3464	3483	3502	3522	3541	3560	3579	3598	2	4	6	8	10	12	14	15	17
23	0·3617	3636	3655	3674	3692	3711	3729	3747	3766	3784	2	4	6	7	9	11	13	15	17
24	0·3802	3820	3838	3856	3874	3892	3909	3927	3945	3962	2	4	5	7	9	11	12	14	16
25	0·3979	3997	4014	4031	4048	4065	4082	4099	4116	4133	2	3	5	7	9	10	12	14	15
26	0·4150	4166	4183	4200	4216	4232	4249	4265	4281	4298	2	3	5	7	8	10	11	13	15
27	0·4314	4330	4346	4362	4378	4393	4409	4425	4440	4456	2	3	5	6	8	9	11	13	14
28	0·4472	4487	4502	4518	4533	4548	4564	4579	4594	4609	2	3	5	6	8	9	11	12	14
29	0·4624	4639	4654	4669	4683	4698	4713	4728	4742	4757	1	3	4	6	7	9	10	12	13
30	0·4771	4786	4800	4814	4829	4843	4857	4871	4886	4900	1	3	4	6	7	9	10	11	13
31	0·4914	4928	4942	4955	4969	4983	4997	5011	5024	5038	1	3	4	6	7	8	10	11	12
32	0·5051	5065	5079	5092	5105	5119	5132	5145	5159	5172	1	3	4	5	7	8	9	11	12
33	0·5185	5198	5211	5224	5237	5250	5263	5276	5289	5302	1	3	4	5	6	8	9	10	12
34	0·5315	5328	5340	5353	5366	5378	5391	5403	5416	5428	1	3	4	5	6	8	9	10	11
35	0·5441	5453	5465	5478	5490	5502	5514	5527	5539	5551	1	2	4	5	6	7	9	10	11
36	0·5563	5575	5587	5599	5611	5623	5635	5647	5658	5670	1	2	4	5	6	7	8	10	11
37	0·5682	5694	5705	5717	5729	5740	5752	5763	5775	5786	1	2	3	5	6	7	8	9	10
38	0·5798	5809	5821	5832	5843	5855	5866	5877	5888	5899	1	2	3	5	6	7	8	9	10
39	0·5911	5922	5933	5944	5955	5966	5977	5988	5999	6010	1	2	3	4	5	7	8	9	10
40	0·6021	6031	6042	6053	6064	6075	6085	6096	6107	6117	1	2	3	4	5	6	8	9	10
41	0·6128	6138	6149	6160	6170	6180	6191	6201	6212	6222	1	2	3	4	5	6	7	8	9
42	0·6232	6243	6253	6263	6274	6284	6294	6304	6314	6325	1	2	3	4	5	6	7	8	9
43	0·6335	6345	6355	6365	6375	6385	6395	6405	6415	6425	1	2	3	4	5	6	7	8	9
44	0·6435	6444	6454	6464	6474	6484	6493	6503	6513	6522	1	2	3	4	5	6	7	8	9
45	0·6532	6542	6551	6561	6571	6580	6590	6599	6609	6618	1	2	3	4	5	6	7	8	9
46	0·6628	6637	6646	6656	6665	6675	6684	6693	6702	6712	1	2	3	4	5	6	7	7	8
47	0·6721	6730	6739	6749	6758	6767	6776	6785	6794	6803	1	2	3	4	5	5	6	7	8
48	0·6812	6821	6830	6839	6848	6857	6866	6875	6884	6893	1	2	3	4	4	5	6	7	8
49	0·6902	6911	6920	6928	6937	6946	6955	6964	6972	6981	1	2	3	4	4	5	6	7	8
50	0·6990	6998	7007	7016	7024	7033	7042	7050	7059	7067	1	2	3	3	4	5	6	7	8
51	0·7076	7084	7093	7101	7110	7118	7126	7135	7143	7152	1	2	3	3	4	5	6	7	8
52	0·7160	7168	7177	7185	7193	7202	7210	7218	7226	7235	1	2	2	3	4	5	6	7	7
53	0·7243	7251	7259	7267	7275	7284	7292	7300	7308	7316	1	2	2	3	4	5	6	6	7
54	0·7324	7332	7340	7348	7356	7364	7372	7380	7388	7396	1	2	2	3	4	5	6	6	7
55	0·7404	7412	7419	7427	7435	7443	7451	7459	7466	7474	1	2	2	3	4	5	5	6	7

COMMON LOGARITHMS
(Logarithms to Base 10)

	0	1	2	3	4	5	6	7	8	9	Add: 1	2	3	4	5	6	7	8	9
55	0·7404	7412	7419	7427	7435	7443	7451	7459	7466	7474	1	2	2	3	4	5	5	6	7
56	0·7482	7490	7497	7505	7513	7520	7528	7536	7543	7551	1	2	2	3	4	5	5	6	7
57	0·7559	7566	7574	7582	7589	7597	7604	7612	7619	7627	1	2	2	3	4	5	5	6	7
58	0·7634	7642	7649	7657	7664	7672	7679	7686	7694	7701	1	1	2	3	4	4	5	6	7
59	0·7709	7716	7723	7731	7738	7745	7752	7760	7767	7774	1	1	2	3	4	4	5	6	7
60	0·7782	7789	7796	7803	7810	7818	7825	7832	7839	7846	1	1	2	3	4	4	5	6	6
61	0·7853	7860	7868	7875	7882	7889	7896	7903	7910	7917	1	1	2	3	4	4	5	6	6
62	0·7924	7931	7938	7945	7952	7959	7966	7973	7980	7987	1	1	2	3	3	4	5	6	6
63	0·7993	8000	8007	8014	8021	8028	8035	8041	8048	8055	1	1	2	3	3	4	5	5	6
64	0·8062	8069	8075	8082	8089	8096	8102	8109	8116	8122	1	1	2	3	3	4	5	5	6
65	0·8129	8136	8142	8149	8156	8162	8169	8176	8182	8189	1	1	2	3	3	4	5	5	6
66	0·8195	8202	8209	8215	8222	8228	8235	8241	8248	8254	1	1	2	3	3	4	5	5	6
67	0·8261	8267	8274	8280	8287	8293	8299	8306	8312	8319	1	1	2	3	3	4	5	5	6
68	0·8325	8331	8338	8344	8351	8357	8363	8370	8376	8382	1	1	2	3	3	4	4	5	6
69	0·8388	8395	8401	8407	8414	8420	8426	8432	8439	8445	1	1	2	2	3	4	4	5	6
70	0·8451	8457	8463	8470	8476	8482	8488	8494	8500	8506	1	1	2	2	3	4	4	5	6
71	0·8513	8519	8525	8531	8537	8543	8549	8555	8561	8567	1	1	2	2	3	4	4	5	5
72	0·8573	8579	8585	8591	8597	8603	8609	8615	8621	8627	1	1	2	2	3	4	4	5	5
73	0·8633	8639	8645	8651	8657	8663	8669	8675	8681	8686	1	1	2	2	3	4	4	5	5
74	0·8692	8698	8704	8710	8716	8722	8727	8733	8739	8745	1	1	2	2	3	4	4	5	5
75	0·8751	8756	8762	8768	8774	8779	8785	8791	8797	8802	1	1	2	2	3	3	4	5	5
76	0·8808	8814	8820	8825	8831	8837	8842	8848	8854	8859	1	1	2	2	3	3	4	5	5
77	0·8865	8871	8876	8882	8887	8893	8899	8904	8910	8915	1	1	2	2	3	3	4	4	5
78	0·8921	8927	8932	8938	8943	8949	8954	8960	8965	8971	1	1	2	2	3	3	4	4	5
79	0·8976	8982	8987	8993	8998	9004	9009	9015	9020	9025	1	1	2	2	3	3	4	4	5
80	0·9031	9036	9042	9047	9053	9058	9063	9069	9074	9079	1	1	2	2	3	3	4	4	5
81	0·9085	9090	9096	9101	9106	9112	9117	9122	9128	9133	1	1	2	2	3	3	4	4	5
82	0·9138	9143	9149	9154	9159	9165	9170	9175	9180	9186	1	1	2	2	3	3	4	4	5
83	0·9191	9196	9201	9206	9212	9217	9222	9227	9232	9238	1	1	2	2	3	3	4	4	5
84	0·9243	9248	9253	9258	9263	9269	9274	9279	9284	9289	1	1	2	2	3	3	4	4	5
85	0·9294	9299	9304	9309	9315	9320	9325	9330	9335	9430	1	1	2	2	3	3	4	4	5
86	0·9345	9350	9355	9360	9365	9370	9375	9380	9385	9390	1	1	2	2	3	3	4	4	5
87	0·9395	9400	9405	9410	9415	9420	9425	9430	9435	9440	0	1	1	2	2	3	3	4	4
88	0·9445	9450	9455	9460	9465	9469	9474	9479	9484	9489	0	1	1	2	2	3	3	4	4
89	0·9494	9499	9504	9509	9513	9518	9523	9528	9533	9538	0	1	1	2	2	3	3	4	4
90	0·9542	9547	9552	9557	9562	9566	9571	9576	9581	9586	0	1	1	2	2	3	3	4	4
91	0·9590	9595	9600	9605	9609	9614	9619	9624	9628	9633	0	1	1	2	2	3	3	4	4
92	0·9638	9643	9647	9652	9657	9661	9666	9671	9675	9680	0	1	1	2	2	3	3	4	4
93	0·9685	9689	9694	9699	9703	9708	9713	9717	9722	9727	0	1	1	2	2	3	3	4	4
94	0·9731	9736	9741	9745	9750	9754	9759	9763	9768	9773	0	1	1	2	2	3	3	4	4
95	0·9777	9782	9786	9791	9795	9800	9805	9809	9814	9818	0	1	1	2	2	3	3	4	4
96	0·9823	9827	9832	9836	9841	9845	9850	9854	9859	9863	0	1	1	2	2	3	3	4	4
97	0·9868	9872	9877	9881	9886	9890	9894	9899	9903	9908	0	1	1	2	2	3	3	4	4
98	0·9912	9917	9921	9926	9930	9934	9939	9943	9948	9952	0	1	1	2	2	3	3	3	4
99	0·9956	9961	9965	9969	9974	9978	9983	9987	9991	9996	0	1	1	2	2	3	3	3	4
100	0·0000	0004	0009	0013	0017	0022	0026	0030	0035	0039	0	1	1	2	2	3	3	3	4

INDEX

dead cells

Corneal layer

multiplying cells

Malpygian layer

blood vessels
sweat glands